GENE
TRANSFER

GENE TRANSFER

Edited by
Raju Kucherlapati

University of Illinois College of Medicine
Chicago, Illinois

PLENUM PRESS • **New York and London**

Library of Congress Cataloging in Publication Data

Gene transfer.

Includes bibliographical references and index.
1. Genetic engineering. 2. Cell hybridization. 3. Recombinant DNA. 4. Genetic regulation. 5. Gene expression. 6. Mammals—Genetics. 7. Mammals—Cytology. I. Kucherlapati, Raju.
QH442.G439 1986   615′.39   86-18743
ISBN-13: 978-1-4684-5169-6   e-ISBN-13: 978-1-4684-5167-2
DOI: 10.1007/978-1-4684-5167-2

© 1986 Plenum Press, New York
Softcover reprint of the hardcover 1st edition 1986
A Division of Plenum Publishing Corporation
233 Spring Street, New York, N.Y. 10013

CONTRIBUTORS

Vijay R. Baichwal   McArdle Laboratory for Cancer Research, University of Wisconsin, Madison, Wisconsin 53706

John W. Belmont   Institute for Molecular Genetics and Howard Hughes Medical Institute, Baylor College of Medicine, Houston, Texas 77030

Michele P. Calos   Department of Genetics, Stanford University School of Medicine, Stanford, California 94305

S. Carroll   Gene Expression Laboratory, The Salk Institute, La Jolla, California 92037

C. Thomas Caskey   Institute for Molecular Genetics and Howard Hughes Medical Institute, Baylor College of Medicine, Houston, Texas 77030

Moses V. Chao   Department of Cell Biology and Anatomy and Hematology/Oncology Division, Cornell University Medical College, New York, New York 10021

R. E. K. Fournier   Department of Microbiology and the Comprehensive Cancer Center, University of Southern California School of Medicine, Los Angeles, California 90033

P. Gaudray   Gene Expression Laboratory, The Salk Institute, La Jolla, California 92037

Stephen P. Goff   Department of Biochemistry and Molecular Biophysics, College of Physicians and Surgeons, Columbia University, New York, New York 10032

David E. Housman   Center for Cancer Research, Department of Biology, Massachusetts Institute of Technology, Cambridge, Massachusetts 02139

Raju Kucherlapati   Center for Genetics, University of Illinois College of Medicine, Chicago, Illinois 60612

Anthea Letsou   Departments of Therapeutic Radiology and Human Genetics, Yale University School of Medicine, New Haven, Connecticut 06510

Arnold J. Levine   Department of Molecular Biology, Princeton University, Princeton, New Jersey 08544

R. Michael Liskay   Departments of Therapeutic Radiology and Human Genetics, Yale University School of Medicine, New Haven, Connecticut 06510

Tracy G. Lugo   Department of Microbiology and the Comprehensive Cancer Center, University of Southern California School of Medicine, Los Angeles, California 90033

J. Meinkoth   Gene Expression Laboratory, The Salk Institute, La Jolla, California 92037

David L. Nelson   Laboratory of Molecular Genetics, National Institute of Neurological and Communicative Disorders and Stroke, National Institutes of Health, Bethesda, Maryland 20892

Angel Pellicer   Department of Pathology and Kaplan Cancer Center, New York University Medical Center, New York, New York 10016

Frank H. Ruddle   Department of Biology, Yale University, New Haven, Connecticut 06510

J. Ruiz   Gene Expression Laboratory, The Salk Institute, La Jolla, California 92037

Thomas B. Shows   Department of Human Genetics, Roswell Park Memorial Institute, New York State Department of Health, Buffalo, New York 14263

Bill Sugden   McArdle Laboratory for Cancer Research, University of Wisconsin, Madison, Wisconsin 53706

Howard M. Temin   McArdle Laboratory for Cancer Research, University of Wisconsin, Madison, Wisconsin 53706

Shirley M. Tilghman   Institute for Cancer Research, Philadelphia, Pennsylvania 19111

G. M. Wahl   Gene Expression Laboratory, The Salk Institute, La Jolla, California 92037

PREFACE

Genetic analysis of microbial systems provided us with the foundation for understanding gene structure, expression, and regulation. It was long felt that the ability to generate mutants and conduct genetic studies in mammalian systems would prove to be equally useful. However, genetic analysis based on sexual systems is difficult in mammals because of the long generation times and the inability to perform controlled matings. As a result, genetic analysis of mammalian systems had to await the development of parasexual systems. This book is an attempt to bring together descriptions of a number of these parasexual systems.

A common theme of all the parasexual systems is the transfer of genetic information from a defined source into a specific cell type. This volume deals with a number of methods of gene transfer into mammalian cells. The early methods of gene transfer involved transfer of relatively large amounts of genetic information. These include somatic cell hybridization, microcell fusion, and chromosome transfer, which constitute the first part of this book. Each of these methods has already proven to be of enormous value in arriving at a genetic understanding of the mammalian genome. Development of recombinant DNA methods, and the ability to introduce purified DNA into mammalian cells, has had a significant impact on our ability to dissect important aspects of mammalian gene expression and regulation. The second part of this book deals with gene transfer systems involving defined nucleic acid sequences.

This volume is intended for use by a wide audience. It is expected to serve students of mammalian genetics at various levels, as a source of information about the various aspects of gene transfer into mammalian cells.

I am very grateful to my colleague Richard Davidson, who encouraged me to put together this volume. The book is the result of dedicated efforts by the authors and I thank them for their timely contributions.

Raju Kucherlapati

Chicago, Illinois

CONTENTS

Chapter 14

Anthea Letsou and R. Michael Liskay

Chapter 15

John W. Belmont and C. Thomas Caskey

GENE TRANSFER: A PERSPECTIVE

Frank H. Ruddle

Somatic-cell genetics is a large, multifaceted field that is not easily defined. This book addresses one subfield of this unruly subject—that concerned with parasexuality. Tissue culture had its origins in the experiments of Ross Harrison (1907) early in this century, but a consideration of sexuality in cultured cell populations did not arise until the 1950s. This way of thinking was motivated by the demonstration of parasexuality in eukaryotes, particularly by Stern (1936) and Pontecorvo (1962), and by sexuality in prokaryotes, as demonstrated by Lederberg (1958). In a sense, the work of Stern dealing with mitotic recombination was the more relevant because it dealt with somatic cells, albeit *in vivo*. However, the demonstration of sexuality in free-living bacteria was more dramatic and, although only analogous, powerfully influenced the development of parasexual systems in cultured prokaryotic cells *in vitro*.

The early tissue-culture experiments on parasexuality were inspired by the quantitative thinking of the lambdologists and microbiologists. The cloning experiments of Puck (1972) and the drug-selection experiments of Harris (M. Harris and Ruddle, 1961), Szybalska and Szybalski (1962), and Littlefield (1964) are examples of these influences. Eukaryotic somatic genetics found its identity with the advent of its first parasexual system—cell hybridization—introduced independently by Okada (1958) and Barski (Barski *et al.*, 1960) and developed as an experimental system by both Ephrussi (Ephrussi and Weiss, 1965) and Harris (H. Harris and Watkins, 1965). The important use of somatic-cell hybrids for gene mapping followed through the application of chromosome segregation (Weiss and Green, 1967), genetic marker systems (Ruddle, 1968), and chromosome identification by banding (Caspersson *et al.*, 1968). More than 1000 entries in the human gene map have been made possible by this parasexual system (see Chapter 2). Parasexual analysis was extended by the introduction of additional techniques, such as microcell transfer (Fournier and Ruddle, 1977) and chromosome-mediated gene transfer (McBride and Ozer, 1973).

Animal-cell virology has been intimately involved in the development of

Frank H. Ruddle • Department of Biology, Yale University, New Haven, Connecticut 06510.

eukaryotic parasexual systems. The genetic complementation of thymidine kinase (TK) deficiency by herpesvirus infection (Munyon *et al.*, 1971) and the incorporation and expression of the herpes *TK* gene in mammalian recipient cells (Smiley *et al.*, 1978) strongly motivated the development of DNA transformation systems. It was the introduction of the herpes *TK* gene (purified but not cloned) using a calcium phosphate transfer procedure developed first in adenovirus studies (Graham and van der Eb, 1973) that first established a workable DNA-mediated gene-transfer procedure (Wigler *et al.*, 1977; Bacchetti and Graham, 1977; Maitland and McDougall, 1977).

DNA-mediated gene transfer and DNA cloning arose at about the same time and together have provided a means of understanding the structural basis of gene function. The mechanisms of promoter-mediated transcription, enhancement, and silencing have all been elucidated to some extent by gene-transfer coupled with gene-rearrangement techniques. Our understanding of gene function has been further enhanced by the introduction of procedures that allow the introduction of genes into germ cells of higher organisms, e.g., *Nicotiana* (Chilton *et al.*, 1977), yeast (Hinnen *et al.*, 1978), *Drosophila* (Rubin and Spradling, 1982), and the mouse (Gordon *et al.*, 1980). These transgenic systems promise to tell us much about the mechanisms that govern mammalian development. On a more practical level, these procedures are also likely to lead to genetically improved livestock and crops, new pharmaceuticals, and the introduction of gene therapies. Such goals will be particularly advanced by recombination of donor genes into homologous recipient sites at high efficiency levels (Smithies *et al.*, 1985).

This volume provides a useful and welcome summary of the current state of gene-transfer methodologies in eukaryotes and of the impact these experiments are making on our understanding of genetic and developmental mechanisms. The contributions attest to the intense activity and excitement that characterize parasexual studies today. It is impossible to foresee with certainty where these activities will lead us. Certainly our understanding of the development as it relates to cytodifferentiation and to morphogenesis will be greatly advanced. Such an outcome would be entirely fitting, since Ross Harrison originated the tissue-culture technique originally as a means to solve problems of this type.

REFERENCES

Bacchetti, S., and Graham, F., 1977, Transfer of the gene for thymidine kinase to thymidine kinase deficient human cells by means of purified herpes simplex viral DNA, *Proc. Natl. Acad. Sci. U.S.A.* **74:**1590–1594.
Barski, G., Sorieul, S., and Cornefert, F., 1960, Production dans des cultures *in vitro* de deux souches cellulaires en association de cellules de caractere "hybrid," *C. R. Acad. Sci.* **251:**1825–1830.

Caspersson, T., Farber, S., Foley, G. E., Kudynoski, J., Modest, E. J., Simonsson, E., Wagh, U., and Zeck, L., 1968, Chemical differentiation along metaphase chromosomes, *Exp. Cell Res.* **49**:219–222.

Chilton, M.-D., Drummond, M. H., Merlo, D. J., Sciaky, D., Montoya, A., Gordon, M. P., and Nester, E. W., 1977, Stable incorporation of plasmid DNA into higher plant cells: The molecular basis of crown gall tumorigenesis, *Cell* **11**:263–271.

Ephrussi, B., and Weiss, M. C., 1965, Interspecific hybridization of somatic cells, *Proc. Natl. Acad. Sci. U.S.A.* **53**:1040–1042.

Fournier, R. E. K., and Ruddle, F. H., 1977, Microcell mediated transfer of murine chromosomes into mouse, Chinese hamster, and human somatic cells, *Proc. Natl. Acad. Sci. U.S.A.* **74**:319–323.

Gordon, J. W., Scangos, G., Plotkin, D., Barbosa, J., and Ruddle, F. H., 1980, Genetic transformation of mouse embryos by microinjection of purified DNA, *Proc. Natl. Acad. Sci. U.S.A.* **77**:7380–7384.

Graham, F. L., and van der Eb, A. J., 1973, A new technique for the assay of infectivity of human adenovirus 5 DNA, *Virology* **52**:456–467.

Harris, H., and Watkins, J. F., 1965, Hybrid cells derived from mouse and man: Artificial heterokaryons of mammalian cells from different species, *Nature (London)* **205**:640–646.

Harris, M., and Ruddle, F. H., 1961, Clone strains of pig kidney cells with drug resistance and chromosomal markers, *J. Natl. Cancer Inst.* **26**:1405–1411.

Harrison, R. G., 1907, Observations on the living developing nerve fiber, *Anat. Rec.* **1**:116–118.

Hinnen, A., Hicks, J. B., and Fink, G. R., 1978, Transformation of yeast, *Proc. Natl. Acad. Sci. U.S.A.* **75**:1929–1933.

Lederberg, J., 1958, Genetic approaches to somatic cell variation: Summary comment, *J. Cell. Physiol.* **52**:383–401.

Littlefield, J., 1964, Selection of hybrids from matings of fibroblasts *in vitro* and their presumed recombinants, *Science* **145**:709–710.

Maitland, H., and McDougall, J., 1977, Biochemical transformation of mouse cells by fragments of herpes simplex virus DNA, *Cell* **11**:233–241.

McBride, O. W., and Ozer, H. L., 1973, Transfer of genetic information by purified metaphase chromosomes, *Proc. Natl. Acad. Sci. U.S.A.* **70**:1258–1262.

Munyon, Y., Kraiselburd, E., Davis, D., and Mann, J., 1971, Transfer of thymidine kinase to thymidine kinaseless L cells by infection with ultraviolet-irradiated herpes simplex virus, *J. Virol.* **7**:813–820.

Okada, Y., 1958, The fusion of Ehrlich's tumor cells caused by HVJ virus *in vitro.*, *Biken J.* **1**:103–110.

Pontecorvo, G., 1962, Methods in microbiological genetics in an approach to human genetics, *Br. Med. Bull.* **18**:18–84.

Puck, T. T., 1972, *The Mammalian Cell as a Microorganism: Genetic and Biochemical Studies in Vitro,* Holden-Day, San Francisco, 1972.

Rubin, G. M., and Spradling, A. C., 1982, Genetic transformation of *Drosophila* using transposable element vectors, *Science* **218**:348–353.

Ruddle, F. H., 1968, Isozymic variants as genetic markers in somatic cell populations *in vitro, Natl. Cancer Inst. Monogr.* **29**:9–13.

Smiley, J. R., Steege, D. A., Juricek, D. K., Summers, W., and Ruddle, F. H., 1978, Herpes simplex virus 1 integration site in the mouse genome defined by somatic cell genetic analysis, *Cell* **15**:455–468.

Smithies, O., Gregg, R. G., Boggs, S. S., Koralewski, M. A., and Kucherlapati, R. S., 1985, Insertion of DNA sequences into the human chromosomal β-globin locus by homologous recombination, *Nature (London)* **317**:230–234.

Stern, C., 1936, Somatic crossing over and segregation in *Drosophila melanogaster, Genetics* **21**:625–730.

Szybalska, E. H., and Szybalski, W., 1962, Genetics of human cell lines. IV. DNA mediated heritable transformation of a biochemical trait, *Proc. Natl. Acad. Sci. U.S.A.* **48**:2026–2034.

Weiss, M. C., and Green, H., 1967, Human–mouse hybrid cell lines containing partial complements of human chromosomes and functioning human genes, *Proc. Natl. Acad. Sci. U.S.A.* **58**:1104–1111.

Wigler, M., Silverstein S., Lee, L.-S., Pellicer, A., Cheng, T., and Axel, R., 1977, Transfer of purified herpes virus thymidine kinase gene to cultured mouse cells, *Cell* **11**:223–232.

2

CELL HYBRIDIZATION AND THE 24 HUMAN GENE MAPS

Thomas B. Shows

1. INTRODUCTION

The transfer of genes is an ancient phenomenon in the biology and evolution of organisms, as is evidenced by a variety of microorganisms. In bacteria during their sexual reproductive phase, cell-to-cell transfer of single-stranded DNA has been demonstrated (Lederberg, 1947; Wollman *et al.*, 1956). In addition, DNA isolated from one genotype could stably change the phenotype of another bacterium (Avery *et al.*, 1944). Similarly, the infection of bacteria by a certain bacteriophage involves transfer of phage DNA into a bacterium and integration into the bacterial genome. When phage replication takes place, the phage genome sometimes acquires a bacterial gene that was closely linked to the inserted phage genome. Subsequent bacterial infections by this phage can transfer the bacterial gene to another bacterial genome (Zinder and Lederberg, 1952; Lennox, 1955). In the life cycle of fungi, the entire haploid genome is transferred by the formation of heterokaryons after fusion of hyphae from different genomes (Fincham, 1966). In eukaryotes, fertilization of an egg could be considered a form of gene transfer, since an entire haploid genome is transferred. Thus, there are biological mechanisms for cells to accept new genetic material and propagate it through cell division. Understanding the principles of gene transfer in mammals, especially *Homo sapiens,* will teach us much about genetic mechanisms, gene expression, cell growth and differentiation, gene mapping, mutagenesis, and both normal and abnormal human biology. This information promises to be essential for understanding and possibly treating disease.

Cell fusion resulting in multinucleated mammalian cells, termed "polykaryocytes," was first observed over 100 years ago after viral infection and was thought to be the first step in tumorigenesis (Langhans, 1868). This fusion

Thomas B. Shows • Department of Human Genetics, Roswell Park Memorial Institute, New York State Department of Health, Buffalo, New York 14263.

represents the transfer of whole genomes into a single cell. It was Barski and Ephrussi and their colleagues who demonstrated, by fusing certain rodent cells in culture, that full complements of two genomes were represented in a single cell (cf. Ephrussi, 1972). Weiss and Green (1967) later showed that in a fusion of human and mouse cells in culture, only a partial complement of human chromosomes was retained after hybrid cells were cloned. This was the key observation that revolutionized and stimulated interest in human gene mapping. By the coupling of these findings with current technology in cytogenetics, enzymology, immunology, and recombinant DNA methodology, detailed human gene maps have been generated for each of the 24 human chromosomes (see Table I, p. 36).

The ability to transfer mammalian genes parasexually has opened new possibilities for gene mapping, fine-structure mapping, and genetic dissection of disease. The DNA transferred has ranged from whole genomes to single genes and smaller segments of DNA. The transfer of whole genomes by cell fusion produces cell hybrids and has promoted the extensive mapping of human and mouse genes. Transfer by cell fusion of rearranged chromosomes has contributed significantly to determining close linkage and the assignment of genes to specific, small chromosomal regions. Transfer of single chromosomes has been achieved utilizing microcells fused to recipient cells. Metaphase chromosomes have been isolated and used to transfer single-to-multigenic DNA segments. DNA-mediated gene transfer, simulating bacterial transformation, has achieved transfer of single-copy genes. Single genes and DNA segments have been cloned into eukaryotic and prokaryotic vectors for transfer into mammalian cells, their fate and chromosome mapping being examined at the molecular level using sequence-specific probes. In fact, recombinant libraries in which entire mammalian genomes are represented collectively are an extensive source of transferable genes. Thus, virtually all gene-transfer methodology lends itself to mapping genes. This review will focus on the transfer of whole mammalian genomes and rearranged chromosomes through cell fusion and the mapping of the human genome stimulated by this technology and other complementing technology. Other chapters in this book will describe the transfer of single chromosomes, segments of chromosomes, cloned genes, and defined fragments of DNA, all of which contribute to the mapping and linear order of genes and to fine-structure gene mapping.

2. SOMATIC-CELL HYBRIDIZATION AND THE TRANSFER OF WHOLE GENOMES

Entire mammalian genomes are transferred routinely by fusing individual cells. A large array of mammalian cells and cell types have been fused principally by two fusing agents: Sendai virus (Okada and Tadokaro, 1963; Klebe *et al.,*

1970) and polyethylene glycol (PEG) (Davidson and Gerald, 1976). Cultured somatic cells growing as monolayers or in suspension, single cells from living organisms (e.g., leukocytes, erythrocytes, and macrophages), and cancer cells in suspension are examples of cells that have been used for cell fusion (Ephrussi, 1972; Sell and Krooth, 1973; Ringertz and Savage, 1976; Shay, 1982). The genetics, state of differentiation, and chromosomal composition of parental cells participating in cell hybridization have been diverse; for example, diploid, sub-diploid, or heteroploid cells and cells with gene mutations, differentiated functions, and chromosome translocations and deletions have been used (Ephrussi, 1972; Sell and Krooth, 1973; Davidson, 1974; Ringertz and Savage, 1976; Shows, 1979; Shows et al., 1982). When growing cell hybrids are desired, at least one of the parental cells must be able to proliferate in cell-culture conditions, and an appropriate selection system must be available to isolate the hybrid cells from parental cells. Most cells can be fused using either inactivated Sendai virus or PEG of different molecular weights (Klebe et al., 1970; Davidson and Gerald, 1976). Homokaryons or heterokaryons constitute the multinucleated-cell hybrid population that is formed at concentrations depending on several criteria, such as concentration of the fusing agent, time, pH, temperature, ions, cell type, and cell cycle (Klebe et al., 1970; Poste, 1970; Croce et al., 1972; Davidson and Gerald, 1976; Ringertz and Savage, 1976; Shay, 1982; Mueller et al., 1983). Heterokaryons can be isolated mechanically, usually by size (Karig Hohmann and Shows, 1979), or by growth as clones in a selection medium that prohibits propagation of parental cells with enzyme deficiencies, drug sensitivities, or temperature dependencies (Littlefield, 1964; Shows, 1972; Ringertz and Savage, 1976; Shay, 1982). Heterokaryons, the parental cells of which have selectable markers, survive after nuclear fusion in a selection medium by complementation of the genetic defects of one or both parental cells (Ruddle, 1972; Shay, 1982; Shows et al., 1982). Thus, a hybrid cell survives that has retained chromosome complements from both genomes. Independent hybrid clones are isolated from proliferating cell hybrids by established cell-cloning techniques (Puck et al., 1956). Details of whole-cell hybridization have been described (Ringertz and Savage, 1976; Shay, 1982).

Two genomes in the same cell were initially used for studying control of expression of differentiated products and for investigating genetic complementation when the two parental cells are deficient for the same phenotype but have mutations at different genes (Fincham, 1966; Davidson, 1974; Karig Hohmann and Shows, 1979; Honey et al., 1982; Mueller et al., 1983). However, use of these genome-transfer techniques for gene-mapping purposes proved to be an additional advantage. Generally, cell hybrids formed between parental cells of the same species or two established tissue-culture lines do not lose large numbers of chromosomes (Ephrussi, 1972; Ringertz and Savage, 1976). However, it was observed that in some combinations of cell hybrids, interspecific hybrids in particular, large numbers of chromosomes from one cell type are lost by unknown mechanisms. This makes it possible to map those genes of the parental type that loses chromosomes; the genes must be distinguishable between parental types. The observation of chromosome loss from only one of the parental cell types, described first by Weiss and Green (1967), has evolved into the dogma that in interspecific cell fusions, chromosomes from one of the species will be lost when the parental cells of that species are not derived from an established, infinitely growing cell line. In these combinations, chromosomes from the established cell line will be retained, but chromosomes from the other parental cell obtained from cells *in vivo* or from a finite growing cell-culture line will be lost. The coupling of gene-transfer and gene-mapping techniques has made it possible to map the genome of the species, the chromosomes of which are lost in cell hybrids. This parasexual strategy for mapping genes in mammals approximates classic parasexual procedures described for microorganisms.

3. SOMATIC-CELL HYBRIDS FOR GENE MAPPING

Procedures for obtaining man–rodent cell hybrids, the mapping of human genes, and the genetic dissection of complex enzyme and protein systems involved in metabolic disease and essential biological functions have been described (Ruddle, 1972; Shows, 1978, 1983a; Shows *et al.*, 1982). Normal or abnormal human cells such as fibroblasts, leukocytes, lymphoblastoid cells, cancer cells in suspension, or transformed cells are fused to rodent cells that possess enzyme deficiencies or other biochemical growth deficiencies that can be selected against. Human cells are fused with mouse or Chinese hamster monolayer cells using inactivated Sendai virus or polyethylene glycol fusing agents (Klebe *et al.*, 1970; Davidson and Gerald, 1976) and are plated for monolayer cell culture. The fused cells (heterokaryons) are subjected to a selection medium that prevents the growth of rodent cells. If the rodent parental cells are deficient, for example, for thymidine kinase (TK) or hypoxanthine phosphoribosyltransferase (HPRT), which function in DNA–RNA salvage pathways, they will not grow in hypoxanthine–aminopterin–thymidine (HAT) medium (Littlefield, 1964). Human cells are selected against, since human cells in suspension are removed with media changes, and human fibroblasts as a monolayer replicate slowly in low dilution. Hybrid cells are able to proliferate if nuclear fusion occurs and the genes for human HPRT or TK are retained to compensate for the HPRT or TK enzyme deficiencies of the rodent cells. Proliferating hybrid colonies derived from independent cell fusions are cloned from separate plates and expanded to large quantities of hybrid cells for analysis. The important characteristic of man–rodent cell hybrids for human genetic studies is that human but not rodent chromosomes are lost, which results in hybrid clones that have an infinite life span and retain different numbers and combinations of functioning human chromosomes (Weiss and Green, 1967). Loss of human chromosomes in man–mouse cell hybrids provides a parasexual system of chromosome "segregation" necessary to map human genes to specific chromosomes, to genetically dissect complex physiological systems, and to determine the organization of genes, which leads to an understanding of the human genome.

When the cells of an individual are investigated by hybridization studies and several independent clones are isolated, it is possible to obtain a population of primary hybrid clones in which the total human chromosome complement of an individual is represented but distributed in reduced numbers of chromosomes among the hybrid clones. These characteristics present a methodology to analyze genetically the somatic cells of an individual and dissect the human genome and metabolic disease (Shows, 1977; Shows *et al.*, 1982).

Development of this somatic-cell-hybrid strategy for mapping human genes has resulted in an exponential increase in the number of human genes assigned to specific chromosomes and regions of chromosomes. Any human gene or product that is present in cultured cells and that can be distinguished from a rodent homologous gene or product can be assigned to a specific human chromosome and syntenic group (Shows *et al.*, 1982, 1983a). This is accomplished by correlating the presence of a human gene product with the retention of a specific human chromosome. Using cells with human-chromosome rearrangements "segregating" in cell hybrids permits regional chromosome assignments when a gene marker and a chromosome region are concordantly retained or lost (Shows *et al.*, 1979, 1982). This capability for mapping human genes parasexually is the result of a very timely culmination of four technologies: growth and selection of mammalian cells in culture, gene-marker identification such as isozyme analyses, availability of cloned genes and DNA segments, and chromosome-identification and -banding techniques to distinguish chromosomes within and between species. Chromosome and regional assignment is further optimized by assembling chromosome-assignment panels of cell hybrids (Creagan and Ruddle, 1975) and, likewise, region-assignment panels. The chromosome-assignment panels consist of a minimum number of cell hybrids, each of which has a unique combination of human chromosomes. In such a panel, only a single chromosome correlates unambiguously with the distribution of the new marker in the hybrid panel. Thus, a gene is assigned to a specific chromosome. For a panel to be successful, a gene marker must be tested on samples correlated on the same cell passage with karyotype and gene-marker analyses. A logical extension to this approach is to construct and isolate regional assignment panels in which each cell hybrid in a set contains a different region of a specific chromosome. Twenty-four such panels of cell hybrids would contribute to the linear ordering of genes and to fine-structure mapping (Shows *et al.*, 1982). The chromosome rearrangements necessary for these panels are being obtained from the human population as translocations or deletions, often involving selectable markers, or from human cultured cells that have been irradiated (Goss and Harris, 1977a,b). (A variety of chromosomal rearrangements in cultured cells can be obtained from the Mutant Human Cell Repository in Camden, New Jersey.)

4. REGIONAL GENE MAPPING WITH CELL HYBRIDS

If parental cells with chromosome translocations or deletions are used in cell-hybridization studies, then genes can be assigned to specific chromosome regions that are "segregating" in cell hybrids (Shows et al., 1982). For regionally mapping human genes, it is possible to obtain several chromosomal abnormalities for each human chromosome from the human population. By obtaining a series of human cells with different lengths of a specific chromosome, human genes can be regionally mapped and linearly ordered using the somatic-cell strategy. Four human X-chromosome genes coding for glucose-6-phosphate dehydrogenase (G6PD), hypoxanthine phosphoribosyltransferase (HPRT), phosphoglycerate kinase (PGK), and α-galactosidase (GLA) have been used extensively for regional mapping studies and have been localized to small regions of the X chromosome. This was accomplished using X-autosome translocations in which regions of the X chromosome are translocated to different autosomes. Since HPRT is a selectable marker in the hypoxanthine–aminopterin–thymidine selection system, the region that encodes HPRT can be selected for and against. This enables regional mapping not only of the X but also of the autosome involved (Shows, 1979). A different method of determining close synteny and gene order of human genes reported by Goss and Harris (1977a,b) utilizes X-rays to induce chromosome breaks in human cells before fusion with mouse cells. This reduces the likelihood that any two syntenic or linked genes will be cotransferred, the probability being inversely proportional to the distance separating the genes. When irradiated human cells are fused with mouse cells having a growth requirement and cell hybrids are isolated in selection medium, different-sized fragments that encode the human selectable marker are retained. The cotransfer of syntenic genes gives results for linear order in good agreement with those obtained by linkage analysis.

The ability to transfer small amounts of DNA to recipient somatic cells in culture facilitates structural mapping of genes closely linked on chromosomes. Chromosome-mediated gene transfer (CMGT) using isolated metaphase chromosomes (McBride and Ozer, 1973) and isolated interphase DNA for DNA-mediated gene transfer (DMGT) (Wigler et al., 1979) can be used in transferring a selectable gene and closely linked genes to cultured somatic cells with a growth requirement (see Chapters 3, 4, and 13). With either technique, CMGT or DMGT, only a small portion of donor genetic material containing at least the selectable gene is ultimately stably retained in recipient cells with a growth requirement (Willecke et al., 1976). The closer the two genes are, the greater is the likelihood that they will both be transferred and stably retained in a recipient cell. The selectable marker thymidine kinase (TK) is very close to galactokinase (GALK) on human chromosome 17 in the q21–q22 region (Elsevier et al., 1974). When CMGT is used to transfer and select for TK, a number of recipient cells also retain the close marker GALK (Wullems et al., 1977). Since the band on chromosome 17 probably contains less than 5000 kilobases (kb) (Ruddle and

McBridge, 1977), the *TK* and *GALK* genes are estimated to be no more than 5000 kb apart. Mouse (*HPRT⁻*) recipient cells isolated after CMGT were found to contain human-donor *HPRT* and a cytologically detectable fragment of the X chromosome in different recipient cells, an unusual occurrence, since donor genetic material usually has not been physically detectable after CMGT (Miller and Ruddle, 1978; Klobutcher and Ruddle, 1979). These X-chromosome fragments were used to regionally localize the X-linked markers *HPRT, G6PD,* and *PGK*. These studies give a semiquantitative measure of physical distances between linked genes and allow gene order to be determined.

5. MARKERS FOR MAPPING THE HUMAN GENOME

The paucity of markers for mapping the human genome just 15 years ago has been transformed into a galaxy of markers that now promises to encompass the entire genome. For previous Mendelian genetic studies, it was necessary to rely on a relatively small number of polymorphic enzyme phenotypes, fairly common diseases, blood-group antigens, morphological traits, and sex-linked phenotypes (Renwick, 1971). Using these markers and chromosomal abnormalities, it was possible to assign only genes responsible for phenotypes known to be X-linked or associated with a particular chromosomal rearrangement to specific chromosomes; the latter represented a very small number. An equally small number of genes were demonstrated to be closely linked by statistical family studies and occasional molecular studies showing duplication of related genes and therefore close linkage. With the advent of gene transfer observed in man–rodent somatic-cell hybrids, the usable gene markers for mapping human genes increased exponentially because of the more than 80 million years of evolutionary divergence between man and mouse or Chinese hamster (Shows, 1977; Shows *et al.*, 1982). Virtually every gene product that can be identified in cultured somatic cells can be distinguished as being human or rodent because of structural or other differences. As a result of this large number of markers, and the cell-hybrid gene-transfer system to dissect and reconstruct the human genome, about 1300 human genes have been mapped (see Table I). Gene markers that have been used for human mapping studies represent a wide variety of gene functions, depending on whether a gene is identified by, for example, somatic-cell hybridization, biochemical–molecular, or recombinant-DNA methodologies. A large number of markers has been described, since man is the species most observed and studied. Many of the gene markers studied in families can be utilized in somatic-cell hybrids. The notable exceptions include blood groups, various diseases, and morphological traits; yet many unique markers are expressed in cultured somatic cells and cell hybrids.

The regional assignments of the β-globin locus in 1979 (Gusella *et al.*, 1979; Jeffreys *et al.*, 1979) and the insulin (Owerbach *et al.*, 1980a), growth hormone (Owerbach *et al.*, 1980b), and chorionic somatomammotropin (Owerbach *et al.*, 1980b) genes in 1980 were the result of combining restriction-endonuclease, recombinant-DNA, and somatic-cell hybrid methodologies. This heralded a new era in human gene mapping and made possible an enormous reservoir of gene markers. In fact, by joining the methodologies of recombinant DNA and somatic-cell hybridization with the isolation of single human chromosomes and family studies for identifying gene linkages, it will be possible to map virtually the majority of the human genome.

Currently, the most common genes assigned to human chromosomes are those that code for enzymes or structural polypeptides (Shows, 1983a), or cloned genes and DNA segments (Shows *et al.*, 1983a). Mapping these genes has been accomplished principally using the cell-hybrid methodology, although these markers can be mapped using other procedures. Gene-linkage studies using only family studies have utilized markers that comprise the blood groups, isozymes, serum proteins, clotting factors, sex-linked traits, diseases, and, most recently, DNA polymorphisms (Botstein *et al.*, 1980). To expand this capability, it is often necessary to combine several genetic methodologies to make an assignment; for example, blood groups not recognizable in cultured cells but known to be linked to enzyme markers or DNA polymorphisms from family studies can be assigned to chromosomes when the enzyme gene or DNA polymorphism is present and mapped in cell hybrids. Thus, most types of gene markers recognized in man have been mapped to a specific chromosome, demonstrating the efficiency and versatility of chromosome-assignment techniques.

5.1. Isozyme Markers

The somatic-cell genetics of a specific gene depends on whether or not it is expressed in tissue-culture cells and has species or allelic differences. A large number of proteins, generally of ubiquitous tissue distribution and involved in basic cell function and maintenance, are expressed in cultured cells, whereas relatively few proteins associated with a differentiated function are expressed (Davidson, 1974). Because of the amino acid differences in homologous proteins that occur among species, human enzymes that differ electrophoretically, antigenically, or kinetically from homologous mouse enzymes become instant markers in man–mouse cell hybrids (Shows, 1977). As long as enzymes can be identified among species by electrophoresis or by specific antigenic or kinetic differences, genetic characterization and mapping of genes that code for enzymes and their expression are possible for man, mouse, and Chinese hamster and any other species the chromosome "segregation" characteristics of which conform to the cell-hybrid strategy. In addition to the study of human genetics, mouse–Chinese hamster cell hybrids that "lose" either mouse or Chinese hamster chromosomes have been isolated (Minna *et al.*, 1976). Thus, the genetic mapping and genetic dissection of mouse and Chinese hamster enzymes can be determined in a parasexual way.

Electrophoretic procedures to separate enzymes (isozymes) have played a significant role in supplying a large number of markers for cell hybrids (Shows, 1983a). For example, after approximately 80×10^6 hybrid cells are grown and then homogenized or sonicated in 1 ml of a buffer, between 40 and 60 different enzymes can be determined on the same cell passage using starch gel or cellogel electrophoresis and specific histochemical staining techniques. Over 100 enzymes have been shown to have different electrophoretic mobilities when comparing man and mouse or man and Chinese hamster (Harris and Hopkinson, 1976; Shows, 1977, 1983a). Enzymes that differ represent markers suitable for genetic studies in cell hybrids. If a human isozyme is identified in a cell hybrid, then it is reasoned that the human gene that codes for this enzyme has been retained in the cell hybrid. The concordant segregation of an enzyme marker and a specific chromosome demonstrates assignment of the enzyme gene to that chromosome. When the presence or absence of a human enzyme in cell hybrids correlates with the appearance of other segregating human enzyme markers, their genes are considered to be syntenic (encoded) on the same chromosome. Following this strategy, enzyme genes have been assigned to each human chromosome (Shows *et al.*, 1982, 1984; Shows, 1983a) (see Table I).

Isozyme markers represent reliable markers for each chromosome that are easily tested by standard electrophoretic techniques, and the chromosome assignments for most have been confirmed by several laboratories (Shows, 1983a). When a new marker is being studied, its presence in cell hybrids can be compared with the presence of enzymes from a test marker panel, the genes of which have previously been mapped. Concordant segregation of a new gene marker and a test marker determines the chromosome assignment. Since human and rodent chromosomes can be distinguished, chromosome studies on cell hybrids identify the human chromosomes retained, and concordant segregation of a gene marker and a specific chromosome confirms the assignment predicted by cosegregation with the enzyme marker test panel. If parental cells that possess chromosomal rearrangements are used, then genes can be assigned to specific chromosomal regions using the same cosegregation strategy.

Mapping enzyme loci to specific chromosomal sites has made possible several observations about the genomic organization of human enzyme genes. In general, genes for enzymes that have similar function, subunit structure, cellular location, or clinical pathology or that are present in the same metabolic pathway have not as yet been found to be tightly clustered on specific chromosomes. For example, nuclear genes that code for soluble and mitochondrial forms of enzymes have not been localized on the same chromosome. Additionally, no chromosome has been identified as encoding a cluster of nuclear-controlled mitochondrial genes. Genes for mitochondrial enzymes assigned to the same chromosomes are far enough apart to exclude the possibility of a mitochondrial gene cluster (Shows, 1983a). Similarly, genes for enzymes that are localized in another cellular organelle, the lysosome, are not located on the same chromosome (Shows, 1977; Shows *et al.*, 1982). Genes for enzymes that act sequentially in a metabolic pathway are not clustered on the same chromosome. This has been observed, for example, for the glycolytic, citric acid, and complex glycolipid pathways, and for galactose metabolism and the urea cycle (Shows, 1977; Shows *et al.*, 1982). Genes that code for enzymes classified as, for example, kinases, transferases, dehydrogenases, or phosphatases have not been clustered on an individual chromosome, nor have genes for enzymes with a similar subunit structure (e.g., monomers, dimers, trimers, tetramers) or genes that code to nonidentical subunits of a specific enzyme (Shows, 1977). If there is an organization in the human genome for enzyme function, structure, or cellular location, the organization and control have not yet become clear; perhaps it is at a different level than at the chromosomal location of the functional structuring gene.

5.2. Molecular Markers

There is a great deal of optimism that the enormous complexity of the human genome will be dissected. This is highly feasible using a combination of several experimental approaches, including gene transfer. The most promising approaches are recombinant-DNA techniques that allow isolation of specific genes and DNA sequences, DNA sequencing to provide a precise nucleotide map, and chromosome-mapping techniques to determine specific-site gene location. All these approaches will complement each other in constructing a molecular or physical map of man (Shows *et al.*, 1982). It is no longer fanciful to speculate that soon the nucleotide sequence of large portions of human chromosomes, perhaps even whole chromosomes, will be determined using DNA-sequencing strategies. Human geneticists will eventually have molecular information and sequence data about the approximately 5000 kb of DNA that resides in a chromosome band. This could provide a molecular understanding of chromosome regions encompassed by a chromosome band or associated with neoplasia and inherited disease. In fact, this information will be obtained in most cases before the functional significance of a DNA region is determined. Gene function will continue to be investigated in cultured cells in conjunction with gene-transfer techniques.

Somatic-cell hybrids have been used to chromosomally assign genes by following the segregation of gene function or products. However, mapping with cloned probes overcomes a major drawback to this method of gene assignment: Gene expression is required to map a protein, but the gene itself can be assigned directly using a cloned DNA probe. Mapping structural gene sequences directly overcomes an uncertainty that a regulatory gene is being assigned in hybrids. Many genes that code for proteins characteristic of differentiated cells are not expressed in hybrid cells. These same genes can be chromosomally assigned using cloned probes (cf. Shows *et al.*, 1982). Over 250 genes and 550 DNA segments have been cloned into plasmids and mapped to specific chromosomal regions (Shows, 1983b; Shows et al., 1983a; Willard et al., 1985).

The methodology for mapping cloned genes by cell-hybrid gene-transfer techniques is reviewed in Shows et al. (1982). Somatic-cell hybrids are constructed by fusing rodent cells with human leukocytes, fibroblasts, or other cell types as presented earlier. DNA is isolated (Naylor et al., 1983) from the same cell-hybrid passage as are chromosome preparations and enzyme homogenates. The DNAs are digested with a restriction endonuclease, and the fragments are separated by electrophoresis in agarose (Owerbach et al., 1980a). The restriction endonuclease chosen is one that produces at least one human DNA fragment with a unique mobility compared to the mouse DNA fragments that hybridize with the probe (Shows et al., 1982). This requirement is usually observed, since there has been considerable divergence in nucleotide sequence in man and mouse in and around homologous genes. Of course, if the human gene probe does not hybridize to mouse DNA, this is no longer a consideration. It should be mentioned that human genes can be mapped using probes derived from other species as long as there is sufficient sequence homology to allow detection by molecular hybridization (Shows et al., 1982; Lally and McKusick, 1985).

To test a cloned probe for mapping purposes, several restriction enzymes (usually EcoRI, BamHI, HindIII, MspI, and Tac) are screened for differences in the size of human and mouse DNA fragments, principally because these enzymes have yielded usable patterns on Southern blots. Once a suitable restriction enzyme is chosen, DNAs from a panel of hybrids are analyzed by Southern blotting using a gene probe that has been radioactively labeled with ^{32}P by nick translation (cf. Shows et al., 1982). The cell hybrids are chosen so that collectively all the human chromosomes are represented among the cell hybrids. The hybridized blots are exposed to X-ray film, which is developed to reveal specific bands that represent fragments of DNA that have hybridized to the probe. A correlation between the presence of a specific hybridization band and a single human chromosome or isozyme marker assigns the gene to a specific chromosome.

The function of the vast majority of human DNA is still undefined. Therefore, most of the clones that comprise a recombinant-DNA library of the human genome are "anonymous." These clones can be used as genetic markers that when mapped serve as genetic landmarks for a specific chromosomal region. Cloned probes of undetermined coding capacity have been isolated and used to identify specific segments on each human chromosome (see Table I) (Willard et al., 1985). Over 550 of these anonymous DNA segments have been mapped in man and serve as excellent chromosomal markers (Table I). Determining the presence of a human undefined DNA segment by hybridization to cell-hybrid chromosomal DNA does not require a gene product to be expressed. Thus, that the coding capacity of a single-copy probe is not known does not diminish its importance as a gene marker. Once an undefined probe has been mapped to a specific chromosome, it becomes a useful marker for genetic studies.

6. HIGH-RESOLUTION CHROMOSOME MAPPING

A precise fine-structure human gene map is a prerequisite for understanding the molecular genetics of human biology and disease. There are 24 different chromosomes and, thus, 24 nuclear gene maps. Eventually, all these maps will be deciphered, an accomplishment that will be of inestimable value in understanding the molecular biology of the human condition. To achieve this end, it is essential to map as many human genetic markers as possible and to assign them to a specific site on a chromosome and, ultimately, to a specific nucleotide sequence on a chromosome.

The advent of recombinant-DNA technology has given the human geneticist a new set of molecular markers that theoretically span the entire genome. These markers are the genes themselves, as well as undefined DNA segments that have been cloned; they represent very large numbers of markers for genetic studies. In fact, about one half the markers currently mapped in man are a result of cloned-DNA probes (see Table I). In addition to providing this enormous and profitable set of new markers, molecular technology has provided ways of using these new markers for high-resolution gene mapping. In these initial phases of study, the degree of resolution is not as fine as will ultimately be required. However, the technology centering around *in situ* hybridization is capable of identifying sites at which genes are encoded in bands and subbands of stained human chromosomes observed at the light-microscopic level (Harper and Saunders, 1981; Shows *et al.*, 1983b; Zabel *et al.*, 1983). Specific-site chromosome assignment depends on having cloned probes without repetitive sequences of specific genes and undefined DNA segments. The radiolabeled probes are hybridized to homologous sequences on metaphase chromosomes; after exposure to autoradiographic emulsion, silver grains identify the chromosomal site (cf. Shows *et al.*, 1983b). In addition to chromosomal *in situ* hybridization, mapping of genes and gene products to chromosomal bands and subbands is made possible by the use of a region-assignment panel of human–mouse cell hybrids, in which each hybrid retains only a segment of a specific human chromosome. At the current level of high-resolution mapping, there are principally these two strategies: (1) smallest chromosomal segment obtained usually by gene-transfer methods utilizing cell hybrids and chromosome rearrangements, and (2) *in situ* chromosomal hybridization. For each method, the smallest chromosomal site for mapping is about the size of a chromosomal band, or about 5000 kb.

6.1. Smallest Chromosomal Segment

Of the techniques available for assigning a human gene to the smallest chromosomal segment (SCS), the method of rearranged chromosomes segregating in cell hybrids and the transfer of genes on a defined chromosomal segment into recipient cells have yielded the most information. In human–rodent cell-hybridization studies, if parental human cells have a chromosomal translocation or deletion, then human genes can be assigned to that chromosomal region, as discussed above. For determining the SCS, a series of independent cell hybrids with different lengths of a specific chromosome are tested for the marker, and in positive hybrids, the smallest segment of chromosomal overlap determines the SCS (Shows, 1979; Shows *et al.*, 1982, 1983b). The most-documented fine-resolution mapping of human genes by the SCS methodology are *PGM1* (phosphoglucomutase) localized to the p22.1 site (Cook and Hamerton, 1982) and the X-linked *G6PD* (glucose-6-phosphate dehydrogenase) gene localized to the p28 site (Shows and Sakaguchi, 1980). Because insufficiently large numbers of chromosomal rearrangements are known for many chromosomes, a host of genes have been mapped only to large segments. This method of mapping to the SCS is thus limited to the availability of large numbers of rearrangements for each human chromosome.

Using microcell-mediated gene transfer, chromosome-mediated gene transfer, gene transfer with defined segments of chromosomes and DNA, and vector-mediated gene transfer, the SCSs of a few genes have been determined (see Chapters 3–5 and 13). These systems have relied on using selectable markers and markers on a region of a chromosome that has been translocated to a chromosome that encodes a selectable marker. Currently, these methodologies are limited to the few selectable markers known and their closely syntenic markers.

6.2. *In Situ* Hybridization

Recent advances in *in situ* hybridization techniques have made it possible to map single-copy DNA sequences to specific chromosomal sites directly on fixed human metaphase chromosomes (Harper and Saunders, 1981; Shows *et al.*, 1983b; Zabel *et al.*, 1983, 1985) (the procedure has been diagrammed in Shows *et al.*, 1983b). The procedure increases the sensitivity of fine-structure gene mapping by combining advances in molecular techniques with high-resolution chromosomal banding to identify the chromosomal site that encodes a gene. Features that have increased the resolving power include using large prometaphase chromosomes from synchronized lymphocyte cultures, high-specific-activity DNA probes, dextran sulfate to promote formation of probe networks, and high-resolution chromosome-banding techniques through the emulsion. Utilizing this current methodology, about 100 genes have been mapped to specific chromosomal sites, making this methodology a major contributor to mapping and identifying genes after gene transfer and chromosomal insertion.

We have found that the procedure detailed in Zabel *et al.* (1983) and Shows *et al.* (1983b) increases our resolution of *in situ* hybridization and specific-site mapping. This procedure employs two methods that are different from that described by Harper and Saunders (1981). One method is to obtain longer prometaphase chromosomes with a distinct chromosomal banding pattern of up to 1000 bands, which is essential for the precise localization of a DNA probe.

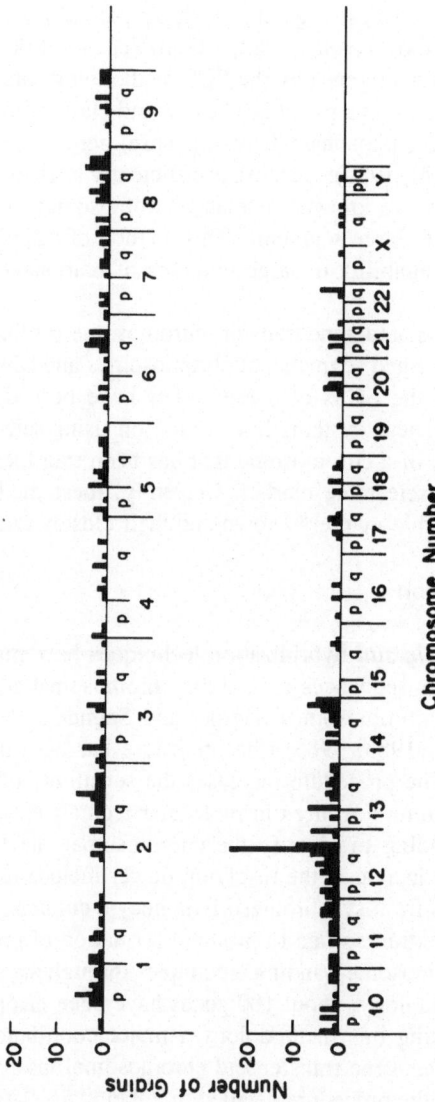

FIGURE 1. Diagram showing the grain distribution in 189 metaphases for the *IGF1* probe. Abscissa: relative size of the chromosomes; ordinate: number of silver grains (see the text).

Large prometaphase chromosomes from leukocytes were obtained using bromodeoxyuridine as the cell-synchronizing agent. This was coupled with the second modification, which uses a reliable Hoechst 33258–Giemsa chromosome staining method at the end of the *in situ* hybridization procedure to produce a high-resolution G-banding pattern on long prometaphase chromosomes. The preparation of chromosomes, radiolabeling of DNA probes, *in situ* hybridization of DNA probes with chromosomes, autoradiography, and staining for G-banding have been described (Zabel *et al.*, 1983, 1985).

In situ hybridization, previously limited to the detection of repetitive gene families, now allows the identification of single-copy genes and the dissection and mapping of genes within a family. For example, the genes that encode the related insulinlike growth factors (IGF) I and II have been mapped to specific sites on human chromosomes 12 and 11, respectively (cf. Tricoli *et al.*, 1984; Morton *et al.*, 1986). IGF I and IGF II are single-chain serum proteins of 70 and 67 amino acids that possess 62% sequence identity and are probably required for normal growth and development. These proteins also share a 47% homology with insulin in the *A* and *B* regions. We and others have previously chromosomally assigned these genes using somatic-cell hybrids (cf. Tricoli *et al.*, 1984; Morton *et al.*, 1986). Knowing the specific chromosomal site of genes that code for growth factors is important for understanding their role in growth control. Their proximity to other growth-control genes, such as protooncogenes or sites involved in chromosomal rearrangements, could suggest mechanisms for abnormal expression involved in establishing and maintaining the malignant state.

Using complementary DNA probes for IGF I and IGF II and the *in situ* technology discussed above, *IGF1* was assigned to human chromosome 12 because a large proportion of metaphases had silver grains located on the long arm of this chromosome. It can be seen in Fig. 1 that the largest number of silver

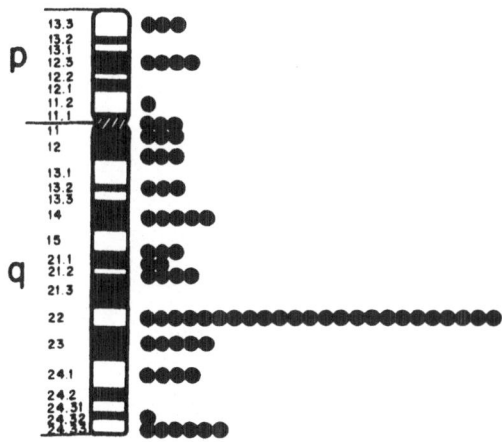

FIGURE 2. Idiogram of human chromosome 12 showing localization of silver grains to q22 on metaphase chromosomes following hybridization of the *IGF1* probe (see the text).

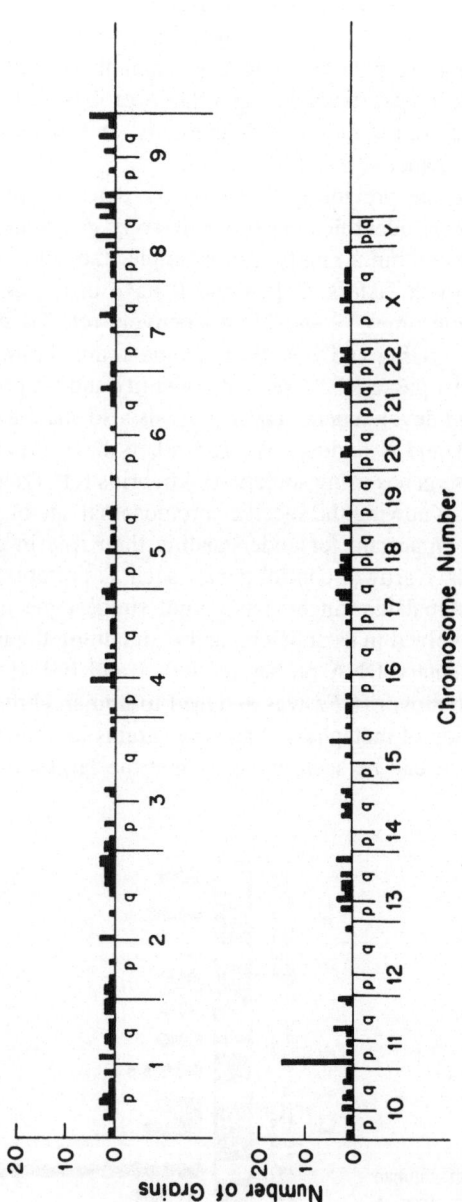

FIGURE 3. Diagram showing grain distribution in 109 metaphases for the *IGF2* probe (see the text).

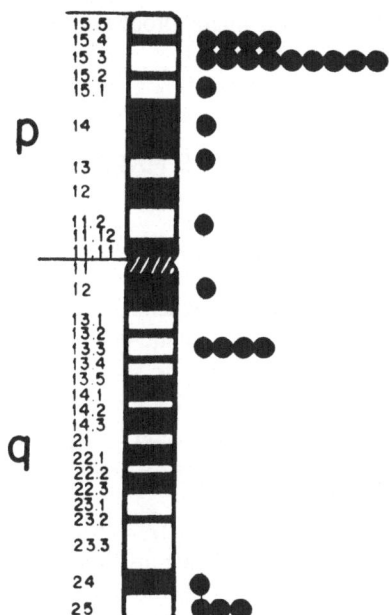

FIGURE 4. Idiogram of human chromosome 11 showing localization of *IGF2* to the 11p15.5–p15.3 region (see the text).

grains were located over the long arm (q) of chromosome 12. This confirms the somatic-cell-hybrid, gene-assignment results. The grain distribution on chromosome 12 locates *IGF1* at q22 (Fig. 2). Of 189 metaphases, 13% had a grain at 12q22, 5% of the total number of grains were located at this band, and 32% of chromosome 12 grains were at q22; no other sites were observed above the background level. Therefore, *IGF1* is located at the q22 band on human chromosome 12. In a similar fashion, *IGF2* was assigned to the short arm (p) of chromosome 11, as observed from the grain distribution displayed in Fig. 3. This confirms previous somatic-cell-hybrid mapping results. The grain distribution over chromosome 11 revealed *IGF2* to be encoded in the p15 short-arm region and, more precisely, in the p15.5–p15.3 subband region (Fig. 4). Of 109 metaphases, 12% were observed with a grain at 11p15.5–p15.3, 5% of the total grains were located at 11p15.5–p15.3, and 50% of grains on chromosome 11 were at p15.5–p15.3. No other sites were observed above the level of background. Thus, *IGF2* is located at 11p15 in the p15.5–p15.3 subband region. The gene for the related insulin gene has been located in the p15 band (cf. Zabel *et al.*, 1985). These results demonstrate not only that single-copy genes can be mapped using this technology but also that highly related genes can be distinguished and mapped to their respective chromosomal sites. This technology represents a major procedure for mapping genes and the sites of insertion for transferred genes.

7. THE 24 GENE MAPS

There are three major strategies for assigning genes to human chromosomes and to specific sites on chromosomes: (1) close genetic linkage to an already mapped gene (Conneally and Rivas, 1980), (2) the parasexual approach using human–rodent somatic-cell hybrids, and (3) *in situ* hybridization using DNA probes to identify the specific chromosomal site. These are the principal technologies that have been employed to generate the human nuclear map in Table I; there are a few additional biochemical–molecular strategies that represent specialized procedures, such as cotransfer of genes and nucleotide sequencing for close linkage.

The more than 1300 markers assigned to the 24 human chromosomes using these methodologies are listed by chromosome in Table I, pp. 36–74. About 700 genes with specific functions have been assigned to specific chromosomes, and approximately 600 additional loci representing cloned human anonymous DNA sequences, many identifying DNA polymorphisms, have been mapped and are also listed in Table I. Of the 600 anonymous DNA sequences, about 300 represent arbitrary DNA fragments that have been chromosomally assigned but do not identify a gene or DNA polymorphism. Although only a small fraction of the total number of human genes have been identified to date, the majority of these genes have been mapped within the last ten years as a result of gene-transfer, recombinant-DNA, and gene-linkage methodologies. With these totals, it is estimated that about 1% of all human functional genes have been identified and mapped and that, including the anonymous DNA fragments, about 2% of the human genome now has markers for mapping. In contrast, the entire human mitochondrial genome has been mapped. There are 37 genes encoded in a circular DNA molecule with several unique features that distinguish it from nuclear chromosomes (Anderson *et al.,* 1981). It will be necessary to understand the human nuclear genome at near this level of resolution to begin to understand human biology and disease.

The chromosome assignments and specific locations of human genes presented in Table I are derived from eight Human Gene Mapping Workshops (Human Gene Mapping Workshop 8, 1985). The original reference for each gene assignment and regional localization has been cited in the workshop reports. The genes are arranged by chromosome and are listed in a tentative linear order from the short-arm terminus (pter) to the long-arm terminus (qter). The map location is based on the most consistent smallest region of overlap using rearranged chromosomes that segregate in cell hybrids or *in situ* hybridization. The chromosome-banding pattern that identifies the map location follows the *International System for Human Cytogenetic Nomenclature* (Harnden and Klinger, 1985). Gene symbols have been standardized in accordance with gene-nomenclature guidelines proposed by the Human Gene Nomenclature Committee and were accepted by Human Gene Mapping Workshop 6 (Shows *et al.*, 1979). This standardization provides a uniform system for human gene nomenclature and is referred to as the International System for Human Gene Nomenclature (1979), or ISGN (1979) (Shows *et al.*, 1979). Amendments to this system have been published in Human Gene Mapping Workshops 6, 7, and 8 (Shows and Mc-Alpine, 1982; Shows *et al.*, 1984; McAlpine *et al.*, 1985). The genes and DNA segments that have been cloned are identified. Those DNA segments that are listed as polymorphic identify loci with allelic variants in the population for gene-linkage studies. Other loci that are polymorphic can be found in Willard *et al.*, (1985), Shows *et al.* (1982), and chromosome committee reports (Human Gene Mapping Workshop 8, 1985). The mode in Table I indicates how the gene was mapped: (S) somatic-cell hybridization; (F) Mendelian family studies; (RE) recombinant methods, usually in conjunction with somatic-cell hybridization; (A) annealing of homologous sequences as in *in situ* hybridization; (D) assignment by gene dosage; (CH) assignment by cytogenetic analysis. The majority of human genes have been mapped utilizing the parasexual strategy of cell hybridization and gene transfer. The status indicates the strength of the assignment—(C) confirmed, (P) provisional, (I) inconsistent, and (T) tentative—but this rating does not necessarily relate to the status of the regional location on a particular chromosome. A confirmed assignment is indicated when two or more independent reports assign a gene to the same chromosome and region; a provisional assignment acknowledges an assignment from a single report; an inconsistent assignment occurs when assignment results disagree. For a listing of human genes alphabetically by symbol and gene name, see McAlpine *et al.* (1985). This last report includes the E.C. number of enzymes and the McKusick number (McKusick, 1982), which leads one to a description of the gene, disease association, and references. A few chromosome markers have not been included in Table I. They are the fragile sites and chromosomal breakpoints associated with cancer, a list of which can be found in Berger *et al.* (1985). Also not listed are a few DNA sequences that recognize repetitive DNA segments and small undefined families of homologous sequences found on multiple chromosomes. These DNA markers are listed by Willard *et al.* (1985).

The contribution of cell hybridization to the mapping of the human genome is considerable. The cell-hybrid parasexual gene-mapping strategy revolutionized human genetics and provided the stimulation to identify the genes on each chromosome and their order. Coupled with the human–rodent cell-hybrid chromosomal characteristics that provided the rapid construction of gene maps was the recombinant-DNA methodology that identified genes and DNA sequences without first recognizing the function of the gene. These technologies could then be combined with classic genetics, including family studies, gene dosage, and genetic polymorphisms, to develop the gene map into a dynamic compilation of genetic information that is leading to an understanding of the organization of the human genome, human biology, and disease.

7.1. Chromosomal Landmarks

It is clear that knowing where important genes are located and how they function will play a significant role in gene transfer. There are certain genes that have proven to be particularly useful in human genetic studies and that serve as benchmarks for each chromosome. These markers and certain characteristics will be identified and briefly discussed for each chromosome. The most markers, over 300, have been assigned to the X chromosome; these represent about a fourth of the mapped markers. The chromosome with the second largest number of markers is 11, with 91 markers assigned. The number of markers per chromosome ranges from 18 to 321, and each chromosome encodes significant genes and polymorphic variants suitable for family, somatic-cell genetic, and gene-transfer studies.

7.1.1. Chromosome 1

There is a total of 80 gene markers on chromosome 1. Of these markers, 45 are located on the short arm and 24 on the long arm. Eleven markers have not been assigned to either arm as yet. These gene markers represent a large variety of functions. They include blood group, enzyme, protein, blood clotting, antibody, viral, protooncogene, cloned gene, DNA segment, growth factor, hormone, pseudogene, and several different types of disease markers. Each chromosome in the genome encodes the same broad array of gene functions. Polymorphic markers exist at several locations on both long and short arms, providing useful markers for gene-linkage studies. These allelic markers are likely to be segregating in families with an unknown inherited disease. Such genes that are variant in the population are *AMY1, AMY2, AT3, CAE, D1S2, D1S4, EL1, ENO1, FUCA1, FY, GALE, GDH, PEPC, PGD, PGM1, RD, RH, SC,* and *UMPK*. With this array of markers segregating in families, extensive genetic studies at the Mendelian, somatic-cell, and molecular levels are being carried out and illustrate the usefulness of a detailed map of each chromosome. Such information contributes to the molecular organization of the genome, genetic counseling, prenatal diagnosis, and possible gene therapy.

For cultured somatic and hybrid cells, *AK2, ENO1, FH, FUCA1, GALE, PEPC, PGD, PGM1,* and *UMPK* are excellent functional markers. Of course, all cloned-gene and DNA-fragment probes are appropriate markers when a functional assay is not required. Molecular markers are on both chromosome arms, many are polymorphic, and they are very useful and essential for somatic-cell genetic and cell-hybrid molecular studies.

This chromosome encodes several important growth control genes; they are nerve growth factor (*NGFB*) and the protooncogenes neuroblastoma RAS viral oncogene homologue (*NRAS*), avian myelocytomatosis viral oncogene homologue (*LMYC*), and avian sarcoma viral oncogene homologue (*SKI*). The extensive genetic information on this chromosome has allowed a probable gene order that will provide excellent benchmarks for genetic studies. Additional information on chromosome 1 and references have been detailed by Povey *et al.* (1985).

7.1.2. Chromosome 2

A total of 49 gene markers have been assigned to the second largest of the human chromosomes: 20 on the short arm, 15 on the long arm, and 14 not yet assigned to either arm. Few polymorphic markers have been assigned to this chromosome. Examples are *ACP1* on the short arm; the DNA polymorphisms *D2S1, D2S3, D2S5,* and *D2S6; IGK,* a κ-chain marker; and the Kidd and Colton blood groups. Many genes assigned to this chromosome are not yet useful in Mendelian genetic studies; however, several cloned-gene probes have been assigned for such important genes as *POMC, APOB, NMYC, TGFA, IGK, IL1,* collagens, *FN1, GCG,* and *REL*. A detailed gene order is developing for this chromosome through cell-hybrid regional mapping and *in situ* hybridization. There are not a large number of functional gene markers available for somatic-cell genetic studies; however, several excellent cloned probes for significant genes are available. Additional details have been reported by Povey *et al.* (1985).

7.1.3. Chromosome 3

There are 32 gene markers assigned to this chromosome: 6 are short-arm markers, 12 are long-arm markers, and 14 have not been assigned to either arm. Several important genes have been assigned to this chromosome; they are β-galactosidase, aminoacylase, the *RAF1* protooncogene, glutathione peroxidase 1, transferrin and its receptor, somatostatin, cholinesterase 1, and ceruloplasmin. The genes *ACY1* and *GLB1;* the DNA polymorphisms *D3S1, D3S2, D3S3, D3S4, D3S5, D3S6, D3S8, D3S9,* and *D3S10;* and *GPX1, TF, SST, CHE1,* and *CP* are the most useful on this chromosome for mapping studies in both family and cell genetic studies. Deletions on the short arm have been associated with small-cell lung carcinoma. It is of interest that the genes for transferrin and its receptor are both located on this chromosome in the same long-arm region. Other mapping information can be found in Kidd and Gusella (1985).

7.1.4. Chromosome 4

There are 68 gene markers that have been assigned to this chromosome; 15 are likely on the short arm, 33 have been assigned to the long arm, and 3 markers are near the centromere without the arm being known. Another 17 markers have not been localized. Recently, the gene that codes for Huntington's disease was assigned to the short arm on the basis of its close linkage with *D4S10* in family studies. The search for the location of this gene has stimulated the mapping of new markers on this chromosome, the localization of old markers, and the identification of DNA polymorphisms for family studies. A considerable number of DNA probes that recognize restriction-fragment-length polymorphisms have been isolated for these studies. Several important genes have been mapped to this chromosome; they include *AFP, ALB, HPAFP, GC, PGM2, PEPS, ADH, AGA, EGF, IL2*, the fibrinogen polypeptides, the *MNS* blood group, and the *RAF2* oncogene. Several cloned genes have been mapped, but there are few genes as yet available for studies at the functional level in cultured cells or cell hybrids. On the other hand, there are numerous DNA markers that are polymorphic and that are excellent markers for gene-linkage family studies. Additional mapping formation has been published (Kidd and Gusella, 1985).

7.1.5. Chromosome 5

There are 28 gene markers on this chromosome: 6 on the p arm, 11 on the q arm, and 11 not localized. Two rare, inherited lysosomal storage disorders are located on this chromosome; they are β-hexosaminidase-B (*HEXB*), which is deficient in Sandhoff–Jatzkewitz disease, and arylsulfatase-B (*ARSB*), which is deficient in Maroteaux–Lamy syndrome. The gene that codes for dihydrofolate reductase and is amplified in transformed cells is located on this chromosome, as is the gene for the *FMS* protooncogene. Several genes on this chromosome have been cloned, and a few DNA segments that recognize polymorphisms are suitable for family studies. There are two receptor genes and two transfer RNA (tRNA) synthetase genes. The diphtheria toxin sensitivity locus is a good marker in somatic-cell studies, and its function can be utilized in a selection system against chromosome 5. See Lamm and Olaisen (1985) for mapping details.

7.1.6. Chromosome 6

A total of 64 gene markers have been assigned to this chromosome. Twenty-nine markers have been assigned to the short arm and another 6 are probably located on this arm, while only 7 markers have been assigned to the long arm with another 2 markers probably being located on the q arm. Twenty additional markers have not yet been assigned to either arm. The markers are concentrated on the short arm of this chromosome. The predominant group of markers on 6p is the major histocompatibility complex, which includes the *HLA* loci, the complement components, glyoxalase 1 (*GLO1*), and the properdin factor B (*BF*). These markers are highly polymorphic, which increases the possibility of finding heterozygosity in families segregating rare Mendelian traits, hence the preponderance of markers on the short arm. Rare traits that have been linked, or provisionally linked, to the *HLA* loci by family studies are congenital adrenal hyperplasia (*CAH*) and hemochromatosis (*HFE*). This degree of polymorphism has allowed linear ordering of the genes by recombination studies in families and chromosome localization by chromosomal rearrangements in cell-hybrid studies. Other polymorphic loci on chromosome 6 are phosphoglucomutase-3 (*PGM3*), the P blood group, and several RFLPs—D6S2-4 and D6S6-11. Reliable markers for somatic-cell functional genetic studies are *GLO* and *HLA* loci on the short arm and *PGM3, ME1* (malic enzyme-1), and *SOD2* (superoxide dismutase-2) on the long arm. Several DNA segments and cloned genes including growth-control genes such as oncogenes and hormones have been assigned using somatic-cell hybrids, the recombinant-DNA strategy, or *in situ* hybridization. A further detailed account of the chromosome 6 genes and fragile sites has been published (Lamm and Olaisen, 1985).

7.1.7. Chromosome 7

There are 50 gene markers that have been assigned to this chromosome: 9 have been localized on the short arm and 20 on the long arm; another 21 need further localization. This chromosome encodes genes with diverse functions representative of many of the different activities of a cell. Markers encoded on 7 are involved in amino acid metabolism, nucleic acid metabolism, blood groups, structural proteins, membrane proteins, chromosomal proteins, hormones, mitochondria, surface antigens, the lysosome, and developmental markers. Recently, several important genes have been assigned to this chromosome using a combination of cell hybrids and cloned DNA probes. These genes code for cystic fibrosis, T-cell receptors, neuropeptide Y, epidermal growth factor receptor, and P glycoprotein. Several DNA segments that recognized DNA polymorphisms have been isolated that are suitable for family studies. For somatic-cell gene-function studies, *GUSB, PSP, UP, MDH2,* and *AS* are reliable markers that identify this chromosome. Additional information on map details has been reported (Smith and Spence, 1985).

7.1.8. Chromosome 8

Only 18 human gene markers have been assigned to this chromosome. Four of these are located on the p arm, 7 on the q arm, and the remaining 7 markers need to be localized. Although chromosomes 8, 9, 10, 11, and 12 are similar in size, fewer markers have been assigned to chromosome 8, whereas the other chromosomes have 2–3 times the number of markers assigned. Two important protooncogenes are encoded on this chromosome, *MYC* and *MOS,* along with the genes for the carbonic anhydrase gene family and plasminogen activator and thyroglobulin. This chromosome does carry the polymorphic markers glutathione reductase (*GSR*) and carbonic anhydrase 2 (*CA2*), which are informative in family studies. A fibronectin marker (*FNZ*) appears to control the retention of *FN* on the cell surface, which may be important in metastasizing cells, some of which do not express the *FN* structural gene encoded on chromosome 2. Additional information has been reported (Smith and Spence, 1985).

7.1.9. Chromosome 9

A total of 33 gene markers have been assigned to this chromosome; 8 have been located on the short arm, 14 on the long arm, and another 11 have not been localized to either arm. The important ABO blood group marker is located at the q terminus (q34) of this chromosome. For family studies, this marker and adenylate kinase-1 (*AK1*) are the most useful polymorphic markers on the long arm, while *ACO1* (aconitase-1) and *GALT* (galactose-1-phosphate uridylyltransferase) are polymorphic in the population and encoded on the short arm. *D9S1, D9S3,* and *D9S4* are DNA probes that recognize polymorphisms. It is of evolutionary interest that both fibroblast and leukocyte interferon structural genes are located on this chromosome. These and the *ACO, ASS, AK1,* and *AK3* markers are very useful for cultured-cell studies. Further mapping information has been reported (Smith and Spence, 1985).

7.1.10. Chromosome 10

A total of 25 gene markers have been assigned to chromosome 10. Three have been localized to the p arm and 6 to the q arm; 16 markers are not yet regionally assigned. Reliable functional markers that are used to identify the chromosome in cell genetic studies are *GOT1, NEU, HK1, ADK, GSAS, OAT, LIPA,* and *PP*. For somatic-cell genetic studies, there is a curious nonviral gene called *FUSE* involved in the important property of fusing cells. Important cloned genes assigned to this chromosome are *IL2R, PLAU, GLUD,* and *VIM*. There are few polymorphic loci suitable for family studies; they are *GOT1, D10S1,* and *D10S3*. Further details have been reported by Grzeschik and Kazazian (1985).

7.1.11. Chromosome 11

There are 91 loci that have been assigned to chromosome 11; only the X chromosome has had more loci assigned. There are 45 markers mapped to regions on the short arm, and 26 loci are mapped on the long arm. There are 6 markers that have been mapped near the centromere, but the arm on which they are located is not known. Fourteen additional markers have been assigned to the chromosome, but regional mapping is not as yet known. The chromosome is being mapped extensively because a number of significant genes and diseases have been mapped to this chromosome. The mapping studies have generated a large number of markers on the short arm. This mapping has been stimulated by the localization of the β-globin gene family (*HBB, HBBP1, HBD, HBE1, HBG1, HBG2*), the insulin gene family (*INS, IGF2*), the deletion at p13 associated with Wilms' tumor, Beckwith–Wiedemann syndrome, the hormone genes *FSHB* and *PTH*, with the Harvey *RAS* protooncogene to the distal portion of the short arm. To isolate markers to study the various disorders associated with these genes, a large number of DNA-fragment probes have been isolated from the short arm. Many of these probes recognize fragment-length polymorphisms and are important for family studies and for prenatal diagnostic studies. While the long arm has not had as many markers assigned to it, it does encode important genes such as the apolipoprotein cluster (*APOA1, APOA4, APOC3*), the pepsinogen cluster (*PGA3, PGA4, PGA5*), the protooncogene *ETS1*, alpha glucosidase, and uroporphyrinogen I synthetase. Very few polymorphic markers exist for linkage studies on the long arm. A large number of markers exist on this chromosome that are expressed in cultured cells for cell genetic studies. It is interesting that there are two sets (*LDHA, HRAS1* and *IGF2, TH*) of genes on the short arm of 11, one of which (*LDHA, HRAS1*) has homologues on the short arm of chromosome 12 (*LDHB, KRAS2*) and the other of which (*IGF2, TH*) has homologues on the long arm of chromosome 12 (*IGF1, PAH*). These examples of conserved evolution point to a common ancestor for chromosomes 11 and 12 and a pericentric inversion involving chromosome 12. Further information has been reported (Grzeschik and Kazazian, 1985).

7.1.12. Chromosome 12

There have been 49 gene markers assigned to this chromosome. Sixteen have been mapped to short-arm regions and 17 to long-arm regions; 16 are yet to be localized. Important genes that have been mapped to this chromosome are *PAH* deficient in phenylketonuria, the coagulation factor (*F8*), *TPI, LDHB, IFNG, A2M*, and the growth-control genes *KRAS2* and insulinlike growth factor 1 (*IGF1*). Although genetic variants have been described for a few of the loci encoded on this chromosome, e.g., *PEPB, TPI, PAH*, and *LDHB*, none is at a level of polymorphism that has played an extensive role in gene-linkage studies. There are several polymorphic loci that are suitable markers for gene-linkage studies; they are *KRAS2, IGF1*, and several DNA polymorphisms. On this chromosome are such loci as *LDHB, PEPB, GAPD*, and *TPI*, which are excellent markers for cultured cells. Further mapping details have been reported by Grzeschik and Kazazian (1985).

7.1.13. Chromosome 13

Forty gene markers have been assigned to this acrocentric chromosome. Only 2 markers have been mapped to the short arm, while 30 markers have been assigned to the long arm; 8 markers are yet to be localized. Retinoblastoma has been assigned to the q14.1 region and is closely linked to *ESD*. A sizable number of DNA-segment probes have been generated for this chromosome, stimulated by the goal to find markers that are closely linked to retinoblastoma. These DNA polymorphisms have been utilized to generate a linkage map and linear order of DNA-sequence probes on the long arm. Markers that are useful for linkage studies are *ESD, RB1, D13S1, D13S2, D13S3, D13S4, D13S5,* and *D13S10*. A large number of cloned probes exist for this chromosome, but there are few functional genes such as *ESD* available for cultured-cell studies. Some important additional genes that have been mapped are a collagen gene, two coagulation factors, osteosarcoma, and Wilson's disease. Additional genetic information has been reported (Cox and Gedde-Dahl, 1985).

7.1.14. Chromosome 14

On this acrocentric chromosome reside 41 gene markers: 1 located on p and 25 located on q. Fifteen markers have been assigned to this chromosome but not yet located to a region. The ribosomal DNA (rDNA) cluster of repeated genes is located on the short arm, and there are several important genes that have been located on the long arm, e.g., the immunoglobulin heavy-chain family, two oncogenes (*FOS* and *AKT1*), a tRNA synthetase (*WARS*), nucleoside phosphorylase, and several DNA polymorphisms. Several genes are excellent markers for population and family studies, such as *NP, PI, IGHG, CKB,* and *D14S1*. The reliable *NP* marker has been used extensively to identify this chromosome in cultured cells. Although several markers are now assigned to this chromosome, their linear order is not well defined. See Cox and Gedde-Dahl (1985) for further details regarding this chromosome.

7.1.15. Chromosome 15

There are 31 gene markers assigned to this chromosome: 1 on the short arm, 19 on the long arm, and 11 as yet not regionally localized. The rDNA cluster of genes is encoded on the short arm, with several useful markers on the long arm. Of these, *HEXA*, altered in Tay–Sachs disease, β_2-microglobulin, α-mannosidase, and the *FES* protooncogene are important markers for investigating molecular disease on the long arm of this chromosome. The Prader–Willi syndrome has been assigned to the long arm (q11) as well. *D15S1, D15S2*, and *D15S4* are polymorphic markers useful in gene-linkage studies. An interesting biological marker that determines human coronavirus 229E sensitivity in human–rodent hybrids has been located on the long arm of chromosome 15. Heavily used cell genetic markers for this chromosome include *HEXA, MPI*, and *PKM2*, all on the long arm. Further mapping information on this chromosome has been summarized by Cox and Gedde-Dahl (1985).

7.1.16. Chromosome 16

To this chromosome have been assigned 28 gene markers: 6 on the p arm and 8 on the q arm; 14 markers have not been regionally determined. The important hemoglobin α gene family (*HBA*), the haptoglobin locus (*HP*), and *GOT2* are polymorphic loci and have been used extensively in human family studies. The cloned chymotrypsinogen B gene (*CTRB*) was recently found to recognize a DNA polymorphism along with probes for DNA polymorphisms that should be useful in family studies with other markers such as *PGP* and *PKD1*. Several important cloned genes have been assigned, including *APRT* and the metallothionein gene family. In cell-culture studies, the *APRT* locus has been most frequently used, and separate selection systems are available to select for and against this gene and chromosome. Excellent markers are available for both arms of this chromosome. Cox and Gedde-Dahl (1985) further describe the genes on this chromosome.

7.1.17. Chromosome 17

There are 30 gene markers that have been assigned to chromosome 17. Six gene markers have been assigned on p, 19 have been assigned on q, and 5 have not been assigned to a region. This chromosome encodes the significant genes for thymidine kinase (*TK*), growth hormone (*GH*), chorionic somatomammotropin hormone (*CSH*), the skeletal muscle myosin heavy-chain family, the p53 tumor protein, pancreatic polypeptide, the *ERBA1* oncogene, a homeobox region, gastrin, and alpha-1 type 1 collagen. Polymorphic loci that are available for family studies are *GAA* (acid α-glucosidase) and *D17S1–S3*. The closely related and linked *GH* and *CSH* genes and those of muscle myosin arose by gene duplication. In cell genetic studies, the *TK* gene can be selected for and against, and the closely syntenic locus *GALK* (galactokinase) is widely used to identify

17. See Naylor *et al.* (1985) for details regarding the gene map of this chromosome.

7.1.18. Chromosome 18

A total of 16 genes have been assigned to this chromosome, the lowest number of genes assigned to a human chromosome. One marker has been mapped on p, 3 markers have been mapped on q, and 12 markers have not been localized. Significant genes have been assigned to this chromosome; they are thymidylate synthetase, the *YES1* oncogene, chorionic gonadotropin, the receptor for gamma interferon, asparaginyl-tRNA synthetase, and prealbumin. Peptidase-A has been the most useful marker in some population studies and in somatic genetic studies. Markers that are useful in population studies are *D18S1*, *D18SD3*, and *PEPA*. For further details regarding the gene map, see Naylor *et al.* (1985).

7.1.19. Chromosome 19

On this chromosome, 46 gene markers have been mapped. Seven are located on the short arm and 13 on the long arm; 26 have not yet been localized. Although this chromosome is smaller than chromosome 18, a larger number of genes have been mapped on 19. For Mendelian genetic studies, *C3* (complement component 3), proline dipeptidase (*PEPD*), low-density lipoprotein receptor (*LDLR*), the blood groups *LE, H, LU, SE,* and *LW,* the apolipoproteins *APOC2* and *APOE,* dystrophia myotonia (*DM*), and the DNA segments *D19S4–S9* and *D19S11* are very useful markers. For somatic-cell genetic studies, *GPI* (glucose phosphate isomerase), lysosomal α-mannosidase-B (*MANB*) associated with mannosidosis, *PEPD,* and cell-surface markers are important markers. It is interesting to note that a gene that controls sensitivity of human cells to poliovirus infection is encoded on this chromosome. Other important genes include insulin receptor, transforming growth factor-B, *CGB, LHB,* a ferritin light chain, and a *PGK* autosomal gene. Further details of chromosome 19 mapping have been reported by Naylor *et al.* (1985).

7.1.20. Chromosome 20

Only 18 gene markers have been assigned to this chromosome: 4 markers on the short arm and 5 markers on the long arm; 9 remain to be localized. One of the most useful polymorphic markers for population studies is red-cell adenosine deaminase (*ADA*); also of use in family studies are *D20S4–S6* and *AHCY*. *ADA, ITPA, AHCY, ARVP,* and *DCE* identify this chromosome in cell genetic studies. Other important markers on this chromosome are the oncogene *SRC1,* a gene involved in galactosialidosis, high L-leucine transport, and growth-hormone-releasing factor. See Tippett and Kaplan (1985) for gene-mapping details.

7.1.21. Chromosome 21

A total of 82 gene markers have been assigned to this chromosome. Three markers have been assigned to the p arm and 54 markers to the q arm; 25 markers have not been localized to a small region. Of these markers, 66 are DNA segments, many of which recognize DNA polymorphisms. The isolation of these numerous probes for undefined DNA segments has been stimulated by the association of Down's syndrome with this chromosome. Only 16 of the markers on this chromosome recognize functional genes. For gene-linkage studies in families, superoxide dismutase 1 (*SOD1*) has been an important marker; now, numerous probes recognize DNA polymorphisms suitable for these studies. *SOD1,* the interferon receptor, and surface antigens are frequently used markers for somatic-cell genetic studies when a functional gene is necessary. Other important genes on this chromosome are receptors for interferon alpha and beta, cystathionine-β-synthase, an estrogen-inducible sequence observed in breast cancer, 5-hydroxytryptamine oxygenase regulator, several antigens, and a glycoprotein. For additional mapping details regarding this chromosome, see Tippett and Kaplan (1985).

7.1.22. Chromosome 22

On this chromosome, 27 gene markers have been assigned. Only 1 marker, the rDNA cluster of genes, is encoded on the short arm; 18 markers have been mapped to the long arm; 8 markers have not been localized to either arm or small region. Significant markers on this chromosome are the immunoglobulin lambda-chain family, myoglobin, the breakpoint cluster region associated with the Philadelphia chromosome, the *SIS* protooncogene, alpha-L-iduronidase, diaphorase 1, and aldolase A. Genes affected in the DiGeorge syndrome and metachromatic leukodystrophy (arylsulfastase A) are encoded on this chromosome. Polymorphic markers that have been used in family studies for gene linkage are the immunoglobulin loci, *SIS, MB, PI,* and several DNA polymorphisms. For further gene-mapping details and references, see Tippett and Kaplan (1985).

7.1.23. X Chromosome

More loci have been assigned to this chromosome than to any other human chromosome. In fact, of the over 1300 gene markers that have been assigned in the human genome, about one fourth have been mapped on the X chromosome. To date, there have been 321 gene markers assigned to the X chromosome. This intense mapping activity has been stimulated for several reasons: (1) Sex-linked traits are easily recognized in families; (2) the search for genetic mechanisms accounting for X-inactivation; (3) the search for genes closely linked to the many X-linked disorders; and (4) the search for defective genes themselves has generated the isolation of many undefined DNA sequences that serve as markers as well. There have been 73 markers assigned to regions on the short arm, 3 likely to be on the short arm, 160 markers assigned to regions on the long arm, and 7 markers have been mapped near the centromere but the arm has not been identified, and 12 markers are likely to be located on the long arm. Another 66 X-chromosome loci have not as yet been localized. Of the 321 mapped markers, 211 of them are anonymous cloned DNA segments (fragments), many of which recognize DNA polymorphisms; those that do not are listed as arbitrary DNA segments. The names, availability, and polymorphic nature of these probes, and references, have been tabulated by Willard *et al.* (1985). Recently, sequences have been identified that are encoded on both the X and Y chromosomes (see Goodfellow *et al.*, 1985; Willard *et al.*, 1985). These are designated in Table I by *DXYS* and the number of the probe. There are a few markers that seem to recognize homologous DNA sequences on the X and an autosome; these markers have not been included in Table I because not enough is known regarding their homologies (see Willard *et al.*, 1985). Fifty-eight well-known X-linked disorders are included in the table, with their map locations when known. Additional rare X-linked disorders are listed in McKusick (1982). Some important genes and disorders encoded on the X chromosome are steroid sulfatase (*STS*), Xg blood group, ocular albinism (*OA*), the protooncogene *HRAS2*, alpha DNA polymerase (*POLA*), Becker muscular dystrophy (*BMD*), Duchenne muscular dystrophy (*DMD*), ornithine carbamoyltransferase (*OTC*), classic retinitis pigmentosa (*RP2*), phosphoglycerate kinase 1 (*PGK1*), Hunter syndrome (*SIDS*), coagulation factors VIII and IX (*F8, F9*), hypoxanthine phosphoribosyltransferase (*HPRT*), mental retardation (*FRAXA*), glucose-6-phosphate dehydrogenase (*G6PD*), color blindness (*CBB*), albinism–deafness syndrome (*ADS*), cleft palate (*CLPA*), forms of deafness (*DFN1, DFN2, DFN3*), and forms of immunodeficiency (*IMD*). There is a rich supply of polymorphic markers for family studies and genes expressed in cultured cells for somatic-cell genetic studies. The large numbers of markers provide an abundant source of loci for prenatal diagnosis and population studies. Refer to Goodfellow *et al.* (1985) for X-chromosome gene-mapping details.

7.1.24. Y Chromosome

This male-determining chromosome has had a total of 53 markers assigned to it. Of these, 23 loci have been assigned to the short arm and 12 to the long arm; 17 loci have not been assigned to either arm or a small region. Of the 53 markers assigned, 45 are DNA sequences, and of these, 23 share sequences on both the Y and X chromosomes. These latter, shared sequences are designated in Table I by *DXYS* and the number of the probe. The wealth of DNA probes has been generated as a result of the molecular characterization of the Y chromosome. Some of the DNA segments identify DNA polymorphisms for gene-linkage studies. There are few defined gene loci that have been described for the Y chromosome; a few important genes are the testis-determining factor (*TDF*) and Y-chromosome growth control (*GCY*). The homologies being described at the molecular level for the X and Y chromosomes represent a major recent advance in understanding the sex chromosomes. See Goodfellow *et al.* (1985) for genetic details regarding the Y-gene map.

8. CONCLUSIONS

The ability to transfer genes in mammals has provided the stimulus and technology for fine-resolution gene mapping in humans. This report has focused on the somatic-cell hybridization form of gene transfer for mapping human genes. These techniques have been the foundation of somatic-cell genetics and gene mapping, since about three quarters of the human genes and markers chromosomally assigned in man have been determined utilizing these strategies. The technology has provided the ability to transfer DNA ranging from an entire genome, as discussed here, to nucleotide sequences. Later chapters will discuss the different sizes of DNA, including chromosomes, that can be transferred. However, it is now essential to refine the techniques, optimize the delivery, and broaden the range of organisms. Optimization of these properties has the potential to increase significantly the ability to carry out fine-structure gene mapping, genetic manipulation of cells, cell differentiation and gene expression, and molecular organization and evolution of mammalian cells. Results obtained from these studies will have direct application to correction of disease and to livestock and agricultural production. Certainly it is more realistic to transfer a gene than to carry out the breeding experiments necessary to change or generate the phenotype desired. The human gene maps presented in Table I have constituted a rich source of chromosomally assigned human genes defining the normal genome and locations of genes that may be involved in disease. Understanding the organization of the human gene maps will contribute to our understanding of normal and abnormal human biology and the correction of molecular disease.

TABLE I. The 24 Human Gene Maps

Map location	Symbol	Marker name	Mode[a]	Status[b]
		Human chromosome 1 map		
pter–p36.13	*EN01*	Enolase-1	F, S	C
pter–p36.13	*GDH*	Glucose dehydrogenase	F, S	C
pter–p34	*EL1*	Elliptocytosis 1	F	C
pter–p21	*F3*	Coagulation factor III	S	P
pter–p32	*GALE*	UDP-galactose-4-epimerase	S	C
pter–p22	*MSK31*	Antigen identified by monoclonal antibody SR75	S	P
pter–p22.1	*RD*	Radin blood group	F	C
p36.3	*RNU1*	U1 small nuclear RNA 1[c]	A, S, RE	C
p36.2–p22.1	*C8A*	Complement component 8, alpha polypeptide	F	C
p36.2–p36.13	*PGD*	Phosphogluconate dehydrogenase	F, S	C
p36.2–p34	*RH*	Rhesus blood group	F	C
p36.2–p22.1	*SC*	Scianna blood group	F	C
p36	*A12M2*	Adenovirus 12 chromosome modification site 1A	S	P
p36	*PND*	Pronatriodilatin[c]	A, S, RE	P
p36–p34	*SRC2*	Avian sarcoma viral (v-src) oncogene homologue 2[c]	A	C
p34	*AK2*	Adenylate kinase-2	S	C
p34	*FUCA1*	Fucosidase, alpha-1-tissue[c]	F, S, RE	C
p34	*UROD*	Uroporphyrinogen decarboxylase[c]	A, S	C
p32	*BLYM*	Avian-lymphoma-virus-derived transforming sequence[c]	A	P
p32	*LMYC*	Avian myelocytomatosis viral (v-myc) oncogene homologue, lung-carcinoma-derived[c]	A, S, RE	P
p32	*UMPK*	Uridine monophosphate	F, S	C
p32–qter	*PYHG7*	Protein spot in 2-dimensional (2-D) gels (MW 82K)	S	P
p32–qter	*PYHG9*	Protein spot in 2-D gels (MW 79K)	S	P
p32–qter	*PYHG21*	Protein spot in 2-D gels (MW 10K)	S	P
p32–qter	*PYHG22*	Protein spot in 2-D gels (MW <10K)	S	P
p31–p22	*FTHL1*	Ferritin, heavy-polypeptide-like 1[c]	A	P
p22.1	*NGFB*	Nerve growth factor, beta polypeptide[c]	A, S, RE	C
p22.1	*PGM1*	Phosphoglucomutase 1	D, F, S	C
p22.1–qter	*SDH*	Succinate dehydrogenase (one of two polypeptides)	S	P
p22	*MSK1*	Antigen identified by monoclonal antibody AJ9	S	P

Footnotes for the table appear on p. 74.

TABLE I. The 24 Human Gene Maps—(Continued)

Map location	Symbol	Marker name	Mode[a]	Status[b]
		Human chromosome 1 map		
p22–qter	MSK32	Antigen identified by monoclonal antibody K66	S	P
p22–qter	MTL2	Metallothionein-like 2	S, RE	P
p22 and/or p11–p12	NRAS	Neuroblastoma RAS viral (v-ras) oncogene homologue[c]	A, S, RE	C P
p22	TSHB	Thyroid-stimulating hormone beta polypeptide[c]	S, RE	P
p21–qter	ACTA	Actin, alpha; skeletal muscle[c]	S, RE	C
p21	AMY1	Amylase, alpha; salivary[c]	A, F, S, RE	C
p21	AMY2	Amylase, alpha; pancreatic[c]	A, F, S, RE	C
p21–qter	APOA2	Apolipoprotein A-II[c]	S, RE	C
p21–q23	CAE	Zonular pulverulent cataract	F	C
p21	COLL6	Collagen-like 6[c]	A, S, RE	P
p21–q23	FY	Duffy blood group	F	C
p21–qter	REN	Renin[c]	S, RE	P
p	C1QB	Complement component 1, q subcomponent, beta polypeptide[c]	S, RE	P
p	EKV	Erythrokeratodermia variabilis	F	P
p	MTL1	Metallothionein-like 1[c]	S, RE	P
cen–q32	PFKM	Phosphofructokinase, muscle type	S	P
q11–q22	RNU1P1	U1 small nuclear RNA pseudogene 1[c]	A	P
q11–q22	RNU1P2	U1 small nuclear RNA pseudogene 2[c]	A	P
q11–q22	RNU1P3	U1 small nuclear RNA psuedogene 3[c]	A	P
q11–q22	RNU1P4	U1 small nuclear RNA pseudogene 4[c]	A	P
q12–q23	APCS	Amyloid P component; serum[c]	S, RE	P
q12–q23	CRP	C-reactive protein[c]	A, S, RE	C
q12–qter	SKI	Avian sarcoma viral (v-ski) oncogene homologue[c]	S, RE	P
q21	A12M3	Adenovirus-12 chromosome modification site 1B	S	P
q21 or q31	GBA	Glucosidase, beta; acid[c]	A, D, S, RE	C
q21	H4F2	H4 histone family 2[c]	S, RE	P
q21–q22	UGP1	UDP glucose pyrophosphorylase 1	S	C
q22–q25	SPTA	Spectrin, alpha[c]	A, S, RE	P
q23–q25	AT3	Antithrombin III[c]	A, D, F, S, RE	C
q25 or q42	PEPC	Peptidase C	S	C
q32–q42	FTHL2	Ferritin, heavy-polypeptide-like 2[c]	A	P
q32–q42	GUK1	Guanylate kinase 1	F, S	C
q32–qter	MSK2	Antigen identified by monoclonal antibody T87	S	P

(Continued)

TABLE I. The 24 Human Gene Maps—(Continued)

Map location	Symbol	Marker name	Mode[a]	Status[b]
		Human chromosome 1 map		
q42.1	FH	Fumarate hydratase	S	C
q42–q43	A12M1	Adenovirus-12 chromosome modification site 1C	S	P
q42–q43	RN5S	5 S RNA	A	C
q	CMT1	Charcot–Marie–Tooth neuropathy 1	F	C
q	ERCC2	Excision repair complementing defective repair in Chinese hamster	S	P
q	XPAC	Fast kinetic complementation DNA repair in xeroderma pigmentosum, group A	S	P
	ALPL	Alkaline phosphatase, liver/bone/kidney isozyme	S	P
	C8B	Complement component 8, beta polypeptide	F	P
	CFAG	Cystic-fibrosis-associated antigen	S	P
	CMM	Cutaneous malignant melanoma/dysplastic nevus	F	T
	D1S2	DNA segment, polymorphic[c]	S, RE	P
	D1S4	DNA segment, polymorphic[c]	S, RE	P
	D1S5	DNA segment, arbitrary	S, RE	P
	GUK2	Guanylate kinase 2	S	C
	H3F2	H3 histone family 2[c]	S, RE	P
	MIC10	Antigen identified by monoclonal antibody TRA-2-10	S	P
	MTR	Tetrahydropteroylglutamate methyltransferase	S	P
		Human chromosome 2 map		
pter–q31	AHH	Aryl hydrocarbon hydroxylase	S	C
pter–p12	CD8	Antigen leu 2/T8[c]	A, S, RE	P
p25 or p23	ACP1	Acid phosphatase 1, soluble	D, F, S	C
p25–cen	ASSP1	Argininosuccinate synthetase pseudogene 1[c]	A, S, RE	P
p25	D2S1	DNA segment, polymorphic[c]	A, S, RE, F	C
p24	APOB	Apolipoprotein B[c]	S, RE	C
p24–p25	D2S9	DNA segment, arbitrary[c]	A, S, RE	P
p24–p25	D2S10	DNA segment, arbitrary[c]	A, S, RE	P
p24–p23	NMYC	Avian-myelocytomatosis (v-myc)-related oncogene, neuroblastoma-derived[c]	A, S, RE	C
p23-qter	ACTL3	Actin-like sequence 3[c]	S, RE	P
p23-qter	IFNB3	Interferon, beta 3; fibroblast	S, RE	P
p23	MDH1	Malate dehydrogenase, NAD (soluble)	S	C

TABLE I. The 24 Human Gene Maps—(Continued)

Map location	Symbol	Marker name	Mode[a]	Status[b]
		Human chromosome 2 map		
p23	POMC	Proopiomelanocortin (adrenocorticotropin/ beta-lipotropin)[c]	S, RE	C
p22–p24	D2S7	DNA segment, arbitrary[c]	A, S, RE	P
p16–p14	FNL1	Fibronectin-like 1	A	P
p15–p16	D2S5	DNA segment, polymorphic[c]	S, RE	P
p13	D2S11	DNA segment, arbitrary[c]	S, RE, A	P
p13	TGFA	Transforming growth factor, type alpha[c]	A, S, RE	C
p12	IGK	Immunoglobulin kappa polypeptide[c]	A, S, RE	C
p	CPS1	Carbamoyl phosphate synthetase 1	A, S, RE	C
q13–q21	IL1	Interleukin 1[c]	A, S, RE	P
q31	COL3A1	Collagen, type III, alpha 1[c]	S, RE	C
q31	COL5A2	Collagen, type V, alpha 2[c]	S, RE	C
q32–q36	D2S6	DNA segment, polymorphic[c]	S, RE, A	P
q32–q33	FTHL3	Ferritin, heavy-polypeptide-like 3[c]	A	P
q32–qter	IDH1	Isocitrate dehydrogenase 1 soluble	S	C
q32–qter	RPE	Ribulose-5-phosphate-3-epimerase	S	C
q33–q35	CRYG1	Crystallin, gamma polypeptide 1[c]	S, RE, A	C
q33–q35	CRYG2	Crystallin, gamma polypeptide 2[c]	S, RE, A	C
q33–q35	CRYG3	Crystallin, gamma polypeptide 3[c]	S, RE, A	C
q33–q35	CRYG4	Crystallin, gamma polypeptide 4[c]	S, RE, A	C
q33–q35	CRYGP1	Crystallin, gamma polypeptide pseudogene 1	A	C
q34–q36	FN1	Fibronectin 1[c]	A, S, RE	C
q35–q37	D2S3	DNA segment, polymorphic[c]	S, RE	P
q	TUBA1	Tubulin, alpha (testis-specific)[c]	S, RE	P
	ADCP2	Adenosine deaminase complexing protein 2	S	C
	CO	Colton blood group	F	P
	CRYGP2	Crystallin, gamma polypeptide pseudogene 2[c]	A	C
	D2S2	DNA segment, arbitrary[c]	S, RE	P
	DES	Desimin[c]	S, RE	P
	GCG	Glucagon[c]	S, RE	P
	GLAT	Galactose enzyme activator	S	P
	JK	Kidd blood group	F	C
	MSK8	Antigen identified by monoclonal antibody L230	S	P
	PCH1	Protein spot in 2-dimensional gels (molecular weight 250K; pI 7.0)	S	P
	PROC	Protein C, inactivator of coagulation factors Va and VIIIa[c]	S, RE	P
	RACH	Acetylcholinesterase derepressor	D	P
	REL	Avian reticuloendotheliosis viral (v-rel) oncogene homologue[c]	S, RE	P
	UGP2	UDP glucose pyrophosphorylase 2	S	P

TABLE I. The 24 Human Gene Maps—(Continued)

Map location	Symbol	Marker name	Mode[a]	Status[b]
		Human chromosome 3 map		
pter–q21	ACTL4	Actin-like sequence 4[c]	S, RE	P
pter–q21	D3S4	DNA segment, polymorphic[c]	S, RE	P
pter–q21	D3S6	DNA segment, polymorphic[c]	S, RE	P
p25–p24	RAF1	Murine leukemia viral (v-raf-1) oncogene homologue 1[c]	S, RE	P
p21	ACY1	Aminoacylase 1	S	C
p21–cen	GLB1	Galactosidase, beta 1	S	C
p14–p21	D3S2	DNA segment, polymorphic[c]	S, RE	P
p13–q12	GPX1	Glutathione peroxidase 1	S	C
p	D3S3	DNA segment, polymorphic[c]	S, RE	C
cen–q21	UMPS	Uridine monophosphate synthetase (orotate phosphoribosyl transferase and orotidine-5′-decarboxylase)	S	P
q12	D3S1	DNA segment, polymorphic[c]	A, S, RE	C
q12–qter	ASSP7	Argininosuccinate synthetase pseudogene 7[c]	S, RE	P
q13.3–q22	PCCB	Propionyl CoA carboxylase, beta polypeptide[c]	A, S, RE	C
q21–q23	FTHL4	Ferritin, heavy-polypeptide-like 4[c]	A	P
q21–q26.1	TF	Transferrin[c]	A, S, RE	C
q21–qter	D3S5	DNA segment, polymorphic[c]	S, R	P
q26.2–qter	TFRC	Transferrin receptor[c]	A, S	C
q28	SST	Somatostatin[c]	A, S, RE	P
q	AHSG	Glycoprotein, alpha$_2$-HS	F, S	C
q	CHE1	Cholinesterase (serum) 1	F	C
q	CP	Ceruloplasmin[c]	S, RE	C
	AF8T	AF8 temperature sensitivity complementing	S	P
	DHFRP4	Dihydrofolate reductase pseudogene 4[c]	S, RE	C
	D3S8	DNA segment, polymorphic[c]	RE	P
	D3S9	DNA segment, polymorphic[c]	RE	P
	D3S10	DNA segment, polymorphic[c]	RE	P
	ERCM1	Excision repair compensating in mouse cells	S	P
	HPRTP1	Hypoxanthine phosphoribosyl-transferase pseudogene 1[c]	S, RE	P
3 or 11	HV1S	Herpes simplex virus type 1 sensitivity	S	I
	MF12	Antigen p97 (melanoma-associated) identified by monoclonal antibodies 133.2 and 96.5	S	P
	MSK9	Antigen identified by monoclonal antibody K15	S	P
	MSK10	Antigen identified by monoclonal antibody AJ425	S	P

TABLE I. The 24 Human Gene Maps—(Continued)

Map location	Symbol	Marker name	Mode[a]	Status[b]
		Human chromosome 4 map		
pter–q12	D4S15	DNA segment, polymorphic[c]	S, RE	P
pter–q12	D4S16	DNA segment, polymorphic[c]	S, RE	P
pter–q12	D4S17	DNA segment, polymorphic[c]	S, RE	P
pter–q12	D4S18	DNA segment, polymorphic[c]	S, RE	P
pter–q12	D4S19	DNA segment, polymorphic[c]	S, RE	P
pter–q12	D4S20	DNA segment, polymorphic[c]	S, RE	P
pter–q12	D4S21	DNA segment, polymorphic[c]	S, RE	P
pter–q12	D4S22	DNA segment, polymorphic[c]	S, RE	P
pter–q12	D4S23	DNA segment, polymorphic[c]	S, RE	P
pter–q12	D4S24	DNA segment, polymorphic[c]	S, RE	P
pter–q12	D4S35	DNA segment, polymorphic[c]	S, RE	P
pter–q21	PPAT	Phosphoribosyl pyrophosphate amidotransferase	S	P
pter–q26	D4S1	DNA segment, polymorphic[c]	S, RE	P
pter–q26	D4S12	DNA segment, polymoprhic[c]	S, RE	P
pter–q26	D4S13	DNA segment, polymorphic[c]	S, RE	P
p16.1	D4S10	DNA segment, polymorphic[c]	A, S, RE, D	C
Near p16.1	HD	Huntington's disease	F	C
p14–q12	PGM2	Phosphoglucomutase 2	S	C
p11–q12	PEPS	Peptidase S	S	C
p11–q21	MT2P1	Metallothionein 2 processed pseudogene 1[c]	A, S, RE	C
p	D4S2	DNA segment, polymorphic[c]	S, RE	P
q11–q13	AFP	Alpha-fetoprotein[c]	A, RE	C
q11–q13	ALB	Albumin[c]	A, R, S, RE	C
q12–q13	GC	Group-specific component[c]	A, F, S, RE, D	C
q12–qter	D4S25	DNA segment, polymorphic[c]	F, RE	P
q12–qter	D4S26	DNA segment, polymorphic[c]	F, RE	P
q12–qter	D4S27	DNA segment, polymorphic[c]	F, RE	P
q12–qter	D4S28	DNA segment, polymorphic[c]	F, RE	P
q12–qter	D4S29	DNA segment, polymorphic[c]	F, RE	P
q12–qter	D4S30	DNA segment, polymorphic[c]	F, RE	P
q12–qter	D4S31	DNA segment, polymorphic[c]	F, RE	P
q12–qter	D4S32	DNA segment, polymorphic[c]	F, RE	P
q12–qter	D4S33	DNA segment, polymorphic[c]	F, RE	P
q12–qter	D4S34	DNA segment, polymorphic[c]	F, RE	P
q12–qter	D4S36	DNA segment, polymorphic[c]	F, RE	P
q12–qter	D4S37	DNA segment, polymorphic[c]	F, RE	P
q12–qter	D4S38	DNA segment, polymorphic[c]	F, RE	P
q21–qter	AGA	Aspartylglucosaminidase	S	P
q21–qter	ASSP8	Argininosuccinate synthetase pseudogene 8[c]	S, RE	P
q21–q25	ADH1	Alcohol dehydrogenase (class I), alpha polypeptide[c]	S, RE	P

(Continued)

TABLE I. The 24 Human Gene Maps—(Continued)

Map location	Symbol	Marker name	Mode[a]	Status[b]
		Human chromosome 4 map		
q21–q25	ADH2	Alcohol dehydrogenase (class I), beta polypeptide[c]	S, RE	P
q21–q25	ADH3	Alcohol dehydrogenase (class I), gamma polypeptide[c]	S, RE	P
q21–q25	ADH5	Alcohol dehydrogenase (class III), chi polypeptide[c]	S, RE	P
q21–q25	FDH	Formaldehyde dehydrogenase	S	C
q21–qter	AGA	Aspartylglucosaminidase	S	P
q25–q27	EGF	Epidermal growth factor[c]	A, S, RE	C
q26–q28	IL2	Interleukin 2[c]	A, S, RE	C
q26–q28	FGA	Fibrinogen, alpha polypeptide[a]	A, S, RE	C
q26–q28	FGB	Fibrinogen, beta polypeptide[c]	A, S, RE	C
q26–q28	FGG	Fibrinogen, gamma polypeptide[c]	A, S, RE	C
q26–qter	D4S14	DNA segment, polymorphic[c]	S, RE	P
q28–q31	MNS	MNS blood group	D, F	C
q	D4S3	DNA segment, arbitrary[c]	S, RE	P
q	D4S9	DNA segment, polymorphic[c]	S, RE	P
	CD2	Leukocyte antigen 5 (CD2p50, identified by monoclonal antibodies 9.6, T1Leu5, and 35.1)	S	P
	DGI1	Dentinogenesis imperfecta	F	C
	D4S4	DNA segment, arbitrary[c]	S, RE	P
	D4S5	DNA segment, arbitrary[c]	S, RE	P
	D4S6	DNA segment, arbitrary[c]	S, RE	P
	D4S7	DNA segment, arbitrary[c]	S, RE	P
	D4S8	DNA segment, arbitrary[c]	S, RE	P
	HPAFP	Hereditary persistence of alpha-fetoprotein	F	P
	MDF1	Antigen identified by monoclonal antibody A-3A4	S	P
	MIC16	Antigen identified by monoclonal antibody OKT10/p45	S	P
	QDPR	Quinoid dihydropteridine reductase	S	P
	RAF2	Murine leukemia viral (v-raf-2) oncogene homologue 2[c]	S, RE	P
	SF	Stoltzfus blood group	F	P
	TYS	Sclerotylosis	F	P
		Human chromosome 5 map		
pter–q11	LARS	Leucyl-tRNA synthetase	S	C
pter–q35	D5S2	DNA segment, polymorphic[c]	S, RE	P
p13–q11	HPRTP2	Hypoxanthine phosphoribosyl-transferase pseudogene 2[c]	S, RE	P
p11–q13	ARSB	Arylsulfatase B	S	C

TABLE I. The 24 Human Gene Maps—(Continued)

Map location	Symbol	Marker name	Mode[a]	Status[b]
		Human chromosome 5 map		
p	AVRR	Antiviral state repressor regulator	D	P
p	IFN2	Interferon 2; fibroblast	S	P
cen–q12	ASSP9	Argininosuccinate synthetase pseudogene 9[c]	S, RE	C
q11–q13	GRL	Glucocorticoid receptor[c]	S, RE	C
q11.1–q13.2 or q23	DHFR	Dihydrofolate reductase[c]	S, RE	C
q13	HEXB	Hexosaminidase B (beta polypeptide)[c]	S	C
q13–q23	ASSP10	Argininosuccinate synthetase pseudogene 10[c]	S, RE	C
q13.3–q14	HMGCR	3-Hydroxy-methylglutaryl coenzyme A reductase[c]	A, S, RE	C
q15–qter	DTS	Diphtheria toxin sensitivity	S	C
q23–q31	COLL4	Collagen-like sequence 4[c]	A, S, RE	P
q31–q35	RPS14	Ribosomal protein S14[c]	S	P
q34	FMS	McDonough feline sarcoma viral (v-fms) oncogene homologue[c]	S, RE	C
q35	CHR	Chromate resistance; sulfate transport	S	P
	ACTBP2	Actin, beta pseudogene 2[c]	S, RE	P
	ACTBP4	Actin, beta pseudogene 4[c]	S, RE	P
	ADRBR	Beta-adrenergic receptor, surface	S	P
	DHLAG	HLA class II gamma chain polypeptide	S, RE	P
	D5S1	DNA segment, polymorphic[c]	S, RE	P
	D5S3	DNA segment, polymorphic[c]	S, RE	P
	D5S4	DNA segment, polymorphic[c]	S, RE	P
	GM2A	Ganglioside GM2 activator protein	S	P
	HARS	Histidyl-tRNA synthetase	S	P
	MSK11	Antigen defined by monoclonal antigen SR84	S	P
	PCH2	Protein spot in 2-dimensional gels (MW 250K; pI 8.3)	S	P
		Human chromosome 6 map		
pter–p21	TUBB	Tubulin, beta polypeptide[c]	S, RE	P
pter–p21	PYHG5	Protein spot in 2-dimensional gels (MW 90K)	S	P
pter–q14	D6S11	DNA segment, arbitrary[c]	S, RE	P
p24–q12	D6S5	DNA segment, insulin-related[c]	S, RE	P
p23	F12	Coagulation factor XII (Hageman)	D, F	P
p23–qter	F13A	Coagulation factor XIII, A polypeptide	D, F	C
p23–q12	KRAS1P	Kirsten rat sarcoma 1 viral (v-Ki-ras1) oncogene homologue, processed pseudogene	S, RE	C

(Continued)

TABLE I. The 24 Human Gene Maps—(Continued)

Map location	Symbol	Marker name	Mode[a]	Status[b]
		Human chromosome 6 map		
p23–q12	PRL	Prolactin[c]	S, RE	P
p23–q12	TRM1	Methionine tRNA initiator 1[c]	S, RE	P
p23–q12	TRM2	Methionine tRNA initiator 2[c]	S, RE	P
p21.3	BF	Properdin factor B[c]	F, RE	C
p21.3	C2	Complement component 2[c]	F, RE	C
p21.3	C3BR	Complement component 3b receptor	S	C
p21.3	C3DR	Complement component 3d receptor	S	P
p21.3	C4A	Complement component 4A[c]	F, RE	C
p21.3	C4B	Complement component 4B[c]	F, RE	C
p21.3	CA21HA	Congenital adrenal hyperplasia (21-hydroxylase deficiency) A[c]	F, RE	C
p21.3	CA21HB	Congenital adrenal hyperplasia (21-hydroxylase deficiency) B[c]	F, RE	C
p21.3	HFE	Hemochromatosis	F	P
p21.3	HLA-A	Major histocompatibility complex, class I antigens[c]	F, S	C
p21.3	HLA-B	Major histocompatibility complex, class I antigens[c]	F, S	C
p21.3	HLA-C	Major histocompatibility complex, class I antigens	F, S	C
p21.3	HLA-DP	Major histocompatibility complex, class II antigens	A, D, F, S, RE	C
p21.3	HLA-DQ	Major histocompatibility complex, class II antigens	A, D, F, S, RE	C
p21.3	HLA-DR	Major histocompatibility complex, class II antigens[c]	A, D, F, S, RE	C
p21.3	HLA-DZ	Major histocompatibility complex, class II antigens	S, RE	P
p21.3	ISSCW	Immune suppression to streptococcal cell-wall antigen	F	P
p21.3	RPL32P	Ribosomal protein L32 pseudogene	S, RE	P
p21.31–p21.1	GLO1	Glyoxalase I	D, F, S, RE	C
p21.3–p12	FTHL5	Ferritin, heavy-polypeptide-like 5[c]	A	P
p21–p12	PGK1P2	Phosphoglycerate kinase 1, pseudogene 2[c]	S, RE	C
p21–qter	D6S2	DNA segment, polymorphic[c]	S, RE	C
p11–cen	D6S4	DNA segment, polymorphic[c]	S, RE	P
p	D6S7	DNA segment, polymorphic[c]	S, RE	P
p	D6S8	DNA segment, polymorphic[c]	S, RE	P
p	D6S9	DNA segment, polymorphic[c]	S, RE	P
p	D6S10	DNA segment, polymorphic[c]	S, RE	P
q12	ME1	Malic enzyme 1, soluble	S	C
q12	PGM3	Phosphoglucomutase 3	F, S	C

TABLE I. The 24 Human Gene Maps—(Continued)

Map location	Symbol	Marker name	Mode[a]	Status[b]
		Human chromosome 6 map		
q12–q21	CGA	Glycoprotein hormone, alpha polypeptide[c]	S, RE	P
q15–q24	MYB	Avian myeloblastosis viral (v-myb) oncogene homologue[c]	S, RE	C
q21	SOD2	Superoxide dismutase 2; mitochondrial	S	C
q21–qter	D6S1	DNA segment, arbitrary[c]	S, RE	P
q25–qter	PLG	Plasminogen[c]	A, S, RE	P
	ADCP1	Adenosine deaminase complexing protein 1	S	P
	ARG1	Arginase; liver[c]	S, RE	P
	ASSP2	Argininosuccinate synthetase pseudogene 2[c]	S, RE	C
	BEV1	Baboon M7 virus integration site	S	C
	DHFRP2	Dihydrofolate reductase pseudogene 2[c]	A, S	C
	D6S3	DNA segment, polymorphic[c]	S, RE	P
	D6S6	DNA segment, polymorphic[c]	S	P
	FUCA2	Fucosidase, alpha-L-2; plasma	F	P
	LQT	Long QT (Romano–Ward syndrome)	F	P
	ME2	Malic enzyme 2; mitochondrial	F	P
	MRBC	Monkey red blood cell activator	S	P
	MSK28	Antigen identified by monoclonal antibody MG2	S	P
	MSK29	Antigen identified by monoclonal antibody A42	S	P
	P	P blood group (globoside)	S	T
	PUT1	Polypeptide m44p in lymphocyte membrane	F	P
	S5	Surface antigen 5 (MW 45K)	S	P
	SCA1	Spinal cerebellar ataxia (olivopontocerebellar ataxia)	F	C
	SRBC	Sheep red blood cell receptor	S	P
	TCP1	T-complex 1[c]	S, RE	P
	YES2	Yamaguchi sarcoma viral (v-yes-2) oncogene homologue 2[c]	S, RE	P
		Human chromosome 7 map		
pter–p15	COLL5	Collagen-like 5	A	P
pter–p14	GCTG	Gamma-glutamyl-cyclotransferase	S	P
pter–p14	D7S10	DNA segment, polymorphic[c]	S	P

(Continued)

TABLE I. The 24 Human Gene Maps—(Continued)

Map location	Symbol	Marker name	Mode[a]	Status[b]
		Human chromosome 7 map		
pter–q22	NPY	Neuropeptide Y[c]	S, RE	P
pter–q22	ACTB	Actin, beta[c]	S, RE	P
pter–q22	PSP	Phosphoserine phosphatase	S	C
pter–q22	D7S11	DNA segment, arbitrary[c]	S	P
pter–q22	D7S12	DNA segment, arbitrary[c]	S, RE	C
p21–q22	ASL	Argininosuccinate lyase	S	P
p15	ASSP11	Argininosuccinate synthetase pseudogene 11[c]	S, RE	P
p15	TCRG	T-cell receptor (rearranging), gamma polypeptide[c]	S, RE	C
p14–p12	ERBB	Avian erythroblastic leukemia viral (v-erb-b) oncogene homologue[c]	S, RE	C
p14–cen	BLVR	Biliverdin reductase	S	C
p13–p11	EGFR	Epidermal growth factor receptor	S	C
p13–q22	MDH2	Malate dehydrogenase, NAD (mitochondrial)	S	C
p11–q11	ASNS	Asparagine synthetase	S	P
cen–q11.2	MYH5	Myosin heavy polypeptide; skeletal muscle, adult	A	P
cen–q22	D7S6	DNA segment, polymorphic[c]	S, RE	P
cen–q22	D7S8	DNA segment, polymorphic[c]	S, RE	P
cen–q22	GUSB	Glucuronidase, beta[c]	S	C
cen–q22	D7S13	DNA segment, arbitrary[c]	S, RE	P
cen–q22	D7S14	DNA segment, arbitrary[c]	S, RE	P
q21.3–q22.1	OI4	Osteogenesis imperfecta type IV	F	P
q21.3–q22.1	COL1A2	Collagen, type I, alpha 2[c]	S, RE	C
	CF	Cystic fibrosis	S, RE, F	C
q22	MET	Met oncogene[c]	S, RE	P
q22 or q32–q36	H1	H1 histone[c]	A	C
q22 or q32–q36	H2A	H2A histone[c]	A	C
q22 or q32–q36	H2B	H2B histone[c]	A	C
q22 or q32–q36	H3F1	H3 histone family 1[c]	A	C
q22 or q32–q36	H4F1	H4 histone family 1[c]	A	C
q22–qter	ACTBP5	Actin, beta pseudogene 5	S, RE	P
q22–qter	NM	Neutrophil migration	D	P
q22–qter	TRY1	Trypsin 1	S, RE	P
q32 or q35	TCRB	T-cell receptor (rearranging), beta polypeptide[c]	S	C

TABLE I. The 24 Human Gene Maps—(Continued)

Map location	Symbol	Marker name	Mode[a]	Status[b]
		Human chromosome 7 map		
q36	PGY1	P glycoprotein 1	S, RE	P
	CPA	Carboxypeptidase A	S, RE	P
	D7S1	DNA segment, polymorphic[c]	S, RE	P
	D7S2	DNA segment, arbitrary[c]	S, RE	P
	D7S3	DNA segment, arbitrary[c]	S, RE	P
	D7S4	DNA segment, arbitrary[c]	S, RE	P
	D7S5	DNA segment, arbitrary[c]	S, RE	P
	GCF1	Growth control factor 1	S	P
	HADH	Hydroxyacyl-CoA dehydrogenase	S	P
	NHCP2	Nonhistone chromosome protein 2	S	P
	PO4DB	Proline, 2-oxoglutarate dioxygenase, beta polypeptide	S	T
	PYHG3	Protein spot in 2-dimensional (2-D) gels (MW 106K)	S	P
	PYHG8	Protein spot in 2-D gels (MW 80K)	S	P
	S7	Surface antigen (chromosome 7) 2	S	P
	UP	Uridine phosphorylase	S	C
		Human chromosome 8 map		
p21.1	GSR	Glutathione reductase	S, D	C
p12	PLAT	Plasminogen activator, tissue[c]	S, RE	C
p11.2–p21	LHRH	Luteinizing-hormone-releasing-hormone precursor[c]	A, S	P
p1	SPH1	Spherocytosis	F	C
q11–q22	MOS	Moloney murine sarcoma viral (v-mos) oncogene homologue[c]	A, RE	C
q13–qter or 16p	GPT	Glutamic-pyruvate transaminase	S	I
q21.1–qter	GLYB	Glycine B complementing	S	P
q22–q24	LGS	Langer–Giedion syndrome	D	C
q24	MYC	Avian myelocytomatosis viral (v-myc) oncogene homologue[c]	A, RE	C
q24	TG	Thyroglobulin[c]	S, RE	C
q24	POLB	Polymerase (DNA), beta polypeptide[c]	A, RE	P
	CA1	Carbonic anhydrase I[c]	S, RE	P
	CA2	Carbonic anhydrase II[c]	S, RE	P
	CA3	Carbonic anhydrase III, muscle-specific[c]	S, RE	P
	F7R	Coagulation factor VII regulator	D	P
	FNZ	Fibronectin; influences presence on cell surface	S	C
	MYLL1	Myosin light polypeptide, cardiac-like[c]	S, RE	P
	TUBBP1	Tubulin, beta polypeptide pseudogene 1[c]	S, RE	P

(Continued)

TABLE I. The 24 Human Gene Maps—(Continued)

Map location	Symbol	Marker name	Mode[a]	Status[b]
		Human chromosome 9 map		
pter–p24	D9S3	DNA segment, polymorphic[c]	S, RE	P
pter–p13	IFNA	Interferon, alpha (leukocyte)[c]	S, RE, A	C
pter–q11	D9S1	DNA segment, polymorphic	S, RE	P
pter–q12	RLN1	Relaxin 1 (H1)[c]	S, RE	P
pter–q12	RLN2	Relaxin 2 (H2)[c]	S, RE	P
p24–p13	IFNB	Interferon, beta (fibroblast)[c]	S, RE	C
p24–p13	AK3	Adenylate kinase 3	S	C
p22–p13	ACO1	Aconitase 1, soluble	S	C
p13	GALT	Galactose-1-phosphate uridylyltransferase	S, D	C
p13–q11	ASSP12	Argininosuccinate synthetase pseudogene 12[c]	S, RE	P
cen–q34	FPGS	Folylpolyglutamate synthetase	S	P
q3	NPS1	Nail patella syndrome 1	F	C
q11–q22	ASSP3	Argininosuccinate synthetase pseudogene 3[c]	S, RE	P
q12	DNCM	DNA associated with cytoplasmic membrane[c]	A	P
q21.3–q32	ALDOB	Aldolase B		P
q34	ABL	Abelson murine leukemia viral (v-abl) oncogene homologue[c]	S, RE	C
q34	AK1	Adenylate kinase 1	S, D	C
q34	ABO	ABO blood group	F	C
q34	WS1	Waardenburg syndrome, type 1	F	T
q34–qter	ASS	Argininosuccinate synthetase	S	C
q	ALAD	Delta-aminolevulinate dehydratase	F	C
q	ALDH1	Aldehyde dehydrogenase I, soluble	S	P
q	GRP78	Glucose-regulated protein [molecular weight (MW) 78K]	S, RE	P
q	ORM	Orosomucoid	F	C
	CPO	Coproporphyrinogen oxidase	S	P
	D9S2	DNA segment, arbitrary[c]	S, RE	P
	D9S4	DNA segment, polymorphic[c]	S, RE	P
	IGHEP2	Immunoglobulin epsilon polypeptide pseudogene 2[c]	S, RE	P
	MTAP	Methylthioadenosine phosphorylase	S	P
	PYHG13	Protein spot in 2-dimensional (2-D) gels (MW 40K)	S	P
	PYHG14	Protein spot in 2-D gels (MW 37K)	S	P
	PYHG15	Protein spot in 2-D gels (MW 35K)	S	P
	PYHG16	Protein spot in 2-D gels (MW 38K)	S	P
		Human chromosome 10 map		
pter–q23	NEU	Neuroaminidase	S	P
pter–p11.1	PFKP	Phosphofructokinase, platelet type	S	C
p15–p14	IL2R	Interleukin 2 receptor[c]	A, RE	P

TABLE I. The 24 Human Gene Maps—(Continued)

Map location	Symbol	Marker name	Mode[a]	Status[b]
		Human chromosome 10 map		
cen–q24	ADK	Adenosine kinase	S	C
q11.1–q24	PP	Pyrophosphatase (inorganic)	S	C
q23–q24	GLUD	Glutamate dehydrogenase[c]	S	P
q24–qter	PLAU	Plasminogen activator, urokinase[c]	S, RE	C
q25.3	GOT1	Glutamic-oxaloacetic transaminase 1, soluble	S, D	C
q25.3	PGAMA	Phosphoglyceratemutase A	D, S	P
	ATPM	ATPase, mitochondrial	S	P
	D10S1	DNA segment, polymorphic[c]	S, RE	P
	D10S2	DNA segment, arbitrary[c]	S, RE	P
	D10S3	DNA segment, polymorphic[c]	S, RE	P
	FUSE	Polykaryocytosis promoter	S	P
	GSAS	Glutamate gamma-semialdehyde synthetase	S	P
	HK1	Hexokinase 1	S	C
	LIPA	Lipase A	S	C
	M130	External membrane protein (MW 130K)	S	P
	MDF2	Antigen identified by monoclonal antibody A-1A15	S	P
	MSK12	Antigen identified by monoclonal antibody AJ2	S	P
	OAT	Ornithine aminotransferase	S	P
	PYHG1	Protein spot in 2-dimensional (2-D) gels (MW 218K)	S	P
	PYHG18	Protein spot in 2-D gels (MW 31K)	S	P
	SAP1	Sphingolipid activator protein 1	S	P
	VIM	Vimentin[c]	S, RE	P
		Human chromosome 11 map		
pter–p15.5	HRAS1	Harvey rat sarcoma 1 viral (v-Ha-ras1) oncogene homologue[c]	A, D, S, RE	C
pter–p15.4	BWS	Beckwith–Wiedemann syndrome	D	C
pter–p15	MER2	Antigen identified by monoclonal antibodies 1D12 and 2F7	S	C
pter–p15	D11S19	DNA segment, polymorphic[c]	S, RE	P
pter–p13	FSHB	Follicle-stimulating hormone beta polypeptide[c]	D, S	C
pter–p13	MER1	Antigen identified by monoclonal antibodies W6/34, 5CL, etc. (formerly lethal antigen 1)	D, S	C
pter–p13	D11S17	DNA segment, polymorphic[c]	RE, D	P
pter–p13	D11S18	DNA segment, polymorphic[c]	RE, D	P
pter–p13	D11S20	DNA segment, polymorphic[c]	RE	P
pter–p12	CALC2	Calcitonin-gene-related peptide 2[c]	D	P

(Continued)

TABLE I. The 24 Human Gene Maps—(Continued)

Map location	Symbol	Marker name	Mode[a]	Status[b]
		Human chromosome 11 map		
pter–p11	D11S25	DNA segment, arbitrary[c]	D	P
pter–p11	D11S26	DNA segment, arbitrary[c]	D	P
pter–p11	D11S21	DNA segment, polymorphic[c]	RE	P
p15.5	IGF2	Insulinlike growth factor 2[c]	A, D	C
p15.5	INS	Insulin[c]	A, D, S	C
p15.5–p15.3	TH	Tyrosine hydroxylase[c]	A, D, S	C
p15.4–p15.1	CALC1	Calcitonin/calcitonin-gene-related peptide 1[c]	A, D, RE	C
p15.4–p15.1	PTH	Parathyroid hormone[c]	A, S, RE	C
p15	CPSD	Cathepsin D[c]	S	P
p15	D11S12	DNA segment, polymorphic[c]	RE, D	C
p15	HBB	Hemoglobin, beta[c]	A, S, RE	C
p15	HBBP1	Hemoglobin, beta pseudogene 1[c]	A, S, RE	C
p15	HBD	Hemoglobin, delta[c]	A, S, RE	C
p15	HBE1	Hemoglobin, epsilon 1[c]	A, S, RE	C
p15	HBG1	Hemoglobin, gamma A[c]	A, S, RE	C
p15	HBG2	Hemoglobin, gamma G[c]	A, S, RE	C
p14–p12	LDHA	Lactate dehydrogenase A[c]	S, RE	C
p14–p13	MIC4	Antigen identified by monoclonal antibodies F10.44.2 and Mab4	S, D	C
p14–p13	MIC11	Antigen identified by monoclonal antibody 16.3A5 (formerly lethal antigen 3)	S	P
p13	AN2	Aniridia without Wilms' tumor, genitourinary abnormalities, and mental retardation	CH, F	C
p13	WAGR	Wilms' tumor—aniridia, genitourinary abnormalities, and mental retardation triad	D	C
p13	CAT	Catalase[c]	S, D, RE	C
p13–p12	D11S3	DNA segment, arbitrary[c]	D	P
p13–p12	D11S6	DNA segment, arbitrary[c]	D	P
p13–p12	D11S8	DNA segment, arbitary[c]	D	P
p13–p12	D11S10	DNA segment, arbitrary[c]	D	P
p13–q23	D11S23	DNA segment, polymorphic[c]	S, D	P
p13–q23	D11S24	DNA segment, polymorphic[c]	S, D	P
p12–p11	D11S9	DNA segment, arbitrary[c]	D	P
p12–p11	D11S11	DNA segment, arbitrary[c]	D	P
p12–p11	D11S14	DNA segment, arbitrary[c]	D	P
p12–p11	MER3	Antigen identified by monoclonal antibody 3C5	D, S	P
p12–p11	MER4	Antigen identified by monoclonal antibody 6C12	D, S	P
p12–cen	ACP2	Acid phosphatase 2; lysosomal	S	C
p11–q12	PGA3	Pepsinogen 3, group I (PGA)[c]	S	P
p11–q12	PGA4	Pepsinogen 4, group I (PGA)[c]	S	P

TABLE I. The 24 Human Gene Maps—(Continued)

Map location	Symbol	Marker name	Mode[a]	Status[b]
		Human chromosome 11 map		
p11–q12	PGA5	Pepsinogen 5, group I (PGA)[c]	S	P
p11–q13	D11S1	DNA segment, arbitrary[c]	D	P
p11–q13	D11S4	DNA segment, arbitrary[c]	D	P
p11–q13	D11S5	DNA segment, arbitrary[c]	D	P
p11–q13	D11S6	DNA segment, arbitrary[c]	D	P
p	MDU2	Antigen identified by monoclonal antibody A1G3	S	P
p	MDU3	Antigen identified by monoclonal antibody A3D8	S	P
p	RRM1	Ribonucleotide reductase, M1 polypeptide	S	P
q12–qter	MEI1	Antigen CDw15 identified by monoclonal antibodies RIB-19, S4.7.13, etc.	S	P
q12.1–q13.5	FNL2	Fibronectin-like 2[c]	A, S	P
q13	APOC3	Apolipoprotein C-III[c]	F, RE, S	C
q13	FTHL6	Ferritin, heavy-polypeptidelike 6[c]	A	P
q13–qter	GANAB	Glucosidase, alpha; neutral AB	S	P
q13–qter	GST3	Glutathione S-transferase, isozyme 3	S	C
q13–qter	MSK13	Antigen identified by monoclonal antibody Q14	S	P
q13–qter	APOA1	Apolipoprotein A-I[c]	S, RE, F	C
q13–qter	APOA4	Apolipoprotein A-IV	F	P
q13.3	BCL1	B cell CLL/lymphoma 1	D	P
q22.3	THY1	Thy-1 cell-surface antigen	A	C
q23–q24	ETS1	E26 acute avian LV oncogene homologue 1[c]	S, RE	P
q23	NCAM	Cell-adhesion molecule, neural	A	P
q23–qter	T3D	TiT3 complex, delta polypeptide (CD3) (p19–p29)[c]	S, RE	P
q23.2–qter	UPS	Uroporphyrinogen I synthetase	S	C
q	ASSP13	Argininosuccinate synthetase pseudogene 13[c]	S, RE	P
q	ESA4	Esterase A4	S	C
q	HPRTP3	Hypoxanthine phosphoribosyltransferase pseudogene 3[c]	S, RE	P
q	HPRTP4	Hypoxanthine phosphoribosyltransferase pseudogene 4[c]	S, RE	P
q	MDU1	Antigen identified by monoclonal antibodies 4F2, TRA1.10, and TROP4	S	P
q	MIC9	Antigen identified by monoclonal antibody 4D12	S	P

(Continued)

TABLE I. The 24 Human Gene Maps—(Continued)

Map location	Symbol	Marker name	Mode[a]	Status[b]
		Human chromosome 11 map		
q	*PC*	Pyruvate carboxylase[c]	D	P
q	*PYGM*	Phosphorylase, glycogen, muscle (McArdle syndrome, glycogen storage disease type I)[c]	D, RE	P
	CD5	Antigen CD5p56-62 identified by monoclonal antibodies A50, 10.2, T1, T101, and Leu1	S	P
	CLG	Collagenase, epidermolysis bullosa, recessive dystrophic	S	P
	D11S15	DNA segment, polymorphic[c]	RE	P
	D11S16	DNA segment, polymorphic[c]	RE	P
	LEU1	Leukocyte antigen 1	S	P
	LEU7	Leukocyte antigen 7	S	P
	MSK14	Antigen identified by monoclonal antibody JF23	S	P
	MSK21	Antigen identified by monoclonal antibody G344	S	P
	MSK22	Antigen identified by monoclonal antibody A124	S	P
	MSK23	Antigen identified by monoclonal antibody T43	S	P
	MSK24	Antigen identified by monoclonal antibody NP13	S	P
	MSK25	Antigen identified by monoclonal antibody MC139	S	P
	MSK26	Antigen identified by monoclonal antibody K117	S	P
	PYHG2	Protein spot in 2-dimensional gels (MW 178K)	S	P
		Human chromosome 12 map		
pter–q12	*BCT1*	Branched-chain amino-transferase 1	S	C
pter–p12	*F8VWF*	Coagulation factor VIII VWF (von Willebrand factor)[c]	A, S, RE	C
pter–q14	*INT1*	Murine mammary tumor virus integration site (v-int 1) oncogene homologue[c]	S	P
pter–q12	*MIC3*	Antigen (CD9p24) identified by monoclonal antibodies 602-29, BA-2, ALB6, and Du-ALL	S	C
p13.2	*PRB1*	Proline-rich protein *Bst*NI subfamily 1[c]	S	P
p13.2	*PRB2*	Proline-rich protein *Bst*NI subfamily 2[c]	S	P
p13.2	*PRB3*	Proline-rich protein *Bst*NI subfamily 3[c]	S	P
p13.2	*PRB4*	Proline-rich protein *Bst*NI subfamily 4[c]	S	P
p13.2	*PRH1*	Proline-rich protein *Hae*III subfamily 1	S	P
p13.2	*PRH2*	Proline-rich protein *Hae*III subfamily 2	S	P

TABLE I. The 24 Human Gene Maps—(Continued)

Map location	Symbol	Marker name	Mode[a]	Status[b]
		Human chromosome 12 map		
p13	GAPD	Glyceraldehyde-3-phosphate dehydrogenase[c]	S, D, RE	C
p13	TPI1	Triosephosphate isomerase 1[c]	D, S, RE	C
p12.2–p12.1	LDHB	Lactate dehydrogenase B[c]	S, RE	C
p12.1	KRAS2	Kirsten rat sarcoma 2 viral (v-Ki-ras2) oncogene homologue[c]	S, RE, A	C
p11–qter	CS	Citrate synthase	S	C
p11–qter	ENO2	Enolase 2	S	C
p	KAR	Alpha-keto acid (aromatic) reductase	S	P
p	MIC17	Antigen identified by monoclonal antibody BB1	S	P
p	MSK3	Antigen identified by monoclonal antibody M68	S	P
cen–q13	D12S6	DNA segment, polymorphic[c]	S, RE	P
cen–q14	MIP	Major intrinsic protein of lens fiber[c]	S	P
cen–q13	MSK4	Antigen identified by monoclonal antibody A123/A127	S	P
cen–qter	MSK27	Antigen identified by monoclonal antibody MG6	S	P
q12–q14	SHMT	Serine hydroxymethyl transferase	S, RE	C
q13–qter	MSK7	Antigen identified by monoclonal antibody VI	S	P
q13.1–q14.3	COL2A1	Collagen, type II, alpha 1[c]	S, RE	C
q14–qter	D12S8	DNA segment, polymorphic[c]	S	C
q14.3–qter	D12S7	DNA segment, polymorphic[c]	S	C
q21–q22	D12S5	DNA segment, arbitrary[c]	S	C
q21	PEPB	Peptidase B	S	C
q22–q24.1	IGF1	Insulin-like growth factor 1[c]	A, S, RE	C
q22–q24.2	PAH	Phenylalanine hydroxylase[c]	S, RE	C
q24.1	IFNG	Interferon, gamma[c]	A, S, RE	C
q	D12S2	DNA segment, polymorphic[c]	S, RE	C
	A2M	Alpha-2-macroglobulin[c]	S	C
	ALDH2	Aldehyde dehydrogenase 2; mitochondrial[c]	S, RE	C
	ASSP14	Argininosuccinate synthetase pseudogene 14[c]	S, RE	P
	CD4	Antigen CD4p55 identified by monoclonal antibodies T4, Leu3a, and 9106	S	P
	D12S1	DNA segment, arbitrary[c]	S, RE	C
	D12S3	DNA segment, polymorphic[c]	S, RE	P
	D12S4	DNA segment, polymorphic[c]	S, RE	C
	ELA1	Elastase 1	S, RE	P
	GPD1	Glycerol-3-phosphate dehydrogenase	S	P
	MARS	Methionine tRNA synthetase	S	P

(Continued)

TABLE I. The 24 Human Gene Maps—(Continued)

Map location	Symbol	Marker name	Mode[a]	Status[b]
		Human chromosome 12 map		
	PYHG6	Protein spot in 2-dimensional (2-D) gels (MW 85K)	S	P
	PYHG10	Protein spot in 2-D gels (MW 77K)	S	P
	PYHG19	Protein spot in 2-D gels (MW 12K)	S	P
	S8	Surface antigen (chromosome 12) 1	S	P
	TPI2	Triosephosphate isomerase 2	S	T
		Human chromosome 13 map		
pter–q12	D13S27	DNA segment, arbitrary[c]	RE, S	P
p12	RNR1	Ribosomal RNA 1[c]	A	C
q11–q22	D13S10	DNA segment, polymorphic[c]	RE, S, F	C
q12–q24	COLL1	Collagen-like 1[c]	A	P
q12–q14	D13S1	DNA segment, polymorphic[c]	RE, S, D, F	C
q12–q22	D13S5	DNA segment, polymorphic[c]	RE, S, D, F	C
q12–q22	D13S6	DNA segment, polymorphic[c]	RE, S, D, F	C
q12–q22	D13S7	DNA segment, polymorphic[c]	RE, S, D, F	C
q12–q22	D13S9	DNA segment, arbitrary[c]	RE, S	P
q12–q13	D13S11	DNA segment, polymorphic[c]	S, RE	P
q12–q13	D13S19	DNA segment, arbitrary[c]	RE, S	P
q12–q13	D13S28	DNA segment, arbitrary[c]	RE, S	P
q12–q14	D13S29	DNA segment, arbitrary[c]	RE, S	P
q12	FTHL7	Ferritin, heavy-polypeptide-like 7[c]	A	P
q13	COL4A1	Collagen, type IV, alpha 1[c]	S, RE	P
q14	D13S21	DNA segment, arbitrary[c]	RE, S, D	C
q14	D13S22	DNA segment, arbitrary[c]	RE, S, D	C
q14	D13S25	DNA segment, arbitrary[c]	RE, S	P
q14	D13S30	DNA segment, arbitrary[c]	RE, S	P
q14.1	ESD	Esterase D	S, F, D	C
q14.1–q14.3	LCP1	Lymphocyte cytosolic protein 1	D, F	P
q14.1	RB1	Retinoblastoma 1	F, D	C
q21–qter	D13S12	DNA segment, polymorphic[c]	RE, S	P
q21–q22	D13S13	DNA segment, arbitrary[c]	RE, S	P
q21–qter	D13S16	DNA segment, arbitrary[c]	RE, S	P
q21–qter	D13S24	DNA segment, arbitrary[c]	RE, S	P
q21–q22	D13S26	DNA segment, arbitrary[c]	RE, S	P
q22	D13S2	DNA segment, polymorphic[c]	RE, S, D, F	C
q22–qter	D13S3	DNA segment, polymorphic[c]	RE, S, D, F	C
q22–qter	D13S4	DNA segment, polymorphic[c]	RE, S, D, F	C
q34	F7	Coagulation factor VII	D	C
q34	F10	Coagulation factor X[c]	D, S, RE	C
	D13S15	DNA segment, polymorphic[c]	RE, S	P
	D13S17	DNA segment, arbitrary[c]	RE, S	P
	D13S18	DNA segment, arbitrary[c]	RE, S	P
	ERCM2	Excision repair complementing defective repair in mouse cells	S	P
	OSRC	Osteosarcoma	D	P

TABLE I. The 24 Human Gene Maps—(Continued)

Map location	Symbol	Marker name	Mode[a]	Status[b]
		Human chromosome 13 map		
	PCCA	Propionyl coenzyme A carboxylase, alpha polypeptide[c]	S, RE	P
	TUBBP2	Tubulin, beta polypeptide pseudogene 2	S, RE	P
	WND	Wilson's disease	F	P
		Human chromosome 14 map		
pter–q21	TCRA	T-cell receptor (rearranging), alpha polypeptide[c]	S, RE	C
p12	RNR2	Ribosomal RNA 2[c]	A, RE, S	C
q11	D14S7	DNA segment, arbitrary[c]	S, RE	P
q13.1	NP	Nucleoside phosphorylase[c]	S, D	C
q21–q31	FOS	Murine FBJ osteosarcoma viral (v-fos) oncogene homologue[c]	A, S	C
q21–qter	WARS	Tryptophanyl-tRNA synthetase	S	C
q22–qter	PGFT	Phosphoribosylglycinamide formyltransferase	S	P
q31–qter	AACT	Alpha-1-antichymotrypsin[c]	A	P
q32	AKT1	Murine thymoma viral (v-akt) oncogene homologue[c]	A, S, RE	P
q32	CKBB	Creatine kinase, brain form[c]	S, RE	C
q32.1–q32.2	D14S1	DNA segment, polymorphic[c]	S, RE, A	C
q32.1	PI	Alpha-1-antitrypsin (protease inhibitor)[c]	D, F, S	C
q32.3	IGD1	Immunoglobulin heavy chain diversity region 1[c]	D, S, RE	C
q32.3	IGHA1	Immunoglobulin alpha 1 polypeptide[c]	S, RE	C
q32.3	IGHA2	Immunoglobulin alpha 2 polypeptide (A2m marker)[c]	F, S, RE	C
q32.3	IGHD	Immunoglobulin delta polypeptide[c]	S, RE	C
q32.3	IGHE	Immunoglobulin epsilon polypeptide[c]	S, RE	C
q32.3	IGHEP1	Immunoglobulin epsilon polypeptide pseudogene 1[c]	S, RE	C
q32.3	IGHG1	Immunoglobulin gamma 1 polypeptide (Gm marker)[c]	D, F, S, RE	C
q32.3	IGHG2	Immunoglobulin gamma 2 polypeptide (Gm marker)[c]	D, F, S, RE	C
q32.3	IGHG3	Immunoglobulin gamma 3 polypeptide (Gm marker)[c]	D, F, S, RE	C
q32.3	IGHG4	Immunoglobulin gamma 4 polypeptide (Gm marker)[c]	D, F, S, RE	C
q32.3	IGHGP1	Immunoglobulin gamma polypeptide pseudogene 1[c]	D, RE	P
q32.3	IGHJ	Immunoglobulin heavy polypeptide joining region[c]	S, RE	C

(Continued)

TABLE I. The 24 Human Gene Maps—(Continued)

Map location	Symbol	Marker name	Mode[a]	Status[b]
		Human chromosome 14 map		
q32.3	IGHM	Immunoglobulin mu polypeptide[c]	D, S, RE	C
q32.3	IGHV	Immunoglobulin heavy polypeptide variable region (polypeptide not specified)[c]	F, S, RE	C
	COLL2	Collagen-like 2[c]	S, RE	P
	D14S2	DNA segment, arbitrary[c]	RE, S	P
	D14S3	DNA segment, polymorphic[c]	RE, S	P
	D14S4	DNA segment, polymorphic[c]	RE, S	P
	D14S5	DNA segment, arbitrary[c]	RE, S	P
	D14S6	DNA segment, arbitrary[c]	RE, S	P
	D14S8	DNA segment, arbitrary[c]	RE	P
	D14S9	DNA segment, arbitrary[c]	RE	P
	D14S10	DNA segment, arbitrary[c]	RE	P
	D14S11	DNA segment, arbitrary[c]	RE	P
	ESAT	Esterase activator	S	P
	LCH	Lentil agglutinin binding	S	P
	M195	External membrane protein (MW 195K)	S	P
	MSK30	Antigen identified by monoclonal antibody A42	S	P
	PFGS	Phosphoribosylformylglycinamidine synthetase (formylglycinamide ribotide)	S	P
		Human chromosome 15 map		
pter–q21	SORD	Sorbitol dehydrogenase	S	P
p12	RNR3	Ribosomal RNA 3[c]	A, RE, S	C
q11–qter	ACTC	Actin, alpha; cardiac muscle[c]	S, RE	P
q11–qter	HCVS	Human coronavirus sensitivity	S	P
q11–q12	IGD2	Immunoglobulin heavy chain diversity region 2[c]	A	P
q11–qter	MANA	Mannosidase, alpha[c]	S	P
q11–q22	MIC7	Antigen identified by monoclonal antibody 28.3.7	S	C
q11–q22	MIC12	Antigen identified by monoclonal antibody 30.2A8	S	P
q11	PWS	Prader–Willi syndrome	D	C
q12–q24	D15S4	DNA segment, polymorphic[c]	RE, S, A	C
q13–q15	B2MR	Beta-2-microglobulin regulator	D	C
q14–q21	D15S1	DNA segment, polymorphic[c]	RE, S, A, F	C
q15–q22	D15S2	DNA segment, polymorphic[c]	RE, S, A, F	C
q15–q24	D15S6	DNA segment, arbitrary[c]	RE, S, A	P
q21–q22	B2M	Beta-2-microglobulin[c]	S, RE	C
q21–q23	COLL3	Collagen-like 3[c]	A	P
q21–qter	IDH2	Isocitrate dehydrogenase 2; mitochondrial	S	C

TABLE I. The 24 Human Gene Maps—(Continued)

Map location	Symbol	Marker name	Mode[a]	Status[b]
		Human chromosome 15 map		
q22–q25.1	HEXA	Hexosaminidase A (alpha polypeptide)[c]	S, RE	C
q22–qter	MPI	Mannose phosphate isomerase	S	C
q22–qter	PKM2	Pyruvate kinase	S	C
q25–q26	FES	Feline sarcoma viral (v-fes) oncogene; Fujinami avian sarcoma viral (v-fps) oncogene homologue[c]	A, S, RE	C
	CYP2	Cytochrome P450, dioxin-inducible 1[c]	S, RE	P
	D15S3	DNA segment, polymorphic[c]	RE	P
	GANC	Glucosidase, alpha; neutral C	S	P
	MSK15	Antigen identified by monoclonal antibody SV13	S	P
	MSK16	Antigen identified by monclonal antibody A0122	S	P
	MSK17	Antigen identified by monoclonal antibody F23	S	P
	PYHG11	Protein spot in 2-dimensional (2-D) gels (MW 74K)	S	P
	PYHG20	Protein spot in (2-D) gels (MW 10K)	S	P
	RPL32	Ribosomal protein L32	RE	P
	SRD1	Specific reading disability 1	S	T
		Human chromosome 16 map		
pter–p12	HBA1	Hemoglobin, alpha 1[c]	A, S	C
pter–p12	HBA2	Hemoglobin, alpha 2[c]	A, S	C
pter–p12	HBAP1	Hemoglobin, alpha pseudogene 1[c]	A, S	C
pter–p12	HBZ1	Hemoglobin, zeta 1[c]	A, S	C
pter–p12	HBZ2	Hemoglobin, zeta 2[c]	A, S	C
pter–p11	PGP	Phosphoglycollate phosphatase	S, F	C
p12–q22	GOT2	Glutamic-oxaloacetic transaminase 2; mitochondrial	F, S	C
q12–q22	DIA4	Diaphorase (NADH/NADPH) (cytochrome b_5 reductase)	F, S	C
q22	APRT	Adenine phosphoribosyltransferase[c]	S, D	C
q22	LCAT	Lecithin-cholesterol acyltransferase	A, F, S	C
q22	MT1	Metallothionein 1[c]	S, RE	C
q22	MT2	Metallothionein 2[c]	S, RE	C
q22.1	HP	Haptoglobin[c]	A, CH, F, D	C

(Continued)

TABLE I. The 24 Human Gene Maps—(Continued)

Map location	Symbol	Marker name	Mode[a]	Status[b]
		Human chromosome 16 map		
q22.1	HPR	Haptoglobin-like[c]	RE	C
	AVR	Antiviral state regulator	S	P
	CTH	Cystathionase	S	P
	CTRB	Chymotrypsinogen B[c]	S, RE	P
	D16S1	DNA segment, polymorphic[c]	S, RE	P
	D16S2	DNA segment, arbitrary[c]	S, RE	P
	ESB3	Esterase B3	S	P
	HAGH	Hydroxyacyl glutathione hydrolase	S	P
	IFNR	Interferon production regulator	S	P
	LIPB	Lipase B	S	P
	NHCP1	Nonhistone chromosome protein 1	S	P
	PKD1	Polycystic kidney disease, adult 1	F	P
	PNI2	Protein (serum) spot in 2-dimensional (2-D) gels (MW 33KD)		
	TK2	Thymidine kinase (mitochondrial)[c]	S	P
	VDI	Vesicular stomatitis virus defective interfering particle suppressor	S	P
		Human chromosome 17 map		
pter–p13	D17S1	DNA segment, polymorphic[c]	S, F, RE	C
pter–p11	MYH1	Myosin heavy polypeptide; skeletal muscle, adult 1[c]	S, RE	C
pter–p11	MYH2	Myosin heavy polypeptide; skeletal muscle; adult 2[c]	S, F	C
pter–p11	MYH3	Myosin heavy polypeptide; skeletal muscle, embryonic[c]	S, RE	C
pter–p11	MYH4	Myosin heavy polypeptide; skeletal muscle[c]	S, RE	P
p13	TP53	Tumor protein p53[c]	S, RE	C
p11.1–qter	PPY	Pancreatic polypeptide[c]	S, RE	P
q11–q12.21	ERBA1	Avian erythroblastic leukemia viral (v-erb-a) oncogene homologue 1[c]	S, RE	C
q11–q22	HOX2	Homeobox region 1[c]	S, RE	C
q21–q22	GALK	Galactokinase	S	C
q21–q22	TK1	Thymidine kinase, soluble[c]	S, RE	C
q21–q22	A12M4	Adenovirus-12 chromosome modification site 17	S	P
q21–q22	RNU2	U1 small nuclear RNA 2[c]	A	P

TABLE I. The 24 Human Gene Maps—(Continued)

Map location	Symbol	Marker name	Mode[a]	Status[b]
		Human chromosome 17 map		
q21–q22	NGL	Neuro-glioblastoma derived proto-oncogene homologue, rat-derived[c]	S, A	P
q21–qter	MIC6	Antigen identified by monoclonal antibody H267	S	P
q21.31–q22	COL1A1	Collagen, type I, alpha 1[c]	S, RE, A	C
q22–q24	GH1	Growth hormone 1[c]	RE, S, A	C
q22–q24	GH2	Growth hormone 2[c]	RE, S	C
q22–q24	CSH1	Chorionic somatomammotropin hormone 1[c]	RE, S, A	C
q22–q24	CSH2	Chorionic somatomammotropin hormone 2[c]	RE, S	C
q22–q24	CSHP1	Chorionic somatomammotropin pseudogene 1[c]	RE, S	C
q23	GAA	Glucosidase, alpha, acid[c]	S, A, RE	C
q23–qter	PEPE	Peptidase E	S	P
q	CRYB1	Crystallin, beta polypeptide 1[c]	S, RE	P
q	GAS	Gastrin[c]	S, RE	P
	CD7	Antigen CD7p41 identified by monoclonal antibodies 3A1, 4A, CL1.3, and TR41	S	P
	D17S2	DNA segment, polymorphic[c]	RE	
	D17S3	DNA segment, polymorphic[c]	RE	
	MSK18	Antigen identified by monoclonal antibody J143	S	P
	S9	Surface antigen (chromosome 17) 1	S	P
		Human chromosome 18 map		
pter–q12	TS	Thymidylate synthase[c]	S	P
p11.3	D18S3	DNA segment, polymorphic[c]	S, A	C
q21.3	YES1	Yamaguchi sarcoma viral (v-yes-1) oncogene homologue 1[c]	S, RE	P
q21.3	BCL2	B cell CLL/lymphoma[c]	CH, RE	P
q23	PEPA	Peptidase A	S, D	C
	ACTBP3	Actin, beta pseudogene 3[c]	S, RE	P
	CGH	Chorionic gonadotropin	S	P
	D18S1	DNA segment, polymorphic[c]	S, RE	P
	D18S4	DNA segment, arbitrary[c]	S, RE	P
	DHFRP1	Dihydrofolate reductase pseudogene 1[c]	S, RE	P
	ERV1	Endogenous retroviral sequence 1	S	P
	GRP	Gastrin-releasing peptide[c]	S, RE	P
	IFNGR	Interferon, gamma; receptor for	S	P
	MTL3	Metallothionein-like 3[c]	S, RE	P
	NARS	Asparaginyl-tRNA synthetase	S	P
	PALB	Prealbumin	S	P

(Continued)

TABLE I. The 24 Human Gene Maps—(Continued)

Map location	Symbol	Marker name	Mode[a]	Status[b]
		Human chromosome 19 map		
p13.3–p13.2	C3	Complement component 3[c]	S, A	C
p13.3–p13.2	INSR	Insulin receptor[c]	S, A	P
p13.2–p13.1	LDLR	Low-density-lipoprotein receptor (familial hypercholesterolemia)[c]	S, A	C
p13.2–cen	D19S8	DNA segment, polymorphic[c]	S, RE	P
p13.2–cen	D19S11	DNA segment, polymorphic[c]	S, A	C
p13.2–cen	PEPD	Peptidase D	S	C
p13.2–q13.2	MANB	Mannosidase, alpha, B	S	C
p13.2–q13.2	D19S7	DNA segment, polymorphic[c]	S, RE	P
p13.2–q13.2	D19S9	DNA segment, polymorphic[c]	S, RE	P
p13.2–q13.4	DNL	Lysosomal DNAase	S	C
p13.1–cen	D19S5	DNA segment, polymorphic[c]	S, A	P
cen–q13.2	APOC1	Apolipoprotein CI[c]	S	C
cen–q13.2	APOC2	Apolipoprotein CII[c]	F, S	C
cen–q13.2	APOE	Apolipoprotein E[c]	S, F	C
cen–q13.2	GPI	Glucose phosphate isomerase	S, D	C
q13–qter	PVS	Polio virus sensitivity	S	C
q13.1–q13.3	CYP1	Cytochrome P450, phenobarbital-inducible 1[c]	S, A	C
q13.1–q13.3	TGFB	Transforming growth factor, type beta[c]	S, A	P
q13.2–q13.3	ERCC1	Excision repair complementing defective repair in Chinese hamster cells 1[c]	S	C
q13.3	CGB	Chorionic gonadotropin, beta polypeptide[c]	S	C
q13.3	MER5	Cell-surface antigen identified by monoclonal antibody 208	S	P
q13.3–q13.4	FTL	Ferritin, light polypeptide[c]	S, A	C
q13.3–qter	D19S6	DNA segment, polymorphic[c]	S, A	P
q	E11S	ECHO-11 virus sensitivity	S	C
	BCT2	Branched-chain aminotransferase 2	S	P
	CKMM	Creatine kinase, muscle form[c]	S, RE	P
	D19S1	DNA segment, arbitrary[c]	S, RE	P
	D19S2	DNA segment, arbitrary[c]	S, RE	P
	D19S4	DNA segment, polymorphic[c]	S	P
	DM	Dystrophia myotonia	F	C
	EF2	Elongation factor-2	S	P
	GUSM	Mouse beta-glucuronidase modifier	S	P
	H	H blood group	F	C
	LE	Lewis blood group	F	C
	LHB	Luteinizing hormone, beta subunit[c]	S, RE	C
	LU	Lutheran blood group	F	C
	LW	Landsteiner–Wiener blood group	F	P
	M7V1	Baboon virus M7 receptor	S	P
	MSK19	Cell-surface antigen identified by monoclonal antibody F8	S	P

TABLE I. The 24 Human Gene Maps—(Continued)

Map location	Symbol	Marker name	Mode[a]	Status[b]
		Human chromosome 19 map		
	MSK20	Cell-surface antigen identified by monoclonal antibody F10	S	P
	NF1	Neurofibromatosis 1 (von Recklinghausen's disease)	F	I
	PGK2	Phosphoglycerate kinase-2[c]	S, RE	C
	PNI1	Protein spot NC22 in 2-dimensional (2-D) gels (MW 50K; pI 5.00)	F	P
	RCC	Repair complementing defective repair in Chinese hamster cells	S	P
	RDRC	RD114 virus receptor	S	P
	SE	ABH secretion	F	C
		Human chromosome 20 map		
p12	D20S5	DNA segment, polymorphic[c]	A	P
p12.2	MEN2A	Multiple endocrine neoplasia type II A	D, F	I
p	D20S6	DNA segment, polymorphic[c]	S, F	P
p	ITPA	Inosine triphosphatase (nucleoside triphosphate pyrophosphatase)	S	C
cen–q13.1	AHCY	S-Adenosylhomocysteine hydrolase	S	C
q12–q13	SRC1	Avian sarcoma viral (v-src) oncogene homologue 1[c]	S, RE, A	C
q12–q13.3	FTLL1	Ferritin light polypeptide-like 1[c]	A	P
q13.2	D20S4	DNA segment, polymorphic[c]	S, RE	C
q13.2–qter	ADA	Adenosine deaminase[c]	S, RE	C
	ARVP	Arginine vasopressin	S	P
	DCE	Desmosterol-to-cholesterol enzyme	S	P
	D20S1	DNA segment, arbitrary[c]	S, RE	P
	D20S2	DNA segment, arbitrary[c]	S, RE	P
	D20S3	DNA segment, arbitrary[c]	S, RE	P
	GHRF	Growth-hormone-releasing factor[c]	S, RE	C
	GSL	Galactosialidosis	S	P
	HTL	High L-leucine transport	S	P
	MTL4	Metallothionein-like 4[c]	S, RE	T
		Human chromosome 21 map		
pter–q21.1	D21S13	DNA segment, arbitrary[c]	S, RE	P
pter–q21.2	D21S26	DNA segment, polymorphic[c]	S, RE	P
pter–q21	D21S16	DNA segment, polymorphic[c]	S, RE	P
pter–q22	D21S46	DNA segment, arbitrary[c]	S, RE	P
pter–q22	D21S48	DNA segment, arbitrary[c]	S, RE	P
p11.2	D21S5	DNA segment, polymorphic[c]	S, RE, F	P
p12	RNR4	Ribosomal RNA 4[c]	A	C
p	D21S24	DNA segment, polymorphic[c]	S, RE	P
cen–q22.2	D21S4	DNA segment, polymorphic[c]	S, RE, F	P

(Continued)

TABLE I. The 24 Human Gene Maps—(Continued)

Map location	Symbol	Marker name	Mode[a]	Status[b]
		Human chromosome 21 map		
cen–q22.2	*D21S12*	DNA segment, polymorphic[c]	S, RE, F	P
cen–q22.2	*D21S54*	DNA segment, polymorphic[c]	S, RE	P
cen–q22.2	*D21S58*	DNA segment, polymorphic[c]	S, RE	P
cen–q22.2	*D21S59*	DNA segment, polymorphic[c]	S, RE	P
cen–q22	*D21S27*	DNA segment, arbitrary[c]	S, RE	P
q11.2–q21	*D21S1*	DNA segment, polymorphic[c]	S, F, A, D	C
q11.2–q21	*D21S11*	DNA segment, polymorphic[c]	S, F, A, D	C
q21	*D21S8*	DNA segment, polymorphic[c]	S, F, A, D	C
q21–q22.1	*CBS*	Cystathionine beta-synthase[c]	S, A, D	C
q21–qter	*IFNAR*	Interferon, alpha; receptor for	S, D	C
q21–qter	*IFNBR*	Interferon, beta; receptor for	S, RE	P
q21.2–qter	*D21S15*	DNA segment, polymorphic[c]	S, RE	P
q21.2–qter	*D21S17*	DNA segment, polymorphic[c]	S, RE	P
q21.2–qter	*D21S18*	DNA segment, polymorphic[c]	S, RE	P
q21.2–qter	*D21S19*	DNA segment, polymorphic[c]	S, RE	P
q22	*D21S33*	DNA segment, arbitrary[c]	S, RE	P
q22	*D21S35*	DNA segment, arbitrary[c]	S, RE	P
q22	*D21S39*	DNA segment, arbitrary[c]	S, RE	P
q22	*D21S40*	DNA segment, arbitrary[c]	S, RE	P
q22	*D21S41*	DNA segment, arbitrary[c]	S, RE	P
q22	*D21S42*	DNA segment, arbitrary[c]	S, RE	P
q22	*D21S43*	DNA segment, arbitrary[c]	S, RE	P
q22	*D21S44*	DNA segment, arbitrary[c]	S, RE	P
q22	*D21S45*	DNA segment, arbitrary[c]	S, RE	P
q22	*D21S47*	DNA segment, arbitrary[c]	S, RE	P
q22	*D21S49*	DNA segment, arbitrary[c]	S, RE	P
q22	*D21S50*	DNA segment, arbitrary[c]	S, RE	P
q22	*D21S51*	DNA segment, arbitrary[c]	S, RE	P
q22	*PFKL*	Phosphofructokinase, liver type	S, D	C
q22.1	*PRGS*	Phosphoribosylglycinamide synthetase	S	C
q22.1	*SOD1*	Superoxide dismutase 1, soluble[c]	S, D, F, RE	C
q22.1–q22.2	*D21S25*	DNA segment, polymorphic[c]	S, RE	P
q22.3	*BCEI*	Estrogen-inducible sequence, expressed in breast cancer	S, A	C
q22.3	*D21S3*	DNA segment, polymorphic[c]	S, F, RE	P
q22.3	*D21S23*	DNA segment, polymorphic[c]	F, RE	P
q22.3	*D21S52*	DNA segment, polymorphic[c]	S, RE	P
q22.3	*D21S53*	DNA segment, polymorphic[c]	S, RE	P
q22.3	*D21S55*	DNA segment, polymorphic[c]	S, RE	P
q22.3	*D21S56*	DNA segment, polymorphic[c]	S, RE	P
q22.3	*D21S57*	DNA segment, polymorphic[c]	S, RE	P
q22.3	*D21S60*	DNA segment, arbitrary[c]	S, RE	P
q22.3	*D21S64*	DNA segment, arbitrary[c]	S, RE	P
q22.3	*D21S70*	DNA segment, arbitrary[c]	S, RE	P
q22.3	*D21S71*	DNA segment, arbitrary[c]	S, RE	P

TABLE I. The 24 Human Gene Maps—(Continued)

Map location	Symbol	Marker name	Mode[a]	Status[b]
		Human chromosome 21 map		
q	*D21S59*	DNA segment, polymorphic[c]	S, RE	P
q	*D21S61*	DNA segment, arbitrary[c]	S, RE	P
q	*D21S62*	DNA segment, arbitrary[c]	S, RE	P
q	*D21S63*	DNA segment, arbitrary[c]	S, RE	P
q	*D21S65*	DNA segment, arbitrary[c]	S, RE	P
q	*D21S66*	DNA segment, arbitrary[c]	S, RE	P
q	*D21S67*	DNA segment, arbitrary[c]	S, RE	P
q	*D21S68*	DNA segment, arbitrary[c]	S, RE	P
q	*D21S69*	DNA segment, arbitrary[c]	S, RE	P
	AABT	Beta-amino acid transport[c]	S, RE	P
	CRYA1	Crystallin, alpha-A2 polypeptide[c]	S, RE	P
	D21S2	DNA segment, arbitrary[c]	S, RE	P
	D21S6	DNA segment, polymorphic[c]	S, RE	P
	D21S7	DNA segment, polymorphic[c]	S, RE	P
	D21S9	DNA segment, arbitrary[c]	S, RE	P
	D21S10	DNA segment, arbitrary[c]	S, RE	P
	D21S20	DNA segment, arbitrary[c]	S, RE	P
	D21S22	DNA segment, polymorphic[c]	S, RE	P
	D21S30	DNA segment, polymorphic[c]	S, RE	P
	D21S31	DNA segment, arbitrary[c]	S, RE	P
	D21S36	DNA segment, arbitrary[c]	S, RE	P
	D21S37	DNA segment, arbitrary[c]	S, RE	P
	D21S38	DNA segment, arbitrary[c]	S, RE	P
	HTOR	5-Hydroxytryptamine oxygenase regulator	D	P
	MFI3	Antigen (glycoprotein; MW 86K) identified by monoclonal antibody 2B2	S	P
	MFI4	Antigen (glycoprotein; MW 145K) identified by monoclonal antibody 2B2	S	P
	MFI7	Glycoprotein (MW 90K) identified by monoclonal antibody 60.3; possible leukocyte cell-adhesion molecule	S	P
	PAIS	Phosphoribosylaminoimidazole synthetase	S	P
	S14	Surface antigen (chromosome 21)	S	P
		Human chromosome 22 map		
pter–q11	*D22S3*	DNA segment, polymorphic[c]	S, RE	P
pter–q11	*IDUA*	Iduronidase, alpha-L	S, D	P
p12	*RNR5*	Ribosomal RNA 5[c]	A	C
q11	*DGS*	DiGeorge syndrome	D	C

(Continued)

TABLE I. The 24 Human Gene Maps—(Continued)

Map location	Symbol	Marker name	Mode[a]	Status[b]
		Human chromosome 22 map		
q11	*IGLV*	Immunoglobulin lambda polypeptide, variable region[c]	S, F	C
q11	*IGLC*	Immunoglobulin lambda polypeptide, constant region[c]	S, A,F	C
q11	*BCR*	Breakpoint cluster region[c]	RE	C
q11	*D22S9*	DNA segment, polymorphic[c]	A	C
q11–qter	*D22S5*	DNA segment, polymorphic[c]	S, RE	P
q11.2–q13.31	*ACO2*	Aconitase 2; mitochondrial	S	C
q11–q13	*D22S1*	DNA segment, polymorphic[c]	S, A, F	C
q11.2–q13	*MB*	Myoglobin[c]	S, F, RE	C
q11.2–qter	*D22S4*	DNA segment, arbitrary[c]	S, RE	P
q11.2–qter	*D22S6*	DNA segment, arbitrary[c]	S, RE	P
q11.2–qter	*D22S7*	DNA segment, arbitrary[c]	S, RE	P
q11.2–qter	*D22S8*	DNA segment, arbitrary[c]	S, RE	P
q11.2–qter	*P1*	P blood group (P1 antigen)	F	P
q12.3–q13.1	*SIS*	Simian sarcoma viral (v-sis) oncogene homologue (structural gene for platelet-derived growth factor)[c]	S, A, F	C
q13–qter	*NAGA*	Acetylgalactosaminidase; alpha-N[c]	S	C
q13.31–qter	*ARSA*	Arylsulfatase A	S	C
q13.31–qter	*DIA1*	Diaphorase 1 (NADH-cytochrome b_5 reductase)	S	C
	ALDOA	Aldolase A	S	I
	D22S2	DNA segment, polymorphic[c]	S, RE	P
	D22S10	DNA segment, polymorphic[c]	S, RE	P
	GLB2	Galactosidase; beta 2	S	P
	PPGB	Stabilizing protein for beta-galactosidase	S	P
	S13	Surface antigen (chromosome 22)	S	P
		Human X chromosome map		
pter–p22.32	*CDPX*	Chondrodysplasia punctata	F, D	P
pter–p22.32	*MIC2* (*MIC2X*)	Antigen identified by monoclonal antibodies 12E7, F21, and 013[c]	S, RE, D	P
pter–p22.32	*STS*	Steroid sulfatase (microsomal)	F, S, D	C
pter–p22.3	*DXS31*	DNA segment, polymorphic[c]	RE	P
pter–p22.3	*XG*	Xg blood group	F, D	C
pter–p22	*DXS89*	DNA segment, polymorphic[c]	RE	P
pter–p22	*OA*	Ocular albinism	F	C
pter–p21	*RS*	Retinoschisis	F	C
pter–p21	*XK*	Kell blood group precursor	F, D	C
pter–q13	*PSF2*	Protein spot in 2-dimensional (2-D) gels (MW 27K)	S	P
pter–q13	*PSF3*	Protein spot in 2-D gels (MW 37K)	S	P

TABLE I. The 24 Human Gene Maps—(Continued)

Map location	Symbol	Marker name	Mode[a]	Status[b]
		Human X chromosome map		
pter–q13	PSF4	Protein spot in 2-D gels (MW 40K)	S	P
pter–q28	HRAS2	Harvey rat sarcoma 2 viral (v-Haras2) oncogene homologue	S	P
pter–p21	CGD	Chronic granulomatous disease	F, D	C
p22.3	DXS69	DNA segment, arbitrary[c]	RE	P
p22.3	DXS22	DNA segment, arbitrary[c]	RE	P
p22.3	DSX70	DNA segment, arbitrary[c]	RE	P
p22.3	DXYS14[d]	DNA segment, polymorphic[c]	RE	P
p22.3	DXYS15[d]	DNA segment, polymorphic[c]	RE	P
p22.3	DXYS17[d]	DNA segment, polymorphic[c]	RE	P
p22.3–p22.2	DXS85	DNA segment, polymorphic[c]	RE, F	C
p22.3–p22.1	DXS123	DNA segment, polymorphic[c]	RE	P
p22.3–p22.1	DXS149	DNA segment, arbitrary[c]	RE	P
p22.3–p21	DXS68	DNA segment, arbitrary[c]	RE	C
p22.3–p21.3	DXS103	DNA segment, polymorphic[c]	RE	P
p22.3–p21	DXS104	DNA segment, polymorphic[c]	RE	P
p22.3–p21	DXS126	DNA segment, arbitrary[c]	RE	P
p22.3–p21	DXS158	DNA segment, arbitrary[c]	RE	P
p22.3–p21	DXS154	DNA segment, arbitrary[c]	RE	P
p22.3–p21.2	FTLL2	Ferritin, light- polypeptide-like 2[c]	A	P
p22.2–p22.1	DXS41	DNA segment, polymorphic[c]	RE, F	C
p22.2–p22.1	DXS43	DNA segment, polymorphic[c]	RE, F	C
p22.1–p21.3	POLA	Polymerase (DNA), alpha polypeptide	S	C
p22.1–p11	DXS44	DNA segment, arbitrary[c]	RE	P
p22–p21	DXS183	DNA segment, arbitrary[c]	RE	P
p22–p21	DXS189	DNA segment, arbitrary[c]	RE	P
p22–p21	DXS2	DNA segment, polymorphic[c]	RE	P
p22–p21	DXS39	DNA segment, arbitrary[c]	RE	P
p22–p11	DXS32	DNA segment, arbitrary[c]	RE	C
p22–p11	DXS27	DNA segment, arbitrary[c]	RE	P
p22–p11	DXS187	DNA segment, arbitrary[c]	RE	P
p22–q26	DXS176	DNA segment, arbitrary[c]	RE	P
p22	AIC	Aicardi syndrome	F, CH	P
p22	DXS9	DNA segment, polymorphic[c]	RE, F	C
p22	DXS16	DNA segment, polymorphic[c]	RE, F	C
p22	DXS89	DNA segment, polymorphic[c]	RE	P
p22	DXS143	DNA segment, polymorphic[c]	RE	P
p21.3	DXS28	DNA segment, polymorphic[c]	RE, F	C
p21.3	DXS67	DNA segment, polymorphic[c]	RE	C
p21.2–p21.1	DXS84	DNA segment, polymorphic[c]	RE, F	C
p21.2–p21.3	AHC	Adrenal hypoplasia, congenital	F, D, RE	C

(Continued)

TABLE I. The 24 Human Gene Maps—(Continued)

Map location	Symbol	Marker name	Mode[a]	Status[b]
		Human X chromosome map		
p21.2–p21.3	GK	Glycerol kinase	F, D, RE	C
p21.1	DXS196	DNA segment, arbitrary[c]	RE	C
p21–p11	DXS90	DNA segment, polymorphic[c]	RE	P
p21–p11	DXS117	DNA segment, arbitrary[c]	RE	P
p21–p11	DXS150	DNA segment, arbitrary[c]	RE	P
p21–p11	DXS166	DNA segment, arbitrary[c]	RE	P
p21–p11	GAPDP1	Glyceraldehyde-3-phosphate dehydrogenase pseudogene 1[c]	S, RE	C
p21	DXS140	DNA segment, arbitrary[c]	RE	P
p21	DXS141	DNA segment, arbitrary[c]	RE	C
p21	DXS142	DNA segment, arbitrary[c]	RE	C
p21	DXS148	DNA segment, arbitrary[c]	RE	P
p21	DXS164	DNA segment, polymorphic[c]	RE	C
p21	BMD	Muscular dystrophy, Becker type	F	C
p21	DMD	Muscular dystrophy, Duchenne type	F, D	C
p21	OTC	Ornithine carbamoyltransferase[c]	F, S, RE, A	C
p21–cen	DXS38	DNA segment, arbitrary[c]	RE	P
p21–cen	DXS66	DNA segment, arbitrary[c]	RE	P
p21–cen	DXS71	DNA segment, arbitrary[c]	RE	P
p21–cen	NDP	Norrie's disease (pseudoglioma)	F	C
p21–cen	RP2	Retinitis pigmentosa 2, classic	F, D	C
p11.4–q13	DXS77	DNA segment, arbitrary[c]	RE, F	C
p11.3	DXS7	DNA segment, polymorphic[c]	RE	P
p11	DXS34	DNA segment, arbitrary[c]	RE	P
p11	DXS184	DNA segment, arbitrary[c]	RE	P
p11	IP	Incontinentia pigmenti	F, CH	P
p11–cen	DXS14	DNA segment, polymorphic[c]	RE	P
p11–q11	DHTR	Dihydrotestosterone receptor (testicular feminization; androgen-receptor deficiency)	F, S	C
p11–q11	MNK	Menkes' syndrome	F	C
p11–q12	ACTL1	Actin-like sequence 1[c]	RE, S	P
p11–q12	DXS62	DNA segment, polymorphic[c]	RE	P
p11–q12	DXS146	DNA segment, polymorphic[c]	RE	P
p11–q13	DXS168	DNA segment, polymorphic[c]	RE	P
p11–q21.2	DXS113	DNA segment, arbitrary[c]	RE	P
p11–q21.2	DXS121	DNA segment, arbitrary[c]	RE	P
p11–q21.3	DXS106	DNA segment, polymorphic[c]	RE	P
p11–q21.3	DXS110	DNA segment, arbitrary[c]	RE	P
p11 q21.3	DXS125	DNA segment, arbitrary[c]	RE	P

TABLE I. The 24 Human Gene Maps—(Continued)

Map location	Symbol	Marker name	Mode[a]	Status[b]
		Human X chromosome map		
p11–q21.3	*DXS127*	DNA segment, arbitrary[c]	RE	P
p11–q21.3	*DXS128*	DNA segment, arbitrary[c]	RE	P
p11–q21.3	*DXS131*	DNA segment, arbitrary[c]	RE	P
p11–q21.3	*DXS132*	DNA segment, arbitrary[c]	RE	P
p11–q21.3	*DXS135*	DNA segment, arbitrary[c]	RE	P
p11–q21.3	*DXS153*	DNA segment, arbitrary[c]	RE	P
p11–q21.3	*DXS159*	DNA segment, arbitrary[c]	RE	P
p11–q22	*DXS133*	DNA segment, arbitrary[c]	RE	P
p	*DXYS20[d]*	DNA segment, polymorphic[c]	RE	P
cen–q11	*DXS65*	DNA segment, arbitrary[c]	RE	P
cen–q12	*DXS18*	DNA segment, polymorphic[c]	RE	P
cen–q12	*DXS136*	DNA segment, arbitrary[c]	RE	P
cen–q13	*DXS171*	DNA segment, arbitrary[c]	RE	P
cen–q22	*DXS191*	DNA segment, arbitrary[c]	RE	P
cen–q22	*DXS194*	DNA segment, arbitrary[c]	RE	P
cen–q22	*DXYS22[d]*	DNA segment, arbitrary[c]	RE	P
q11–q12	*DXS61*	DNA segment, arbitrary[c]	RE	P
q11–q13	*DXS1*	DNA segment, polymorphic[c]	RE	P
q11–q13	*DXS91*	DNA segment, polymorphic[c]	RE	P
q11–q13	*DXS94*	DNA segment, polymorphic[c]	RE	P
q11–q13	*DXS96*	DNA segment, polymorphic[c]	RE	P
q11–q13	*PGK1P1*	Phosphoglycerate kinase 1, pseudogene 1[c]	S, RE	C
q11–q21	*DXS88*	DNA segment, polymorphic[c]	RE	P
q11–q21	*DXS188*	DNA segment, arbitrary[c]	RE	P
q11–q22	*DXS26*	DNA segment, arbitrary[c]	RE	P
q11–qter	*DXS178*	DNA segment, polymorphic[c]	RE	P
q12–q13	*DXS55*	DNA segment, arbitrary[c]	RE	P
q12–q21.3	*DXS56*	DNA segment, arbitrary[c]	RE	P
q12–q21.3	*DXS165*	DNA segment, arbitrary[c]	RE	P
q12–q22	*DXYS4[d]*	DNA segment, arbitrary[c]	RE	P
q12–q22	*DXYS5[d]*	DNA segment, arbitrary[c]	RE	P
q12–q22	*DXYS6[d]*	DNA segment, arbitrary[c]	RE	P
q12–q22	*DXYS7[d]*	DNA segment, arbitrary[c]	RE	P
q12–q22	*DXYS8[d]*	DNA segment, arbitrary[c]	RE	P
q12–q22	*DXYS9[d]*	DNA segment, arbitrary[c]	RE	P
q12–q28	*PSF1*	Protein spot in 2-D gels (MW 24K)	S	P
q13	*FGDY*	Faciogenital dysplasia (Aarskog syndrome)	F, CH	C
q13	*PGK1*	Phosphoglycerate kinase 1[c]	S, RE	C
q13–q21	*DXYS25[d]*	DNA segment, arbitrary[c]	RE	P
q13–q21.1	*DXYS1[d]*	DNA segment, polymorphic[c]	RE	P
q13–q21.1	*DXS169*	DNA segment, arbitrary[c]	RE	P
q13–q22	*DXS54*	DNA segment, arbitrary[c]	RE	P

(Continued)

TABLE I. The 24 Human Gene Maps—(Continued)

Map location	Symbol	Marker name	Mode[a]	Status[b]
		Human X chromosome map		
q13–q22	DXS24	DNA segment, arbitrary[c]	RE	P
q13–q22	DXS72	DNA segment, polymorphic[c]	RE	P
q13–q22	DXS73	DNA segment, arbitrary[c]	RE	P
q13–q22	PLP	Proteolipid protein[c]	S	P
q13–q22	ACTBP1	Actin, beta pseudogene 1[c]	S, RE	P
q13–q22	DXYS2[d]	DNA segment, polymorphic[c]	RE	P
q13–q22	DXYS12[d]	DNA segment, polymorphic[c]	RE	P
q13–q24	DXYS21[d]	DNA segment, arbitrary[c]	RE	P
q13–q26	DXS78	DNA segment, arbitrary[c]	RE	P
q13–q27	BA2R	BALB/c 3T3 ts2 temperature-sensitivity complementing	S	P
q13–qter	MIC5	Antigen identified by monoclonal antibody R1	S	P
q21	DXS93	DNA segment, polymorphic[c]	RE	P
q21	DXS95	DNA segment, polymorphic[c]	RE	P
q21	DXS190	DNA segment, arbitrary[c]	RE	P
q21	DXYS27[d]	DNA segment, arbitrary[c]	RE	P
q21–q22	GLA	Galactosidase, alpha[c]	F, S	C
q21–q24	DXS87	DNA segment, polymorphic[c]	RE	P
q21–q26	DXS74	DNA segment, arbitrary[c]	RE	P
q21–q27	PRPS	Phosphoribosyl pyrophosphate synthetase	F, S	C
q21.2–q24	DXS112	DNA segment, arbitrary[c]	RE	P
q21.3–q22	DXS3	DNA segment, polymorphic[c]	RE	P
q21.3–q22	DXS17	DNA segment, polymorphic[c]	RE	P
q21.3–q22	DXS83	DNA segment, arbitrary[c]	RE	P
q21.3–q22	DXS101	DNA segment, polymorphic[c]	RE	P
q21.3–q22	DXS114	DNA segment, arbitrary[c]	RE	P
q21.3–q22	DXS147	DNA segment, arbitrary[c]	RE	P
q21.3–q22	DXS156	DNA segment, arbitrary[c]	RE	P
q21.3–q24	DXS118	DNA segment, arbitrary[c]	RE	P
q22–q23	DXS48	DNA segment, arbitrary[c]	RE	P
q22–q24	DXS4	DNA segment, arbitrary[c]	RE	P
q22–q24	DXS5	DNA segment, arbitrary[c]	RE	P
q22–q24	DXS36	DNA segment, arbitrary[c]	RE	P
q22–q24	DXS82	DNA segment, arbitrary[c]	RE	P
q22–q24	DXS46	DNA segment, arbitrary[c]	RE	P
q22–q24	DXS111	DNA segment, arbitrary[c]	RE	P
q22–q24	DXS124	DNA segment, arbitrary[c]	RE	P
q22–q24	DXS139	DNA segment, polymorphic[c]	RE	P
q22–q26	DXYS13[d]	DNA segment, polymorphic[c]	RE	P
q22–q28	DXS13	DNA segment, arbitrary[c]	RE	P
q22–q28	DXYS10[d]	DNA segment, arbitrary[c]	RE	P
q22–q28	DXYS11[d]	DNA segment, arbitrary[c]	RE	P
q22–qter	DXS192	DNA segment, arbitrary[c]	RE	P
q22–qter	DXS193	DNA segment, arbitrary[c]	RE	P

TABLE I. The 24 Human Gene Maps—(Continued)

Map location	Symbol	Marker name	Mode[a]	Status[b]
		Human X chromosome map		
q23–q25	DXS75	DNA segment, arbitrary[c]	RE	P
q23–q26	DXS30	DNA segment, arbitrary[c]	RE	P
q23–qter	DXS182	DNA segment, arbitrary[c]	RE	P
q23–qter	DXS195	DNA segment, arbitrary[c]	RE	P
q24–q26	DXS8	DNA segment, polymorphic[c]	RE	P
q24–q26	DXS57	DNA segment, arbitrary[c]	RE	P
q24–q26	DXS58	DNA segment, arbitrary[c]	RE	P
q24–q26	DXS138	DNA segment, arbitrary[c]	RE	P
q24–q26	DXS145	DNA segment, arbitrary[c]	RE	P
q24–q26	DXS11	DNA segment, polymorphic[c]	RE	P
q24–q26	GLUDP1	Glutamate dehydrogenase pseudogene 1[c]	S, RE	P
q24–q27	DXS109	DNA segment, arbitrary[c]	RE	P
q24–q27	DXS129	DNA segment, arbitrary[c]	RE	P
q24–q27	DXS151	DNA segment, arbitrary[c]	RE	P
q24–q28	DXS35	DNA segment, arbitrary[c]	RE	P
q24–qter	DXS6	DNA segment, arbitrary[c]	RE	P
q24–qter	DXS12	DNA segment, arbitrary[c]	RE	P
q24–qter	DXS25	DNA segment, arbitrary[c]	RE	P
q24–qter	DXS42	DNA segment, polymorphic[c]	RE	P
q24–qter	DXS63	DNA segment, arbitrary[c]	RE	P
q24–qter	DXS64	DNA segment, arbitrary[c]	RE	P
q26	DXS10	DNA segment, polymorphic[c]	RE	P
q26	DXS53	DNA segment, arbitrary[c]	RE	P
q26	DXS177	DNA segment, polymorphic[c]	RE	P
q26–q27	DXS23	DNA segment, arbitrary[c]	RE	P
q26–q27	DXS37	DNA segment, polymorphic[c]	RE, F	C
q26–q27	DXS59	DNA segment, arbitrary[c]	RE	P
q26–q27	DXS79	DNA segment, polymorphic[c]	RE	P
q26–q27	DXS86	DNA segment, polymorphic[c]	RE	P
q26–q27	DXS92	DNA segment, polymorphic[c]	RE	P
q26–q27	DXS99	DNA segment, polymorphic[c]	RE, F	C
q26–q27	DXS100	DNA segment, polymorphic[c]	RE, F	C
q26–q27	DXS172	DNA segment, arbitrary[c]	RE	P
q26–q27	DXS173	DNA segment, arbitrary[c]	RE	P
q26–q27	DXS174	DNA segment, arbitrary[c]	RE	P
q26–q27	DXS175	DNA segment, arbitrary[c]	RE	P
q26–q27	SIDS	Sulfoiduronate sulfatase (Hunter's syndrome)	F, CH	C
q26–q27.3	DXS19	DNA segment, polymorphic[c]	RE	P
q26–q27.3	F9	Coagulation factor IX (plasma thromboplastic component; Christmas disease; hemophilia B)[c]	F, A, S, RE	C

(Continued)

TABLE I. The 24 Human Gene Maps—(Continued)

Map location	Symbol	Marker name	Mode[a]	Status[b]
		Human X chromosome map		
q26–q27.3	HPRT	Hypoxanthine phosphoribosyltransferase[c]	F, S, RE	C
q26–q28	DXS40	DNA segment, arbitrary[c]	RE	P
q26–q28	DXS98	DNA segment, polymorphic[c]	RE	P
q26–q28	DXS144	DNA segment, polymorphic[c]	RE	P
q26–q28	FTHL8	Ferritin, heavy-polypeptide-like 8[c]	A, RE	
q26–q28	S10	Surface antigen (X-linked) 1	S	P
q26–q28	S11	Surface antigen (X-linked) 2	S	P
q26–qter	DXS181	DNA segment, arbitrary[c]	RE	P
q26–qter	DXS185	DNA segment, arbitrary[c]	RE	P
q27	DXS15	DNA segment, polymorphic[c]	RE, F	C
q27–q28	DXS167	DNA segment, arbitrary[c]	RE	P
q27–q28	DXS80	DNA segment, arbitrary[c]	RE	P
q27–qter	DXS29	DNA segment, arbitrary[c]	RE	P
q27–qter	DXS33	DNA segment, arbitrary[c]	RE	P
q27–qter	DXS51	DNA segment, polymorphic[c]	RE, F	C
q27–qter	DXS81	DNA segment, arbitrary[c]	RE	P
q27–qter	DXS102	DNA segment, polymorphic[c]	RE, F	C
q27–qter	DXS105	DNA segment, polymorphic[c]	RE	P
q27–qter	DXS107	DNA segment, polymorphic[c]	RE	P
q27–qter	DXS115	DNA segment, arbitrary[c]	RE	P
q27–qter	DXS119	DNA segment, arbitrary[c]	RE	P
q27–qter	DXS122	DNA segment, arbitrary[c]	RE	P
q27–qter	DXS130	DNA segment, arbitrary[c]	RE	P
q27–qter	DXS134	DNA segment, arbitrary[c]	RE	P
q27–qter	DXS137	DNA segment, arbitrary[c]	RE	P
q27–qter	DXS152	DNA segment, arbitrary[c]	RE	P
q27–qter	DXS155	DNA segment, arbitrary[c]	RE	P
q27–qter	DXS157	DNA segment, arbitrary[c]	RE	P
q27–qter	DXS160	DNA segment, arbitrary[c]	RE	P
q27–qter	DXS161	DNA segment, arbitrary[c]	RE	P
q27–qter	DXS162	DNA segment, arbitrary[c]	RE	P
q27.3–qter	CBD	Color blindness, deutan[c]	F, RE, S	C
q27.3–qter	CBP	Color blindness, protan[c]	F, RE, S	C
q27.3	FRAXA	Fragile site, folic acid type fra(X) (q27.3) (macroorchidism, mental retardation)	CH	C
q28	ALD	Adrenoleukodystrophy	F	C
q28	DXS49	DNA segment, arbitrary[c]	RE	P
q28	DXS52	DNA segment, polymorphic[c]	RE, F	C
q28	DXS60	DNA segment, arbitrary[c]	RE	P
q28	DXS180	DNA segment, arbitrary[c]	RE	P
q28	F8C	Coagulation factor VIIIc, procoagulant component (hemophilia A)[c]	F, A, S, RE	C

TABLE I. The 24 Human Gene Maps—(Continued)

Map location	Symbol	Marker name	Mode[a]	Status[b]
		Human X chromosome map		
q28	G6PD	Glucose-6-phosphate dehydrogenase[c]	F, S, A, RE	C
q28–qter	TKC	Torticollis, keloids, cryptorchidism, and renal dysplasia	F, CH	P
q	CBBM	Color blindness, blue monochromatic	F	P
q	CMT2	Charcot–Marie–Tooth neuropathy 2	F	P
q	TCD	Choroideremia (progressive tapeto-choroidal dystrophy)	F	C
	ACAD	Acyl-CoA dehydrogenase, multiple; glutaric aciduria II	F	P
	ADS	Albinism–deafness syndrome	F	C
	AIH	Amelogenesis imperfecta, hypomaturation or hypoplastic	F	P
	ALAS	Delta-aminolevulinate synthetase	F	P
	ASB	Anemia, sideroblastic/ hypochromic	F	C
	ASLHN	Alport-syndrome-like hereditary nephritis	F	P
	ASSP4	Argininosuccinate synthetase pseudogene 4[c]	S, RE	P
	ASSP5	Argininosuccinate synthetase pseudogene 5[c]	S, RE	P
	BORJ	Borjeson syndrome	F	P
	CCT	Congenital cataract, total	F	C
	C1HR	C1AGOH temperature sensitivity complementing	S	P
	CLA2	Cerebellar ataxia 2	F	C
	CLPA	Cleft palate	F	P
	CLS	Coffin–Lowry syndrome	F	P
	DFN1	Deafness, progressive	F	P
	DFN2	Deafness, perceptive, congenital	F	C
	DFN3	Deafness, conductive, with fixed stapes	F	C
	DIN	Diabetes insipidus, neurohypophyseal type	F	C
	DIR	Diabetes insipidus, renal	F	C
	DKC	Dyskeratosis, congenital	F	C
	DXS97	DNA segment, polymorphic[c]	RE	P
	DXS108	DNA segment, polymorphic[c]	RE	P

(Continued)

TABLE I. The 24 Human Gene Maps—(Continued)

Map location	Symbol	Marker name	Mode[a]	Status[b]
		Human X chromosome map		
	DXS116	DNA segment, arbitrary[c]	RE	P
	DXS179	DNA segment, arbitrary[c]	RE	P
	DXS186	DNA segment, arbitrary[c]	RE	P
	DXYS18[d]	DNA segment, arbitrary[c]	RE	P
	DXYS19[d]	DNA segment, arbitrary[c]	RE	P
	DXYS23[d]	DNA segment, polymorphic[c]	RE	P
	EBM	Epidermolysis bullosa, macular type	S	P
	EDA	Ectodermal dysplasia, anhydrotic	F	C
	EFE	Endocardial fibroelastosis	F	P
	EMD	Muscular dystrophy, Emery–Dreifus type	F	C
	HPDR	Hypophosphatemia, vitamin-D-resistant rickets	F	C
	HPT	Hypoparathyroidism	F	P
	HSAS	Hydrocephalus, stenosis of the aqueduct of Sylvius	F	C
	HSH	Hypomagnesemia, secondary hypocalcemia	F	C
	IHG	Iris hypoplasia with glaucoma	F	P
	IMD1	Immunodeficiency 1, combined, progressive (Bruton agammaglobulinemia)	F	P
	IMD2	Immunodeficiency 2 (Wiskott–Aldrich syndrome)	F	C
	IMD3	Immunodeficiency 3, with increased IgM	F	P
	IMD4	Immunodeficiency 4, thymic epithelial hypoplasia	F	C
	KAL	Kallmann syndrome	F	P
	LOX	Lysyl oxidase, cutis laxa-X, Ehlers–Danlos V	F	C
	MAA	Microphthalmia or anophthalmia, associated anomalies	F	C
	MAOA	Monoamine oxidase	S	P
	MF4	Metacarpal 4–5 fusion	F	C
	MSD	Microcephaly with spastic diplegia (Paine syndrome)	F	P
	NKRS	Natural killer cell radiation sensitivity	F	P
	NYS	Nystagmus	F	C
	OCRL	Oculocerebrorenal syndrome of Lowe	F	C

TABLE I. The 24 Human Gene Maps—(Continued)

Map location	Symbol	Marker name	Mode[a]	Status[b]
		Human chromosome X map		
	OPD	Otopalatal digital syndrome	F	P
	OPEM	Ophthalmoplegia, external, with myopia	F	P
	PHP	Panhypopituitarism	F	P
	PMD	Pelizaeus–Merzbacher disease	F	C
	PYHG4	Protein spot in 2-D gels (MW 104K)	S	P
	PYHG12	Protein spot in 2-D gels (MW 55K)	S	P
	PYHG17	Protein spot in 2-D gels (MW 32K)	S	P
	PYK	Phosphoryl kinase, liver glycogen storage disease VIII	F	C
	SBMA	Spinal and bulbar muscular atrophy	F	C
	SEDL	Spondylolpiphyseal dysplasia, late	F	P
	S12	Surface antigen (X-linked) 3	S	P
	TBG	Thyroxin-binding globulin	F	C
	TDD	Testicular 17,20-desmolase deficiency	F	C
	XS	Suppressor of Lu antigens	F	P
		Human Y chromosome map		
pter–p11.2	*TDF*	Testis-determining factor	F	P
p11–cen	*DXYS21[d]*	DNA segment, arbitrary[c]	RE	P
p	*DYS2*	DNA segment, arbitrary[c]	RE	P
p	*DYS3*	DNA segment, arbitrary[c]	RE	P
p	*DYS4*	DNA segment, arbitrary[c]	RE	P
p	*DYS5*	DNA segment, arbitrary[c]	RE	P
P	*DYS6*	DNA segment, arbitrary[c]	RE	P
p	*DYS7*	DNA segment, arbitrary[c]	RE	P
p	*DYS8*	DNA segment, arbitrary[c]	RE	P
p	*DYS9*	DNA segment, arbitrary[c]	RE	P
p	*DYS13*	DNA segment, arbitrary[c]	RE	P
p	*DXYS1[d]*	DNA segment, polymorphic[c]	RE	P
p	*DXYS5[d]*	DNA segment, arbitrary[c]	RE	P
p	*DXYS6[d]*	DNA segment, arbitrary[c]	RE	P
p	*DXYS7[d]*	DNA segment, arbitrary[c]	RE	P
p	*DXYS14[d]*	DNA segment, arbitrary[c]	RE	P
p	*DXYS15[d]*	DNA segment, polymorphic[c]	RE	P
p	*DXYS17[d]*	DNA segment, polymorphic[c]	RE	P
p	*DXYS20[d]*	DNA segment, polymorphic[c]	RE	P
p	*DXYS25[d]*	DNA segment, arbitrary[c]	RE	P

(Continued)

TABLE I. The 24 Human Gene Maps—(Continued)

Map location	Symbol	Marker name	Mode[a]	Status[b]
		Human Y chromosome map		
p	DXYS27[d]	DNA segment, arbitrary[c]	RE	P
p	MIC2 (MIC2Y)	Antigen identified by monoclonal antibodies 12E7, F21, and 013[c]	S, RE, D	P
p	YG	Yg (controls expression of antigen identified by monoclonal antibody 12E7 on erythrocytes)	F	P
q11	ACTL2	Actin-like sequence 2[c]	S	P
q11	AZF	Azoospermia		P
q11	DYS1	DNA segment, polymorphic[c]	RE	P
q11	DYS14	DNA segment, arbitrary[c]	RE	P
q11	DYS15	DNA segment, arbitrary[c]	RE	P
q11	DYS16	DNA segment, arbitrary[c]	RE	P
q11	DYS22	DNA segment, arbitrary[c]	RE	P
q11	DXYS18[d]	DNA segment, arbitrary[c]	RE	P
q11	DXYS19[d]	DNA segment, arbitrary[c]	RE	P
q11–qter	DYS12	DNA segment, arbitrary[c]	RE	P
q11–qter	DYS20	DNA segment, arbitrary[c]	RE	P
q11–qter	DYS21	DNA segment, arbitrary[c]	RE	P
	ASSP6	Argininosuccinate synthetase pseudogene 6[c]	S, RE	P
	GCY	Growth control, Y-chromosome-influenced	F	
	DYS10	DNA segment, arbitrary[c]	RE	P
	DYS11	DNA segment, polymorphic[c]	RE	P
	DYS17	DNA segment, arbitrary[c]	RE	P
	DYS18	DNA segment, arbitrary[c]	RE	P
	DYS19	DNA segment, arbitrary[c]	RE	P
	DXYS2[d]	DNA segment, polymorphic[c]	RE	P
	DXYS4[d]	DNA segment, arbitrary[c]	RE	P
	DXYS8[d]	DNA segment, arbitrary[c]	RE	P
	DXYS9[d]	DNA segment, arbitrary[c]	RE	P
	DXYS10[d]	DNA segment, arbitrary[c]	RE	P
	DXYS11[d]	DNA segment, arbitrary[c]	RE	P
	DXYS12[d]	DNA segment, polymorphic[c]	RE	P
	DXYS13[d]	DNA segment, polymorphic[c]	RE	P
	DXYS22[d]	DNA segment, arbitrary[c]	RE	P
	DXYS23[d]	DNA segment, arbitrary[c]	RE	P

[a] Mode: (A) annealing of homologous sequences as in *in situ* hybridization; (CH) cytogenetic analysis; (D) gene dosage; (F) Mendelian family studies; (H) (RE) recombinant methods, usually in conjunction with somatic-cell hybridization; (S) somatic-cell hybridization.
[b] Status: (C) confirmed; (I) inconsistent; (P) provisional; (T) tentative.
[c] Has been cloned.
[d] (DXYS) Denotes homologous gene sequences on both X and Y chromosomes.

ACKNOWLEDGMENTS. In our cell-hybridization and gene-mapping efforts, the contributions of many colleagues, present and past, are very much appreciated. They include G. Bell, M. Byers, M. Champion, R. Eddy, L. Haley, W. Henry, N. Honey, P. Lalley, O. Mueller, H. Nakai, S. Naylor, D. Owerbach, A. Sakaguchi, J. Tricoli, and B. Zabel. The expert assistance of C. Young, S. Shows, and J. Young in the preparation of the manuscript is gratefully acknowledged. This work has been supported in part by NIH Grants GM 20454 and HD 05196, ACS Grant CD-62, and March of Dimes Grant 1-935.

REFERENCES

Anderson, S., Bankier, A. T., Barrell, B. G., deBruijn, M. H. L., Coulson, A. R., Drouin, J., Eperon, I. C., Nierlich, D. P., Roe, B. A., Sanger, E., Schreier, P. H., Smith, A. J. H., Staden, R., and Young, I. G., 1981, Sequence and organization of the human mitochondrial genome, *Nature (London)* **290:**457–465.

Avery, O. T., MacLeod, C. M., and McCarty, M., 1944, Studies on the chemical nature of the substance inducing transformation of pneumococcal types: Induction of transformation by a deoxyribonucleic acid fraction isolated from pneumococcus type III, *J. Exp. Med.* **79:**137–158.

Berger, R., Bloomfield, C. D., and Sutherland, G. R., 1985, Report of the committee on chromosome rearrangements in neoplasia and on fragile sites, *Cytogenet. Cell Genet.* **40:**490–535.

Botstein, D., White, R. L., Skolnick, M., and Davis, R. W., 1980, Construction of genetic linkage map in man using restriction fragment length polymorphisms, *Am. J. Hum. Genet.* **32:**314–331.

Conneally, P. M., and Rivas, M. L., 1980, Linkage analysis in man, in: *Advances in Human Genetics,* Vol. 10 (H. Harris and K. Hirschhorn, eds.), pp. 209–266, Plenum Press, New York.

Cook, P. J. L., and Hamerton, J. L., 1982, Report of the committee on the genetic constitution of chromosome 1, *Cytogenet. Cell Genet.* **32:**111–120.

Cox, D. R., and Gedde-Dahl, T., Jr., 1985, Report of the committee on the genetic constitution of chromosomes 13, 14, 15 and 16, *Cytogenet. Cell Genet.* **40:**206–241.

Creagan, R. P., and Ruddle, F. H., 1975, The clone panel: A systematic approach to gene mapping using interspecific somatic cell hybrids, *Cytogenet. Cell Genet.* **14:**282–286.

Croce, C. M., Koprowski, H., and Eagle, H., 1972, Effect of environmental pH on the efficiency of cellular hybridization, *Proc. Natl. Acad. Sci. U.S.A.* **69:**1953–1956.

Davidson, R. L., 1974, Gene expression in somatic cell hybrids, *Annu. Rev. Genet.* **8:**195–218.

Davidson, R. L., and Gerald, P. S., 1976, Improved techniques for the induction of mammalian cell hybridization by polyethylene glycol, *Somat. Cell Genet.* **2:**165–176.

Elsevier, S. M., Kucherlapati, R. S., Nichols, E. A., Creagan, R. P., Giles, R. E., Ruddle, F. H., Willecke, K., and McDougall, J. K., 1974, Assignment of the gene for galactokinase to human chromosome 17 and its regional localization to band q21–22, *Nature (London)* **251:**633–636.

Ephrussi, B., 1972, *Hybridization of Somatic Cells,* Princeton University Press, Princeton, New Jersey, 175 pp.

Fincham, J. R. S., 1966, *Genetic Complementation,* pp. 3–9, W. A. Benjamin, New York.

Goodfellow, P. N., Davies, K. E., and Ropers, H.-H., 1985, Report of the committee on the genetic constitution of the X and Y chromosomes, *Cytogenet. Cell Genet.* **40:**296–352.

Goss, S. J., and Harris, H., 1977a, Gene transfer by means of cell fusion I, *J. Cell Sci.* **25:**17–37.

Goss, S. J., and Harris, H., 1977b, Gene transfer by means of cell fusion II, *J. Cell Sci.* **25:**39–58.

Grzeschik, K.-H., and Kazazian, H. H., 1985, Report of the committee on the genetic constitution of chromosomes 10, 11 and 12 *Cytogenet. Cell Genet.* **40:**179–205.

Gusella, J., Varsanyi-Breiner, A., Kao, F.-T., Jones, C., Puck, T. T., Keys, C., Orkin, S., and Housman, D., 1979, Precise localization of human beta-globin gene complex on chromosome 11, *Proc. Natl. Acad. Sci. U.S.A.* **76**:5239–5243.

Harnden D. G., and Klinger, H. P. (eds.), 1985, *ISCN (1985): An International System for Human Cytogenetic Nomenclature*, published in collaboration with *Cytogenetics and Cell Genetics*, Karger, Basel; also in *Birth Defects: Original Article Series* **21** (1), March of Dimes Birth Defects Foundation, New York (1985).

Harper, M.E., and Saunders, G.F., 1981, Localization of single copy DNA sequences on G-banded human chromosomes by *in situ* hybridization, *Chromosoma* **83**:431–439

Harris, H., and Hopkinson, D.A., 1976, *Handbook of Enzyme Electrophoresis in Human Genetics*, North-Holland, Amsterdam.

Honey, N.K., Mueller, O.T., Little, L.E., Miller, A.L., and Shows, T.B., 1982, Mucolipidosis III is genetically heterogeneous, *Proc. Natl. Acad. Sci. U.S.A.* **79**:7420–7424.

Human Gene Mapping Workshop 8, 1985, Eighth International Workshop on Human Gene Mapping, *Cyogenet. Cell Genet.* **40**:1-4.

Jeffreys, A. J., Craig, I. W., and Francke, U., 1979, Localization of the G-gamma, A-gamma, delta-, and beta-globin genes on the short arm of human chromosome 11, *Nature (London)* **281**:606–608.

Karig Hohmann, L., and Shows, T. B., 1979, Complementation of genetic disease: A velocity sedimentation procedure for the enrichment of heterokaryons, *Somat. Cell Genet.* **5**:1013–1029.

Kidd, K. K., and Gusella, J., 1985, Report of the committee on the genetic constitution of chromosomes 3 and 4, *Cytogenet. Cell Genet.* **40**:107–127.

Klebe, R. J., Chen, T. R., and Ruddle, F. H., 1970, Controlled production of proliferating somatic cell hybrids, *J. Cell Biol.* **45**:74–82.

Klobutcher, L. A., and Ruddle, F. H., 1979, Phenotype stabilization and integration of transferred material in chromosome mediated gene transfer, *Nature (London)* **280**:657–660.

Lalley, P. A., and McKusick, V. A., 1985, Report of the committee on comparative mapping, *Cytogenet. Cell Genet.* **40**:536–566.

Lamm, L. U., and Olaisen, B., 1985, Report of the committee on the genetic constitution of chromosomes 5 and 6, *Cytogenet. Cell Genet.* **40**:128–155.

Langhans, T., 1868, Uber Riesenzellen mit wandständigen Kernen in Tuberkeln an die fibrose Form des Tuberkels, *Arch. Pathol. Anat. Physiol. Klin. Med.* **42**:382–404.

Lederberg, J., 1947, Gene recombination and linked segregations in *Escherichia coli, Genetics* **32**:505–525.

Lennox, E. S., 1955, Transduction of linked genetic characters of the host by bacteriophage P1, *Virology* **1**:190–206.

Littlefield, J. W., 1964, Selection of hybrids from matings of fibroblasts *in vitro* and their presumed recombinants, *Science* **145**:709–710.

McAlpine, P. J., Shows, T. B., Miller, R. L., and Pakstis, A. J., 1985, The 1985 catalog of mapped genes and report of the nomenclature committee, *Cytogenet. Cell Genet.* **40**:8–66.

McBride, O. W., and Ozer, H. L., 1973, Transfer of genetic information by purified chromosomes, *Proc. Natl. Acad. Sci. U.S.A.* **70**:1258–1262.

McKusick, V. A., 1982, *Mendelian Inheritance in Man*, 6th ed., Johns Hopkins University Press, Baltimore.

Miller, C. L., and Ruddle, F. H., 1978, Cotransfer of human X-linked markers into murine somatic cells via isolated metaphase chromosomes, *Proc. Natl. Acad. Sci. U.S.A.* **75**:3346–3350.

Minna, J. D., Lalley, P. A., and Francke, U., 1976, Comparative mapping using somatic cell hybrids, *In Vitro* **12**:726–733.

Morton, C. C., Byers, M. G., Nakai, H., Bell, G. I., and Shows, T. B., 1986, Human genes for insulin-like growth factors I and II and epidermal growth factor are located on 12q22→q24.1, 11p15, and 4q25→q27, respectively *Cytogenet. Cell Genet.* **41**:245–249.

Mueller, O. T., Honey, N. K., Little, L. E., Miller, A. L., and Shows, T. B., 1983, Mucolipidosis II and III: The genetic relationships between two disorders of lysosomal enzyme biosynthesis, *J. Clin. Invest.* **72:**1016–1023.

Naylor, S. L., Sakaguchi, A. Y., Shows, T. B., Law, M. L., Goeddel, D. V., and Gray, P. W., 1983, Human immune interferon gene is located on chromosome 12, *J. Exp. Med.* **57:**1020–1027.

Naylor, S., Lalouel, J.-M., and Shaw, D. J., 1985, Report of the committee on the genetic constitution of chromosomes 17, 18 and 19, *Cytogenet. Cell Genet.* **40:**242–267.

Okada, Y., and Tadokaro, J., 1963, The distribution of cell fusion capacity among cell strains or cells caused by HVJ, *Exp. Cell Res.* **32:**417–423.

Owerbach, D., Bell, G. I., Rutter, W. J., and Shows, T. B., 1980a, The insulin gene is located on chromosome 11 in humans, *Nature (London)* **286:**82–84.

Owerbach, D., Rutter, W. J., Martial, J. A., Baxter, J. D., and Shows, T. B., 1980b, Genes for growth hormone, chorionic somatomammotropin, and growth hormone-like gene on chromosome 17 in humans, *Science* **209:**289–292.

Poste, G., 1970, Virus-induced polykaryocytosis and the mechanism of cell fusion, *Adv. Virus Res.* **16:**303–356.

Povey, S., Morton, N. E., and Sherman, S. L., 1985, Report of the committee on the genetic constitution of chromosomes 1 and 2, *Cytogenet. Cell Genet.* **40:**67–106.

Puck, T. T., Marcus, P. T., and Cieciura, S. J., 1956, Clonal growth of mammalian cells *in vitro:* Growth characteristics of colonies from single HeLa cells with and without a "feeder" layer, *J. Exp. Med.* **103:**653–666.

Renwick, J. H., 1971, The mapping of human chromosomes, *Annu. Rev. Genet.* **5:**81–120.

Ringertz, N. R., and Savage, R. E., 1976, *Cell Hybrids,* Academic Press, New York, 366 pp.

Ruddle, F. H., 1972, Linkage analysis using somatic cell hybrids, in: *Advances in Human Genetics,* Vol. 3 (H. Harris and K. Hirschhorn, eds.), pp. 173–235, Plenum Press, New York.

Ruddle, F. H., and McBride, O. W., 1977, New approaches to cell genetics: Cotransfer of linked genetic markers by chromosome mediated gene transfer, in: *The Molecular Biology of the Mammalian Genetics Apparatus* (P. Tso, ed.), pp. 163–169, Elsevier/North-Holland, Amsterdam.

Sell, E. K., and Krooth, R. S., 1973, Tabulation of somatic cell hybrids formed between lines of cultured cells, *J. Cell Physiol.* **80:**453–462.

Shay, J. W., 1982, *Techniques in Somatic Cell Genetics,* Plenum Press, New York.

Shows, T. B., 1972, Genetics of human–mouse somatic cell hybrids: Linkage of human genes for lactate dehydrogenase-*A* and esterase-*A₄*, *Proc. Natl. Acad. Sci. U.S.A.* **69:**348–352.

Shows, T. B., 1977, Genetic and structural dissection of human enzymes and enzyme defects using somatic cell hybrids, in: *Isozymes: Current Topics in Biological and Medical Research,* Vol. 2 (M. C. Rattazzi, J. G. Scandalios, and G. S. Whitt, eds.), pp. 107–158, Alan R. Liss, New York.

Shows, T. B., 1978, Mapping the human genome and metabolic diseases, in: *Birth Defects* (J. W. Littlefield and J. DeGrouchy, eds.), pp. 66–84, Excerpta Medica, Amsterdam.

Shows, T. B., 1979, The X chromosome gene map, in: *Genetic Mechanisms of Sexual Development* (H. L. Vallet and I. H. Porter, eds.), pp. 253–269, Academic Press, New York.

Shows, T. B., 1983a, Human genome organization of enzyme loci and metabolic diseases, in: *Isozymes: Current Topics in Biological and Medical Research,* Vol. 10 (M. C. Rattazzi, J. G. Scandalios, and G. S. Whitt, eds.), pp. 323–339, Alan R. Liss, New York.

Shows, T. B., 1983b, The human molecular map of cloned genes and DNA polymorphisms, *Banbury Rep.* **14:**347–356 (Cold Spring Harbor Laboratory publication).

Shows, T. B., and McAlpine, P. J., 1982, The 1981 catalogue of assigned human genetic markers and report of the nomenclature committee, *Cytogenet. Cell Genet.* **8:**667–675.

Shows, T. B., and Sakaguchi, A. Y., 1980, Gene transfer and gene mapping in mammalian cells in culture, *In Vitro* **16:**55–76.

Shows, T. B., Alper, C. A., Bootsma, D., Dorf, M., Douglas, T., Huisman, T., Kit, S., Klinger, H. P., Kozak, C., Lalley, P. A., Lindsley, D., McAlpine, P. J., McDougall, J. K., Meera Khan, P., Meisler, M., Morton, N. E., Opitz, J. M., Partridge, C. W., Payne, R., Roderick, T. H., Rubinstein, P., Ruddle, F. H., Shaw, M., Spranger, J. W., and Weiss, K., 1979, International System for Human Gene Nomenclature (1979), *Cytogenet. Cell Genet.* **25**:96–116.

Shows, T. B., Sakaguchi, A. Y., and Naylor, S. L., 1982, Mapping the human genome, cloned genes, DNA polymorphisms, and inherited disease, in: *Advances in Human Genetics*, Vol. 12 (H. Harris and K. Hirschhorn, eds.), pp. 341–452, Plenum Press, New York.

Shows, T. B., Naylor, S. L., Sakaguchi, A. Y., Zabel, B. U., and Tricoli, J. V., 1983a, Chromosome mapping of cloned genes and DNA polymorphisms to study human disease, *Banbury Rep.* **14**:347–356 (Cold Spring Harbor Laboratory publication).

Shows, T. B., Zabel, B. U., and Tricoli, J. V., 1983b, High resolution chromosome mapping of cloned genes and DNA polymorphisms, in: *Recombinant DNA and Medical Genetics* (A. Messer and I. Porter, eds.), pp. 79–99, Academic Press, New York.

Shows, T. B., McAlpine, P. J., and Miller, R. L., 1984, The 1983 catalogue of mapped human genetic markers and report of the nomenclature committee, *Cytogenet. Cell Genet.* **37**:340–393.

Smith, M., and Spence, M. A., 1985, Report of the committee on the genetic constitution of chromosomes 7, 8 and 9, *Cytogenet. Cell Genet.* **40**:156–178.

Tippett, P., and Kaplan, J.-C., 1985, Report of the committee on the genetic constitution of chromosomes 20, 21 and 22, *Cytogenet. Cell Genet.* **40**:268–295.

Tricoli, J. V., Rall, L. B., Scott, J., Bell, G. I., and Shows, T. B., 1984, Localization of insulin-like growth factor genes to human chromosomes 11 and 12, *Nature (London)* **310**:784–786.

Weiss, M. C., and Green, H., 1967, Human–mouse hybrid cell lines containing partial complements of human chromosomes and functioning human genes, *Proc. Natl. Acad. Sci. U.S.A.* **58**:1104–1111.

Wigler, M., Pellicer, A., Silverstein, S., Axel, R., Urlaub, G., and Chasin, L., 1979, DNA-mediated transfer of the adenine phosphoribosyltransferase locus into mammalian cells, *Proc. Natl. Acad. Sci. U.S.A.* **76**:1373–1376.

Willard, H. F., Skolnick, M. H., Pearson, P. L., and Mandel, J.-L., 1985, Report of the committee on human gene mapping by recombinant DNA techniques, *Cytogenet. Cell Genet.* **40**:360–489.

Willecke, K., Lange, R., Kruger, A., and Reber, T., 1976, Cotransfer of two linked human genes into cultured mouse cells, *Proc. Natl. Acad. Sci. U.S.A.* **73**:1274–1278.

Wollman, E. L., Jacob, F., and Hayes, H., 1956, Conjugation and genetic recombination in *Escherichia coli* K-12, *Cold Spring Harbor Symp. Quant. Biol.* **21**:141–162.

Wullems, G. J., van der Horst, J., and Bootsma, D., 1977, Transfer of the human gene coding for thymidine kinase and galactokinase to Chinese hamster cells and human–Chinese hamster cell hybrids, *Somat. Cell Genet.* **3**:281–293.

Zabel, B. U., Naylor, S. L., Sakaguchi, A. Y., Bell, G. I., and Shows, T. B., 1983, High-resolution chromosomal localization of human genes for amylase, proopiomelanocortin, somatostatin, and a DNA fragment (*D3S1*) by *in situ* hybridization, *Proc. Natl. Acad. Sci. U.S.A.* **80**:6932–6936.

Zabel, B. U., Kronenberg, H. M., Bell, G. I., and Shows, T. B., 1985, Chromosome mapping of genes on the short arm of human chromosome 11: Parathyroid hormone gene is at 11p15 together with the genes for insulin, c-Harvey-*ras* 1, and β-hemoglobin, *Cytogenet. Cell Genet.* **11**:505–509.

Zinder, N. D., and Lederberg, J., 1952, Genetic exchange in *Salmonella*, *J. Bacteriol.* **64**:679–699.

3

MICROCELL FUSION AND MAMMALIAN GENE TRANSFER

Tracy G. Lugo and R. E. K. Fournier

1. INTRODUCTION

Microcell-mediated chromosome transfer (Fournier and Ruddle, 1977a; Fournier, 1981) is a parasexual genetic technique that can be used to transfer single, intact chromosomes from one mammalian cell to another. As such, microcell fusion is not a gene-transfer method in the usual sense of the term; rather, it is an approach for the construction of simple hybrid cells with precisely defined genotypes. Microcell hybrids have proved to be valuable tools for mammalian gene mapping, for genetic analyses of complex cellular phenotypes, and, in conjunction with other gene-transfer methods, for genetic manipulation of specific chromosomes and chromosomal regions.

1.1. Overview of the Method

Microcell fusions generally involve the steps that are illustrated diagramatically in Fig. 1 (for a detailed discussion of experimental methods, see Fournier, 1982). First, exponentially growing populations of donor cells are subjected to conditions that induce micronucleation. The micronucleation method that has been most commonly used is prolonged mitotic arrest (Phillips and Phillips, 1969; Ege and Ringertz, 1974; Fournier and Ruddle, 1977a). An alternate method is to plate populations of mitotic cells in the presence of agents that perturb normal nuclear division (Fournier, 1981). In either case, progression

Tracy G. Lugo and R. E. K. Fournier • Department of Microbiology and the Comprehensive Cancer Center, University of Southern California School of Medicine, Los Angeles, California 90033.

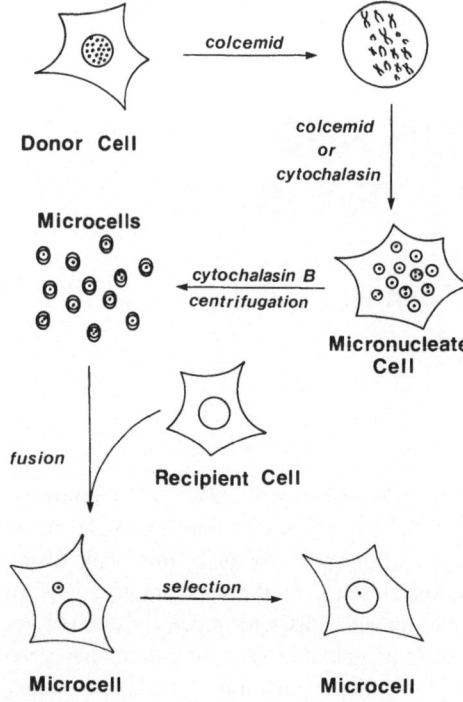

FIGURE 1. Diagrammatic representation of the steps involved in the preparation of microcell hybrids.

of the mitotic cells into interphase is accompanied by the formation of many small micronuclei, each containing a subset of the donor cell's chromosome complement. Thus, micronucleation partitions the genome into discrete subnuclear packets.

The next step in the procedure is to physically isolate these packets from the cells. This is accomplished by centrifugal enucleation in the presence of cytochalasin B. When micronucleate cells are enucleated, each micronucleus is recovered as a discrete particle. These particles, called "microcells," consist of a single micronucleus surrounded by a thin rim of cytoplasm and an intact plasma membrane.

Isolated microcells, with or without subsequent purification, can be fused with intact recipients using standard cell-fusion methods. Since each microcell is enveloped by a surface membrane, the product of fusion is a recipient cell into which a single donor micronucleus has been introduced—a microcell heterokaryon. These heterokaryons can be generated at frequencies of 10^{-1}–10^{-2}.

Some of the microcell heterokaryons will subsequently proliferate, ultimately generating microcell-hybrid clones that contain the donor chromosome(s) originally present in the introduced micronucleus. The frequencies with which proliferating hybrids are recovered in microcell fusions are typically 10^{-4}–10^{-5}; thus, some strategy for their selective isolation is required. Metabolic selection for expression of a donor-encoded trait, i.e., intergenic complementation, is the usual means to achieve this goal.

1.2. Properties of Microcell Hybrids

Not all micronuclei contain single chromosomes, and not all microcell hybrids are monochromosomal. In a given experiment, monochromosomal hybrids may represent as few as 20% or as many as 90% of the clones that survive selection. This fraction is influenced not only by the chromosome content [i.e., physical size (Sekiguchi et al., 1978)] of the microcells used for fusion, but also by the propensity of the recipient cells to retain or to segregate chromosomes. Depending on the experiment, polychromosomal microcell hybrids retaining 2, 3, 4, or more donor chromosomes may be either useful or a nuisance. True microcell hybrids retaining more than 5 or 6 donor chromosomes are rarely observed. The microcell-hybrid clones that survive selection usually contain an intact copy of the donor chromosome that encodes the selected marker. However, some clones may retain that chromosome in deleted or translocated form, or in both forms. Again, this property seems to depend primarily on the recipient cells used for fusion. A few cell lines that rearrange donor chromosomes at high frequency have been identified (Fournier and Moran, 1983; Landolph and Fournier, 1983); other recipient cells do not show this tendency (Fournier, 1982; Killary and Fournier, 1984; Peterson et al., 1985).

2. MAMMALIAN GENE MAPPING USING MICROCELL HYBRIDS

The karyotypic simplicity of microcell hybrids makes them particularly useful tools for mammalian gene mapping. The most straightforward cases are those in which expression of the donor locus of interest can be selectively fixed in the cultured recipient cells. This situation is realized in experiments designed to map genes that complement recessive, conditional–lethal mutations or genes that confer dominant drug-resistance phenotypes. Collections of polychromosomal microcell hybrids have also been useful for mapping genes to particular chromosomes by the usual methods of synteny and assignment testing (Ruddle and Creagan, 1975).

2.1. Complementation Mapping in Microcell Hybrids

The basic strategy of gene mapping by complementation is to fuse cells carrying a recessive, conditional–lethal mutation with microcells derived from wild-type cells of a different species and to select hybrids in which the mutant phenotype has been complemented. The chromosome that carries the complementing gene is selectively fixed in every hybrid clone. Since microcell hybrids retain only one or a few donor chromosomes, unambiguous assignments can usually be made by analyzing a very limited number of independent primary clones.

The assignment of the murine gene that encodes folylpolyglutamyl synthetase (FPGS) to mouse chromosome 2 (Fournier and Moran, 1983) illustrates the utility of this approach. AuxB1 Chinese hamster ovary (CHO) cells are deficient in FPGS and are consequently auxotrophic for glycine, adenosine, and thymidine (gat$^-$ phenotype). Chromosomes from diploid mouse fibroblasts were transferred into AuxB1 recipients, and 20 gat$^+$ microcell-hybrid clones were selected in medium lacking adenosine and thymidine. All these hybrids expressed FPGS at levels similar to that of wild-type CHO cells.

Of the 20 hybrid clones, 7 retained one or a few intact donor-derived mouse chromosomes. Karyotypic analysis showed that mouse chromosome 2 was present at high frequency in all 7 clones, and no other mouse chromosome was common among them. Two clones were monochromosomal hybrids that retained only mouse chromosome 2. All 7 clones expressed the murine form of adenylate kinase-1 (AK-1), a marker assigned to mouse chromosome 2 (Francke *et al.*, 1977). When the hybrids were grown under nonselective conditions to permit segregation of the *Fpgs* gene, loss of the gat$^+$ phenotype was invariably accompanied by loss of murine AK-1 expression and loss of mouse chromosome 2.

The remaining gat$^+$ hybrid clones carried fragments of mouse chromosome 2, which permitted regional localization of the *Fpgs* gene. Eight clones retained a centric fragment that appeared to have arisen from mouse chromosome 2 by terminal deletion; all 8 expressed murine AK-1. The centric fragment segregated concordantly with FPGS and murine AK-1 expression, indicating that both markers resided in that region of mouse chromosome 2, identified as 2(cen–C1).

Five of the hybrid clones did not contain mouse chromosomes visible in karyotypes. However, all 5 expressed murine AK-1, indicating that they did contain a fragment of chromosome 2. The gat$^+$ phenotype of these clones was stable even under nonselective conditions. It is likely that these hybrids retained a small fragment of mouse chromosome 2 that had been translocated to one of the recipient hamster chromosomes.

More than half the gat$^+$ microcell hybrids isolated in the experiment described above retained murine chromosomal fragments. This fraction is higher than that observed in most microcell fusions, but is typical of crosses in which CHO recipient cells have been employed (Fournier and Frelinger, 1982; Fournier and Moran, 1983). Even in these cases, however, a fraction of hybrids with intact donor chromosomes sufficiently large to permit an unambiguous chromosome assignment is almost always recovered. Furthermore, as the mapping of *Fpgs* illustrates, rearrangements that can be identified can be used to provide regional mapping data.

Complementation mapping is a powerful technique, limited only by the availability of mutant recipient cell lines. It is particularly useful in instances in which the trait to be mapped is nonpolymorphic or difficult to assay in cultured cells. The use of microcell hybrids enhances its efficiency, since only a limited number of independent clones need be analyzed. Monochromosomal hybrids recovered by direct selection can provide a chromosome assignment in a single step and become permanently useful tools for subsequent genetic studies.

2.2. Mapping Dominant Mutations

As described above, genes that complement recessive mutations can be mapped by transferring single chromosomes from wild-type donor cells to mutant recipients and selecting complemented hybrids with the wild-type phenotype. Dominant alleles are mapped in precisely the converse manner; i.e., chromosomes from mutant donors are transferred to wild-type recipients, and hybrids with the (dominant) mutant phenotype are directly selected. For example, the murine gene that encodes ouabain resistance (*Oua-1*) was assigned to mouse chromosome 3 using human × mouse microcell hybrids (Kozak *et al.*, 1979). In this case, the murine allele confers a resistance phenotype that is dominant to the ouabain sensitivity of its human counterpart. Following transfer of chromosomes from diploid mouse cells to human recipients, microcell hybrids that expressed the level of ouabain resistance characteristic of the mouse donor cells were isolated. Only mouse chromosome 3 was retained by five independent ouabain-resistant hybrids, and segregation analyses showed that loss of murine *Oua-1* was concordant with loss of murine chromosome 3. Furthermore, mutagen-induced ouabain-resistance alleles apparently map to the same locus (Landolph and Fournier, 1983). This general approach can be used to map any mutation with dominant phenotypic expression.

2.3. Gene Mapping Using Polychromosomal Microcell Hybrids

Microcell hybrids generated as described above can also be used for mapping by concordance in the traditional manner. For example, the collection of hybrids constructed for *Fpgs* mapping was subsequently used to provide chromosome assignments for mouse genes that encode phosphoenolpyruvate carboxykinase (PEPCK) (Lem and Fournier, 1985) and tyrosine aminotransferase (TAT) (Peterson *et al.*, 1985). Cloned complementary DNAs corresponding to the rat genes for PEPCK and TAT were used to screen hybrids for DNA restriction fragments characteristic of the homologous mouse genes. All hybrids that retained an intact mouse chromosome 2 also retained the murine *PEPCK* gene. This chromosome assignment was then confirmed using specially constructed monochromosomal microcell hybrids (see Section 3.1). The murine *PEPCK* gene was absent from hybrid clones that retained only a centric fragment of mouse chromosome 2, localizing it to the region 2(C1–ter) (Lem and Fournier, 1985).

On the other hand, polychromosomal hybrids from this collection did not contain the murine *TAT* gene. Screening these microcell hybrids and three other highly segregated hybrid clones reduced the number of candidate chromosomes to six. Monochromosomal microcell hybrids were then used to provide a definitive assignment of the murine *TAT* gene to one of those chromosomes, mouse chromosome 8 (Peterson *et al.*, 1985). These examples illustrate a generally useful and extremely straightforward approach for mammalian gene mapping in which 3 or 4 hybrid clones that retain 6–8 donor chromosomes are initially assayed for the marker of interest and monochromosomal hybrids are then used to screen the candidate chromosomes one by one. Unambiguous assignments can frequently be obtained by screening as few as 6–8 hybrid and microcell-hybrid clones.

3. MONOCHROMOSOMAL HYBRID PANELS

The major advantage of complementation mapping is that hybrids that contain the chromosome of interest are selected directly. The traditional strategy for localizing genes the expression of which is not selectable relies on the tendency of interspecific hybrids to segregate chromosomes of one of the parental cells. Using hybrids that have lost all but a subset of one parent's chromosomes, it is possible to correlate the expression of phenotypic traits with the presence of particular chromosomes (Ruddle and Creagan, 1975). However, chromosome segregation in hybrids is uncontrolled and not completely random. Typically, large panels of hybrid clones are required to cover the entire chromosome complement, and the karyotypes of the hybrids are unstable and must be analyzed frequently. Microcell fusion can be used to create panels of monochromosomal hybrids in which single, specific chromosomes are maintained by direct selective pressure (Fournier and Ruddle, 1977b; Fournier and Frelinger, 1982). The microcell hybrid clones that make up such a monochromosomal hybrid panel (MHP) display karyotypes that are simple, stable, and homogeneous.

The construction of MHPs requires that (1) single chromosomes be transferred from one mammalian cell to another and (2) each chromosome of the donor's complement be selectively fixed in individual hybrid clones. Microcell fusion provides a means to achieve the first goal. With respect to the latter, certain chromosomes of any species can be maintained selectively in hybrids because they encode markers that result in drug-resistance phenotypes. Some well-known examples of selectable markers include thymidine kinase (*TK*), the adenine and hypoxanthine phosphoribosyltransferases (*APRT* and *HPRT*), and the genes for ouabain (*Oua^r*) and methotrexate resistance (*Mtx^r*). *Mtx^r* and *Oua^r* are dominant alleles that can be fixed in wild-type recipients, but these markers suffer the disadvantage that no direct selection against their expression is available. *TK, APRT,* and *HPRT* expression can be selected in either direction, but mutant recipients (*TK^−, APRT^−,* or *HPRT^−*) are required. In any case, selectable markers are not available for every chromosome of any mammalian species.

There are at least two approaches that can be employed to overcome this limitation. The first is to utilize defined, naturally occurring translocations between chromosomes that encode selectable markers and other chromosomes. The second method employs chromosomes into which an exogenous selectable gene has integrated after gene transfer. In either case, the marked chromosomes are transferred to recipient cells by microcell fusion and selectively fixed in every hybrid clone.

3.1. Robertsonian Translocations and Monochromosomal Hybrid Panels

Translocations between chromosomes that encode selectable markers and other chromosomes of the diploid complement can be used to fix both types of chromosomes in microcell hybrids by direct selective pressure (Fournier and Frelinger, 1982). This approach is particularly advantageous in the case of the murine genome. The diploid karyotype of *Mus musculus* consists of 40 telocentric chromosomes. Robertsonian (Rb) translocations are metacentric chromosomes formed by centric fusion of two chromosomes of the normal diploid complement. Rb translocations are particularly useful in the context of this discussion, since (1) whole chromosomes are translocated, (2) ambiguities concerning translocation breakpoints are avoided, and (3) Rb translocations involving chromosomes that encode selectable markers and virtually all other autosomes are available. Rb-translocation chromosomes can be transferred intact via microcells, and they are stable in microcell hybrids (Fournier and Frelinger, 1982).

As an example, Rb translocations have been used to confirm the assignment of the *PEPCK* gene to mouse chromosome 2 (Lem and Fournier, 1985). Microcell hybrids that contained an Rb translocation between mouse chromosomes 2 and 8 were constructed by selecting for chromosome-8-encoded *APRT* and were compared with analogous hybrids containing chromosome 8 alone. The clones that contained the Rb(2.8) translocation retained the mouse *PEPCK* gene, while those that retained mouse chromosome 8 did not. Back-selection against *APRT* expression resulted in concordant loss of the *PEPCK* gene and the Rb(2.8) translocation chromosome. Monochromosomal microcell hybrids that retain many different Rb translocations have been isolated and characterized in an exactly analogous manner (Fournier and Frelinger, 1982).

3.2. Recovery of Chromosomes That Contain Exogenous Selectable Loci

A second approach to constructing MHPs involves inserting exogenous selectable genes into mammalian chromosomes after some form of gene transfer and isolating individual marked chromosomes in monochromosomal hybrids by microcell fusion. This strategy is based on the fact that exogenous gene sequences introduced into cells by chromosome-, DNA-, or vector-mediated gene transfer become integrated into host-cell chromosomes in a manner that appears at least quasi-random. Microcell hybrids that selectively retain such marked chromosomes can be readily isolated. The feasibility of this approach was demonstrated, and its potential utility first discussed, in 1977 (Fournier and Ruddle, 1977b). Since that time, several laboratories have constructed microcell hybrids of this type (Athwal *et al.*, 1985; Lugo *et al.*, 1986; Saxon *et al.*, 1985; Tunnacliffe *et al.*, 1983).

Cloned genes the expression of which can be selectively fixed in wild-type recipient cells have been widely employed in these experiments. These have included the bacterial *Neo^r* (Colbere-Garapin *et al.*, 1981) and *Xprt* (Mulligan and Berg, 1981) genes as well as a mutant mammalian *Dhfr* gene (Simonsen and Levinson, 1983). Most studies reported to date have concerned the isolation of microcell hybrids that carry chromosomes into which selectable loci have been introduced after gene transfer by DNA–calcium phosphate coprecipitation. Such hybrids may contain many copies of the inserted sequences in large tandem arrays, and the marked chromosomes have been derived from heteroploid donor cells. Both potential problems can be avoided.

For the construction of microcell hybrids to be used for genetic analyses, diploid cells are the best donor materials because they are free of the chromosomal aberrations found in virtually all established cell lines. Microcells can be produced efficiently from both rodent and human diploid fibroblasts (Fournier and Ruddle, 1977a; McNeill and Brown, 1980). However, the introduction of exogenous selectable genes into primary cells has been constrained by the limited proliferative potential of these cells *in vitro* and by the low efficiency of DNA-calcium phosphate-mediated gene transfer. Furthermore, gene transfer by calcium phosphate coprecipitation generally yields cells that contain multiple copies of the transferred gene integrated in tandem arrays (Robins *et al.*, 1981). Other gene-transfer techniques, such as electroporation, result in fewer tandem integrations, but do not sufficiently increase the yield of independent integration events (Potter *et al.*, 1984). For primary cells, an efficient means of achieving integration at many different chromosomal sites is required.

High-frequency gene transfer into primary cells can be achieved using retroviral vectors. Transducing retroviruses are replication-defective and require helper functions from an intact virus in order to grow. The recent development of packaging cell lines that can provide these functions in *trans* has permitted the isolation of stocks of defective transducing virus free of contaminating helper virus (Mann *et al.*, 1983). Vectors that can infect both rodent and human cells are available (Miller *et al.*, 1985; Cone and Mulligan, 1984). The integration of these defective viruses is quasi-random, and since the efficiency of infection can approach 100%, many independent integration events can be recovered in a single step (Weis *et al.*, 1984).

Microcell hybrids in which human chromosomes 14 and 20 are selectively maintained in a mouse background have been produced in this manner (Lugo *et al.*, 1986). Diploid human fibroblasts were infected with pZIP-NeoSV(X)1, an amphotropic, defective retrovirus that carries the *G418^r* gene (Cepko *et al.*, 1984; Cone and Mulligan, 1984). Infected cells from approximately 75 independent events were pooled and used as donors for microcell transfer into mouse 3T6 cells. The G418-resistant microcell hybrids were screened by staining the chromosomes with alkaline Giemsa, which differentiates human and rodent chromosomes (Friend *et al.*, 1976). Two hybrids that each retained a single human chromosome were recovered. The chromosomes were identified as 14 and 20 by chromosome banding and by testing the cells for the presence of human isozymes and polymorphic DNA-restriction fragments assigned to these chromosomes. The sites of vector insertion were confirmed by *in situ* hybridization.

Hybrids of this kind can be used as starting materials for the regional localization of genes by any of several methods that produce chromosome fragments. They can be used as donors for metaphase-chromosome transfer (Weis *et al.*, 1984). They can also be gamma-irradiated and rescued by fusion with a nonirradiated partner (Goss and Harris, 1975; Cirullo *et al.*, 1983). Both these techniques produce cells in which subchromosomal fragments are retained by virtue of the introduced selectable marker. Another advantage of retroviral vectors is that since the ends of the integrated sequences are precisely defined, it is straightforward to recover the flanking sequences by molecular cloning.

The ability to introduce selectable markers into any mammalian chromosome has implications for genetic analysis far beyond gene mapping. By combining different selective schemes, it is possible to construct hybrid cells with two or more desired chromosomes, and it should soon be possible to create much more complex combinations. This will permit the investigation of functional interactions among multiple asyntenic genes in a wide variety of backgrounds.

4. GENETIC ANALYSIS OF COMPLEX PHENOTYPES

Somatic-cell hybrids have been useful for relatively simple genetic experiments such as gene mapping. However, the genotypic complexity, heterogeneity, and instability of traditional cell × cell hybrids have severely limited their utility in experiments designed to investigate the genetic basis of more complex cellular phenotypes. Microcell-hybrid clones containing single, specific donor chromosomes are well suited to this task.

4.1. Regulation of Differentiated Functions

One of the most intriguing phenotypic alterations that occurs in hybrid cells is the phenomenon termed "extinction." Extinction was first documented by Davidson *et al.* (1966), who reported that hybrids between pigmented melanoma cells and nonpigmented fibroblasts failed to produce the pigment characteristic of the differentiated melanoma parent. In fact, extinction of tissue-specific functions in intertypic hybrids is a general phenomenon: Hybrids between different cell types almost invariably fail to express the differentiated products of either parent (for reviews, see Davis and Adelberg, 1973; Davidson, 1974; Ringertz and Savage, 1976). It is important to note that extinction affects the expression of tissue-specific products exclusively; the housekeeping functions of both parental cells are coexpressed in the hybrids. Furthermore, if interspecific hybrids have been prepared such that chromosomes of one parent are segregated during serial passage, the tissue-specific traits of the other parent may subsequently be reexpressed (Weiss and Chaplain, 1971; Sparkes and Weiss, 1973; Weiss *et al.*, 1975).

The alterations in expression of differentiated products that occur in intertypic hybrid cells constitute one example of what might be considered a complex change in the cellular phenotype. Before we consider how microcell hybrids have been useful in defining the genetic basis of extinction, it is perhaps worthwhile to summarize some important features of this phenomenon. The first point concerns the generality of extinction. Extinction is general in two respects: (1) It occurs in virtually all hybrid crosses in which the parents are distinctly different cell types and (2) the vast majority of tissue-specific products are extinguished in the hybrids. With respect to the latter point, a recent survey demonstrated that 16 of 17 liver-specific genes assayed were extinguished in hepatoma × fibroblast hybrids (A. C. Chin and R. E. K. Fournier, unpublished findings). An increasing body of evidence indicates that extinction of tissue-specific products in hybrids is primarily the result of regulation at the level of gene transcription. Thus, intertypic hybrids fail to accumulate specific messenger RNAs (mRNAs) that encode differentiated products (Orkin *et al.*, 1975; Cassio *et al.*, 1981), largely due to diminished transcription rates of the corresponding genes. Extinction is also a gene regulatory phenomenon that occurs in *trans*. Indeed, extinction has been observed in intertypic heterokaryons within hours of fusion, times at which the parental nuclei remain separate (Thompson and Gelehrter, 1971). Transient extinction of serum albumin production has also been documented in fusions between intact rat hepatoma cells and enucleated fibroblast cytoplasms (Kahn *et al.*, 1981).

These observations are consistent with the hypothesis that extinction of tissue-specific gene expression in intertypic hybrids is mediated by factors that are encoded by the heterologous parental genome, factors the action of which is *trans*-dominant and the final effects on gene expression of which are negative. The critical assumption of this simple hypothesis is that *specific* factors are involved—an assumption that, until recently, remained untested.

Using hepatoma × fibroblast hybrids as a model system, we have performed experiments to explore the genetic basis of extinction. Answers to the following questions have been sought: (1) Does extinction of liver-specific gene expression in hepatoma × fibroblast hybrids have a specific genetic basis; i.e., does extinction of a given hepatic trait require the presence of specific fibroblast loci? (2) If it does, are single or multiple loci required for extinction of a given hepatic trait? (3) Are sets of liver-specific genes coordinately extinguished and reexpressed? (4) What are the linkage relationships between liver-specific structural genes and presumptive loci required for their extinction? Intertypic microcell hybrids have been useful in approaching these questions, as summarized below.

In a manner analogous to that described in Section 2.3, which was concerned with gene mapping, polychromosomal hybrids were used initially to attempt to identify candidate chromosomes involved in extinction. From a large collection of rat hepatoma × mouse fibroblast hybrids, several clones with highly segregated karyotypes were identified. Three of these clones were extinguished for expression of hepatic-marker traits [e.g., tyrosine aminotransferase (*TAT*), alanine aminotransferase (*AAT*), alcohol dehydrogenase (*ADH*), aldehyde dehydrogenase (*AHD*), aldolase B (*Ald B*)]. Karyotypic analyses revealed that five particular fibroblast chromosomes (autosomes 8, 9, 10, 11, 13) were retained at high frequency in each hybrid clone. Thus, extinction of a particular group of hepatic markers was correlated with the retention of five specific fibroblast chromosomes. These data provided the first indication that extinction has a specific genetic basis.

By using the approaches outlined in Section 3.1, monochromosomal microcell hybrids that retain each of the chromosomes implicated in extinction were constructed. These families of microcell hybrids revealed that specific genetic loci localized on single fibroblast chromosomes were both necessary and sufficient for extinction of particular hepatic-marker traits. For example, hepatoma microcell hybrids that selectively retain fibroblast chromosome 11 were specifically extinguished for hepatic *TAT* expression, while expression of *AAT*, *ADH*,

AHD, and *Ald B* was unaffected (Killary and Fournier, 1984). These monochromosomal hybrids lacked *TAT* mRNA, and transcribed the *TAT* structural gene at reduced rates (M. J. Thayer, T. C. Peterson, and R. E. K. Fournier, unpublished findings). Furthermore, removal of fibroblast chromosome 11 from the cells by back-selection resulted in reexpression of *TAT* and *TAT* mRNA to full parental levels. None of these effects was seen in hepatoma hybrids that retained fibroblast chromosomes other than 11 (Peterson *et al.,* 1985).

The ability to introduce single, specific fibroblast chromosomes into hepatoma recipient cells, to maintain the introduced chromosome in the population by direct selective pressure, and to remove the chromosome from the cells by back-selection has allowed the complex phenotypic alteration that occurs in extinguished hybrids to be resolved into its component parts. Through studies similar to those summarized above, specific genetic loci responsible for extinction of hepatic TAT, phosphoenolpyruvate carboxykinase, and serum albumin have been identified. In each case, extinction of a particular hepatic trait required the presence of a single, specific fibroblast chromosome. These data define and map specific genetic loci responsible for extinction. The loci so defined, which we have termed *tissue-specific extinguishers* (*Tse-1, Tse-2,* etc.), exert negative regulatory effects on the expression of unlinked, tissue-specific structural genes in *trans* (Killary and Fournier, 1984; Peterson *et al.,* 1985).

The studies outlined above provide a good example of how a complex cellular phenotype can be genetically dissected using monochromosomal microcell hybrids. Molecular genetic analyses of the individual loci so defined will require the use of several different strategies for mammalian gene transfer.

4.2. Analysis of Other Phenotypes

Extinction is only one example of a complex cellular phenotype the genetic basis of which can be profitably explored using microcell hybrids. Studies like those outlined in the preceding section can be applied generally to determine whether specific phenotypes have a discernible genetic basis, whether that basis is simple or polygenic, and whether individual loci control particular aspects of the overall phenotype. Cellular phenotypes that might be amenable to this approach include the sensitivity or resistance to infection by specific viruses, the ability to respond to environmental signals that control cell growth or differentiation, and the expression of the transformed or malignant state.

5. CONCLUDING REMARKS

Single chromosomes are the smallest subdivisions of the mammalian genome that can be uniquely identified given current technology. For example, while every chromosome of any mammalian complement can be recognized from its banding pattern, subchromosomal fragments are often difficult to identify by any method. Ultimately, mammalian genomes may be so densely populated with genetic markers (especially DNA markers) that any segment might be recognizable. In that case, a gene-mapping approach based entirely on physical linkage to known markers would be possible. For the present, however, microcell hybrids provide a straightforward means to fractionate the donor genome into a discrete number of easily recognizable parts. Panels of microcell hybrids constitute one of the most useful resources currently available for mammalian gene mapping.

Microcell hybrids would have limited utility were it not possible to selectively fix every donor chromosome in an individual microcell-hybrid clone. Using the approaches outlined in this chapter, monochromosomal hybrids that retain each mouse or human chromosome can be constructed, and other species' genomes are susceptible to the same strategies. Thus, monochromosomal hybrids with stable, defined genotypes can be readily obtained. This makes possible a systematic, chromosome-by-chromosome approach to mammalian gene mapping and allows the genetic bases of complex cellular phenotypes to be explored more meaningfully.

Finally, like chromosome-mediated gene transfer, microcell fusion is a technique in which a fraction of the donor genome is introduced into recipient cells as organized chromatin. If heritable chromatin structures with functional consequences do exist, these gene-transfer methods may contribute to their identification and characterization.

ACKNOWLEDGMENTS. We thank D. E. Banker for critically reading the manuscript. Work performed in the authors' laboratory was supported by Grant GM26449 from the National Institute of General Medical Sciences. T. G. L. received postdoctoral fellowship support from the Leukemia Society of America and the Hereditary Disease Foundation. R. E. K. F. is the recipient of a Faculty Research Award from the American Cancer Society.

REFERENCES

Athwal, R. S., Smarsh, M., Searle, B. M., and Deo, S. S., 1985, Integration of a dominant selectable marker into human chromosomes and transfer of marked chromosomes to mouse cells by microcell fusion, *Somat. Cell Mol. Genet.* **11**:177–187.

Cassio, D., Weiss, M. C., Ott, M. O., Sala-Trepat, J., Fries, J., and Erdos, T., 1981, Expression of the albumin gene in rat hepatoma cells and their dedifferentiated variants, *Cell* **27**:351–358.

Cepko, C. L., Roberts, B. E., and Mulligan, R. C., 1984, Construction and applications of a highly transmissible murine retrovirus shuttle vector, *Cell* **37**:1053–1062.

Cirullo, R. E., Dana, S., and Wasmuth, J. J., 1983, Efficient procedure for transferring specific genes into Chinese hamster cell mutants: Interspecific transfer of the human genes encoding leucyl- and asparaginyl-tRNA synthetases, *Mol. Cell. Biol.* **3**:892–902.

Colbere-Garapin, F., Horodniceanu, F., Kourilski, P., and Garapin, A.-C., 1981, A new dominant hybrid selective marker for higher eukaryotic cells, *J. Mol. Biol.* **150**:1–14.

Cone, R. D., and Mulligan, R. C., 1984, High-efficiency gene transfer into mammalian cells: Generation of helper-free recombinant retrovirus with broad mammalian host range, *Proc. Natl. Acad. Sci. U.S.A.* **81**:6349–6353.

Davidson, R. L., 1974, Gene expression in somatic cell hybrids, *Annu. Rev. Genet.* **8**:195–218.

Davidson, R. L., Ephrussi, B., and Yamamoto, K., 1966, Regulation of pigment synthesis in mammalian cells as studied by somatic hybridization, *Proc. Natl. Acad. Sci. U.S.A.* **56**:1437–1440.

Davis, F. M., and Adelberg, E. A., 1973, Use of somatic cell hybrids for analysis of the differentiated state, *Bacteriol. Rev.* **37**:197–214.

Ege, T., and Ringertz, N. R., 1974, Preparation of microcells by enucleation of micronucleate cells, *Exp. Cell Res.* **87**:378–382.

Fournier, R. E. K., 1981, A general high-efficiency procedure for production of microcell hybrids, *Proc. Natl. Acad. Sci. U.S.A.* **78**:6349–6353.

Fournier, R. E. K., 1982, Microcell-mediated chromosome transfer, in: *Techniques in Somatic Cell Genetics* (J. Shay, ed.), pp. 309–327, Plenum Press, New York.

Fournier, R. E. K., and Frelinger, J. A., 1982, Construction of microcell hybrid clones containing specific mouse chromosomes: Application to autosomes 8 and 17, *Mol. Cell. Biol.* **2**:526–534.

Fournier, R. E. K., and Moran, R. G., 1983, Complementation mapping in microcell hybrids: Localization of *Fpgs* and *Ak-1* on *Mus musculus* chromosome 2, *Somat. Cell Genet.* **9**:69–84.

Fournier, R. E. K., and Ruddle, F. H., 1977a, Microcell-mediated transfer of murine chromosomes into mouse, Chinese hamster, and human somatic cells, *Proc. Natl. Acad. Sci. U.S.A.* **74**:319–323.

Fournier, R. E. K., and Ruddle, F. H., 1977b, Stable association of the human transgenome and host murine chromosomes demonstrated with trispecific microcell hybrids, *Proc. Natl. Acad. Sci. U.S.A.* **74**:3937–3941.

Francke, U., Lalley, P. A., Moss, W., Ivy, J., and Minna, J. D., 1977, Gene mapping in *Mus musculus* by interspecific cell hybridization: Assignment of the genes for tripeptidase-1 to chromosome 10, dipeptidase-2 to chromosome 18, acid phosphatase-1 to chromosome 12, and adenylate kinase-1 to chromosome 2, *Cytogenet. Cell Genet.* **19**:57–84.

Friend, K. K., Chen, S., and Ruddle, F. H., 1976, Differential staining of interspecific chromosomes in somatic cell hybrids by alkaline Giemsa stain, *Somat. Cell Genet.* **2**:183–188.

Goss, S. J., and Harris, H., 1975, New method for mapping genes in human chromosomes, *Nature (London)* **255**:680–684.

Kahn, C. R., Bertolotti, R., Ninio, M., and Weiss, M. C., 1981, Short-lived cytoplasmic regulators of gene expression in cell hybrids, *Nature (London)* **290**:717–720.

Killary, A. M., and Fournier, R. E. K., 1984, A genetic analysis of extinction: *Trans*-dominant loci regulate expression of liver-specific traits in hepatoma hybrid cells, *Cell* **38**:523–534.

Kozak, C. A., Fournier, R. E. K., Leinwand, L. A., and Ruddle, F. H., 1979, Assignment of the gene governing cellular ouabain-resistance to *Mus musculus* chromosome 3 using human/mouse microcell hybrids, *Biochem. Genet.* **17**:23–34.

Landolph, J. R., and Fournier, R. E. K., 1983, Microcell-mediated transfer of carcinogen-induced ouabain resistance from C3H/10T1/2 Cl 8 mouse fibroblasts to human cells, *Mutat. Res.* **107**:447–463.

Lem, J., and Fournier, R. E. K., 1985, Assignment of the gene encoding cytosolic phosphoenol-pyruvate carboxykinase (GTP) to *Mus musculus* chromosome 2, *Somat. Cell Mol. Genet.* **11**:633–638.

Lugo, T. G., Leach, R. L., and Fournier, R. E. K., 1986, Parasexual approaches to the study of human genetic disease, *Ann. N.Y. Acad. Sci.* (in press).

Mann, R., Mulligan, R. C., and Baltimore, D., 1983, Construction of a retrovirus packaging mutant and its use to produce helper-free defective retrovirus, *Cell* **33**:153–159.

McNeill, C. A., and Brown, R. L., 1980, Genetic manipulation by means of microcell-mediated transfer of normal human chromosomes into recipient mouse cells, *Proc. Natl. Acad. Sci. U.S.A.* **77**:5394–5398.

Miller, A. D., Law, M.-F., and Verma, I. M., 1985, Generation of helper-free amphotropic retroviruses that transduce a dominant-acting, methotrexate-resistant dihydrofolate reductase gene, *Mol. Cell. Biol.* **5**:431–437.

Mulligan, R. C., and Berg, P., 1981, Selection for animal cells that express the *Escherichia coli* gene coding for xanthine-guanine phosphoribosyltransferase, *Proc. Natl. Acad. Sci. U.S.A.* **78**:2072–2076.

Orkin, S. H., Harosi, F. J., and Leder, P., 1975, Differentiation in erythroleukemia cells and their somatic hybrids, *Proc. Natl. Acad. Sci. U.S.A.* **72**:98–102.

Peterson, T. C., Killary, A. M., and Fournier, R. E. K., 1985, Chromosomal assignment and *trans* regulation of the tyrosine aminotransferase gene in hepatoma hybrid cells, *Mol. Cell. Biol.* **5**:2491–2494.

Phillips, S. G., and Phillips, D. M., 1969, Sites of nucleolus production in cultured Chinese hamster cells, *J. Cell Biol.* **40**:248–268.

Potter, H., Wier, L., and Leder, P., 1984, Enhancer-dependent expression of human kappa immunoglobulin genes introduced into mouse pre-B lymphocytes by electroporation, *Proc. Natl. Acad. Sci. U.S.A.* **81**:7161–7165.

Ringertz, N. R., and Savage, R. E., 1976, *Cell Hybrids*, pp. 196–212, Academic Press, New York.

Robins, D. M., Ripley, S., Henderson, A. S., and Axel, R., 1981, Transforming DNA integrates into the host chromosome, *Cell* **23**:29–39.

Ruddle, F. H., and Creagan, R. P., 1975, Parasexual approaches to the genetics of man, *Annu. Rev. Genet.* **9**:407–486.

Saxson, P. J., Srivatsan, E. S., Leipzig, G. V., Sameshima, J. H., and Stanbridge, E. J., 1985, Selective transfer of individual human chromosomes to recipient cells, *Mol. Cell Genet.* **5**:140–146.

Sekiguchi, T., Shelton, K., and Ringertz, N. R., 1978, DNA content of microcells prepared from rat kangaroo and mouse cells, *Exp. Cell Res.* **113**:247–258.

Simonsen, C. C., and Levinson, A. D., 1983, Isolation and expression of an altered mouse dihydrofolate reductase cDNA, *Proc. Natl. Acad. Sci. U.S.A.* **80**:2495–2499.

Sparkes, R. S., and Weiss, M. C., 1973, Expression of differentiated functions in hepatoma cell hybrids: Alanine aminotransferase, *Proc. Natl. Acad. Sci. U.S.A.* **70**:377–381.

Thompson, E. B., and Gelehrter, T. D., 1971, Expression of tyrosine aminotransferase activity in somatic cell heterokaryons: Evidence for negative control of enzyme expression, *Proc. Natl. Acad. Sci. U.S.A.* **68**:2589–2593.

Tunnacliffe, A., Parkar, M., Povey, S., Bengtsson, B. O., Stanley, K., Solomon, E., and Goodfellow, P., 1983, Integration of *Ecogpt* and SV40 early region sequences into human chromosome 17: A dominant selection system in whole cell and microcell human–mouse hybrids, *EMBO J.* **2**:1577–1584.

Weis, J. H., Nelson, D. L., Przyborski, M. J., Mulligan, R. C., Housman, D. E., and Seidman, J. G., 1984, Eukaryotic chromosome transfer: Linkage of the murine major histocompatibility complex to an inserted dominant selectable marker, *Proc. Natl. Acad. Sci. U.S.A.* **81**:4879–4883.

Weiss, M. C., and Chaplain, M., 1971, Expression of differentiated functions in hepatoma cell hybrids: Reappearance of tyrosine aminotransferase inducibility after the loss of chromosomes, *Proc. Natl. Acad. Sci. U.S.A.* **68**:3026–3030.

Weiss, M. C., Sparkes, R. S., and Bertolotti, R., 1975, Expression of differentiated functions in hepatoma cell hybrids. IX. Extinction and reexpression of liver-specific enzymes in rat hepatoma–Chinese hamster fibroblast hybrids, *Somat. Cell Genet.* **1**:27–40.

4

USE OF METAPHASE-CHROMOSOME TRANSFER FOR MAMMALIAN GENE MAPPING

David E. Housman and David L. Nelson

1. INTRODUCTION

The mapping of mammalian chromosomes can be approached by two basic strategies: genetic linkage and physical mapping. The precision with which genetic mapping can be applied to a given chromosomal region is directly related to the number of polymorphic markers available in that region and the spacing of those markers within the region. Physical mapping techniques are limited by the sizes of DNA fragments that can be manipulated, and until recently, these limits were in the range of tens of kilobase pairs of DNA. Recent technical innovations promise the potential for manipulating mammalian DNA segments hundreds of kilobase pairs in length. It is clear that the power of both genetic-linkage and physical-isolation techniques would be enhanced by the ability to isolate and characterize chromosomal segments 10^5 base pairs or greater in length by a direct genetic technique. Previously, we have suggested a strategy based on metaphase-chromosome transfer to address this issue (Housman *et al.*, 1983). For the past several years, our research efforts have been devoted to development of this strategy. The results presented herein bear on the utility of chromosome transfer both for the physical mapping of chromosomes and for the development of DNA-based markers at intervals of a centimorgan or less within these chromosomal segments.

The detailed technical discussion that follows is focused exclusively on the

David E. Housman • Center for Cancer Research, Department of Biology, Massachusetts Institute of Technology, Cambridge, Massachusetts 02139.   **David L. Nelson** • Laboratory of Molecular Genetics, National Institute of Neurological and Communicative Disorders and Stroke, National Institutes of Health, Bethesda, Maryland 20892.

transfer of isolated metaphase chromosomes. The rationale for the choice of this technique was as follows: Previous workers had demonstrated that genes that can be selected as dominant markers in somatic-cell genetic experiments can be transferred from a donor cell to a recipient cell via a preparation of isolated metaphase chromosomes. In some cases, other markers were transferred to the recipient cell along with the selected marker. The physical distance between the two markers cotransferred in this fashion appeared to be quite considerable, perhaps as much as 10^7 base pairs of DNA. In conjunction with techniques we had previously described for identifying species-specific cloned DNA segments from somatic-cell hybrids, the metaphase-chromosome transfer technique appeared ideally suited to solve the problems posed above. Two major outstanding issues remained: First, selectable markers useful in somatic-cell genetics exist naturally at only a small number of chromosomal sites. The ability to generalize the technique to all regions of a given mammalian genome required development. Second, a detailed test of the technique in a region of a mammalian genome that was already well mapped at the level of physical and genetic precision described above had not been carried out. Without a test of this type, our assumptions regarding the behavior of continuous DNA segments during metaphase-chromosome transfer could not be effectively tested.

To address these two issues, we developed the use of retroviral vectors to introduce selectable genes into the mouse genome at random. We then demonstrated the ability of the introduced vectors to serve as selectable markers during metaphase-chromosome transfer (Nelson et al., 1984). Finally, to initiate a detailed analysis of a series of chromosome transferents representing a well-studied region of the mouse genome, we identified a viral insertion that allowed the transfer of the murine H-2 region of chromosome 17 (Weis et al., 1984). Using transferents derived from this, the best-mapped region of the mouse genome in both genetic and molecular terms (Hood et al., 1983), we have been able to investigate the process of chromosome transfer at a very detailed level of DNA structure.

2. A SHORT HISTORY OF METAPHASE-CHROMOSOME TRANSFER

Of the four major techniques of somatic-cell genetics—fusion, microcell fusion, metaphase-chromosome transfer, and DNA transfer—only metaphase-chromosome transfer provides somatic-cell hybrids containing contiguous stretches of small chromosome segments. Fusion generally results in the retention of several whole chromosomes, while microcell fusion usually reduces the foreign chromosome complement to a single whole chromosome (Fournier, 1983). In the case of DNA transfer from genomic sequences, contiguous regions of a few tens of kilobase pairs can be obtained after two to three rounds of transfer to reduce the amount of adventitiously cotransferred material (Shih and Weinberg, 1982). The attraction of metaphase-chromosome transfer is its capacity to provide small contiguous regions of chromosomes containing up to 1–2% of the genome.

The ability to transfer the gene that encodes hypoxanthine-guanine phosphoribosyltransferase (HGPRT) from Chinese hamster cells to mouse cells mutant for the enzyme via metaphase chromosomes was first described by McBride and Ozer (1973). The frequencies of transfer obtained in these early experiments were 10^{-6}–10^{-7}. Such low frequencies of colony formation require sensitive selection systems, and the hypoxanthine–aminopterin–thymidine (Syzbalski et al., 1962) system for selecting HGPRT is one such system. Improvements in the methodology of transfer have increased transfer efficiencies to the 10^{-4}–10^{-5} range. Willecke and Ruddle (1975) simplified the isolation methods to transfer HGPRT from human cells. Their method for isolating chromosomes to be transferred has been utilized by many groups since. It involves a long (16- to 24-hr) mitotic block of the donor cells using colcemid, colchicine, or vinblastine; physical disruption of the cells in low-ionic-strength buffer; and the partial purification of chromosomes by differential centrifugation. Mitotically blocked cells yield free chromosomes due to the breakdown of the nuclear membrane prior to cell division.

Miller and Ruddle (1978) made a major increase in chromosome-transfer efficiency by borrowing the methodology of Graham and van der Eb (1973) for introducing DNA into cells. This method of introducing chromosomes into recipient cells involved the precipitation of chromosomes with calcium phosphate and the treatment of recipients with 10% dimethyl sulfoxide (DMSO) 4 hr after the addition of chromosomes. With this improvement, efficiencies up to 4×10^{-5} have been observed. Finally, Lewis et al. (1980) studied the factors involved in chromosome transfer using these methods, using transfer of the thymidine kinase (TK) and dihydrofolate reductase (DHFR) genes. They found that efficiencies increase with the amount of chromosomes added and the length of time between addition of precipitate and treatment with DMSO. With these improvements, the efficiencies of metaphase-chromosome transfer approach those of DNA transfer.

The mechanism of uptake of transferred chromosomes has been studied extensively. Using radiolabeled donor chromosomes, Simmons et al. (1978) have followed the uptake and early fate of chromosomes added to recipient cells. They found that chromosome uptake is rapid: After 4 hr of incubation, two thirds of cell-associated radioactivity was inside the cell (measured by resistance to DNase I). Additionally, it appeared that labeled DNA in the nucleus had undergone an increase in length by 6 hr, while in the cytoplasm, label was being degraded. Simmons also reported that between 1 and 4 chromosome equivalents of DNA were associated with each recipient nucleus at 6 hr. These results and others suggest a model of uptake involving phagocytosis by most recipients and lysosomal degradation of chromosomes with frequent escape to the nucleus.

However, if the frequency of successful generation of transferents is compared
to the average number of chromosomes per recipient nucleus, the retention (or
expression) of the introduced chromosome must be quite rare.

The length of chromosome transferred by this technique is variable, but is
generally thought to be between 0.1 and 1% of the genome (3×10^6 to 3×10^7
base pairs). These numbers have been derived from three methods. The earliest
estimates of the size of transferred sequences were made by measuring the
frequency of cotransfer of genes known to be linked. Willecke *et al.* (1976)
found human galactokinase (GALK) expression in two of eight transferents
selected for TK activity. These two genes are thought to be separated by less
than 0.2% of the human genome, since they both map to a single cytogenetic
band on human chromosome 17. Miller and Ruddle (1978) observed expression
of glucose-6-phosphate dehydrogenase (G6PD) in 25% of transferents selected
for the linked *HGPRT* gene. These genes are estimated to be separated by roughly
1% of the genome (Burch and McBride, 1975; Willecke and Ruddle, 1975).
Although the expression of GALK and G6PD is usually seen in somatic-cell
hybrids, the failure to detect the enzymes in these studies does not prove the
absence of the genes. Assuming that these genes are expressed when present,
these cotransfer results suggest transfer of segments up to 20% of the average
chromosome. Cytogenetic analysis of transferents has yielded a range of sizes
of transferred chromosomes. Wullems *et al.* (1976) reported transfer of an intact
X chromosome, but the majority of studies report either no detectable material
or small segments ranging in size to one fourth of a chromosome (Miller and
Ruddle, 1978; Klobutcher and Ruddle, 1979). The third method of analysis of
the size of the transferred segment involves the use of nucleic-acid hybridization.
Olsen *et al.* (1981) utilized differential hybridization to prepare unique-sequence
DNA enriched in X-chromosome sequences for use as a probe in liquid hybridization to transferent DNA. With this procedure, they were able to demonstrate
transfer of a range of sizes from less than 0.3% of the human X chromosome
to nearly 20%. These values translate to a range of 0.015–1% of the genome,
or 4.5×10^5 to 3×10^7 base pairs.

The stability of the transferred phenotype has been investigated by many
groups. Klobutcher *et al.* (1980) defined three stability classes in transferents of
TK and proposed a model to explain each. The first class is as stable as any
host phenotype and is thought to involve incorporation of the transferred chromosome into a host chromosome. In the second class, the phenotype is unstable,
but lost at a relatively slow rate, between 0.5 and 2% of the cells losing the trait
per generation. This rate is close to that seen for whole chromosome somatic-
cell hybrids and is thought to involve a transferred chromosome segment that
has retained its centromere. Such a chromosome is usually segregated properly,
but occasionally fails to assort to both daughters. Data to support this hypothesis
include the frequent observation of visible chromosome segments in "slow-loss"
transferents. The third class of transferents loses its phenotype at a very rapid

rate: from 5 to 10% percent of the cells per generation. This class exhibits no microscopically detectable chromosome and is believed to be a foreign chromosome segment that has no centromere and is unable to be properly segregated at cell division. This class is not yet stabilized by insertion into the host chromosome. If carried in culture for a sufficient length of time, such transferents will give rise to stable populations. This is thought to be accomplished via the insertion of the transferred chromosome into a host chromosome in a cell of the population and subsequent overgrowth by that stable variant. This phenomenon has been observed in several studies (Degnen *et al.*, 1976; Willecke *et al.*, 1976).

One prediction from the improper assortment of unstable chromosome segments at mitosis is the accumulation of chromosome segments in those daughter cells that survive selection. This process will lead to an eventual average of three to four copies of the chromosome per cell. Although this has not been tested carefully, Scangos *et al.* (1979) found an apparent increase in copy number of herpes TK (H-TK) DNA in unstable chromosome transferents by Southern blotting. The H-TK enzyme levels were also greater in the unstable transferents. This increase in copy number could be advantageous for the isolation of recombinant clones from a transferred region. On stabilization, however, the average number of copies reduces to one per cell (half copy).

Little is known about the expression of genes in chromosome transferents. Clearly, the genes selected for are being expressed, and the products of cotransferred isozyme genes have also been detected. Assuming that genes introduced by chromosome transfer are equivalent to those introduced by fusion, then "proper" expression of genes (e.g., in developmental pathways) may be possible with chromosome transfer. This is in contrast to the situation in DNA transfer, in which regulated expression is often incomplete (Willing *et al.*, 1976; Chao *et al.*, 1983; Wright *et al.*, 1983).

One very useful aspect of DNA transfer is the ability to cotransfer unlinked molecules at a frequency much higher than would be predicted by the product of the transfection frequencies. It is of great interest whether chromosome cotransfer is also possible, since this would complicate the analysis of transferred chromosomes. Klobutcher and Ruddle (1979) report data that indicate that cotransfer of asyntenic markers can occur. In an analysis of seven transferents of human TK, two contained isozyme markers from other human chromosomes. By back-selection against the introduced *TK* gene, it was demonstrated that the asyntenic markers were associated with TK in the transferents. Although it was not possible to distinguish whether linkage occurred on transfer or prior to transfer in the parent HeLa cells, the authors favored the former hypothesis. We have also found restructuring of the introduced chromosome segment, and we discuss it in detail below.

A major limitation of standard metaphase-chromosome transfer is the absolute requirement for a selection system. Frequencies of transfer are quite low, and without a means for selecting recipients that contain transferred sequences, the technique is useless. This limitation has restricted the chromosome regions transferred to those few that contain native selectable markers. The genes for TK (Willecke *et al.*, 1976), HGPRT (McBride and Ozer, 1973), adenosine phosphoribosyltransferase (Lugo and Baker, 1983), and DHFR (Haber and Schmike, 1982) have been used widely to transfer human and rodent chromosomes, since powerful selection systems exist for each. Additional selection systems have been developed, and additional chromosome regions have been transferred. These include the "oncogenes" (Shih *et al.*, 1979), which can confer an altered phenotype to certain cell recipients, and argininosuccinate synthetase (Hudson *et al.*, 1980). As further selection systems become available, chromosome regions previously refractory to analysis by chromosome transfer will be studied. Good current candidates include the low-density-lipoprotein receptor (Sege *et al.*, 1984) and certain aminoacyl-transfer RNA transferases (Cirullo *et al.*, 1983). However, for all the selection systems listed (with the exception of DHFR), recipient cell lines are limited to those mutant in the gene to be transferred. This additional limitation on the generality of chromosome transfer, by restricting the range of cell recipients, reduces the variety of experiments possible and further diminishes the utility of the technique.

3. INTRODUCTION OF NOVEL DOMINANT MARKERS

Two developments in somatic-cell genetics have eliminated both the need for native selectable markers in chromosome regions of interest and the requirement for mutant recipients. The advent of bacterial genes the products of which can be directly selected in a dominant fashion in mammalian cells makes it possible to insert novel selectable genes into any genomic region. Additionally, the development of retroviral vectors that can efficiently deliver such genes into a wide variety of mammalian cell types allows large numbers of selectable markers to be introduced into a population of cells, exposing the entire genome of any mammalian species to analysis by chromosome transfer. Furthermore, since the selection systems for these bacterial genes can be generalized to any cell, recipient cells can be of any variety, allowing the retention of chromosome regions in numerous cell backgrounds.

The ability to extend metaphase-chromosome-transfer methods to the entire genome requires the insertion of selectable markers into many regions. Achievement of this goal requires both selectable markers to insert and efficient means of introducing such markers into cells. Two *Escherichia coli* genes were chosen to satisfy the first requirement. These genes, xanthine-guanosine phosphoribosyltransferase (*gpt*) and neomycin acetyltransferase (*neo*), were chosen for their ability to confer a dominantly selectable phenotype on a wide variety of mammalian cell types (Mulligan and Berg, 1980; P. J. Southern and Berg, 1982).

This characteristic allows the introduction of these genes into any cell line, and the transfer of chromosomes containing these genes to any cell background, eliminating the requirement for mutant cell lines and expanding the potential donor–recipient combinations.

There are several methods of introducing genes into mammalian cells. Transfection of purified DNA by the $CaPO_4$ method, first described by Graham and van der Eb (1973) and later refined and extended (Maitland and McDougall, 1977; Pellicer et al., 1978; Wigler et al., 1977, 1978, 1979a–c, 1980) to cloned DNA as well as to total cellular DNA, is relatively inefficient, but suitable for a wide variety of studies. Although there is wide variation in the efficiency of DNA uptake among cell lines, stable transfection can often be found in 1 in 10,000 cells. For the insertion of selectable markers into a few chromosomal locations, such frequencies are sufficient. Variations on this technique have increased frequencies or ease of transfection. These include protoplast fusion (10^{-3}–10^{-2}) (Schaffner, 1980), microinjection (10^{-1}) (Cappecchi, 1980), and phage-mediated gene transfer (10^{-5}) (Ishiura et al., 1982). These methods are also suitable for the insertion of selectable markers into small numbers of chromosomes. However, the recent development of cloned retroviral sequences as vehicles for the introduction of DNA has increased not only the frequency of transfer, but also the range of susceptible cells, the precise insertion of the exogenous sequences, and the ease with which the transfer is accomplished.

Retroviruses are enveloped RNA-containing viruses the unique replication of which via reverse transcription distinguishes them from other RNA-containing viruses (Temin and Baltimore, 1972). Since these viruses replicate through a DNA intermediate, construction of viruses that contain cloned DNAs replacing nonessential sequences followed by transfection into cell lines that contain trans-acting helper functions has allowed the generation of high-titer virus stocks that contain thymidine kinase (Shimotohno and Temin, 1981; Tabin et al., 1982) as well as both the gpt and neo genes (Mulligan, 1983). There are considerable advantages to the use of retroviral infection to deliver selectable markers.

The discovery of the RNA sequences necessary for the packaging of the retroviral genome, along with the finding that proviruses deleted for this region (the ψ sequences) can still provide the trans-acting factors necessary for viral production (Mann et al., 1983), has allowed the production of viral stocks deficient in competent helper virus. This development has added to the utility of virus for introducing selectable markers into the genome by eliminating the "superinfection barrier" found in productively infected cells, allowing multiple infection of a single cell.

Murine retroviruses are classified by their affinities for various cell types. The class of ecotropic viruses has heretofore been found to infect only mouse and rat cell lines. Xenotropic retroviruses make up the majority of murine proviruses, but do not infect mouse cells, preferring rat and other mammalian cells as well as cells form avian species (Oie et al., 1976). The third class of murine retroviruses, the amphotropic viruses found in wild mice, are able to infect a wide variety of mammalian cells, including mouse, human, monkey, rabbit, and guinea pig, but not hamster (Hartley and Rowe, 1976).

Cells that produce retrovirus are refractory to infection by virus of the same tropism ("superinfection") due to the formation of a complex of the budding retrovirus coat proteins with the cell-surface receptor that the virus uses to enter cells. This phenomenon can interfere with the introduction of selectable markers into many mouse cell lines (and somatic-cell hybrids with a mouse background). A method to detour this interference was developed by Rein et al. (1982). The method involves the overnight treatment of producing cells with tunicamycin. Tunicamycin inhibits glycosylation of membrane proteins (including those of the retroviral coat) and disturbs their export to the cell surface. The treatment frees the receptors for incoming virus by stopping the production of internal virus particles. With this methodology and the ability to pseudotype viral constructs in amphotropic coats, it is possible to introduce selectable markers into virtually any common mammalian cell.

One more issue must be addressed in the use of retroviruses to introduce selectable markers into chromosomes: Is the integration of retroviruses random? Clearly, to accomplish the goal of marking all regions of the genome with selectable markers, the markers must integrate into the genome in a reasonably random fashion. Inspection of the flanking DNA sequences of avian and murine proviruses has given no indication of sequence specificity of the integration site (Shimotohno and Temin, 1980; Shoemaker et al., 1980, 1981). However, this result does not preclude a specificity of integration defined by some property of the DNA other than its primary sequence, nor does it preclude a specificity of sequence at a distance from the site of integration. The limited mapping of germline integrations in the mouse genome seems to indicate a large number of integration sites.

Recent reports of homologous recombination mediating site-specific integration of DNA introduced into mammalian cells (see Chapters 13 and 14) validate the concept of targeting a specific region for analysis by introducing a selectable marker attached to DNA homologous to a gene known to be in the region. Although the frequencies reported for this event are small (10^{-2}–10^{-3}), the ability to target integration sites into regions of interest would be quite useful.

4. METAPHASE-CHROMOSOME TRANSFER OF INTRODUCED SELECTABLE MARKERS

We embarked on a series of experiments designed to test the following general approach: Introduce selectable neomycin acetyltransferase (neo) genes into mouse and human cells through the use of retroviral vectors. Transfer chromosomes from infected cells either as populations or as single clones to recipient cells of a different species. Select for cell clones taking up the chromosome segment containing the *neo* gene and analyze them for cotransferred DNA sequences through the use of DNA hybridization. As a model system for the analysis of the technique, we chose to attempt the transfer of the murine major histocompatibility complex (MHC) of chromosome 17, since this region has been extensively analyzed on both the genetic and molecular scales. The details of these studies have been described elsewhere (Weis *et al.*, 1984, 1986; Nelson *et al.*, 1984, 1986). Here, we shall describe the general findings and their importance to the large-scale analysis of mammalian genomes.

4.1. Inserting Selectable Markers

We used both DNA transfection and retroviral infection as methods of inserting selectable *neo* genes into chromosomes. Dominant markers introduced by both procedures function equally well in chromosome-transfer experiments in terms of frequency of transfer. There are two distinct advantages to the use of retroviral vectors: (1) Efficiencies of infection approaching 100% of the target cells are routinely obtained. This allows selectable markers to be introduced at many sites with little effort. Additionally, retroviral infection offers the ability to control the number of introduced markers by adjusting the multiplicity of infection. DNA transfer often results in multiple, occasionally tandem copies of the introduced gene (see Chapters 8 and 10). (2) Retroviral DNA inserts into the chromosome in an orderly fashion, allowing ready determination of the number of integrated copies through Southern hybridization. Since transfected DNA often rearranges on insertion into the chromosome, it is more difficult to determine the number of copies integrated.

The use of helper-deficient packaging lines for the introduction of retroviral vectors into cells to be used as chromosome donors is quite important. Our early efforts to use retroviral vectors to introduce *neo* genes into human cell lines required the use of amphotropic, helper-containing viral stocks. Infection of the human lines was accomplished readily, and chromosome-transfer experiments to hamster and mouse cells seemed to work at a very high frequency. Later analysis of the transferents demonstrated that they contained no human DNA and that the restriction fragment carrying the vector in the transferents bore no relationship to that found in the donor line. On checking the donor line for the production of virus, we found an ample amphotropic titer. We concluded that viral particles sufficient for infection copurify with preparations of chromosomes. "Transferents" derived by infection with virus masked the true transferents. The development by Cone and Mulligan (1984) of an amphotropic packaging line eliminated this difficulty.

4.2. Identifying the Desired Insertion Site

The most difficult aspect of using metaphase-chromosome transfer to isolate a desired region of chromosome in a heterologous cell background is that of identifying a cell containing an inserted selectable marker within the region. Two approaches can be taken. The sites of retrovirally inserted selectable genes can be determined in cell clones prior to their use as donors in chromosome transfer, allowing the identification of insertions in the chromosome or region of interest. Alternatively, chromosomes can be transferred from individual cells, or populations of cells, that harbor introduced selectable markers, and the region transferred can be assayed in the transferents. Since the transfer of chromosomes is the rate-limiting step in this methodology (due to its low frequency), it is preferable to identify insertion sites prior to transfer. However, this approach is not feasible when there is insufficient information regarding the chromosomal position of the desired region.

Identification of an insertion site prior to transfer is the most logical approach to the overall problem of mapping by chromosome transfer. Aside from the possibility of directly selecting for an insertion site through the use of homologous recombination (discussed briefly above), there are four techniques that can be used to demonstrate the site of insertion of the selectable marker. The first method, and the most direct one, is to use *in situ* hybridization to the introduced selectable marker using a DNA probe. Since the introduced *neo* gene is unique in the cell, hybridization will occur only at the site of the introduced sequence. This technique, while technically difficult, can allow the assignment of the insertion site to a small region of chromosome (Gerhard *et al.*, 1981). For this technique to be reliable, it is important that the infected cell have a normal karyotype for its species, so that the chromosome can be identified. This technique is most useful for human chromosomes, since they, unlike mouse chromosomes, can be readily identified. The second most direct technique involves reisolating the introduced gene along with chromosomal DNA flanking it. The flanking DNA can be used as a hybridization probe on Southern blots of DNA from a hybrid cell chromosome mapping panel. Such panels have been used extensively to assign DNA segments to individual chromosomes in mouse and human (Shows *et al.*, 1982). This can assign the flanking DNA (and thus the insertion site) to a particular chromosome. This technique can also be used after transfer. Various methods of isolating inserted retroviral sequences have been described. The most straightforward of these is through the use of standard cloning techniques. Perhaps the simplest method involves the use of the cosfusion rescue protocol developed for the pSVX neo retroviral vector described by Cepko *et al.* (1984). While only a small percentage of DNA clones derived from this method contain flanking sequences, such a large number of clones can be isolated that identifying those that contain flanking DNA is not difficult.

Two somatic-cell genetic approaches can also be used to determine the integration site. The first, cosegregation, requires that the targeted chromosome be contained in a somatic cell-hybrid and be retained by selection. If this chromosome is lost from the hybrid at a reasonable frequency ($>1\%$ per cell generation), the presence of a second selectable marker on the chromosome can be assayed by the simultaneous loss of both phenotypes in a cell clone. If all the aforestated requirements can be met, this technique can be quite fruitful. This method has been used successfully in our laboratory (Glaser and Housman, 1984). Furthermore, once a selectable marker has been introduced into a chromosome, cosegregation can be employed to assay for the insertion of additional markers into the same chromosome. The second technique from somatic-cell genetics is microcell transfer. This technique is covered in detail in Chapter 3. Briefly, the microcell technique generally transfers a single chromosome to a recipient cell. By transferring the chromosome that contains the inserted selectable marker, it is possible to identify the chromosome by karyotype analysis. Microcell transfer often transfers two or more chromosomes, but only that containing the selectable marker should be common to all microcell hybrids. Further reductions in the size of the chromosomal segment can subsequently be achieved by chromosome transfer from the microcell hybrid line.

4.3. Transferring Chromosomes

Metaphase-chromosome transfer is a technically challenging technique. It involves four steps:

1. Chromosome preparation from donor cells
2. Chromosome coprecipitation with calcium phosphate
3. Application to and other treatments of recipient cells
4. Selection of transferents.

On occasion, the frequency of transfer of chromosomes can approach that of DNA; however, it is usually less efficient. Since the transfer of chromosomes is the limiting step in the technology described here, it is important to optimize the technique to as great an extent as possible. The major difference between chromosome and DNA transfer is in step 1. With DNA transfer, the DNA to be applied to the cells need be prepared only once, and can then be used for all subsequent transfections. Chromosomes must be prepared fresh for each transfer, and this probably accounts for much of the variability seen in the frequency of transfer. In general, we have obtained the best chromosomes for transfer from rapidly dividing donor cells. This may simply be due to increased yields of chromosomes produced from greater numbers of cells blocked in mitosis. Steps 2–4 can be optimized through the use of DNA-transfer control studies, since conditions that allow good DNA transfer seem to be close to optimal for chromosome transfer as well. There are clearly differences among recipient cell types, however, and optimization of steps 2–4 is important for each new recipient cell. A detailed transfer protocol can be found in Nelson *et al.* (1984).

4.4. Stability Analysis

We have performed stability tests on a small number of the chromosome transferents derived by this methodology (Nelson *et al.*, 1984). Transferents that exhibit either the totally stable or the quite unstable phenotypes (see Section 2) were found. No transferents of the moderately unstable class were identified. Southern-blot analyses of secondary transferents derived from primary transferents of both the stable and the highly unstable types were consistent with the hypothesis that these types of transferents are integrated into a host chromosome and free in the nucleus respectively (see Section 4.6).

4.5. Nature of the Transferred Segment

Once a chromosome segment has been isolated in a heterologous cell background, differences between the two species can be utilized to identify the transferred region. The simplest of these, karyotypic analysis of the chromosome transferred, is the least reliable, since the transferred segment is only rarely large enough to identify. Assays for isozymic forms of enzymes linked to the region desired can be used when the forms differ between the two species. This mainstay of somatic-cell genetics is less useful for chromosome transferents due to the small likelihood of transfer of an assayable isozyme. Antisera can also be used to identify specific protein products associated with the transferred region. This was the approach we used to identify transferents containing the murine MHC. We used antisera specific for the Class I surface antigens from the mouse. These were readily detectable on the surface of the hamster-cell transferents (Weis *et al.*, 1984). This approach could also be used in conjunction with fluorescence-activated cell sorting for the isolation of cells that express a surface antigen for which an antibody is available, as has been used for DNA transfer (see Chapter 1). DNA probes can also be used to identify transferred regions. In the case of the MHC transferents, we were able to use the numerous clones from the MHC to identify cotransferred sequences. Using single-gene probes to randomly assay transferents for regions of interest would be quite laborious, however.

4.6. DNA-Hybridization Studies

DNA-hybridization technology has allowed better estimates of the amount of DNA transferred through Southern (E. M. Southern, 1975) and dot blotting.

Dot blots containing as little as 100 ng of transferent DNA can give an estimate of the percentage of DNA in the cell derived from the donor when a total donor cell DNA probe is used (Nelson *et al.*, 1984; Murphy *et al.*, 1985). A dilution series of donor DNA mixed into recipient DNA is used as a standard for comparison. While this method relies on repeat DNA being equally distributed in the genome and on the lack of amplification of introduced sequences, it is quite predictive of the relative amounts of a given region in the transferents. The lower limit of detection using this technique appears to be 0.01% of the genome [300,000 base pairs (bp)]. We have obtained transferents that contain as much as 1–2% of the genome as measured with this methodology. However, the typical transferent contains between 0.05 and 0.1% of the donor genome (Nelson *et al.*, 1984).

The capability to visualize large quantities of DNA in hybrid cell lines through the use of Southern blotting and repeat sequence probes was demonstrated for segments of human chromosome 21 by Law *et al.* (1982) and for human chromosome 11 by Gusella *et al.* (1982). We have adapted this methodology for use in analyzing chromosome transferents (Nelson *et al.*, 1984). Using this approach, it is possible to detect 50 or more restriction fragments specific to the introduced chromosome segment. Since the frequency of the repeat sequence chosen as probe determines the number of bands contained within a chromosome segment of a given size, we have used high-frequency repeat probes for analyzing small chromosome segments and relatively low-frequency repeat sequences for larger ones. This has allowed us to compare several transferents derived from a single inserted selectable marker for overlapping sets of repeat DNA-containing restriction fragments (Nelson *et al.*, 1984), to investigate changes in the chromosome segment on secondary chromosome transfer (Nelson *et al.*, 1984), and to follow changes in the introduced chromosome segment during repeated passage of a single transferent (Nelson, unpublished observations).

We have attempted to use cloned repeat-sequence probes to analyze several independent transferents derived from individual *neo* integration sites. Such transferents should contain varying lengths of the single chromosome surrounding the integrated *neo* vector. Repeat DNAs flanking the integration site can be visualized by Southern blotting, with each copy of the repeat sequence generating a restriction fragment of unique size. In principle, a map of the repeat DNAs surrounding the integration site can then be derived, using the frequency of a restriction fragment's appearance among the various transferents as a measure of its distance from the integration site. In practice, no cloned repeat has proven useful for this purpose in either human- or mouse-derived primary transferents. The patterns of restriction fragments on Southern blots of the various transferents have been too complex to be useful in the derivation of a simple map (Nelson *et al.*, 1984). Whether this is due to the limitations of the resolution of Southern blotting or to the transfer of noncontiguous chromosomal segments has not been resolved. However, secondary transferents derived from primary transferents containing the murine MHC in a hamster cell background retain either the entire parental repeat DNA restriction pattern or a subset of it (Nelson *et al.*, 1984). One interpretation of this finding would argue that primary transferents often contain previously unlinked chromosomal segments, confounding analysis of the repeat DNA pattern. The preservation of the complete parental pattern in some secondary transferents would also predict that these segments become linked in the primary transferent.

Analysis of secondary transferents with repeat DNA probes has also allowed us to demonstrate a correlation between stability of the introduced chromosome segment in the primary transferent and the amount of chromosome transferred to secondary transferents derived from it. Two primary transferents containing the MHC were used as donors in chromosome-transfer experiments to hamster and monkey cell backgrounds. Secondary transferents were obtained and their DNA was analyzed by Southern hybridization with cloned repeat DNA probes specific for mouse sequences (Nelson *et al.*, 1984). Several secondary transferents from the line CHMD D2 exhibited a pattern of bands essentially identical to that in the parent line. The CHMD D2 line is a member of the "stable" class of transferents and may therefore represent a chromosome segment embedded in host chromosomal material. It is likely that surrounding hamster chromosome acts as a buffer in secondary chromosome transfer, leading to the preservation of the entire mouse chromosome segment and its restriction pattern. The second primary transferent, CHMD E1, gave rise to secondary transferents containing only small subsets of the restriction pattern found in the parental line. The CHMD E1 line is a member of the "very unstable" class of transferents and probably contains an unattached chromosome segment without a centromere. The reduction in amount of murine chromosomal material on secondary transfer from this line is consistent with this view of the unstable class of transferent and with the notion that the reduction in length of chromosomes on transfer is mediated primarily through attack of the ends of the segment. These findings predict that only primary transferents of the "unstable" class would be useful in serial chromosome-transfer experiments designed to reduce the size of the chromosomal segment retained.

We have recently applied this method of analysis to the study of the CHMD E1 line maintained in culture over a period of 2 months. Since the chromosome segment in this transferent is not attached to host sequences, we wondered whether it would undergo changes over time. DNA was prepared from the line weekly for a total of 8 weeks and analyzed by Southern blotting with a cloned mouse repeat probe. Although the majority of restriction fragments did not change during passage, several did—some appearing, some disappearing, and a few changing in intensity (Nelson, unpublished results). This result directly demonstrates dynamic changes in an unstable chromosome segment. Since this effect was seen in a population of cells over only 16 generations, the effects in individual cells may be quite dramatic. Thus, changes in unstable chromosome transferent lines over time could be quite drastic and may account for some of the anomalies observed among transferents from a single chromosomal region.

5. APPLICATIONS AND PROSPECTS

We embarked on this series of experiments to develop the use of chromosome transfer as a tool for mapping mammalian chromosome segments 10^5–10^7 in length and as a methodology for isolating DNA segments from specific genomic regions. In the preceding sections, we have described the approaches taken and characterized the limitations of the transfer technique. In this section, we will present our findings as they relate to our two objectives and estimate the prospects of this methodology in mammalian genetics.

5.1. Fine-Structure Genetic Mapping

The use of chromosome transferents for fine-structure genetic mapping can be realized in two ways. The first would be to use a panel of many transferents, each containing a different segment of a chromosome (or genome) to assign markers to regions. This approach is much the same as that used for chromosome assignments and sublocalization, but on a much finer scale. Such an approach would take anywhere from 100 to 1000 transferents to cover an entire genome and is not currently feasible, since such a panel of transferents does not exist. The other approach involves the use of a panel of several transferents containing differing segments of the same chromosomal region, all derived from a single integrated selectable marker, to order markers within a single region. We have developed such a panel for the murine major histocompatibility complex (MHC) and used it to order markers within this segment of chromosome 17. This method is similar to deletion mapping, since the presence or absence of a marker is scored for all members of the panel, and the frequency of cotransfer of markers is used as an index of their distance from one another. This allows markers to be grouped and ordered with respect to one another and the inserted selectable marker. In the MHC, with its abundance of cloned DNA sequences, we assigned markers to transferents through the use of Southern blotting, developing a map for each of the 22 primary and 29 secondary transferents that contain this region (D. L. Nelson, unpublished results). The development of such maps also allowed us to identify the integration site of the neomycin acetyltransferase (*neo*) vector in the region. The vector was found to be between the *K* and *I* genes of the MHC by observation of transferents containing *K*, *neo* and *I*, *K* and *neo*, or *I* and *neo*. This prediction was confirmed by cloning the integrated vector and mapping a flanking sequence to the region genetically (Nelson *et al.*, 1986).

The maps of the various transferents have also allowed us to confirm both the order of the regions of the MHC with respect to one another (using breakpoints located between each) and the order of genes within the I region (using breakpoints between the individual genes) predicted by genetics. One transferent, CHMD(B) B1, predicts an order for the genes within the S region ($Bf, Slp, C4$) with respect to the centromere and the other regions (Weis *et al.*, 1986). This is significant, since the S region has never been oriented by genetic means; no crossover events have been detected within the region.

Although the majority of transferents are contiguous for all the markers tested, about 5% are not. For example, some transferents exhibit gaps between genes of the I region, exhibiting no hybridization to genes between the *neo* marker and retained markers. This phenomenon of discontinuity has also been observed for transferents of human chromosome 11 (T. Glaser and D. E. Housman, unpublished results; Porteus and McHastie, personal communication). How such presumed breakage–rejoining events occur is not understood. It is possible that on uptake into recipient cells, chromosomes are broken and re-ligated in a manner similar to that observed in DNA transfection, but there is no direct evidence to support this hypothesis. It seems unlikely that such events could occur during chromosome preparation, since broken ends would not be likely to be rejoined *in vitro*. Deletions occurring either in the recipient cell after transfer or in the donor cell prior to transfer could also account for these discontinuities. Such deletions may be quite similar to those that give rise to mutant phenotypes *in vivo* and may occur with a higher frequency than previously presumed. We have observed one MHC transferent that would support the occurrence of deletions in the recipient cell. This transferent, MDMDA, when analyzed with a probe for the complement gene $C4$, exhibits strong hybridization of the $C4$ gene containing restriction fragment, but reduced hybridization of the fragment representing the cross-hybridizing (and more proximal) Slp gene (Weis *et al.*, 1986). We have interpreted this finding as evidence for deletion of the Slp region in a portion of the cell population carrying this chromosome segment. This may explain the existence of discontinuities, especially if selective pressure allows overgrowth of the cells containing the deleted chromosome.

5.2. Isolating Chromosomal Regions

Perhaps the most promising application of this methodology is its use in isolating mammalian genomic DNA clones solely on the basis of their positions on a chromosome. This approach, which has been employed over the last several years to isolate interesting genes in *Drosophila*, has not yet been established in mammalian systems. The utility of such an approach applied to mammalian genomes should be quite large. Many interesting mutant phenotypes in the mouse have been mapped to single loci. Isolation of the genes involved in these mutations would provide insight into the complex processes of mammalian development. Additionally, through the use of restriction-fragment-length polymorphisms, the locations of loci involved in a number of human genetic disorders have been described (Gusella *et al.*, 1983; Murray *et al.*, 1982). The ability to isolate the genes involved in these disorders not only would allow more precise diagnosis of carriers and affected individuals, but also should provide insight into the nature and potential treatment of the diseases (Gusella *et al.*, 1984). We believe that metaphase-chromosome transfer provides an avenue of approach to cloning by location in mammals.

To demonstrate the feasibility of metaphase-chromosome transfer for the isolation of DNA sequences by their genetic positions, we utilized a transferent from the *H-2* region for the construction of a cosmid library, with the hope of isolating the entire murine region in a relatively small number of clones (Weis *et al.*, 1986). Clones were screened for the presence of mouse DNA through the use of cloned, murine-specific repeat DNA sequence probes. Since a hamster transferent was chosen for cloning, we could not use a total mouse DNA probe due to cross-hybridizing sequences found in hamster DNA. This may have limited the number of clones obtained. Such an approach is feasible when the evolutionary distance between the two species' DNAs is larger (Gusella *et al.*, 1980). We obtained 28 cosmid clones, which fell into eight clusters of overlapping sequence. An additional 4 clones could not be assigned to any cluster. Two of the clusters could be identified. One contained the integrated *neo* vector, while the other represented a portion of the *I* region of the MHC. The remaining clones represent MHC-linked sequences that have not previously been isolated. These sets of cosmid clones represent a small recombinant library specific for this region of the mouse genome. The reduction in complexity is the key point; such a library should readily yield additional genetic markers closely linked to this locus. In the search for sequences involved in genetic disease, such libraries will be extremely valuable, and metaphase-chromosome transfer provides the reduction in complexity necessary for their production.

6. CONCLUSIONS

At the time of the writing of this review, the prospects for applying metaphase-chromosome transfer for the fine-structure mapping of mammalian chromosomes appear good. The detailed evaluation of transferents of the *H-2* region reveals transfer of chromosomal segments between 10^5 and 10^7 in length. The use of retroviral vectors and dominant selectable markers is also well documented in this and other studies allowing extension of the technique to most if not all of the genome. The question of chromosomal rearrangement or deletion remains a significant one. While events of this type can clearly be demonstrated, the degree to which they serve as a limitation on the proposed strategy and the extent to which transferents bearing such rearrangements can be incorporated into the overall strategy remain to be determined. The detailed answers to these questions are likely to result from efforts to apply chromosome transfer to regions of mammalian genomes that have not yet been mapped in fine detail.

REFERENCES

Burch, J. W., and McBride, O. W., 1975, Human gene expression in rodent cells after uptake of isolated metaphase chromosomes, *Proc. Natl. Acad. Sci. U.S.A.* **72:**1797–1801.

Capecchi, M. R., 1980, High efficiency transformation by direct microinjection of DNA into cultured mammalian cells, *Cell* **22:**479–488.

Cepko, C. L., Roberts, B. E., and Mulligan, R. C., 1984, Construction and applications of a highly transmissible murine retrovirus shuttle vector, *Cell* **37:**1053–1062.

Chao, M. V., Mellon, P., Charnay, P., Maniatis, T., and Axel, R., 1983, The regulated expression of beta globin genes introduced into mouse erythroleukemia cells, *Cell* **32:**483–493.

Cirullo, R. E., Dana, S., and Wasmuth, J. J., 1983, Efficient procedure for transferring specific human genes into Chinese hamster cell mutants: Interspecific transfer of the human genes encoding leucyl- and asparaginyl-tRNA synthetases, *Mol. Cell. Biol.* **3:**892–902.

Cone, R. D., and Mulligan, R. C., 1984, High efficiency gene transfer into mammalian cells: Generation of helper-free recombinant retrovirus with broad mammalian host range, *Proc. Natl. Acad. Sci. U.S.A.* **81:**6349–6353.

Degnen, G. E., Miller, I. L., Eisenstadt, J. M., and Adelberg, E. A., 1976, Chromosome-mediated gene transfer between closely related strains of cultured mouse cells, *Proc. Natl. Acad. Sci. U.S.A.* **73:**2838–2842.

Fournier, R. E. K., 1983, Microcell-mediated chromosome transfer, in: *Techniques in Somatic Cell Genetics* (J. M. Shay, ed.), pp. 309–327, Plenum Press, New York.

Gerhard, D. S., Kawasaki, E. S., Bancroft, F. C., and Szabo, P., 1981, Localization of a unique gene by direct hybridization *in situ*, *Proc. Natl. Acad. Sci. U.S.A.* **78:**3755–3759.

Glaser, T., and Housman, D., 1984, Insertion of a selectable marker into various sites on human chromosome #11, *ICSU Short Reports*, Vol. 1, pp. 174–175, Miami Winter Symposium.

Graham, F. L., and van der Eb, A., 1973, A new technique for the assay of infectivity of human adenovirus 5 DNA, *Virology* **52:**456–467.

Gusella, J. F., Keys, C., Varsanyi-Breiner, A., Kao, F.-T., Jones, C., Puck, T. T., and Housman, D., 1980, Isolation and localization of DNA segments from specific human chromosomes, *Proc. Natl. Acad. Sci. U.S.A.* **77:**2829–2833.

Gusella, J. F., Jones, C., Kao, F.-T., Housman, D., and Puck, T. T., 1982, Genetic fine-structure mapping in human chromosome 11 by use of repetitive DNA sequences, *Proc. Natl. Acad. Sci. U.S.A.* **79:**7804–7808.

Gusella, J. F., Wexler, N. S., Conneally, P. M., Naylor, S. L., Anderson, M. A., Tanzi, R. E., Watkins, P. C., Ottina, K., Wallace, M. R., Sakaguchi, A. Y., Young, A. B., Shoulson, I., Bonilla, E., and Martin, J. B., 1983, A polymorphic DNA marker genetically linked to Huntington's disease, *Nature (London)* **306:**234–238.

Gusella, J. F., Tanzi, R. E., Anderson, M. A., Hobbs, W., Gibbons, K., Raschtchian, R., Gilliam, T. C., Wallace, M. R., Wexler, N. S., and Conneally, P. M., 1984, DNA markers for nervous system diseases, *Science* **225:**1320–1326.

Haber, D. A., and Schimke, R. T., 1982, Chromosome-mediated transfer and amplification of an altered mouse dihydrofolate reductase gene, *Somat. Cell Genet.* **8:**499–508.

Hartley, J. W., and Rowe, W. P., 1976, Naturally occurring murine leukemia viruses in wild mice: Characterization of a new "amphotropic" class, *J. Virol.* **19**:19–25.

Hood, L., Steinmetz, M., and Malissen, D., 1983, Genes of the major histocompatibility complex of the mouse, *Annu. Rev. Immunol.* **1**:529–568.

Housman, D., Nelson, D. L., Albritton, L. M., Minden, M., Wieder, S., and Mulligan, R. C., 1983, Gene mapping by metaphase chromosome transfer: Use of an inserted bacterial gene as a selectable marker, in: *Banbury Report No. 14* (C. T. Caskey and R. L. White, eds.), pp. 197–203, Cold Spring Harbor Press, Cold Spring Harbor, New York.

Hudson, L. D., Erbe, R. W., and Jacoby, L. B., 1980, Expression of the human argininosuccinate synthetase gene in hamster transferents, *Proc. Natl. Acad. Sci. U.S.A.* **77**:4234–4238.

Ishiura, M., Hirose, S., Uchida, T., Hamada, Y., Suzuki, Y., and Okada, Y., 1982, Phage particle-mediated gene transfer to cultured mammalian cells, *Mol. Cell. Biol.* **2**:607–616.

Klobutcher, L. A., and Ruddle, F. H., 1979, Phenotype stabilisation and integration of transferred material in chromosome-mediated gene transfer, *Nature (London)* **280**:657–660.

Klobutcher, L. A., Miller, C. L., and Ruddle, F. H., 1980, Chromosome-mediated gene transfer results in two classes of unstable transformants, *Proc. Natl. Acad. Sci. U.S.A.* **77**:3610–3614.

Law, M. L., Davidson, J. N., and Kao, F.-T., 1982, Isolation of a human repetitive sequence and its application to regional chromosome mapping, *Proc. Natl. Acad. Sci. U.S.A.* **79**:7390–7394.

Lewis, W. H., Srinivasan, P. R., Stokoe, N., and Siminovitch, L., 1980, Parameters governing the transfer of the genes for thymidine kinase and dihydrofolate reductase into mouse cells using metaphase chromosomes or DNA, *Somat. Cell Genet.* **6**:333–347.

Lugo, T. G., and Baker, R. M., 1983, Chromosome-mediated gene transfer of HPRT and APRT in an intraspecific human cell system, *Somat. Cell Genet.* **9**:175–188.

Maitland, N. J., and McDougall, J. K., 1977, Biochemical transformation of mouse cells by fragments of herpes simplex virus DNA, *Cell* **11**:233–241.

Mann, R., Mulligan, R. C., and Baltimore, D., 1983, Construction of a retrovirus packaging mutant and its use to produce helper-free defective retrovirus, *Cell* **33**:153–159.

McBride, O. W., and Ozer, H. L., 1973, Transfer of genetic information by purified metaphase chromosomes, *Proc. Natl. Acad. Sci. U.S.A.* **70**:1258–1262.

Miller, C. L., and Ruddle, F. H., 1978, Co-transfer of human X-linked markers into murine somatic cells via isolated metaphase chromosomes, *Proc. Natl. Acad. Sci. U.S.A.* **75**:3987–3990.

Mulligan, R. C., 1983, Construction of highly transmissible mammalian cloning vehicles derived from murine retroviruses, in: *Experimental Manipulation of Gene Expression* (M. Inouye, ed.), pp. 155–173, Academic Press, New York.

Mulligan, R. C., and Berg, P., 1980, Expression of a bacterial gene in mammalian cells, *Science* **209**:1422–1427.

Murphy, P. D., Miller, C. L., and Ruddle, F. H., 1985, Quantitation of the transgenome size in chromosome-mediated gene transfer lines, *Cytogenet. Cell Genet.* **39**:125–133.

Murray, J. M., Davies, K. E., Harper, P. S., Meredith, L., Mueller, C. R., and Williamson, R., 1982, Linkage relationship of a cloned DNA sequence on the short arm of the X chromosome to Duchenne muscular dystrophy, *Nature (London)* **300**:69–71.

Nelson, D. L., Weis, J. H., Przyborski, M. J., Mulligan, R. C., Seidman, J. G., and Housman, D. E., 1984, Metaphase chromosome transfer of introduced selectable markers, *J. Mol. Appl. Genet.* **2**:563–577.

Nelson, D. L., Przyborski, M. J., Housman, D. E., Seidman, J. G., and Weis, J. H., 1986, Insertion of a retroviral vector between the *K* and *I* region of the murine histocompatibility complex (submitted).

Oie, H. K., Russell, E. K., Dotson, J. H., Rhoads, J. M., and Gazdar, A. F., 1976, Host-range properties of murine xenotropic and ecotropic type-C viruses, *J. Natl. Cancer Inst.* **56**:423–426.

Olsen, A. S., McBride, O. W., and Moore, D. E., 1981, Number and size of human X chromosome fragments transferred to mouse cells by chromosome-mediated gene transfer, *Mol. Cell. Biol.* **1**:439–448.

Pellicer, A., Wigler, M., Axel, R., and Silverstein, S., 1978, The transfer and stable integration of the HSV-thymidine kinase gene into mouse cells, *Cell* **14**:133–141.

Rein, A., Schultz, A. M., Bader, J. P., and Bassin, R. H., 1982, Inhibitors of glycosylation reverse retroviral interference, *Virology* **119**:185–192.

Scangos, G. A., Huttner, K. M., Silverstein, S., and Ruddle, F. H., 1979, Molecular analysis of chromosome-mediated gene transfer, *Proc. Natl. Acad. Sci. U.S.A.* **76**:3987–3990.

Schaffner, W., 1980, Direct transfer of cloned genes from bacteria to mammalian cells, *Proc. Natl. Acad. Sci. U.S.A.* **77**:2163–2167.

Sege, R. D., Kozarsky, K., Nelson, D. L., and Krieger, M., 1984, Expression and regulation of human low-density lipoprotein receptors in Chinese hamster ovary cells, *Nature (London)* **307**:742–745.

Shih, C., and Weinberg, R. A., 1982, Isolation of a transforming sequence from a human bladder carcinoma cell line, *Cell* **29**:161–169.

Shih, C., Shilo, B.-Z., Goldfarb, M. P., Dannenberg, A., and Weinberg, R. A., 1979, Passage of phenotypes of chemically transformed cells via transfection of DNA and chromatin, *Proc. Natl. Acad. Sci. U.S.A.* **76**:5714–5718.

Shimotohno, K., and Temin, H. M., 1980, No apparent nucleotide sequence specificity in cellular DNA juxtaposed to retrovirus proviruses, *Proc. Natl. Acad. Sci. U.S.A.* **77**:7357–7361.

Shimotohno, K., and Temin, H. M., 1981, Formation of infectious progeny virus after insertion of herpes simplex thymidine kinase gene into DNA of an avian retrovirus, *Cell* **26**:67–77.

Shoemaker, C. S., Goff, S., Gilboa, E., Paskind, M., Mitra, S. W., and Baltimore, D., 1980, Structure of a cloned circular Moloney murine leukemia virus DNA molecule containing an inverted segment: Implication for retrovirus integration, *Proc. Natl. Acad. Sci. U.S.A.* **77**:3932–3936.

Shoemaker, C., Hoffmann, J., Goff, S. P., and Baltimore, D., 1981, Intramolecular integration within Moloney murine leukemia virus DNA, *J. Virol.* **40**:164–172.

Shows, T. B., Naylor, S. L., and Sakaguchi, A. Y., 1982, Mapping the human genome, cloned genes, DNA polymorphisms, and inherited disease, in: *Advances in Human Genetics*, Vol. 12 (H. Harris and K. Hirschhorn, eds.), pp. 341–452, Plenum Press, New York.

Simmons, T., Lipman, M., and Hodge, L. D., 1978, Uptake and early fate of metaphase chromosomes ingested by the Wi-L2 human lymphoid cell line, *Somat. Cell Genet.* **4**:55–76.

Southern, E. M., 1975, Detection of specific sequences among DNA fragments separated by gel electrophoresis, *J. Mol. Biol.* **98**:503–517.

Southern, P. J., and Berg, P., 1982, Transformation of mammalian cells to antibiotic resistance with a bacterial gene under control of the SV40 early region promoter, *J. Mol. Appl. Genet.* **1**:327–341.

Szybalski, W., Szybalska, E. H., and Ragni, G., 1962, Genetic studies with human cell lines, *Natl. Cancer Inst. Monogr.* **7**:75–89.

Tabin, C. J., Hoffman, J. W., Goff, S. P., and Weinberg, R. A., 1982, Adaptation of a retrovirus as a eukaryotic vector transmitting the herpes simplex virus thymidine kinase gene, *Mol. Cell. Biol.* **2**:426–436.

Temin, H., and Baltimore, D., 1972, RNA-directed DNA synthesis and RNA tumor viruses, *Adv. Virus Res.* **17**.

Weis, J. H., Nelson, D. L., Przyborski, M. J., Mulligan, R. C., Chaplin, D. D., Housman, D. E., and Seidman, J. G., 1984, Eukaryotic chromosome transfer: Linkage of the murine major histocompatibility complex to an inserted dominant selectable marker, *Proc. Natl. Acad. Sci. U.S.A.* **81**:4879–4883.

Weis, J. H., Seidman, J. G., Housman, D. E., and Nelson, D. L., 1986, Eukaryotic chromosome transfer: Production of a murine specific cosmid library from a Neo^r linked fragment of murine chromosome 17, *Mol. Cell. Biol.* **6**:441–451.

Wigler, M., Silverman, S. Lee, L.-S., Pellicer, A., Cheng, Y., and Axel, R., 1977, Transfer of purified herpes virus thymidine kinase genes to cultured mouse cells, *Cell* **14**:725–731.

Wigler, M., Pellicer, S., Silverstein, S., and Axel, R., 1978, Biochemical transfer of single-copy eukaryotic genes using total cellular DNA as donor, *Cell* **14:**725–731.

Wigler, M., Pellicer, S., Silverstein, S., Axel, R., Urlaub, G., and Chasin, L., 1979a, DNA-mediated transfer of the adenine phosphoribosyl transferase locus into mammalian cells, *Proc. Natl. Acad. Sci. U.S.A.* **76:**1373–1376.

Wigler, M., Silverstein, S., Lee, L.-S., Pellicer, A., Cheng, Y., and Axel, R., 1979b, Transfer of purified herpes virus thymidine kinase gene to cultured mouse cells, *Cell* **14:**725–731.

Wigler, M., Sweet, R., Sim, G. K., Wold, B., Pellicer, A., Lacy, E., Maniatis, T., Silverstein, S., and Axel, R., 1979c, Transformation of mammalian cells with genes from prokaryotes and eukaryotes, *Cell* **16:**777–785.

Wigler, M., Perucho, M., Kurtz, K., Dana, S., Pellicer, A., Axel, R., and Silverstein, S., 1980, Transformation of mammalian cells with an amplifiable dominant-acting gene, *Proc. Natl. Acad. Sci. U.S.A.* **77:**3567–3570.

Willecke, K., and Ruddle, F. H., 1975, Transfer of the human gene for hypoxanthine phosphoribosyltransferase via isolated human metaphase chromosomes into mouse L-cells, *Proc. Natl. Acad. Sci. U.S.A.* **72:**1792–1796.

Willecke, K., Lange, R., Kruger, A., and Reber, T., 1976, Cotransfer of two linked human genes into cultured mouse cells, *Proc. Natl. Acad. Sci. U.S.A.* **73:**1274–1278.

Willing, M. C., Nienhuis, A. W., and Anderson, W. F., 1976, Selective activation of human beta but not gamma globin gene in human fibroblast × mouse erythroleukemia cell hybrids, *Nature (London)* **277:**534–538.

Wright, S., deBoer, E., Grosveld, F. G., and Flavell, R. A., 1983, Regulated expression of the human beta globin gene family in murine erythroleukaemia cells, *Nature (London)* **305:**333–336.

Wullems, G. J., van der Horst, J., and Bootsma, D., 1976, Transfer of the human X chromosome to human–Chinese hamster cell hybrids via isolated HeLa metaphase chromosomes, *Somat. Cell Genet.* **2:**359–371.

5

VECTORS FOR GENE TRANSFER DERIVED FROM ANIMAL DNA VIRUSES: TRANSIENT AND STABLE EXPRESSION OF TRANSFERRED GENES

Vijay R. Baichwal and Bill Sugden

1. INTRODUCTION

In this chapter, a "vector" is defined as a DNA molecule that facilitates the intracellular expression of genes inserted into it. Vectors that contain inserted genes will be termed "recombinant vectors." Most eukaryotic vectors to date contain DNA sequences derived from viruses, and those derived from animal DNA viruses are eclectic. They may or may not be infectious; they may or may not replicate autonomously; they may permit the expression of an inserted gene transiently or indefinitely. They have been derived from papova-, parvo-, adeno-, herpes-, and poxviruses. Few have been studied in great detail, and reference to them in the literature is often incidental. We shall try to abstract from recent literature general findings derived from the construction and use of vectors developed from animal DNA viruses. These abstracts will not be complete; however, we shall present examples from each of the cited families of DNA viruses. Many recent experiments have used *cis*-acting elements [e.g., simian virus 40 (SV40) 72-base-pair (bp) repeat or the herpes simplex type 1 thymidine kinase promoter] derived from viruses that affect transcription of linked genes in otherwise nonviral recombinant DNAs. Whether or not these recombinant DNAs should be termed viral vectors is moot. We shall refer only

Vijay R. Baichwal and Bill Sugden ● McArdle Laboratory for Cancer Research, University of Wisconsin, Madison, Wisconsin 53706.

occasionally to the large body of observations in which they were used. We hope that readers who have not used vectors derived from animal DNA viruses can in reading this review identify likely candidates to fulfill their needs and then turn to the cited primary sources for professional help.

Different viruses have different strategies for replication, different host ranges, and different packaging limits, and these properties may influence the use of vectors derived from them. Therefore, vectors derived from each family of viruses will be discussed separately. For each family of viruses, a description of the life cycle in cell culture of a representative member will be given. Following this introduction, first noninfectious and then infectious vectors derived from these viruses will be described. The distinction between noninfectious and infectious vectors is made for the following reasons: Noninfectious vectors cannot be packaged as mature virions and therefore cannot infect cells in culture. The variety of recipient cells and the percentage of cells in a population into which the noninfectious vectors can be introduced are limited by the means of introducing DNAs into cells in culture. On the other hand, infectious vectors can be packaged, have the host range of the virus from which they are derived, and can infect the majority of cells in a susceptible population. General conclusions pertaining to the advantages and limitations of each set of vectors will be drawn. Although we have not found general reviews of this field, wherever possible we have cited references that detail the construction and use of the vectors that we have discussed. Gluzman (1982) has edited a set of early contributions to this field.

Vectors derived from SV40, a papovavirus, will be reviewed first. It is the most thoroughly studied animal DNA virus, and the vectors derived from it are used most frequently. Observations with SV40 vectors will serve as a standard for comparison with vectors derived from other DNA viruses. Vectors constructed from parvo-, adeno-, herpes-, and poxviruses will be reviewed next. These vectors permit transient expression of transferred genes because they usually replicate and kill the host cell. Finally, vectors derived from bovine papilloma virus and Epstein–Barr virus will be described. These vehicles can be maintained as plasmids and have been used to express genes indefinitely in appropriate host cells.

2. VECTORS THAT PERMIT STABLE OR TRANSIENT EXPRESSION OF TRANSFERRED GENES

2.1. Simian Virus 40

Simian virus 40, a papovavirus, was isolated from an Old World monkey (reviewed in Tooze, 1981). Its genome consists of a covalently closed DNA duplex of approximately 5200 bp. In tissue culture, it infects and grows lytically in monkey cells, and it infects and transforms rodent cells. After adsorption, penetration, and uncoating in the lytic host cell, one of the viral DNA strands, termed "early," is transcribed to yield an RNA molecule that is spliced in two ways. The two differently spliced RNAs yield the two known early proteins of SV40, small tumor (small-t) and large tumor (large-T) antigen. Large-T antigen binds to sites around the viral origin of DNA replication, is required for the initiation of viral DNA synthesis, and modulates transcription of both DNA strands. After viral DNA synthesis begins, transcription of the other DNA strand, the "late" strand, can be detected in infected cells, and the viral coat proteins encoded by these late transcripts are synthesized. The late transcripts have heterogeneous 5′ ends and are spliced complexly. Between 10^5 and 10^6 molecules of viral DNA can be synthesized in an infected cell over the course of a 48- to 72-hr infectious cycle, and approximately 1–10% of this DNA is packaged into virus particles. Host macromolecular synthesis is not inhibited in the infected cell, but the cell dies and lyses; the specific contribution SV40 makes to these two events is not known. The expression of the SV40 genome is not regulated in all cells as it is in its lytic host cells. In rodent cells infected by SV40, early viral RNA synthesis occurs, viral DNA synthesis is inhibited partially or completely, and late gene expression is usually not detected. Infected rodent cells that maintain the viral genome so that the viral early genes are expressed display a variety of transformed phenotypes.

The foregoing précis of SV40 provides enough information to outline major features of vectors derived from this virus. In the cells SV40 infects, early viral gene expression occurs in the absence of viral gene products. This finding indicates that the *cis*-acting DNA sequences required for early-strand transcription act in a variety of cells in the absence of viral proteins. We can expect that vectors with early-strand transcriptional signals will express genes without requiring DNA replication. The only viral protein required for viral DNA synthesis is the large-T antigen. Hence, if large-T antigen is provided in *trans* by a helper virus or a helper cell, a vector carrying the SV40 origin of DNA synthesis should replicate autonomously. Infectious vectors will require helper cells or helper virus, since SV40 has a small genome with no large dispensable genes and a narrow packaging limit (≥75% to ≤102% of the wild-type size). Thus, infectious vectors derived from SV40 must be tailored carefully. Finally, we can expect cells in which SV40 vectors replicate to be killed by the vector. All these expectations have been fulfilled by vectors derived from SV40.

2.1.1. Noninfectious Vectors Derived from SV40

Of the vectors constructed from SV40, the most widely used employ just the *cis*-acting signals of the early region to mediate transcription and processing of the inserted gene. These signals include an enhancing element (the 72- and 21-bp repeats), the early promoter, often the splice sites for an intervening sequence, and a poly-adenylic acid addition signal. These vectors are exemplified by the pSV2 series (reviewed by Subramani and Southern, 1983). They permit transient expression of inserted genes in a variety of cells. A selectable marker can be added to such vectors, the vectors introduced into cells, and maintenance and expression of the vectors in the cells made a requirement for cell survival. In this manner, for example, genes transcribed by RNA polymerase II or III have been introduced into mouse and human B lymphoblasts and found to be expressed indefinitely (D. Rice and Baltimore, 1982; Rymo, 1983). The efficiency of isolating cells that express the desired genes varies enormously; it depends on the efficiency of introducing the vector DNA into the cell, on the frequency with which the recombinant vector integrates into the cellular DNA, and the efficiency with which the selectable marker and the transferred gene are expressed once integrated (Luciw et al., 1983; Thillet et al., 1984).

The utility of noninfectious SV40-based vectors has been extended by the introduction of helper cells that express the large-T antigen constitutively. The helper cells, first developed by Gluzman (1981) and termed COS cells, include monkey cells (Gluzman, 1981), human fibroblasts (Boast et al., 1983), and monkey cells expressing a thermolabile large-T antigen (Rio et al., 1985). The SV40 origin of DNA synthesis is embedded within the viral transcriptional signals so that it is present in most SV40-based vectors. Introduction of recombinant vectors with an SV40 origin into these helper cells can result in the amplification of these recombinant vectors up to 10^5 copies per transfected cell (Boast et al., 1983). The increase in gene copy usually leads to an increase in the expression of the messenger RNA (mRNA) and protein product of the inserted gene (Gething and Sambrook, 1981; Mellon et al., 1981; Boast et al., 1983; Semler et al., 1984).

SV40 large-T antigen modulates transcription from both the early and late promoters of the viral DNA. Binding of large-T antigen to its specific binding sites, I and II, inhibits transcription from the early promoter (Rio and Tjian, 1983), while the presence in cells of large-T antigen may enhance expression from the late promoter (Brady et al., 1984; Keller and Alwine, 1984). These observations indicate that genes inserted downstream from the SV40 early promoter may be expressed less efficiently in COS helper cells than those inserted downstream from the late promoter region. However, the late promoter region of SV40 is complex, containing multiple start sites for transcription, and detailed comparisons of noninfectious vectors using the same gene expressed from the early and the late promoters have not been reported. It is clear that recombinant vectors using the late promoter of SV40 that replicate in COS helper cells can synthesize the products of an inserted gene efficiently. In one study, after correction for the fraction of cells (10%) that took up the recombinant vector, approximately 1% of the [^{35}S]methionine label incorporated into protein was found in a transferred vesicular stomatis virus N protein after 24 hr (Sprague *et al.*, 1983).

2.1.2. Infectious Vectors Derived from SV40

Infectious vectors have the advantage that they can be introduced efficiently into an appropriate host cell. The size of a transferred gene is limited in an infectious SV40 vector because the virus has a narrow packaging limit ($\geq$75% to $\leq$102% of the wild-type size). The largest genes inserted to date are less than 2500 bp, since the size of the vector plus insert must be close to 5240 bp, the size of SV40 DNA itself.

Some infectious SV40 vectors have been developed with the inserted genes replacing the early genes of the virus. These constructions can be propagated in the COS helper-cell line. Laub *et al.* (1983) found that such an infectious recombinant vector, the genome of which was 92% of the size of the wild-type genome and contained the gene for the hepatitis B virus surface antigen (HBsAg), replicated in COS helper cells to yield unrearranged viruses with titers of 10^8 plaque-forming units (PFU)/ml. Infection of COS cells with this recombinant vector yielded 10^7 molecules of the HBsAg per cell over a 24-hr period on the 3rd day postinfection, prior to cell degeneration. The yield of the HBsAg from this "early region replacement" vector is approximately 10-fold greater than several "late region replacement" vectors expressing the same gene (Laub *et al.*, 1983). This observation indicates that inhibition of early gene transcription by SV40 large-T antigen in COS cells may not be a major limiting factor for expression of inserted genes in such infectious SV40 vectors.

Infectious vectors in which the late region of SV40 is replaced with an inserted gene require helper virus to be packaged and infectious. The helper viruses used often carry the gene for a thermolabile large-T antigen, or an insertion within the early region, so that both the vector and the helper virus must be complemented for their growth. Stocks of such infectious vectors vary in their content of helper virus, and the cause of this variation is not obvious. When Oda *et al.* (1983) used constructions that varied between 85 and 89% of wild-type genome length, they found that the recombinant vectors that contained the adenovirus type 12 E1A gene constituted only 4–15% of the virus stocks. Will *et al.* (1984) inserted the HBsAg and core antigen genes into the late region of SV40 in vectors that ranged from 85 to 100% of the wild-type genome length. These recombinants comprised 10–50% of the virus stocks. Although such infectious SV40 vectors can be introduced efficiently into appropriate cells, it is clear that the requirement for a helper virus, coupled with the potential for the helper virus to outreplicate the vector, could limit the usefulness of such vectors.

Infectious vectors in which the late region of SV40 has been replaced by foreign genes have been used to express a variety of genes. HBsAg and core antigen (Will *et al.*, 1984), the E1A protein of adenovirus 12 (Oda *et al.*, 1983), and the neuraminidase gene (Davis *et al.*, 1983) and NS gene (Lamb and Lai, 1984) of influenza virus have all been synthesized in monkey cells with helper viruses. The expression of the transferred genes in the infected cells has been sufficient to permit careful analysis of both the structure of the RNA and that of the protein product.

2.1.3. Frequency of Mutations in SV40 Vectors

Recombinant vectors derived from SV40 that contain a marker gene the expression of which can be measured in *Escherichia coli* have been introduced into mammalian cells, isolated from those cells, and shuttled into *E. coli*. When the frequency of mutations introduced into the marker gene in the mammalian cells was measured (Calos *et al.*, 1983; Razzaque *et al.*, 1984), it was found to be on the order of 1%. The cause of this high frequency of mutation is not clear. Razzaque *et al.* (1984) have found that the frequency of mutations at a cellular locus in cells transfected with vector DNAs is not abnormally high. Therefore, transfection *per se* is not mutagenic for a cellular gene. Similarly, frequency of mutations of a cellular gene were not atypical in cells expressing SV40 large-T antigen; therefore, expression of large-T antigen is not mutagenic for a cellular gene. Ashman and Davidson (1984) fused clones of V79 cells (a Chinese hamster cell ovary cell line) into which an SV40-derived vector has been integrated with COS helper cells. When the amplified vectors were rescued and scored in *E. coli*, about 1% of the vectors contained detectable mutations. This observation indicates that the observed mutations need not be generated solely at the step of transfection. Thus, the frequency of mutations for SV40-derived vectors is high, and the potential problems caused by this phenomenon must be considered when these vectors are used. It is not clear whether this high frequency of mutations is related to SV40, to the recipient cells used, or to the method of introducing the vectors into the recipient cells.

2.2. Polyoma Virus

Polyoma virus is a papovavirus that resembles SV40 in its lytic mode of replication (Tooze, 1981). In terms of this review, it differs from SV40 in one major characteristic: It grows lytically in mouse cells. Vectors derived from polyoma can be used to transfect mouse cells and yield high levels of expression of transferred genes, presumably in part as a result of the efficient amplification of the transferred gene. For example, when a recombinant vector that contained polyoma replication functions and carried a mouse immunoglobin α-heavy-chain gene was transfected into a mouse myeloma cell line, it yielded 0.5 μg of secreted immunoglobulin A per 10^6 cells between 48 and 72 hr post-transfection (Deans et al., 1984). A similar construction has been used to map sensitively the effects of cis-acting elements that influence immunoglobulin gene transcription in different cell types (Grosschedl and Baltimore, 1985). A second approach to making noninfectious but versatile vectors, which has worked successfully with SV40, can now be used with polyoma, as well. Hassell and his colleagues (Muller et al., 1984) have isolated polyoma helper cells, termed MOP cells, that are derived from NIH3T3 cells and express each of the three large-T antigens of polyoma. Vectors containing the origin of replication of polyoma are amplified in these cells to yield between 5×10^3 and 2×10^4 molecules per transfected cell (Muller et al., 1984). It appears that replication-competent vectors derived from polyoma will be as convenient to use in mouse cells as are their SV40-derived counterparts in monkey and human cells.

2.3. Adeno-Associated Virus 2

We shall focus on the adeno-associated virus (AAV), which among the parvoviruses has been studied most extensively as a vector for gene transfer. AAV has a single-stranded DNA molecule of approximately 4600 bp as its genome (Salzman, 1978; Berns, 1984a,b). It is a defective virus; a complete lytic cycle requires either adenovirus or herpes simplex virus as a helper. The lytic host range of the virus appears to be restricted only by the host range of the helper virus. Thus, the common human cell lines—HeLa, KB, and 293— and the African green monkey kidney cell lines—BSC-1, Vero, and CV-1—can be lytic hosts for the virus. The kinetics of the viral infectious cycle depend on the time of infection of the helper virus. If, for example, cells are infected with adenovirus 10–12 hr prior to infection with AAV, the latent period (time period in which infectious virions are not detectable) is 4–6 hr. Depending on the exact conditions of infection, approximately 10^5 AAV particles are produced per infected cell.

In the absence of helper virus, AAV establishes a latent infection. The latent infection appears to involve integration of the viral DNA into the cellular genome. In latently infected cells, approximately 1–10 copies of the viral genome are found integrated in head-to-tail tandem repeats with occasional inclusion of additional DNA of uncertain origin. It is not clear whether the latent viral DNA retains all the DNA sequences present in the virion DNA. A characteristic of the structure of the integrated virus is that the junction between the viral and cellular DNA sequences always involves the viral termini. However, the site of integration within the cellular DNA is not specific. The latent viral DNA can be rescued by superinfection of latently infected cells with a helper virus. The rescue process is efficient, and the number of cells positive for AAV antigen is comparable to that obtained with a lytic infection. The integrated latent viral DNA is apparently stable on cell passage and thus shows some features in common with retroviral proviruses (see Chapter 6).

Vectors Derived from AAV2

Studies involving use of AAV as a vector have become possible with the molecular cloning of the AAV genome into prokaryotic plasmids (Samulski *et al.*, 1982) and the demonstration that the virus could be rescued from animal cells carrying such recombinant plasmids by infection with a helper adenovirus (Samulski *et al.*, 1983). Most studies using AAV as a vector have concentrated on unraveling the intriguing molecular biology of the virus. These studies have been reviewed recently (Berns, 1984a,b) and will not be discussed here.

No region larger than 1000 bp of AAV appears to be entirely dispensable for rescue from a recombinant plasmid. Thus, any infectious recombinant vector is doubly defective (i.e., in addition to a helper adenovirus, it requires a helper wild-type AAV). Sequences at the AAV termini are needed in *cis* for viral DNA replication. The requirements, if any, of viral functions needed to establish a latent infection have not been described. Given the small size of the viral genome (approximately 4000 bp) and a strict packaging limit [no greater than 120% of the viral genome (Hermonat and Muzyczka, 1984)], the versatility of AAV vectors is compromised.

Nevertheless, it has been shown that a foreign gene can be introduced into the viral genome, and the infectious recombinant vector can be used to transfer the foreign gene. In one case, the transient expression of a foreign gene from viral promoters was studied. It was found that regulation of expression of the gene was complex (Tratschin *et al.*, 1984), depending on the cell type being infected and whether or not the cells were coinfected with helper virus. In another study (Hermonat and Muzyczka, 1984), the gene that confers resistance to the antibiotic G418 under the control of the SV40 early promoter was substituted for a part of the viral DNA to yield a recombinant virus that was only about 25 bases longer than the wild-type virus. To obtain infectious-virus particles, the recombinant plasmid was transfected into adenovirus-infected cells along with another recombinant plasmid that had the entire wild-type genome. Even though the wild-type viral recombinant plasmid was constructed to have a size of 5800 bp to make it unpackageable, the virus stock obtained contained approximately 20% wild-type virus that was presumably generated by recombination *in vivo*. With this virus stock (10^6 units/ml), 0.4–10% of infected D6 cells (a human cell line) were transformed to G418 resistance, depending on the multiplicity of infection. The transferred DNA was integrated in the genome and could be rescued by infection with adenovirus.

The dependence of virus production on the structure of the AAV constructions introduced into adenovirus-infected cells has been determined (Laughlin *et al.*, 1983). Plasmids in which the wild-type viral DNA was separated from prokaryotic plasmid sequences and recircularized produced approximately 10 times higher virus yields than the untreated plasmid. However, this yield was still only one third of that obtained during a viral infection. Overall, the intricacies of using parvoviruses to construct recombinant vectors, at least as judged by AAV, indicate that at present they may be most powerful in analyzing parvoviruses themselves.

2.4. Adenoviruses 2 and 5

Among the adenoviruses, the human adenovirus type 2 (Ad 2) and type 5 (Ad 5) viruses have been studied most extensively for use as vectors for gene transfer and will be discussed here. These viruses have as their genome a linear DNA duplex of 35,000 bp that contains inverted terminal repeats (Ginsberg, 1984; Tooze, 1982; Wold *et al.*, 1978). DNA sequences near the termini contain the origins of DNA synthesis and a packaging signal (Kelly, 1984; Tibbetts, 1977; Hammerskjöld and Winberg, 1980).

In most cases, the lytic life cycle of Ad 2 and 5 have been studied in the permissive human cell lines, HeLa and KB. Typically, the infection can be divided temporally into two stages, early and late. The dividing point is the onset of viral DNA synthesis, which occurs at 6–7 hr postinfection in the case of Ad-2-infected KB cells. At early stages, less than half the viral genome is transcribed from four viral promoters to yield viral mRNA. At late stages, additional transcription initiates at a single place in the viral genome from the major late promoter (MLP), situated at 17% on the genome. Late transcription reaches a maximum after 15 hr. Both early and late transcripts are spliced in a complex manner. Most late transcripts have at the 5' end a common structure that consists of RNA sequences homologous to three noncontiguous regions of the viral genome. These RNA sequences constitute the tripartite leader. In cases in which individual complementary DNAs (cDNAs) corresponding to late proteins have been sequenced, the tripartite leader appears to be noncoding. In addition to the viral mRNA, the viruses also encode genes for two short (approximately 160-bp) RNAs that are transcribed by RNA polymerase III. These virus-associated RNAs, VA-RNA I and VA-RNA II, accumulate to high levels (approximately 10^5–10^6 copies per cell) late in infection. Late in infection, anywhere from 20 to 40% of the pulse-labeled total RNA is viral, and a typical late mRNA reaches a steady-state level of 5×10^3 to 5×10^4 copies per cell. This high accumulation of late mRNA is aided in part by the presence of 10^3–10^5 copies of viral DNA in the infected cell.

Viral infection eventually alters cellular macromolecular synthesis. However, early after infection, synthesis of cellular DNA, RNA, and protein is essentially unaffected. With the onset of viral DNA synthesis, cellular DNA and protein synthesis is inhibited and ultimately ceases. Thus, late in infection, only viral proteins are synthesized. Eventually, by 48–72 hr postinfection, the cell is lysed. Given that adenoviruses efficiently synthesize their late proteins, it is no surprise that adenovirus-derived infectious vectors hold great promise for overproducing foreign proteins in human cells.

2.4.1. Introduction to Vectors Derived from Adenoviruses 2 and 5

Infectious adenovirus vectors have their antecedents in the naturally occurring adenovirus–SV40 hybrids, which are hybrid viruses in which part of the SV40 virus is inserted within the adenoviral genome. These viruses presumably were selected for their efficient replication in monkey cells, since wild-type human adenoviruses replicate quite inefficiently, if at all, in monkey cells. In most cases, polypeptides that correspond to the inserted SV40 sequences are synthesized. The structure and known properties of adenovirus–SV40 hybrids permit us to predict in part how nonadenoviral genes can be expressed in infectious adenoviral vectors. Indeed, most of the studies involving adenovirus expression vectors use the SV40 large-T antigen as a model gene. A recent review summarizes our knowledge of adenovirus–SV40 hybrids and adenovirus expression vectors (Klessig, 1984). We will focus on factors that need to be considered while constructing infectious adenovirus vectors and describe the two methods that have been used to construct such vectors.

The size limit on DNA that can be packaged into adenovirus virions may be quite close to the size of wild-type DNA. The largest described adenovirus–SV40 hybrids are only 1 or 2% larger than adenovirus itself. Thus, insertion of large genes into infectious adenoviral vectors would require replacement of viral sequences. So far, only one region of the viral genome has been identified as being dispensable for replication of the virus in cell culture. This region is the E3 region, which codes for an early mRNA, and approximately 2000 bp of DNA can be deleted from this region (Berkner and Sharp, 1983). This mutant virus behaves like the wild-type with respect to the number of particles released per cell and kinetics of growth.

The human cell line 293, which was obtained by transfection with sheared Ad 5 DNA, contains the left-end 14% of the virus (Graham *et al.*, 1977) and serves as a helper cell for some mutants. These 293 cells constitutively express the *E1* region gene products of Ad 5 and can thus complement viruses that lack this region. These helper cells permit the generation of a helper-virus-independent, defective virus stock that lacks the *E1* region (Van Doren *et al.*, 1984; Berkner and Sharp, 1984). Hence, the *E1* region of the virus is dispensable in vectors propagated in 293 cells.

Ad 5–SV40 hybrids can contain as few as 7000 bp of adenoviral DNA, corresponding to sequences from the left end of the virus (Gluzman and Van Doren, 1983). The only known *cis*-acting sequences required for replication are the origins of DNA replication and packaging signals (Kelly, 1984; Tibbetts, 1977; Hammerskjöld and Winberg, 1980), which are found within the inverted terminal repeats of the genome. It is conceivable, therefore, that foreign DNA sequences in excess of 30,000 bp could be propagated as a recombinant virus with the aid of helper virus.

Most, if not all, adeno vectors derived to date use the adeno MLP for expression of transferred genes. A common conclusion of all investigations designed to determine which *cis*-acting viral sequences are needed to ensure maximal expression of the transferred gene is that in addition to the MLP, the tripartite leader sequences are required. Kaufman (1985) has constructed non-infectious vectors that include the complete first, and second and two thirds of the third leader exon from the virus at the 5' end of a dihydrofolate reductase (DHFR) gene cDNA. When cells are transfected with this plasmid and subsequently infected with virus, the synthesis of DHFR increases 3- to 10-fold. This increase in synthesis of the DHFR protein is not seen if the second and third leader exon sequences are absent or are present in the opposite orientation. Using similar plasmids in which the expression of human γ-interferon was monitored by a sensitive bioassay, the author demonstrated that the need for adenovirus superinfection could be obviated at least partially if the adenovirus VA genes were present. Although the exact amount of interferon produced depended on the structure of the plasmid tested and whether the VA-RNA genes were provided in *cis* or *trans,* overall there appeared to be a 10- to 50-fold increase in interferon activity when both the leader sequences and VA-RNA were present. In addition, it was reported that a plasmid containing the adenovirus MLP and tripartite leader upstream from the cDNA for γ-interferon, the VA-RNA genes, and the SV40 origin of replication when introduced into COS cells yielded more than 1 μg γ-interferon/10^6 cells, 72 hr posttransfection (Kaufman, 1985).

The function of the tripartite leader sequence has been more closely examined with adenoderived infectious vectors (Logan and Shenk, 1984). This study clearly establishes that the leader sequence serves to enhance translation of the mRNA *in vivo* late after infection. Since the leader sequence does not function at the 3' noncoding end of an mRNA, it probably has a role during initiation of translation. However, an mRNA that has 172 of the 200 nucleotides of the tripartite leader (because of a truncated third leader exon) is translated *in vivo* one fifth to one tenth as efficiently as an authentic late mRNA. Thus, presence of an intact leader sequence and/or nucleotides surrounding the start codon might be important for obtaining maximal expression of a foreign gene. In the absence of leader sequences, a hybrid late mRNA containing the SV40 large-T antigen coding sequences was translated inefficiently *in vivo,* even though it was translated with normal efficiency *in vitro* in a rabbit reticulocyte lysate (Thummel *et al.,* 1983).

2.4.2. Construction of Infectious Vectors Derived from Adenoviruses

Recombinant adenoviruses have generally been made in one of two ways. In one approach, a recombinant plasmid with the SV40 large-T antigen gene downstream from the MLP is made *in vitro* and propagated in *E. coli*. The recombinant plasmid can be appropriately cut and ligated *in vitro* to virion DNA, which has been cut correspondingly. The ligated mixture along with helper wild-type viral DNA is transfected into 293 cells, and a mixed virus stock is obtained. The mixed stock is then passaged through CV-1 monkey cells to amplify selectively those recombinant molecules that contain SV40 large-T antigen gene (Thummel *et al.*, 1983). The virus stock can contain up to 30% recombinant virus. This approach has been used to overproduce proteins other than the SV40 large-T antigen. In such cases, recombinant plasmids were constructed in which the gene for the SV40 large-T antigen was placed downstream from the gene being studied (Mansour *et al.*, 1985). In this case, however, the recombinant virus constituted less than 5% of the viral stock.

In the other approach, infectious adenoviral vectors have been constructed by first inserting a transcription unit, consisting of the foreign gene and adenoviral sequences used for its expression, into the *E1* region of a cloned viral fragment. This recombinant clone contains the entire left end of the genome. It is then linearized at a site within the viral sequence downstream from the inserted gene and ligated with a piece of virion DNA that lacks the left end. The virion DNA fragment lacking the left end is generated by restriction-enzyme digestion followed by separation of the larger right-end fragment from the smaller left-end fragment. The ligated DNA is then transfected into 293 cells, and the recombinant virus is obtained from plaques. DNA molecules that have sequences corresponding to the left and right ends of the genome yield virus. Alternatively, the fragments from the left and right moieties of the genome can be introduced into cells without ligation and infectious progeny formed by recombination *in vivo* (Berkner and Sharp, 1983). In both cases, plasmid DNA sequences present upstream from the viral DNA (with respect to the left end of the viral DNA) are lost in the process of viral replication (Berkner and Sharp, 1983). This method has been used to introduce the DHFR gene (Berkner and Sharp, 1984) and the gene that encodes resistance to the antibiotic G418 (Van Doren *et al.*, 1984) and to reintroduce the *E1a* coding sequences under the control of the MLP (Logan and Shenk, 1984). The advantages to this approach of virus construction are that it gives a helper-free recombinant virus stock and that there is no need for any special selection for the recombinant virus. Its disadvantage is that the recombinant virus is usually defective and replicates only in 293 cells unless it is provided with a wild-type helper virus.

All the hepatitis B virus genome (2800 bp) has been introduced near the right end of Ad 5 by an approach analogous to that used to introduce genes into the *E1* region of the virus. The recipient virus initially carried a 2100-bp deletion in its *E3* region, and the recombinant virus was nondefective and grew in HeLa cells as well as did wild-type Ad 5 (Saito *et al.*, 1985).

Infectious adenoviral vectors have been used to express only a few different genes. In one study, in which the SV40 large-T antigen gene with only 12-bp upstream from the translation initiation codon was fused to the third leader exon, the authors reported that approximately 2 mg of the protein could be purified from 1 liter of HeLa spinner cells infected with the virus (Thummel *et al.*, 1983). In another study, the early region of polyoma was fused to the third leader segment, and in this case approximately 250 μg large-T antigen could be obtained from 1 liter of infected HeLa spinner cells (Mansour *et al.*, 1985). These studies do indicate that infectious adenoviral vectors can be used to synthesize enough of some foreign gene products to permit their careful biochemical study. They do not indicate the range of foreign genes that could be transferred successfully and synthesized efficiently.

2.5. Herpesviruses

The herpes family of viruses contains many members that latently infect different cell types in their natural hosts (Miller, 1985; Gelb, 1985; Rawls, 1985). In humans, herpes simplex virus type 1 (HSV-1) and varicella–zoster virus (VZV) latently infect neurons, and Epstein–Barr virus (EBV) latently infects both B lymphocytes and epithelial cells of the salivary glands. A vaccine strain of VZV is now being used to immunize people, and a vaccine strain of herpes turkey virus has been used to immunize chickens against Marek's disease virus for many years. It seems likely that vectors derived from herpesviruses will be developed with the goal of their being used as a vaccine. Whether or not such vectors will be approved as vaccines for use in people or animals is uncertain. HSV-1 will be our model for a herpesvirus that grows lytically in cell culture; EBV, which infects cells latently in culture, will be considered later.

The HSV genome consists of a single molecule of linear, double-stranded DNA (dsDNA) of approximately 150,000 bp. The vast coding capacity of the virus includes proteins involved in DNA metabolism, in addition to regulatory and structural proteins. The structure of the viral genome is complex. DNA sequences present at the termini (approximately 5000 bp) are present in the reverse orientation in juxtaposition within the molecule. Thus, there are two unique regions of the genome termed "long unique" (U_L) and "short unique" (U_S), which are flanked by inverted repeats. In addition, the sequences at the viral termini (approximately 500 bp) are direct repeats.

The normal replicative cycle of HSV lasts from 15 to 19 hr depending on the cell type used as host and the conditions of infection. During this time, viral gene expression proceeds in a temporally ordered fashion (reviewed in Tooze, 1981; Roizman and Batterson, 1985). Expression of viral genes is regulated in a cascade; those expressed first direct the expression of viral genes expressed later. Replication of viral DNA is initiated several hours after infection and can yield up to 2000 progeny viral particles per infected cell, prior to cell lysis. Infection leads to cessation of host protein and DNA synthesis.

2.5.1. Helper-Independent Infectious Vectors Derived from HSV-1

Two sorts of vectors have been constructed from HSV. In one case, the gene of interest is directly recombined into a conditionally dispensable region of the genome, and the recombinant virus is plaque-purified from permissive cells (reviewed by Roizman and Jenkins, 1985). Specifically, the gene for HbsAg was inserted in the viral thymidine kinase (TK) gene immediately downstream from the promoter in a recombinant plasmid (Shih *et al.*, 1984). The recombinant plasmid was then transfected into cells along with wild-type viral DNA. Homologous recombination between the viral sequences on the plasmid and the virus, which is known to occur with this virus, results in the formation of a recombinant virus (Post *et al.*, 1981; Mocarski *et al.*, 1980; and Smiley, 1980). The recombinant virus is TK-deficient and can be selectively isolated from a mixture of wild-type and recombinant viruses. The expression of the HBsAg protein in the recombinant virus was temporally regulated in a manner analogous to the expression of TK in the wild-type virus. Unfortunately, the absolute amount of protein produced by this recombinant virus was not apparent; hence, the efficiency of expression of the inserted gene in this recombinant vector cannot be compared with that of other viral vectors.

In general, we do not know whether genes with nonviral promoters can be expressed when integrated into HSV-1. In addition, the maximum amount of DNA that can be packaged into an HSV particle is not known. However, up to 7000 bp of DNA have been introduced into the virus (Knipe *et al.*, 1978), and the use of mutant viruses with internal deletions might allow the amount of foreign DNA that can be transferred by the recombinant virus to be increased further.

2.5.2. Helper-Dependent Infectious Vectors Derived from HSV-1

The second approach to constructing vectors from HSV-1 depends on the ability of HSV-1-infected cells to polymerize, amplify, and package fragments of HSV-1 DNA that carry the appropriate *cis*-acting signals. It has been known that serial passage of HSV at high multiplicities of infection leads to accumulation of defective viral genomes within the virus stock (reviewed by Frenkel *et al.*, 1981). These defective molecules are deleted for most of the viral sequences, retaining only those *cis*-acting sequences that are needed for replication and packaging into virion particles. Thus, a defective molecule consists of multiple copies of a repeated sequence, arranged in a head-to-tail tandem array, that is large enough to be packaged. A single molecularly cloned copy of the repeated sequence can be polymerized in an HSV-1-infected cell into defective DNA of a packageable size (Spaete and Frenkel, 1982). Genes to be transferred can be introduced into the single cloned copy prior to its introduction into the infected cell. Although this vector system has been used extensively to explore the molecular biology of HSV-1, it has not been used widely to transfer foreign genes. The initial hope that this approach might be used to introduce large segments of foreign DNA, since the HSV genome is approximately 150,000 bp, has not been realized. Data from two studies indicate that there might be a size limit to the amount of DNA that can be introduced into the vectors, and this limit is well below 150,000 bp (Kwong and Frenkel, 1984; Bear et al., 1984). In one case with a starting recombinant vector of approximately 45,000 bp, the recovered recombinant virus had extensive deletions and rearrangements (Bear et al., 1984). In another study, a starting recombinant vector that was 19,000 bp in size resulted in a virus stock that also had deletions and rearrangements. However, vectors between 11,000 and 15,000 bp in size were polymerized, amplified, and packaged without rearrangement. There might therefore be a size limit of approximately 15,000 bp for the initial recombinant vector (Kwong and Frenkel, 1984).

At this juncture, it is clear that vectors derived from HSV-1 are powerful tools to study this virus. It does appear that they are not as tractable as those derived from the papova- and adenoviruses for producing large amounts of foreign gene products. In addition, they have not yet been tested as vaccines.

2.6. Vaccinia Virus

Vaccinia virus has been used to immunize humans effectively against smallpox. The vaccine did produce complications: Between 1 in 10,000 and 1 in 300,000 recipients developed encephalitis that was often fatal (Fenner, 1985). If the risk of postvaccination encephalitis can now be diminished or is outweighed by the benefit of the immunization, then infectious vectors derived from vaccinia may be used for human immunizations. We shall summarize the work that has been done toward achieving this end. In addition, we shall describe the use of noninfectious vaccinia virus vectors to analyze transcription from viral promoters.

Vaccinia virus contains a dsDNA genome of approximately 186,500 bp, with a covalent linkage of the two strands of DNA at the termini of the molecule (Moss, 1978, 1985). Vaccinia can infect cell lines from many vertebrate species. On infection, the viral genome is transcribed in the cytoplasm by a viral RNA polymerase present in the virion. Viral DNA synthesis ensues 1.5–2 hr postinfection and is essentially completed within 5 hr postinfection. Progeny virions become visible at about this time, and at about 12–15 hr postinfection, there are approximately 10,000 virion particles/infected cell. The course of viral infection is amazingly efficient; viral plaques are detectable within 48 hr of infection. A special feature of the poxvirus life cycle is that the entire infectious cycle— including viral transcription, replication, and maturation—takes place in the cytoplasm. Viral infection also causes inhibition of host synthesis of protein and DNA.

The life cycle of vaccinia virus suggests some limitations inherent in infectious vectors derived from it: (1) All infectious vectors would be necessarily cytocidal. (2) Foreign genes with cellular transcriptional signals would not be expected to be transcribed by the viral RNA polymerase unless they were provided with viral transcriptional signals. (3) Cellular genes with intervening sequences would not be expected to be spliced when inserted into the viral genome. Thus, expression of cellular genes containing intervening sequences in vectors derived from vaccinia would require the use of cDNAs expressed from viral promoters.

2.6.1. A Noninfectious Vector Derived from Vaccinia Virus

The study of promoters from vaccinia virus in transient expression systems with noninfectious vectors is important if we are to understand how best to express foreign genes in infectious vectors. In one study, a plasmid containing the chloramphenicol acetyltransferase (CAT) gene downstream from a viral promoter was transfected into CV-1 monkey cells, which were either infected by vaccinia virus or mock-infected (Cochran et al., 1985). In infected cells, CAT activity was detectable within 4–6 hr after transfection. Expression of CAT was dependent on viral infection, but was not due to replication of the plasmid in the infected cells. The amount of CAT activity seen in infected cells 24 hr posttransfection was approximately 40-fold higher than that seen with a plasmid containing the Rous sarcoma viral long terminal repeat (which is considered to be an efficient promoter) in uninfected cells. This result indicates that this transient expression system might be useful for expression of foreign proteins in animal cells—in particular, proteins toxic to the host cell.

2.6.2. Infectious Vectors Derived from Vaccinia Virus

The approach used to construct infectious vectors is essentially similar to that described for introducing a foreign gene into the HSV genome (reviewed by Moss et al., 1983). The coding sequences for the gene to be transferred are introduced downstream from a viral promoter (Mackett et al., 1982) or inserted into a viral gene (Panicali and Paoletti, 1982), so that some viral sequences flank both sides of the foreign sequences. The plasmid is then transfected into virus-infected cells, and the desired recombinant virus formed by homologous recombination is obtained by selection or screening among the progeny. In one study in which the resultant plaques were screened by nucleic-acid hybridization, approximately 0.4% of the plaques contained recombinant virus (Panicali and Paoletti, 1982). In the case in which the recombinant virus is selected, the selection employed is conversion of a TK$^+$ phenotype to TK$^-$ or vice versa (Mackett *et al.*, 1982; Panicali and Paoletti, 1982). Indeed, plasmids with cloning sites for the facile introduction of genes downstream from a TK promoter, or another viral promoter inserted into the viral TK gene, have been constructed (Mackett *et al.*, 1984).

Viral vectors constructed in this way have been used to express the rabies virus glycoprotein gene (Wiktor *et al.*, 1984), the sporozoite antigen of the malaria parasite *Plasmodium knowlesi* (Smith *et al.*, 1984), the HSV glycoprotein D gene, the HbsAg (Paoletti *et al.*, 1984), the influenza hemagglutinin gene (Panicali *et al.*, 1983), and a cDNA of the 26 S RNA of Sindbis virus (C. M. Rice *et al.*, 1985). In all the cases in which the protein product was studied, proteins of the predicted molecular weight were synthesized and, when examined grossly, showed the expected posttranslational modifications. Whenever tested, the recombinant viruses induced an antibody response against the foreign protein in the infected animals. Animals that were simultaneously infected with a recombinant virus containing HBsAg and the HSV glycoprotein D gene appeared to make antibodies to both foreign proteins; these findings indicate the possibility of using vaccinia vectors to make polyvalent vaccines (Paoletti *et al.*, 1984). Immunization of experimental animals with recombinant viruses containing either the rabies virus glycoprotein gene (Wiktor *et al.*, 1984) or the HSV glycoprotein D gene protected these animals from infection on challenge with the respective viruses (Paoletti *et al.*, 1984). Perkus *et al.* (1985) introduced the genes encoding HBsAg, HSV-1 glycoprotein D, and influenza hemagglutinin into one vaccinia genome. Animals inoculated with this recombinant vector produced antibodies reactive against all three proteins. These results demonstrate the efficacy of these recombinant viruses as live vaccines in experimental animals. However, the amount of the foreign protein produced with this vector system was described in only one instance. Paoletti *et al.* (1984) found that 150–200 ng HBsAg was synthesized in a 24-hr period following infection of 10^6 CV-1 cells at a multiplicity of infection of approximately 2 PFU/cell.

One more attribute of vaccinia virus makes it ideally suited for use as an infectious vector. Smith and Moss (1983) have shown that approximately 25,000 bp of foreign DNA could be inserted into the viral genome and that the recombinant genome was still packaged into the virion. Furthermore, the recombinant virus was stable and infectious, replicated as well, and gave the same yield as did the wild-type virus in tissue culture. In all these examples, the foreign DNA that was inserted into the viral genome was integrated into nonessential regions of the genome. Therefore, the recombinant viruses were not defective and could be propagated as pure recombinant stocks. Thus, in several respects, vectors derived from vaccinia virus appear to be sufficiently versatile to be widely useful.

3. VECTORS THAT PERMIT THE MAINTENANCE OF TRANSFERRED GENES AS PLASMIDS

3.1. Bovine Papillomavirus

In general, host cells that support lytic infection by papillomaviruses *in vitro* have not been identified. *In vivo,* these viruses are found in keratinocytes. The viral DNA can be found in the dividing basal epithelial cells, but no viral capsid antigens are found in these cells (Tooze, 1981). On division, the daughter of an infected basal cell migrates toward the surface and keratinizes, the viral DNA in it is amplified, and viral capsid proteins are synthesized. Therefore, the life cycle of the papillomaviruses is intimately associated with the differentiated state of their host cell.

Bovine papillomavirus (BPV) is a peculiarly tractable member of the papillomavirus family. The virus (Black *et al.,* 1963; Thomas *et al.,* 1964) and its DNA (Dvoretzky *et al.,* 1980) transform some mouse cells in tissue culture. BPV is apparently maintained and expressed in these murine cell lines much as it is in the basal epithelial cells of the infected animal.

Much progress in molecular analysis of the expression of papillomaviruses in infected cells has come recently from the studies of BPV in infected rodent cells. The viral genome, which is a covalently closed duplex DNA of approximately 8000 bp, has been sequenced (Chen *et al.,* 1982). The region of the genome expressed in rodent cells contains eight open reading frames. These open reading frames are expressed to yield different RNAs, some of which are formed by differential splicing (Stendlund *et al.,* 1985; Yang *et al.,* 1985). Several important kinds of *cis*-acting elements have been identified within the BPV genome. Two that are particularly relevant for the use of BPV as a vector are: (1) a transcriptional enhancer (Spalholz *et al.,* 1985) and (2) an origin of plasmid DNA synthesis (Waldeck *et al.,* 1984). The latter element maps close to and is most probably identical with an element termed a plasmid maintenance sequence (PMS) (Lusky and Botchan, 1984). The PMS has been defined as a sequence that permits linked DNAs to be maintained as plasmids in cells that provide BPV functions in *trans.* These two elements are important for BPV-derived vectors. They allow vectors derived from BPV to be maintained as plasmids and, perhaps, facilitate expression of inserted genes. Both the transcriptional enhancer and the PMS require BPV gene products for their function (see below). Functions have been assigned to the translational products of several open reading frames of BPV. At least two (E2 and E6) are necessary for full morphological transformation (Sarver *et al.*, 1984; Schiller *et al.*, 1984), one (E2) is required for transactivation of the enhancer (Spalholz *et al.*, 1985), at least one (E1) is required for plasmid maintenance (Lusky and Botchan, 1985), and one (E7) is required for obtaining a high copy number of plasmids in the transfected cell (Lusky and Botchan, 1985). We can expect that the integrity of the *cis*-acting elements of BPV and the functions of some or all of the virally encoded gene products will be required for vectors derived from BPV to function as intact plasmids in recipient cells.

Noninfectious Vectors Derived from BPV

Vectors derived from BPV can be used to synthesize up to 1 μg of a foreign protein/10^6 cells per day indefinitely (Hsiung *et al.*, 1984). In addition, these vectors can be maintained as plasmids indefinitely in some rodent cells at 20–200 copies/cell without special selection. This latter feature has made them particularly attractive to researchers. However, the requirements for maintaining different vectors containing part or all of BPV as intact plasmids are obscure. The variables that appear to affect maintenance of intact plasmids are: (1) whether or not the vector is rendered linear to remove prokaryotic sequences prior to transfection, (2) the structure or arrangement of DNA sequences within the vector, (3) the method of selection used after transfection of the cells, and (4) the recipient cells used for transfection.

A listing of the observations made with different BPV-derived vectors is given in Table I.

The observations presented in Table I do not allow us to define rules that, if followed, would permit the construction of a reliable vector derived from BPV. We may guess that the different selections require the expression of different viral genes and that the different foreign genes being expressed may affect the expression of viral genes by altering either their transcription or their splicing or both. However, this guess is largely indicative of our ignorance. The use of these vectors must be contemplated as a series of potential trials and errors if the desired goal is to maintain intact, autonomously replicating plasmids in the recipient cell. On the other hand, foreign genes have been expressed efficiently in vectors derived from BPV. BPV has a narrow host range (several rodent and bovine cell lines), which is a further limitation on the use of vectors derived from it, but this problem may have a practical remedy in the near future (see below).

TABLE I. Observations on Vectors Derived from Bovine Papilloma Virus

Reference	Structure of vector[a]	Transferred gene	Vector linearized before transfection?	Selection used	Recipient cells	Structure of intracellular recombinant vector	Physical state of recombinant vector in the cell
Mitrani-Rosenbaum et al. (1983)	0.69 BPV	Human interferon β_1	Yes	Morphological transformation	Mouse C127 and 1D13	Rearranged	Plasmids
Sekiguchi et al. (1983)	0.69 BPV	HSV-1 *TK*	No	HAT	Mouse and hamster TK-	Intact	Both integrated and plasmids
Law et al. (1983)	100% BPV	neo^r	No	G418	Mouse C127	Intact	Plasmids
Giri et al. (1983)	0.69 BPV and 100% BPV	SV2*gpt*	No	MAX	Rat FR3T3	Intact	Large tandems[b]
Wang et al. (1983)	0.69 BPV	HBsAg	No	Morphological transformation	Mouse C127	Rearranged	Plasmids
DiMaio et al. (1984)	0.69 BPV	Human β-globin and *HLA*	No	Morphological transformation	Mouse C127	Rearranged	Integrated
Meneguzzi et al. (1984)	0.69 BPV	neo^r	No	neo^r	Mouse C127, rat FR3T3	Intact	Large tandems
				Morphological transformation	Mouse C127, rat FR3T3	Rearranged	Plasmids
Fukunaga et al. (1984)	0.69 BPV	Human interferon	Yes	Morphological transformation	Mouse C127	Rearranged	13 Clones with plasmids; one with integrants
Lehn and Sauer (1984)	0.69 and 100% BPV	SV40 early region	Yes	Morphological transformation	Mouse C127	Rearranged	Plasmids
Pintel et al. (1984)	100% BPV	Minute virus of mouse	No	Morphological transformation	Mouse C127	Rearranged	Plasmids
Hsiung et al. (1984)	100% BPV	HBsAg	No	Morphological transformation	Mouse C127	Rearranged	Large tandems

[a] 0.69 BPV denotes that region of the BPV that includes all the open reading frames expressed in rodent cells and is sufficient to transform them (Lowy et al., 1980).

[b] Large tandems indicate that the DNA is present as high-molecular-weight DNA in the cells and yields a uniform structure after digestion with an endonuclease that cleaves the recombinant vector once. Often, because of the high copy number, it is not possible to determine whether the tandem molecules of the vector are or are not integrated within the cellular genome.

3.2. Other Papillomaviruses as Sources for Plasmid Vectors

Papillomaviruses have been isolated from many species of mammals, but none has been shown to productively infect cells in culture (Tooze, 1981). LaPorta and Taichman (1982) found that infection of human keratinocyte cultures with human papillomavirus type 1 led to maintenance of viral DNA as a plasmid in this cell strain, but to no phenotypic changes in the infected cell. More recently, Chesters and McCance (1985) have found that when human papillomavirus type 16 is cloned into pSV2-neo and introduced into a monkey cell line, BSC-1, and *neo*[r] clones selected, the vector DNA is present as plasmids. Perhaps, therefore, what is novel to BPV is not that it can be maintained as a plasmid in a cell line derived from a species different from its natural host, but that it can morphologically transform those cell lines. Hence, it seems reasonable to conjecture that by adding a selectable marker to other papillomaviral DNAs and selecting for the marker in different cells and not for morphological transformation, a variety of mammalian cell lines will be identified that would support one or more papillomavirus-derived vectors as plasmids.

3.3. Epstein–Barr Virus

EBV is a herpesvirus that infects human B-lymphoid and some epithelial cells *in vivo* (Tooze, 1981). Unlike the case of herpes simplex virus type 1, there are no cell lines that serve as efficient lytic hosts for EBV. EBV-infected epithelial cells have not been established in culture. In the EBV-infected B-lymphoid cells in culture, the virus displays two relationships with its host cell. In most cells of any clonal population, the viral DNA is maintained as a multicopy plasmid, and only a subset of the viral genes, termed "latent," is expressed. With some frequency, which varies from clone to clone, the resident, latent viral genome undergoes a maturation phase in which it is amplified and additional viral genes are expressed. Mature virus particles are formed, the host cell dies, and virus is released from the cell.

The plasmids that constitute the latent viral genomes are approximately 172,000 bp long and are too large to be manipulated as recombinant vectors. However, the *cis*-acting (Yates *et al.*, 1984; Reisman *et al.*, 1985) and the *trans*-acting (Yates *et al.*, 1985) viral elements required for plasmid replication have been identified. Vectors that carry only the *cis*-acting element of the EBV replicon, termed *ori*P, can be introduced and maintained efficiently under selection in EBV-transformed human B lymphoblasts (Sugden *et al.*, 1985a). The resident EBV genome supplies the only required viral *trans*-acting gene, termed *EBNA-1*, that is necessary for the vectors that contain *ori*P to replicate as plasmids. Thus, all EBV-transformed B lymphoblasts are helper cells for these EBV-derived vectors, and in this respect they are akin to COS cells for simian virus

40 (SV40)-derived, to MOP cells for polyoma-derived, and to 293 cells for adenovirus-derived vectors. However, in striking contrast to these other virus-based constructions, vectors that carry *oriP* from EBV do not kill their helper cells on replication. Rather, they are maintained under selection at a moderate copy number of 5–50 plasmids per cell, and the cells can proliferate indefinitely. In this respect, *oriP* vectors behave similarly to those derived from BPV.

The *cis*- and *trans*-acting elements of the EBV replicon have been combined in a single DNA molecule that replicates as a plasmid in a variety of cells that cannot be infected by EBV (Yates *et al.*, 1984). Such *oriP*, *EBNA-1* containing vectors are maintained stably under selection as a plasmid at 5–50 copies per cell in dog, monkey, and human cell lines from fibroblastic, epithelial, lymphoid, and erythroid lineages (Yates *et al.*, 1984). Again, as with the *oriP* vectors, these EBV vectors are akin to those derived from BPV in that they are maintained as plasmids in the recipient cells in a manner consistent with indefinite survival of the cell. They differ from BPV vectors in having a wider host range, which, however, does not include rodent cell lines. In addition, vectors with *oriP* are lost from cells in the absence of selection.

The use of EBV-derived vectors is in its infancy. However, a variety of selectable markers and the genes for the Leu-2 antigen (our unpublished observations) and for an HLA molecule (Shimizu *et al.*, 1986) have been expressed with these vectors.

One unexpected benefit of the EBV-derived vectors is that they usually appear to be unrearranged in the recipient cells (Sugden *et al.*, 1985b). More interestingly, the frequency of mutations observed in *oriP* vectors rescued from lymphoblastoid cells is 1/100th to 1/1000th of that reported for SV40-derived vectors in COS cells (Drinkwater and Klinedinst, 1986). It is not clear whether the difference in the observed frequency of mutations results from the vectors themselves, the recipient cells used, or the methods used for introducing the vectors into the recipient cells. Vectors derived from EBV and containing marker genes may be useful to study mutations induced in mammalian cells.

The broad host range and low frequency of mutations found with EBV-derived vectors indicate that they may be the vector of choice to maintain foreign genes in the plasmidial state in mammalian cells. However, potential difficulties in working with these vectors may not yet have been identified, since these vectors have been developed only recently.

4. COMPARISONS AND CONCLUSIONS

The advantages and disadvantages of the various vectors derived from animal DNA viruses that have been reviewed can be readily summarized. One caveat must be emphasized in such a summary, however. The construction, analyses, and uses of these vectors have just begun. This area of research will evolve rapidly; conclusions drawn today will certainly require revision tomorrow and may even need to be discarded the day after.

Major features of vectors derived from eight different animal DNA viruses are listed in Table II. The amounts of protein expressed from different recombinant vectors are provided only as representative examples. It is not clear that any of the examples represent the maximum amount attainable with a given recombinant vector in a given cell. It is clear that different host cells are likely to support the expression of different recombinant vectors, and the accumulation of transferred gene products, with differing efficiencies.

Vectors derived from simian virus 40 (SV40) have the unique advantage of being the most thoroughly studied. The simplest approach today for expressing a transferred gene efficiently would be to use a vector derived from SV40 in COS cells or to use its polyoma analogue in MOP cells. For simplicity, the recombinant vectors need not be infectious, so that the packaging limits imposed by virion assembly can be ignored. There are two disadvantages associated with these vectors: Expression is usually transient, although an exception has been reported (Tsui *et al.*, 1982), and the use of a noninfectious recombinant vector necessitates the use of DNA transfection to introduce the recombinant vectors into cells. This method of introduction usually results in approximately 10% of the cells taking up and expressing the added DNAs. Therefore, the recipient cells are heterogeneous with respect to expression of the transferred gene, and the total expression of that gene in the cell population is necessarily less than that which would be obtained with an infectious recombinant vector.

Vectors derived from adeno-associated virus 2 (AAV2) may eventually be used as are retroviral vectors today, because AAV2 integrates colinearly with respect to the virion DNA and can be rescued. However, retroviral vectors themselves seem more tractable today.

For those familiar with human adenoviruses, infectious vectors derived from adenovirus 2 (Ad 2) or Ad 5 would be the instruments of choice to try to synthesize milligram quantities of a desired protein. Derivatives of these viruses have been made that should accommodate approximately 3000 bp in the *E1* region (Van Doren *et al.*, 1984), and in time these derivatives will be modified by deletion of the *E3* region such that they should accommodate up to 5000 bp of exogenous DNA. Such recombinant vectors can be propagated readily on 293 cells. In addition, a productive infection of a large number of cells ($\geq 10^9$) can be easily achieved with HeLa or KB cells in suspension by co-infecting with the recombinant vector and wild-type Ad 2 (Yakov Gluzman, personal communication). The disadvantage of this scheme is that so far, few genes have been tested in infectious adenoviral vectors. We cannot predict how many genes, or which ones, can be inserted into adenoviral vectors and expressed as efficiently as is SV40 large-T antigen.

Post *et al.* (1981) provided a general method for selectively introducing

TABLE II. Comparison of Vectors Derived from Animal DNA Viruses

Virus from which vector is derived	Host cell in which vectors replicate	"Helper" cells[a]	Expression of transferred genes yielded by autonomous vector replication[b]	Largest insert in vector tested to date[c]	Amount of protein expressed from 10^6 cells of a transferred gene[d]
SV40	Monkey and human	Large-T antigen expressing monkey and human, COS	Transient	2,500 bp	0.5 µg HBsAG (Laub et al., 1983)
Polyoma	Mouse	Large-T antigen expressing mouse, MOP	Transient	2,500 bp	?
AAV 2	Human	—	Transient	1,800 bp	?
Ad 2 or 5	Human	Human 293, which expresses E1A and E1B region of Ad 5	Transient	2,800 bp	4 µg SV40 large-T antigen (Thummel et al., 1983)
HSV-1	Human, monkey, mouse	—	Transient	2,800 bp	?

Vaccinia	Human, monkey, mouse	—	Transient	24,700 bp	0.2 μg HBsAg (Paoletti et al., 1984)
BPV	Mouse C127, NIH3T3; rat FR3T3	—	Stable	8,000 bp	10 μg HBsAg (Hsiung et al., 1984)
EBV	Human, monkey, dog	All EBV-transformed lymphoblasts expressing EBNA-1	Stable	13,000 bp	0.5 μg HLA[c] (Shimizu et al., 1986)

[a] "Helper" cells are cell lines that contain and express viral genes required for the replication and expression of recombinant vectors. These cells support in *trans* the functioning of the appropriate recombinant vectors that lack these genes.

[b] Some vectors derived from SV40 and polyoma permit stable expression of transferred genes when autonomous replication is abrogated and the recombinant vectors are maintained in the recipient cells as integrated copies. The line separating the first six rows from those of BPV and EBV underscores the finding that autonomous replication of vectors derived from the first six viruses kills the host cells, while autonomous replication of vectors from BPV and EBV is consistent with host-cell survival.

[c] The size of genes transferred in vectors is given for cases in which recombinant vectors replicated autonomously and, where possible, were infectious. For noninfectious derivatives of SV40 and polyoma and for vectors derived from BPV and EBV, the maximum size limit of DNA that can be inserted is likely to be that which can be handled reasonably with the techniques of recombinant DNA.

[d] The amounts of protein listed for vectors derived from SV40, adenovirus, and vaccinia represent expression in cells destined to be killed by the vectors and therefore approximate the total amount synthesized. The amounts listed for BPV and EBV are per day because the expression of these recombinant vectors in recipient cells is stable and the recipient cells proliferate indefinitely.

[e] This number is obtained by using the fluorescence data cited in Shimizu et al. (1986) and the enumeration of HLA molecules on the parent cell line, 721, presented in Sugden and Metzenberg (1983).

foreign genes into the *TK* locus of herpes simplex virus type 1 (HSV-1). Spaete and Frenkel (1982) have developed vectors for facile introduction of DNA sequences into defective HSV genomes. Both these approaches have been instrumental in studying gene regulation in HSV-1. The potential uses of HSV-1 or of other lytic herpesviruses as vectors are now uncertain. Because attenuated strains of herpesviruses are used currently as vaccines, it is conceivable that vectors derived from lytic herpesviruses will be tested as vaccines in domestic animals and perhaps in man.

Vectors derived from vaccinia virus are versatile and are the ones that have been most thoroughly studied for their application as vaccines. Infectious vaccinia vectors can accommodate up to 25,700 bp of foreign DNA. The work of Moss and Paoletti and their colleagues has provided the protocols for introducing foreign genes into vaccinia and demonstrated that such recombinant vectors elicit specific immune responses to the products of the transferred genes. These are the vectors of choice today for developing live vaccines to be tested in animals. The disadvantage associated with these vectors for this application is that the risks perceived to be associated with their use may be judged to outweigh their potential benefit. Noninfectious derivatives when introduced into cells that are subsequently infected with vaccinia may permit the efficient synthesis of transferred genes the products of which are toxic to the recipient cell.

Finally, if stable expression of transferred genes from a plasmid is desired, then vectors derived from bovine papillomavirus (BPV) or Epstein–Barr virus (EBV) should be employed. These two families of vectors complement each other in that those derived from BPV replicate autonomously only in some rodent cell lines, while those derived from EBV replicate autonomously in cell lines derived from dogs, monkeys, and humans, but not from rodents. One advantage of vectors derived from BPV and EBV is that although they are not infectious, those vectors that carry antibiotic resistance markers can be selected in recipient cells, and the resistant cells can be cloned. These cell clones are then at least as homogeneous with respect to their plasmid vectors as are a population of cells that have been infected with some infectious vector. Another advantage of these plasmid vectors is that the transferred genes are stably maintained, which means that the products of the transferred genes can be harvested over long periods of time by collecting a portion of the proliferating cells. If the product is secreted from the cells, then it can be collected from the supernatant fluid above the cells. In one example, Hsiung *et al.* (1984) found with a BPV-derived vector that 10 µg HBsAg was secreted per 10^6 cells per 24 hr throughout the 85 days of the experiment. A third advantage of vectors that replicate autonomously as plasmids is that they should be able to accommodate large segments of foreign DNA— EBV itself replicates autonomously as a plasmid and is 172,000 bp in size!

Vectors derived from BPV and EBV do have disadvantages, too. Those derived from BPV are sometimes rearranged or are not maintained as plasmids in recipient cells. Those derived from EBV are too new to have been tested widely. This disadvantage applies to most vectors derived from animal DNA viruses and is likely to be eliminated rapidly for many of them.

ACKNOWLEDGMENTS. We thank Kevin Van Doren and Yakov Gluzman for discussing their findings with us. We thank our colleagues, Joyce Knutson, Joan Mecsas, Stan Metzenberg, Dave Reisman, and John Yates, for their constructive criticisms of this review. Finally, we thank Ilse Riegel for her editorial guidance and Kristen Luick for carefully typing our numerous drafts. We were supported by NIH Grants CA-07175 and CA-22443.

REFERENCES

Ashman, C. R., and Davidson, R. L., 1984, High spontaneous mutation frequency in shuttle vector sequences recovered from mammalian cellular DNA, *Mol. Cell. Biol.* **4:**2266–2272.

Bear, S. E., Colberg-Poley, A. M., Court, D. L., Carter, B. J., and Enquist, L. W., 1984, Analysis of two potential shuttle vectors containing herpes simplex virus defective DNA, *J. Mol. Appl. Genet.* **2:**471–484.

Berkner, K. L., and Sharp, P. A., 1983, Generation of adenovirus by transfection of plasmids, *Nucleic Acids Res.* **11:**6003–6020.

Berkner, K. L., and Sharp, P. A., 1984, Expression of dihydrofolate reductase, and of the adjacent E1b region, in an Ad5-dihydrofolate reductase recombinant virus, *Nucleic Acids Res.* **12:**1925–1941.

Berns, K. I., 1984a, Adeno-associated virus, in: *The Adenoviruses* (H. S. Ginsberg, ed.), pp. 563–592, Plenum Press, New York.

Berns, K. I. (ed.), 1984b, *The Parvoviruses,* Plenum Press, New York.

Black, P. H., Rowe, W. P., Turner, H. C., and Huebner, R. J., 1963, Transformation of bovine tissue culture cells by bovine papilloma virus, *Proc. Natl. Acad. Sci. U.S.A.* **50:**1148–1156.

Boast, S., La Mantia, G., Lania, L., and Blasi, F., 1983, High efficiency of replication and expression of foreign genes in SV40-transformed human fibroblasts, *EMBO J.* **2:**2327–2331.

Brady, J., Bolen, J. B., Radonovich, M., Salzman, N., and Khoury, G., 1984, Stimulation of simian virus 40 late gene expression by simian virus 40 tumor antigen, *Proc. Natl. Acad. Sci. U.S.A.* **81:**2040–2044.

Calos, M. P., Lebkowski, J. S., and Botchan, M. R., 1983, High mutation frequency in DNA transfected into mammalian cells, *Proc. Natl. Acad. Sci. U.S.A.* **80:**3015–3019.

Chen, E. Y., Howley, P. M., Levinson, A. D., and Seeburg, P. H., 1982, The primary structure and genetic organization of the bovine papillomavirus type 1 genome, *Nature (London)* **299:**529–534.

Chesters, P. M., and McCance, D. J., 1985, Human papillomavirus type 16 recombinant DNA is maintained as an autonomously replicating episome in monkey kidney cells, *J. Gen. Virol.* **66:**615–620.

Cochran, M. A., Mackett, M., and Moss, B., 1985, Eukaryotic transient expression system dependent on transcription factors and regulatory DNA sequences of vaccinia virus, *Proc. Natl. Acad. Sci. U.S.A.* **82:**19–23.

Davis, A. R., Bos, T. J., and Nayak, D. P., 1983, Active influenza virus neuraminidase is expressed in monkey cells from cDNA cloned in simian virus 40 vectors, *Proc. Natl. Acad. Sci. U.S.A.* **80:**3976–3980.

Deans, R. J., Denis, K. A., Taylor, A., and Wall, R., 1984, Expression of an immunoglobulin heavy chain gene transfected into lymphocytes, *Proc. Natl. Acad. Sci. U.S.A.* **81:**1292–1296.

DiMaio, D., Corbin, V., Sibley, E., and Maniatis, T., 1984, High-level expression of a cloned HLA heavy chain gene introduced into mouse cells on a bovine papillomavirus vector, *Mol. Cell. Biol.* **4**:340–350.

Drinkwater, N. R., and Klinedinst, D. K., 1986, Chemically-induced mutagenesis in a shuttle vector with a low background mutant frequency, *Proc. Natl. Acad. Sci. U.S.A.* (in press).

Dvoretzky, I., Shober, R., Chattopadhyay, S. K., and Lowy, D. R., 1980, A quantitative *in vitro* focus assay for bovine papilloma virus, *Virology* **103**:369–375.

Fenner, R., 1985, Poxviruses, in: *Virology* (B. N. Fields, ed.), pp. 661–684, Raven Press, New York.

Frenkel, N., Locker, H., and Vlanzy, D., 1981, Structure and expression of Class I and II defective interfering HSV genomes, in: *Herpesvirus DNA* (Y. Becker, ed.), pp. 149–184, Martinus Nijhoff, The Hague.

Fukunaga, R., Sokawa, Y., and Nagata, S., 1984, Constitutive production of human interferons by mouse cells with bovine papillomavirus as a vector, *Proc. Natl. Acad. Sci. U.S.A.* **81**:5086–5090.

Gelb, L. D., 1985, Varicella-zoster virus, in: *Virology* (B. N. Fields, ed.)., pp. 591–627, Raven Press, New York.

Gething, M.-J., and Sambrook, J., 1981, Cell-surface expression of influenza hemagglutinin from a cloned DNA copy of the RNA gene, *Nature (London)* **293**:620–625.

Ginsberg, H. S. (ed.), 1984, *The Adenoviruses,* Plenum Press, New York.

Giri, I., Jouanneau, J., and Yaniv, M., 1983, Comparative studies of the expression of linked *Escherichia coli gpt* gene and BPV-1 DNAs in transfected cells, *Virology* **127**:385–396.

Gluzman, Y., 1981, SV40-transformed simian cells support the replication of early SV40 mutants, *Cell* **23**:175–182.

Gluzman, Y. (ed.), 1982, *Eukaryotic Viral Vectors,* Cold Spring Harbor Press, Cold Spring Harbor, New York.

Gluzman, Y., and Van Doren, K., 1983, Palindromic adenovirus type 5-simian virus 40 hybrid, *J. Virol.* **45**:91–103.

Graham, F. L., Smiley, J., Russell, W. C., and Nairn, R., 1977, Characteristics of a human cell line transformed by DNA from human adenovirus type 5, *J. Gen. Virol.* **36**:59–72.

Grosschedl, R., and Baltimore, D., 1985, Cell-type specificity of immunoglobulin gene expression is regulated by at least three DNA sequence elements, *Cell* **41**:885–897.

Hammarskjöld, M.-L., and Winberg, G., 1980, Encapsidation of adenovirus 16 DNA is directed by a small DNA sequence at the left end of the genome, *Cell* **20**:787–795.

Hermonat, P. L., and Muzyczka, N., 1984, Use of adeno-associated virus as a mammalian DNA cloning vector: Transduction of neomycin resistance into mammalian tissue culture cells. *Proc. Natl. Acad. Sci. U.S.A.* **81**:6466–6470.

Hsiung, N., Fitts, R., Wilson, S., Milne, A., and Hamer, D., 1984, Efficient production of hepatitis B surface antigen using a bovine papilloma virus-metallothionein vector, *J. Mol. Appl. Genet.* **2**:497–506.

Kaufman, R. J., 1985, Identification of the components necessary for adenovirus translational control and their utilization in cDNA expression vectors, *Proc. Natl. Acad. Sci. U.S.A.* **82**:689–693.

Keller, J. M., and Alwine, J. C., 1984, Activation of SV40 late promoter: Direct effects of T antigen in the absence of viral DNA replication, *Cell* **36**:381–389.

Kelly, T. J., Jr., 1984, Adenovirus DNA replication, in: *The Adenoviruses* (H. S. Ginsberg, ed.), pp. 271–308, Plenum Press, New York.

Klessig, D. F., 1984, Adenovirus–SV40 interactions, in: *The Adenoviruses* (H. S. Ginsberg, ed.), pp. 399–449, Plenum Press, New York.

Knipe, D. M., Ruyechan, W. T., Roizman, B., and Halliburton, I. W., 1978, Molecular genetics of herpes simplex virus: Demonstration of regions of obligatory and nonobligatory identity within diploid regions of the genome by sequence replacement and insertion, *Proc. Natl. Acad. Sci. U.S.A.* **75**:3896–3900.

Kwong, A. D., and Frenkel, N., 1984, Herpes simplex amplicon: Effect of size on replication of constructed defective genomes containing eucaryotic DNA sequences, *J. Virol.* **51**:595–603.

Lamb, R. A., and Lai, C.-J., 1984, Expression of unspliced NS_1 mRNA, spliced NS_2 mRNA, and a spliced chimera mRNA from cloned influenza virus NS DNA in an SV40 vector, *Virology* **135**:139–147.

LaPorta, R. F., and Taichman, L. B., 1982, Human papilloma viral DNA replicates as a stable episome in cultured epidermal keratinocytes, *Proc. Natl. Acad. Sci. U.S.A.* **79**:3393–3397.

Laub, O., Rall, L. B., Truett, M., Shaul, Y., Standring, D. N., Valenzuela, P., and Rutter, W. J., 1983, Synthesis of hepatitis B surface antigen in mammalian cells: Expression of the entire gene and the coding region, *J. Virol.* **48**:271–280.

Laughlin, C. A., Tratschin, J.-D., Coon, H., and Carter, B. J., 1983, Cloning of infectious adeno-associated virus genomes in bacterial plasmids, *Gene* **23**:65–73.

Law, M.-F., Byrne, J. C., and Howley, P. M., 1983, A stable bovine papilloma virus hybrid plasmid that expresses a dominant selective trait, *Mol. Cell. Biol.* **3**:2110–2115.

Lehn, H., and Sauer, G., 1984, The physical state of hybrid genomes containing simian virus 40 and bovine papilloma virus DNA sequences in transformed clonal cell lines. *J. Gen. Virol.* **65**:1413–1418.

Logan, J., and Shenk, T., 1984, Adenovirus tripartite leader sequence enhances translation of mRNAs late after infection, *Proc. Natl. Acad. Sci. U.S.A.* **81**:3655–3659.

Lowy, D. R., Dvoretzky, I., Shober, R., Law, M.-F., Engel, L., and Howley, P. M., 1980, *In vitro* tumorigenic transformation by a defined sub-genomic fragment of bovine papilloma virus DNA, *Nature (London)* **287**:72–74.

Luciw, P. A., Bishop, J. M., Varmus, H. E., and Capecchi, M. R., 1983, Location and function of retroviral and SV40 sequences that enhance biochemical transformation after microinjection of DNA, *Cell* **33**:705–716.

Lusky, M., and Botchan, M. R., 1984, Characterization of the bovine papilloma virus plasmid maintenance sequences, *Cell* **36**:391–401.

Lusky, M., and Botchan, M. R., 1985, Genetic analysis of bovine papillomavirus type 1 *trans*-acting replication factors, *J. Virol.* **53**:955–965.

Mackett, M., Smith, G. L., and Moss, B., 1982, Vaccinia virus: A selectable eukaryotic cloning and expression vector, *Proc. Natl. Acad. Sci. U.S.A.* **79**:7415–7419.

Mackett, M., Smith, G. L, and Moss, B., 1984, General method for production and selection of infectious vaccinia virus recombinants expressing foreign genes, *J. Virol.* **49**:857–864.

Mansour, S. L., Grodzicker, T., and Tjian, R., 1985, An adenovirus vector system used to express polyoma virus tumor antigens, *Proc. Natl. Acad. Sci. U.S.A.* **82**:1359–1363.

Mellon, P., Parker, V., Gluzman, Y., and Maniatis, T., 1981, Identification of DNA sequences required for transcription of the human α1-globin gene in a new SV40 host–vector system, *Cell* **27**:279–288.

Meneguzzi, G., Binetruy, B., Grisoni, M., and Cuzin, F., 1984, Plasmidial maintenance in rodent fibroblasts of a BPV1-pBR322 shuttle vector without immediately apparent oncogenic trans-formation of the recipient cells, *EMBO J.* **3**:365–371.

Miller, G., 1985, Epstein–Barr virus, in: *Virology* (B. N. Fields, ed.), pp. 563–589, Raven Press, New York.

Mitrani-Rosenbaum, S., Maroteaux, L., Mory, Y., Revel, M., and Howley, P., 1983, Inducible expression of the human interferon β_1 gene linked to a bovine papilloma virus DNA vector and maintained extrachromosomally in mouse cells, *Mol. Cell. Biol.* **3**:233–240.

Mocarski, E. S., Post, L. E., and Roizman, B., 1980, Molecular engineering of the herpes simplex virus genome: Insertion of a second L-S junction into the genome causes additional genome inversions, *Cell* **22**:243–255.

Moss, B., 1978, Poxviruses, in: *The Molecular Biology of Animal Viruses*, Vol. 2 (D. P. Nayak, ed.), pp. 849–890, Marcel Dekker, New York.

Moss, B., 1985, Replication of poxviruses, in: *Virology* (B. N. Fields, ed.), pp. 685–703, Raven Press, New York.

Moss, B., Smith, G. L., and Mackett, M., 1983, Use of vaccinia virus as an infectious molecular cloning and expression vector, in: *Gene Amplification and Analysis*, Vol. 3, *Expression of Cloned Genes in Prokaryotic and Eukaryotic Cells* (T. S. Papas, M. Rosenberg, and J. G. Chirikjian, eds.), pp. 201–213, Elsevier, New York.

Muller, W. J., Naujokas, M. A., and Hassell, J. A., 1984, Isolation of large T antigen-producing mouse cell lines capable of supporting replication of polyomavirus–plasmid recombinants, *Mol. Cell. Biol.* **4:**2406–2412.

Oda, K., Kato, H., Saito, I., Sugano, S., Maruyama, K., Masuda, M., Shiroki, K., and Shimojo, H., 1983, Expression of adenovirus type 12 E1A gene in monkey cells, using a simian virus 40 vector, *J. Virol.* **45:**408–419.

Panicali, D., and Paoletti, E., 1982, Construction of poxviruses as cloning vectors: Insertion of the thymidine kinase gene from herpes simplex virus into the DNA of infectious vaccinia virus, *Proc. Natl. Acad. Sci. U.S.A.* **79:**4927–4931.

Panicali, D., Davis, S. W., Weinberg, R. L., and Paoletti, E., 1983, Construction of live vaccines by using genetically engineered poxviruses: Biological activity of recombinant vaccinia virus expressing influenza virus hemagglutinin, *Proc. Natl. Acad. Sci. U.S.A.* **80:**5364–5368.

Paoletti, E., Lipinskas, B. R., Samsonoff, C., Mercer, S., and Panicali, D., 1984, Construction of live vaccines using genetically engineered poxviruses: Biological activity of vaccinia virus recombinants expressing the hepatitis B virus surface antigen and the herpes simplex glycoprotein D, *Proc. Natl. Acad. Sci. U.S.A.* **81:**193–197.

Perkus, M. E., Piccini, A., Lipinskas, B. R., and Paoletti, E., 1985, Recombinant vaccinia virus: Immunization against multiple pathogens, *Science* **229:**981–984.

Pintel, D., Merchlinsky, M. J., and Ward, D. C., 1984, Expression of minute virus of mice structural proteins in murine cell lines transformed by bovine papillomavirus–minute virus of mice plasmid chimera, *J. Virol.* **52:**320–327.

Post, L. E., Mackem, S., and Roizman, B., 1981, Regulation of α genes of herpes simplex virus: Expression of chimeric genes produced by fusion of thymidine kinase with α gene promoters, *Cell* **24:**555–565.

Rawls, W. E., 1985, Herpes simplex virus, in: *Virology* (B. N. Fields, ed.), pp. 527–561, Raven Press, New York.

Razzaque, A., Chakrabarti, S., Joffee, S., and Seidman, M., 1984, Mutagenesis of a shuttle vector plasmid in mammalian cells, *Mol. Cell. Biol.* **4:**435–441.

Reisman, D., Yates, J., and Sugden, B., 1985, A putative origin of replication of plasmids derived from Epstein–Barr virus is composed of two *cis*-acting components, *Mol. Cell. Biol.* **5:**1822–1832.

Rice, C. M., Franke, C. A., Strauss, J. H., and Hruby, D. E., 1985, Expression of Sindbis virus structural proteins via recombinant vaccinia virus: Synthesis, processing, and incorporation into mature Sindbis virions, *J. Virol.* **56:**227–239.

Rice, D., and Baltimore, D., 1982, Regulated expression of an immunoglobulin K gene introduced into a mouse lymphoid cell line, *Proc. Natl. Acad. Sci. U.S.A.* **79:**7862–7865.

Rio, D. C., and Tjian, R., 1983, SV40 T antigen binding site mutations that affect autoregulation, *Cell* **32:**1227–1240.

Rio, D. C., Clark, S. G., and Tjian, R., 1985, A mammalian host–vector system that regulates expression and amplification of transfected genes by temperature induction, *Science,* **227:**23–28.

Roizman, B., and Batterson, W., 1985, Herpesviruses and their replication, in: *Virology* (B. N. Fields, ed.), pp. 497–526, Raven Press, New York.

Roizman, B., and Jenkins, F. J., 1985, Genetic engineering of novel genomes of large DNA viruses, *Science* **229:**1208–1214.

Rymo, L., 1983, Transfer of the Epstein–Barr virus genes coding for small RNAs to human lymphoid cells with a vector carrying a dominant selectable marker, *EMBO J.* **2:**839–844.

Saito, I., Oya, Y., Yamamoto, K., Yuasa, T., and Shimojo, H., 1985, Construction of nondefective adenovirus type 5 bearing a 2.8-kilobase hepatitis B virus DNA near the right end of its genome, *J. Virol.* **54:**711–719.

Salzman, L. A., 1978, The parvoviruses, in: *The Molecular Biology of Animal Viruses*, Vol. 2 (D. P. Nayak, ed.), pp. 539–587, Marcel Dekker, New York.

Samulski, R. J., Berns, K. I., Tan, M., and Muzyczka, N., 1982, Cloning of adeno-associated virus into pBR322: Rescue of intact virus from the recombinant plasmid in human cells, *Proc. Natl. Acad. Sci. U.S.A.* **79:**2077–2081.

Samulski, R. J., Srivastava, A., Berns, K. I., and Muzyczka, N., 1983, Rescue of adeno-associated virus from recombinant plasmids: Gene correction within the terminal repeats of AAV, *Cell* **33:**135–143.

Sarver, N., Rabson, M. S., Yang, Y.-C., Byrne, J. C., and Howley, P. M., 1984, Localization and analysis of bovine papillomavirus type 1 transforming functions, *J. Virol.* **52:**377–388.

Schiller, J. T., Vass, W. C., and Lowy, D. R., 1984, Identification of a second transforming region in bovine papillomavirus DNA, *Proc. Natl. Acad. Sci. U.S.A.* **81:**7880–7884.

Sekiguchi, T., Nishimoto, T., Kai, R., and Sekiguchi, M., 1983, Recovery of a hybrid vector, derived from bovine papilloma virus DNA, pBR322 and the HSV TK gene, by bacterial transformation with extrachromosomal DNA from transfected rodent cells, *Gene* **21:**267–272.

Semler, B. L., Dorner, A. J., and Wimmer, E., 1984, Production of infectious poliovirus from cloned cDNA is dramatically increased by SV40 transcription and replication signals, *Nucleic Acids Res.* **12:**5123–5141.

Shih, M.-F., Arsenakis, M., Tiollais, P., and Roizman, B., 1984, Expression of hepatitis B virus S gene by herpes simplex virus type 1 vectors carrying α- and β-regulated gene chimeras, *Proc. Natl. Acad. Sci. U.S.A.* **81:**5867–5870.

Shimizu, Y., Koller, B., Geraghty, D., Orr, H., Shaw, S., Kavathas, P., and De Mars, R., 1986, Transfer of cloned human class I MHC genes into HLA mutant human lymphoblastoid cells, *Mol. Cell. Biol.* **6:**1074–1087.

Smiley, J. R., 1980, Construction *in vitro* and rescue of a thymidine kinase-deficient deletion mutation of herpes simplex virus, *Nature (London)* **285:**333–335.

Smith, G. L., and Moss, B., 1983, Infectious poxvirus vectors have capacity for at least 25,000 base pairs of foreign DNA, *Gene* **25:**21–28.

Smith, G. L., Godson, G. N., Nussenzweig, V., Nussenzweig, R. S., Barnwell, J., and Moss, B., 1984, *Plasmodium knowlesi* sporozoite antigen: Expression by infectious recombinant vaccinia virus, *Science* **224:**397–399.

Spaete, R. R., and Frenkel, N., 1982, The herpes simplex virus amplicon: A new eucaryotic defective-virus cloning-amplifying vector, *Cell* **30:**295–304.

Spalholz, B. A., Yang, Y.-C. and Howley, P. M., 1985, Transactivation of a bovine papilloma virus transcriptional regulatory element by the E2 gene product, *Cell* **42:**183–191.

Sprague, J., Condra, J. H., Arnheiter, H., and Lazzarini, R. A., 1983, Expression of a recombinant DNA gene coding for the vesicular stomatitis virus nucleocapsid protein, *J. Virol.* **45:**773–781.

Stenlund, A., Zabielski, J., Ahola, H., Moreno-Lopez, J., and Pettersson, U., 1985, Messenger RNAs from the transforming regions of bovine papillomavirus type I, *J. Mol. Biol.* **182:**541–554.

Subramani, S., and Southern, P. J., 1983, Analysis of gene expression using simian virus 40 vectors, *Anal. Biochem.* **135:**1–15.

Sugden, B., and Metzenberg, S., 1983, Characterization of an antigen whose cell surface expression is induced by infection with Epstein–Barr virus. *J. Virol.* **46:**800–807.

Sugden, B., Marsh, K., and Yates, J., 1985a, A vector that replicates as a plasmid and can be efficiently selected in B-lymphoblasts transformed by Epstein–Barr virus, *Mol. Cell. Biol.* **5:**410–413.

Sugden, B., Yates, J., Reisman, D., and Warren, N., 1985b, Transforming functions of Epstein–Barr virus, in: *UCLA Symposium on Molecular and Cellular Biology*, Vol. 28 (R. P. Gale and D. W. Golde, eds.), pp. 171–186, Alan R. Liss, New York.

Thillet, J., Kunst, F., Lasserre, C., Bucchini, D., Pictet, R., and Jami, J., 1984, Transformation of teratocarcinoma stem cells and fibroblasts with various vectors containing the *Eco.gpt* gene as a selection marker, *Exp. Cell. Res.* **151**:494–501.

Thomas, M., Boiron, M., Tanzer, J., Levy, J. P., and Bernard, J., 1964, *In vitro* transformation of mice cells by bovine papilloma virus, *Nature (London)* **202**:709–710.

Thummel, C., Tjian, R., Hu, S.-L., and Grodzicker, T., 1983, Translational control of SV40 T antigen expressed from the adenovirus late promoter, *Cell* **33**:455–464.

Tibbetts, C., 1977, Viral DNA sequences from incomplete particles of human adenovirus type 7, *Cell* **12**:243–249.

Tooze, J. (ed.), 1981, *DNA Tumor Viruses*, Cold Spring Harbor Press, Cold Spring Harbor, New York.

Tratschin, J.-D., West, M. H. P., Sandbank, T., and Carter, B. J., 1984, A human parvovirus, adeno-associated virus, as a eucaryotic vector: Transient expression and encapsidation of the procaryotic gene for chloramphenicol acetyltransferase, *Mol. Cell. Biol.* **4**:2072–2081.

Tsui, L.-C., Breitman, M. L., Siminovitch, L., and Buchwald, M., 1982, Persistence of freely replicating SV40 recombinant molecules carrying a selectable marker in permissive simian cells, *Cell* **30**:499–508.

Van Doren, K., Hanahan, D., and Gluzman, Y., 1984, Infection of eucaryotic cells by helper-independent recombinant adenoviruses: Early region 1 is not obligatory for integration of viral DNA, *J. Virol.* **50**:606–614.

Waldeck, W., Rösl, F., and Zentgraf, H., 1984, Origin of replication in episomal bovine papilloma virus type 1 DNA isolated from transformed cells, *EMBO, J.* **3**:2173–2178.

Wang, Y., Stratowa, C., Schaefer-Ridder, M., Doehmer, J., and Hofschneider, P. H., 1983, Enhanced production of hepatitis B surface antigen in NIH 3T3 mouse fibroblasts by using extrachromosomally replicating bovine papillomavirus vector. *Mol. Cell. Biol.* **3**:1032–1039.

Wiktor, T. J., Macfarlan, R. I., Reagan, K. J., Dietzschold, B., Curtis, P. J., Wunner, W. H., Kieny, M.-P., Lathe, R., Lecocq, J.-P., Mackett, M., Moss, B., and Koprowski, H., 1984, Protection from rabies by a vaccinia virus recombinant containing the rabies virus glycoprotein gene, *Proc. Natl. Acad. Sci. U.S.A.* **81**:7194–7198.

Will, H., Cattaneo, R., Pfaff, E., Kuhn, C., Roggendorf, M., and Schaller, H., 1984, Expression of hepatitis B antigens with a simian virus 40 vector, *J. Virol.* **50**:335–342.

Wold, W. S. M., Green, M., and Büttner, W., 1978, Adenoviruses, in: *The Molecular Biology of Animal Viruses*, Vol. 2 (D. P. Nayak, ed.), pp. 673–768, Marcel Dekker, New York.

Yang, Y.-C., Okayama, H., and Howley, P. M., 1985, Bovine papillomavirus contains multiple transforming genes, *Proc. Natl. Acad. Sci. U.S.A.* **82**:1030–1034.

Yates, J., Warren, N., Reisman, D., and Sugden, B., 1984, A *cis*-acting element from the Epstein–Barr viral genome that permits stable replication of recombinant plasmids in latently infected cells, *Proc. Natl. Acad. Sci. U.S.A.* **81**:3806–3810.

Yates, J., Warren, N., and Sugden, B., 1985, Stable replication of plasmids derived from Epstein–Barr virus in various mammalian cells, *Nature (London)* **313**:812–815.

6

RETROVIRUS VECTORS FOR GENE TRANSFER: EFFICIENT INTEGRATION INTO AND EXPRESSION OF EXOGENOUS DNA IN VERTEBRATE CELL GENOMES

Howard M. Temin

1. INTRODUCTION

Retroviruses are animal viruses with an RNA genome that replicate through a DNA intermediate. The DNA intermediate, the provirus, is stably integrated into cellular DNA. (Alternatively, retroviruses can be considered the RNA form of cellular movable genetic elements that transpose through an extracellular RNA intermediate.) Because of this efficient integration and other properties to be discussed, retroviruses are considered the most likely vectors for use in human gene therapy (Wyngaarden, 1985).

Retroviruses, like all other viruses, have efficient mechanisms to enter particular host cells. In addition, retroviruses, like all cellular movable genetic elements, have efficient mechanisms to integrate their genomes stably into cellular DNA and to express genes encoded within their genomes. Furthermore, many retroviruses isolated from animals also include sequences from particular cellular genes, the proto-oncogenes. These retroviruses, called "highly oncogenic retroviruses," integrate and express oncogenes, which are modified forms of the proto-oncogenes. Thus, highly oncogenic retroviruses are natural vectors to introduce nonviral sequences into cell DNA.

There are a large number of different retroviruses able to infect many

Howard M. Temin ● McArdle Laboratory for Cancer Research, University of Wisconsin, Madison, Wisconsin 53706.

different types and species of vertebrate cells. All the retroviruses have the appropriate machinery to introduce their genes into vertebrate cells, to have those genes stably integrated into chromosomal DNA, and to have those genes expressed efficiently. Furthermore, retroviruses have a genetic organization that allows other genes to be inserted easily into them without preventing any step in the normal virus life cycle, and usual retrovirus replication does not kill infected cells. The only significant limitations on the use of retrovirus vectors are the lack of integration into a specific chromosomal locus and the possible difficulty of getting a high enough titer of an unmodified helper-free vector stock (see Section 4.2). [Vectors have also been developed from a number of animal DNA viruses (see Chapter 5). None of these vectors has as part of its life cycle a mechanism for efficient and stable integration as retroviruses do. Thus, only retroviruses can stably transform a majority of a cell population.]

In this chapter, I shall first give a brief background description of the relevant features of the retrovirus life cycle and then discuss possible types of retrovirus vectors and their properties, uses, and potential problems.

I shall discuss primarily C-type retroviruses, since they are the only ones that have been used in vector construction and are by far the most studied. The most commonly used C-type retroviruses have been Rous sarcoma virus, murine leukemia virus, and reticuloendotheliosis virus.

Since integration is an obligatory part of the life cycle of most retroviruses, all study of retroviruses could be considered relevant to gene transfer. [Visna virus, a lentivirus, may be an exception, since it may replicate through an unintegrated DNA intermediate (Harris *et al.*, 1984).] However, since the subject matter of this book is not viruses *per se*, but gene transfer, I shall restrict discussion in this chapter to retroviruses modified for purposes of studying transfer of nonviral genes (also see Chapter 12). This distinction between retroviruses and retrovirus vectors is somewhat difficult to make for highly oncogenic retroviruses, since oncogenes are part of these retroviruses, but have evolved from cellular genes. I shall make reference to highly oncogenic retroviruses only to illustrate principles of retrovirus vector construction (for a recent review of oncogenes, see Bishop and Varmus, 1985).

In addition, retrovirus long terminal repeats (see below) have been widely used as transcriptional elements to drive easily assayed genes. These constructions have been used primarily in transient transfection assays to study retrovirus transcriptional controls and will not be discussed in detail in this chapter.

2. RETROVIRUS LIFE CYCLE

The reader is referred to Weiss *et al.* (1985) for an exhaustive review of this subject.

2.1. Retrovirus Genes

Replication-competent retroviruses have three protein-coding genes, *gag*, *pol*, and *env*, which are involved, respectively, in viral encapsidation and assembly, synthesis of viral DNA (reverse transcription) and integration, and viral budding and entrance (Figs. 1, 2, and 3). These genes are *trans*-acting. In addition, retroviruses have *cis*-acting sequences for reverse transcription (r sequences involved in viral DNA synthesis, and primer binding site and polypurine tract sequences involved in priming DNA synthesis), for integration (attachment sequences), and for encapsidation (Fig. 2).

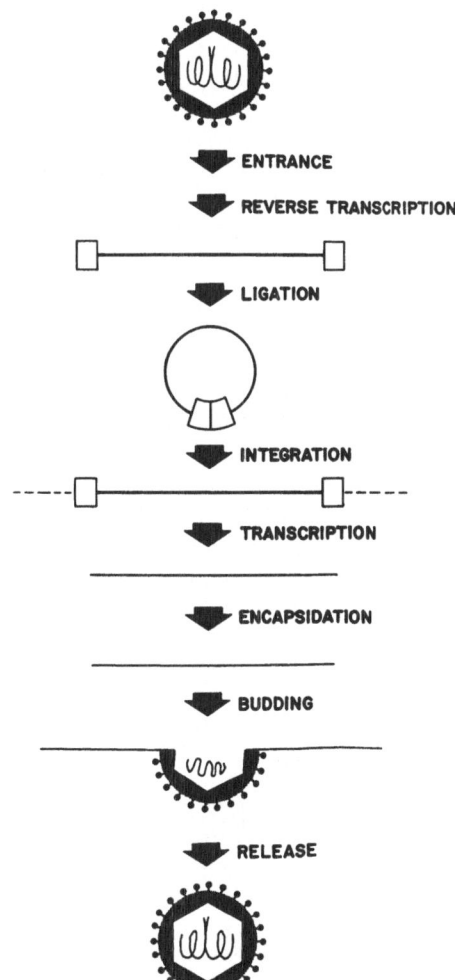

FIGURE 1. Retrovirus life cycle. The virion containing two molecules of single-stranded RNA enters a sensitive cell, and the RNA is reverse-transcribed into a molecule of unintegrated viral DNA with two long terminal repeats (LTRs) (□—□). The unintegrated viral DNA is ligated to form a closed circular molecule that is integrated to cellular chromosomal DNA (– – –) to form the provirus. Viral RNA is transcribed from promoter sequences in the 5' LTR, processed at 3' RNA processing signals in the 3' LTR, and encapsidated in viral proteins. The encapsidated viral RNA buds from the cell to give progeny virus particles.

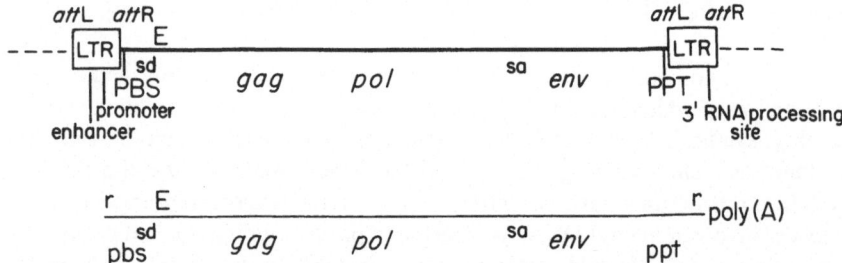

FIGURE 2. DNA and RNA genomes of C-type retroviruses. The viral DNA contains attachment sequences (*att*L, *att*R), encapsidation sequences (E), enhancer and promoter sequences in the long terminal repeats (LTR), a primer binding site (PBS), splice donor sequences (sd), viral coding genes (*gag, pol, env*) splice acceptor sequences (sa), a polypurine tract (PPT), and a 3' RNA processing site in the LTRs. The viral RNA contains a copy of the DNA missing most of one copy of the LTR and having a small repeat (r).

2.2. Entrance and Control of Virus Host Range

Retroviruses have been isolated from many different classes of vertebrates. However, any one species of retrovirus usually has a fairly limited host range. For example, the most common group of mouse retroviruses, ecotropic murine leukemia viruses (MLV), replicate well only in mouse and rat cells, and the most common species of avian retroviruses, avian leukosis viruses [e.g., Rous sarcoma virus (RSV)], replicate well only in chicken and quail cells. But this limited host range is not characteristic of all retroviruses. Another group of mouse retroviruses, amphotropic MLVs, replicate well in cells from several other mammalian species as well as mouse, and another species of avian retroviruses, reticuloendotheliosis virus (Rev), replicates well in all avian cells tested and in dog cells (Embretson and Temin, 1986).

The primary determinant of retrovirus host range appears to be the envelope glycoproteins (Fig. 3). For example, there are several subgroups of avian leukosis

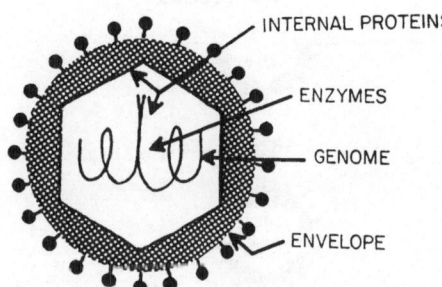

FIGURE 3. Schematic drawing of a retrovirus virion. The outer envelope contains products of the *env* gene. The inner hexagon contains products of the *gag* gene as structural components and products of the *pol* gene as enzymatic components.

viruses with different host ranges on chicken cells of different genotypes. These avian leukosis virus subgroups differ only in their *env* gene, the gene that codes for the virion glycoproteins. Similarly, there are three subgroups of MLVs, eco-, xeno-, and amphotropic, with different host ranges on mammalian and avian cells corresponding to their different *env* genes. However, the envelope-specific control of host range can be abrogated for one virus cycle by fusion of virions to resistant cells using inactivated Sendai virus or polyethylene glycol.

In addition, there are other determinants of retrovirus host range. For example, the transcriptional elements of Revs are inefficient in mouse cells, in which the viruses grow poorly or not at all (Embretson, personal communication). Furthermore, there are tissue-specific controls on the transcriptional activity of some retroviruses (see Section 2.4).

Other controls on retrovirus host range exist, such as the effect of the *Fv-*1 gene on MLV integration and the lack of processing of RSV proteins in mammalian cells. However, these controls are not relevant to retrovirus vectors, since they are easily avoided in vector construction.

2.3. Replication of Retrovirus Nucleic Acids

After entrance into and uncoating in sensitive cells, the viral reverse transcriptase copies the viral genomic RNA into linear unintegrated viral DNA (Fig. 1) (see Fig. 2 for genome structure and Fig. 3 for virion structure). The reverse transcriptase uses a cellular transfer RNA (tRNA) (packaged with viral RNA in the virion) as a primer for minus-strand DNA synthesis [bound at PBS (Fig. 2)] and a section of viral genomic RNA near the 3' end (PPT), as a primer for plus-strand DNA synthesis. The linear unintegrated viral DNA has a long terminal repeat (LTR) at each end, while the viral RNA has a small terminal repeat at each end (r), ignoring the polyadenylic acid [poly (A)] of the viral RNA (Fig. 2). (For more complete descriptions of retrovirus DNA synthesis, see Temin, 1981, 1982; Varmus, 1982.)

The linear unintegrated viral DNA is transported to the cell nucleus, where it is ligated without loss or gain of nucleotides to form closed circular viral DNA. The closed circular viral DNA is the precursor to the integrated viral DNA, the provirus. It is integrated into cellular DNA by means of a virus-coded protein that acts at the nucleotide sequence formed by the ligation of the ends of the unintegrated viral DNA, the attachment sequence (*att*) (Panganiban, 1985). There is no apparent specificity of retrovirus integration with respect to cellular DNA, although there is strict sequence specificity as far as the attachment sequence of viral DNA. Proviruses are sometimes found clustered near certain genes in tumors. However, this clustering appears to be the result of selection for integration at that position, rather than site-specific integration; that is, the tumor results from activation of a proto-oncogene by the integration (also see

Chapter 12). Transcription of the provirus by cellular enzymes using signals in the LTRs results in progeny viral RNA.

2.4. Control of Expression

Retroviruses use several mechanisms to express their genes (Fig. 4). The *gag* gene product is translated into a precursor polyprotein from full-length viral RNA. The *gag* precursor polyprotein is processed after budding by posttranslational cleavage to give the internal virion proteins (and in some viruses the protease used for polyprotein processing). The *pol* gene product is translated as part of a *gag–pol* fusion precursor polyprotein from full-length viral RNA. This polyprotein is processed by posttranslational cleavage to give the viral enzymes: reverse transciptase, integrase, and, in some viruses, the processing protease. The *env* gene product is translated from spliced viral RNA into a precursor polyprotein that is processed by glycosylation and posttranslational cleavage to give the virion envelope glycoprotein and (in different retrovirus species) a transmembrane protein or second glycoprotein. In RSV, the product of a fourth

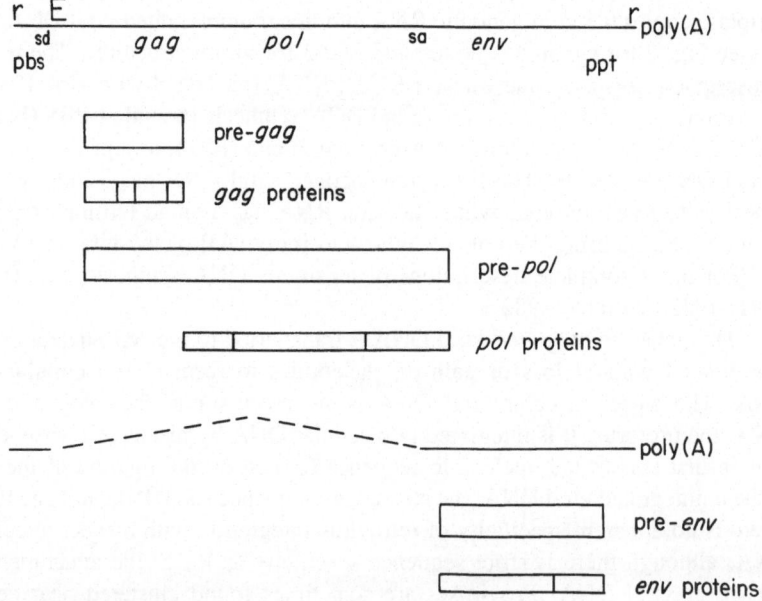

FIGURE 4. Expression of retrovirus RNAs and proteins. Precursor proteins for *gag* and *gag–pol* proteins are translated from full-length viral RNA transcribed from the provirus. Viral proteins are translated from a spliced subgenomic messenger RNA into a precursor protein for *env* proteins. [poly(A)] Polyadenylic acid.

gene, *src* (3' to *env*), is translated from a second smaller spliced viral RNA into a protein responsible for cellular neoplastic transformation.

For retrovirus vectors, the important controls so far studied are those at the level of transcription from the LTR and at the level of splicing. Both processes are affected by the nature of the nucleotide sequences present in the vector.

In cells that produce replication-competent retroviruses, between one half and one tenth of the viral RNA is spliced, depending on the virus species. For avian and mouse leukemia viruses, the value is one half; for Rev's it is one tenth (Quintrell *et al.*, 1980; Mann and Baltimore, 1985; C. K. Miller and Temin, 1986). The replication-defective, highly oncogenic retrovirus Rev strain T expresses its oncogene from a spliced messenger RNA (mRNA). Studies with this virus indicate that sequences in the intron of the spliced mRNA, i.e., in *gag* and *pol,* control the extent of splicing (C. K. Miller and Temin, 1986). Studies with transfection of incomplete MLV genomes suggest that sequences in the intron are needed for splicing to occur (Hwang *et al.*, 1984), but others disagree with this conclusion (Mann and Baltimore, 1985).

The retrovirus LTR contains typical RNA polymerase II promoter sequences. The LTR usually also contains enhancer sequences. The overall level of transcription from a retrovirus provirus is affected by the presence or absence of the enhancer sequences. For example, the avian leukosis virus Rous-associated virus 1 has an enhancer, while Rous-associated virus 0 does not. There is a tenfold difference in transcription between the two otherwise similar viruses, and Rous-associated virus 1 causes cancer, while Rous-associated virus 0 does not (Cullen *et al.*, 1985; Weber and Schaffner, 1985). The tissue specificity of transcription is also apparently controlled by the nature of the enhancer. For example, there are several strains of MLV that have different tissue specificities for transcription and cause different leukemias that can be correlated with the presence of different enhancers (Clark *et al.*, 1985; Holland *et al.*, 1985). Retrovirus enhancers can also be hormone-sensitive; e.g., the mouse mammary tumor virus LTR is stimulated by glucocorticoids. In addition, transcription from the LTR of human T-lymphotropic viruses is stimulated by virus-coded proteins (Felber *et al.*, 1985). Furthermore, MLVs are not expressed in embryonal carcinoma cells as a result of methylation or other processes (see Section 5.1).

Thus, a variety of different controls on expression are available for use in retrovirus vectors.

2.5. Encapsidation and Budding

2.5.1. Encapsidation

Some full-length retroviral RNA is encapsidated by *gag*-gene proteins interacting with a specific region in the viral RNA near the primer binding site. This region is called *E* or *psi* in the case of Rev and MLV, respectively (Watanabe and Temin, 1982; Mann *et al.*, 1983) (spliced mRNA is, however, not encapsidated as a result of deletion of this *E* region, which is present in the intron). In avian leukosis viruses, the location of the *E* region is not quite so clear. It has been variously reported to be in a similar position to the Rev and MLV *E* regions and, additionally in the *gag* gene 3' to the splice donor sequence and also near the 3' LTR (see references in Norton and Coffin, 1985). However, an RSV-based vector lacking the hypothesized *gag*-gene encapsidation sequences appears to be normally packaged (Norton and Coffin, 1985). It also has not been tested whether mammalian retroviruses have additional *E* sequences in the LTR.

2.5.2. Helper Cells

With knowledge of the location of the *E* sequences, helper cells for defective retroviruses have been made for MLV and Rev. These cells constitutively express all virus proteins, but do not produce infectious virus themselves as a result of the absence of the *E* sequences in the viral nucleic acids. When the helper cells are transfected with retrovirus genomes containing appropriate *E* sequences, these genomes are packaged with proteins provided by the helper cell (Fig. 5). The virus produced from the helper cells is called "helper-free," since it contains no replication-competent helper virus. Thus, virus produced from a helper cell can go through only one cycle of virus infection.

Helper cells have been made with ecotropic and amphotropic MLV and with Rev (Mann *et al.*, 1983; Watanabe and Temin, 1983; Sorge *et al.*, 1984b; A. D. Miller *et al.*, 1985).

2.5.3. Budding

Encapsidated RNA is budded from the plasma membrane at locations at which the membrane is modified by the insertion of viral proteins and the exclusion of host-cell proteins. After budding, there is virion maturation involving proteolytic cleavage of certain viral proteins and changes in morphology of the internal virion core.

2.5.4. Phenotypic Mixing (Pseudotype Formation)

The process of budding is not virus-species-specific, even as far as requiring that the envelope glycoproteins are from the retrovirus family. For example, herpes simplex or vesicular stomatitis virus glycoproteins can be used to bud retrovirus virions. In addition, encapsidation is not completely virus-species-specific, although it is more specific than budding. RNA of a mouse cellular movable genetic element, VL30, is efficiently encapsidated by proteins of a variety of mammalian retroviruses, and an Rev-based vector is efficiently encapsidated by MLV proteins (Embretson and Temin, 1986).

Thus, proteins in a retrovirus virion can be encoded by the genome in the virion (replication-competent retrovirus), by a complementing viral genome (helper replication-competent retrovirus or superinfecting retrovirus), or by a helper cell (cell with viral genomes deleted of encapsidation sequences and constitutively expressing virion proteins) (Fig. 5).

In stocks of most retrovirus vectors, all the virion proteins come from other viral genomes [some replication-competent RSV-based vectors are the only exception (Sorge and Hughes, 1982; Foster and Hanafusa, 1983; Iba *et al.*, 1984)]. With such replication-defective vectors, it has been established that *cis*-acting sequences like *att* are not completely virus-specific; e.g., an Rev-based vector encapsidated by MLV proteins forms a provirus efficiently in dog and rat cells (Embretson and Temin, 1986). In this case, the MLV proteins function with heterologous *cis*-acting Rev sequences.

2.6. Genetic Variation

2.6.1. Mutation

Like all viruses with single-stranded nucleic acid genomes, retroviruses have a very high rate of genetic change (Coffin *et al.*, 1980; O'Rear *et al.*, 1980; Hayashida *et al.*, 1985). Thus, retroviruses, when going through viral infection

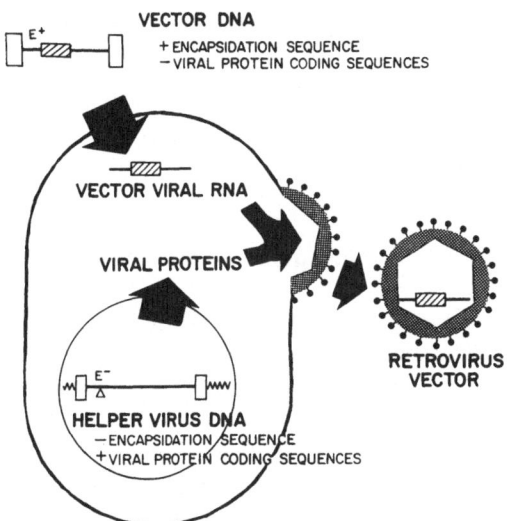

FIGURE 5. Helper cell. Vector virus DNA with an encapsidation sequence and no protein coding sequences is introduced into a helper cell. A helper cell expresses viral proteins, but produces no virus. In the helper cell, the vector DNA is encapsidated and forms infectious virus. However, this virus can undergo only one cycle of infection.

cycles, vary much more frequently than cellular DNA genes. Therefore, when there is selection for a variant, a retrovirus population can change after a few cycles of virus replication (see Section 6.4). The variant viruses can differ by base mutation, deletion, duplication, and insertion. However, retrovirus proviruses do not appear to vary more than cellular genes.

It is not clear whether retroviruses vary more than other single-stranded RNA viruses. The high rate of mutation may reflect only the single-stranded character of the genome in the virus particle. In addition, retroviruses might have an intrinsically higher rate of variation as a result of an unusually error-prone DNA polymerase (Battula and Loeb, 1974; Mizutani and Temin, 1976). In addition, other aspects of the retrovirus life cycle, in particular the proposed movements of the DNA product during synthesis of double-stranded DNA from an RNA template, may result in a high level of mutation (Shimotohno and Temin, 1982b; O'Rear and Temin, 1982).

2.6.2. Recombination

There is a very high rate of recombination during retrovirus replication, especially for different genomes replicating together. In addition, there can also be recombination of infecting virus with resident viral genomes. For example, an infectious virus carrying sequences from an endogenous viral genome was recovered after passage of a replication-defective RSV deleted of its *env* gene (Takeya *et al.*, 1981). The mechanism of these recombinations is not known.

3. NATURAL RETROVIRUS VECTORS

There are several examples of biological systems in which reverse transcription has been used to synthesize DNA that is integrated into chromosomal DNA.

3.1. Highly Oncogenic Retroviruses

Highly oncogenic retroviruses are natural retrovirus vectors that contain viral oncogenes either substituted for or added to the retrovirus genome (for a recent review with numerous references, see Bishop and Varmus, 1985). Most highly oncogenic retroviruses are replication-defective, because of the size limit on infectious retroviruses (see Section 6.2.2) and the nature of the oncogene product, which is often a fusion protein of virus and cell sequences. The viral oncogenes are always expressed from a transcript originating and terminating in the viral long terminal repeats. Viral oncogenes are translated from either full-length or spliced mRNAs, using their own or retrovirus initiation codons. In the latter case, the oncogene product is a fusion protein. The termination codons are derived from the proto-oncogene or the virus sequences (the virus sequences can be out of frame). In the latter case, a fusion protein again results. Viral oncogenes do not contain intervening sequences.

Thus, viral oncogenes contain no transcriptional control elements or other control sequences except possibly an initial splice-acceptor sequence. The cellular proto-oncogenes, from which the viral oncogenes evolved, are typical cellular genes with promoters and 3' RNA processing sites. In addition, most proto-oncogenes contain intervening sequences. All these sequences are lost during the evolution of highly oncogenic retroviruses. As we shall see below, study of retrovirus vectors indicates why there was selection against these sequences in the evolution of highly oncogenic retroviruses.

3.2. Retrotransposons

Although retroviruses were originally the only biological organisms known that use reverse transcription, it is now known that reverse transcription is widely distributed among eukaryotes (for a recent review, see Temin, 1985b). Many eukaryotes, ranging from yeast to *Drosophila* to mouse and to man, contain in their genomes cellular movable genetic elements that use reverse transcription for transposition. These elements have been termed "retrotransposons" (Boeke *et al.*, 1985). Retrotransposons share sequence organization and sometimes sequence similarity with retroviruses, including sequences that apparently code for a reverse transcriptase. However, retrotransposons do not make infectious virus particles and in many cases do not have nucleotide sequences similar to *env* genes, e.g., Tyl and copia (Mount and Rubin, 1985).

The number of copies of retrotransposons in the mammalian genome is large, perhaps 1500 per haploid genome in mouse and in man, representing perhaps 1.5% of these genomes. Thus, integration of reverse transcripts in many locations is not incompatible with organismic survival. However, integration of retrotransposons, like integration of retroviruses, can be mutagenic (Hawley *et al.*, 1982). Thus, during evolution, the formation of new copies of retrotransposons must have led to a large deleterious mutational load (Temin, 1985b).

3.3. Retrotranscripts

Another type of product of reverse transcription is found in the vertebrate genome. These sequences bear some mark of reverse transcription, such as loss of intervening sequences and/or presence of 3' poly(A) sequences as well as some marks of transposition such as different locations in different cells and the presence of surrounding direct repeats (see Temin, 1985b). These sequences include complementary DNA (cDNA) or processed pseudogenes and Alu sequences. They are called "retrotranscripts" to differentiate them from retrotransposons. Retrotranscripts do not have coding sequences for reverse transcription or integration functions.

There are many copies of retrotranscripts in the mammalian genome, at least 500,000 per haploid genome in mouse and man, representing about 10% of these genomes. The presence of retrotranscripts has the same evolutionary implications as the presence of retrotransposons, although retrotranscripts might be less mutagenic than retrotransposons because of the absence of control sequences.

4. TYPES OF RETROVIRUS VECTORS

4.1. Classification

All retrovirus vectors have at a minimum two long terminal repeats (LTRs) containing at least a repeated (R) sequence and attachment sequences (attR, attL), a primary binding site (PBS), polypurine tract (PPT), and an encapsidation (E) sequence (Fig. 6 and Section 2). Thus, they can produce RNA that can be packaged, reverse-transcribed, and integrated. Retrovirus vectors can be classified into types according to the coding and internal controlling sequences they contain. In addition, retrovirus vectors contain different numbers of coding and control sequences organized in different ways (for recent reviews, see Temin, 1985a; Coffin, 1985).

I have classified retrovirus vectors in terms of the number of inserted genes and the number and types of control sequences (Fig. 7). In Fig. 7, all the sequences are oriented in the direction of LTR transcription, i.e., from left to right. All the LTRs contain promoters except for the deleted promoter in the right LTR in vector 8 (the promoters in the left LTR could theoretically be

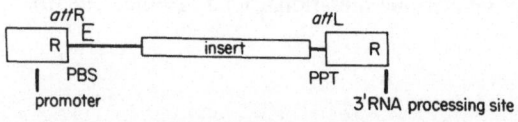

FIGURE 6. Minimal sequences required in a retrovirus vector. The minimal sequences required for a retrovirus vector include a promoter in the 5' long terminal repeat ((LTR), a repeated (R) sequence in both LTRs, a primary binding site (PBS) and polypurine tract (PPT), and one copy of each attachment site (attR, attL), an encapsidation (E) sequence, an insert to bring the virus size above the minimum needed, and a 3' RNA processing site in the 3' LTR.

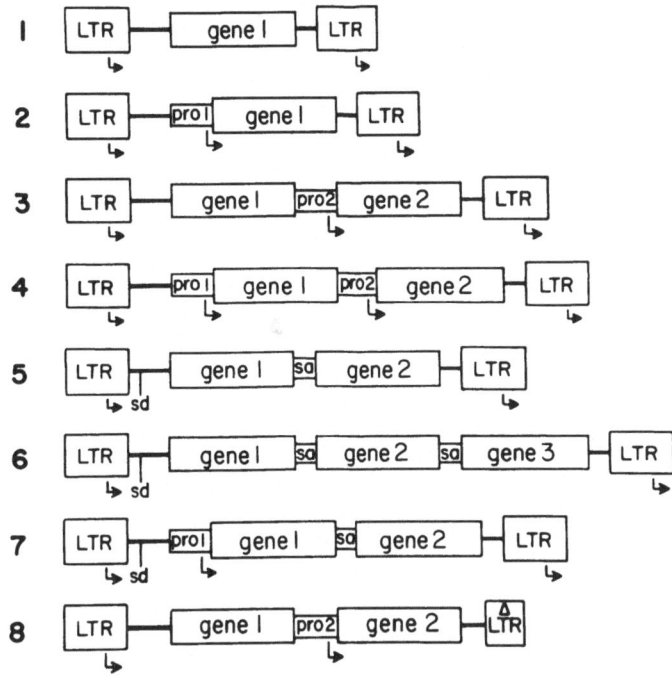

FIGURE 7. Types of retrovirus vectors (+ orientation). (pro) Internal promoter; (sa) splice acceptor sequence; (sd) splice donor sequence.

replaced with heterologous promoters). All the vectors have at least *att*R in the left LTR and *att*L in the right LTR, as well as PBS and E next to the left LTR and PPT next to the right LTR (Fig. 6). The subclassification is as follows:

1. LTR–gene 1–LTR
2. LTR–promoter 1–gene 1–LTR
3. LTR–gene 1–promoter 2–gene 2-LTR
4. LTR–promoter 1–gene 1–promoter 2–gene 2–LTR
5. LTR–splice donor–gene 1–splice acceptor–gene 2–LTR
6. LTR–splice donor–gene 1–splice acceptor–gene 2–splice acceptor–gene 3–LTR
7. LTR–splice donor–promoter 1–gene 1–splice acceptor–gene 2–LTR (and other complex constructions, e.g., with a promoter for gene 2)
8. LTR–gene 1–promoter 2–gene 2–deleted LTR (and other constructions with a deleted 3′ LTR)

All these vectors can be constructed with some of the internal sequences

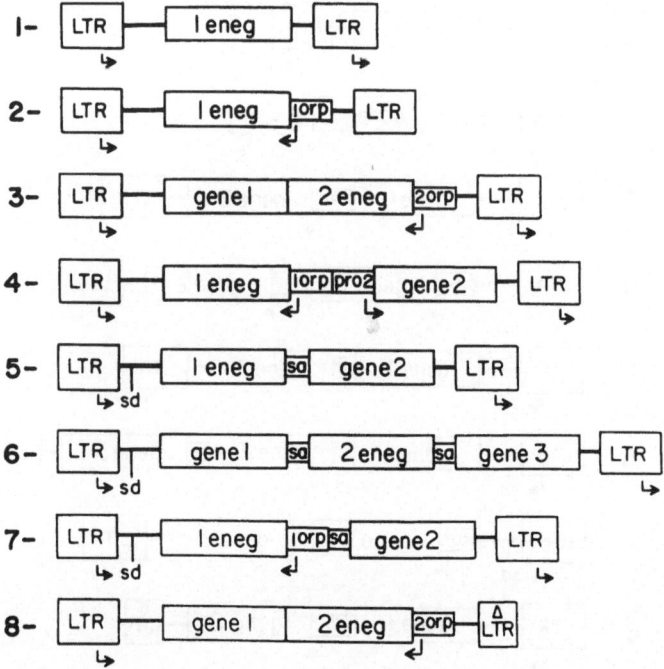

FIGURE 8. Types of retrovirus vectors (− orientation). Some of the possible structures are given. The sequences in opposite orientation are written backwards; e.g., 1 eneg is gene 1 in opposite orientation (cf. Fig. 7 caption).

oriented opposite to the orientation of LTR transcription, i.e., in a minus orientation (Fig. 8 presents some possible examples). They are then called $1-$, $2-$, etc.

A detailed description of each of these types of vectors is given in Section 7 after a general discussion of uses and problems of retrovirus vectors.

4.2. Helper-Free Retrovirus Vectors

With the availability of helper cells, any of these types of vectors can be prepared in the absence of helper virus (see Fig. 5). In such cases, there is no replication-competent retrovirus present, so only one cycle of infection by the vector can occur. It is easier to prepare a stock of helper-free virus when the vector contains a gene that can be selected in the helper cell. However, when no such gene is present in the vector, it can be cotransfected into helper cells with a plasmid containing a selectable marker. Stocks of helper-free virus often have a low titer. However, suggestions have been made of ways to increase the titles (discussed in Section 5.1). Helper-free stocks of spleen-focus-forming virus have been used to transform erythroid cells *in vivo* (Wolff and Rusceltin, 1985; Bestwick *et al.,* 1985; Palacios *et al.,* 1985).

5. USES OF RETROVIRUS VECTORS

5.1. General

There are many possible retrovirus vectors. For example, it is possible to vary the virus species of long terminal repeat (LTR) and encapsidation (E) sequences, the vector type (1–8), the nature of any internal promoters and splice sites, and the species of the virion proteins. The choice of any specific vector depends on the particular use desired. In addition to the vectors already discussed, vectors could be of type 3 with an inducible promoter, could have heterologous E sequences, or could have a tissue-specific enhancer or heterologous promoter in the LTR. The existence of all these possibilities suggests that ultimately a retrovirus vector could be made to target any cell and obtain tissue-specific regulation in that cell.

Since there are many different ways to obtain apparently the same effects with retrovirus vectors, it is probably best when considering the initial use of retrovirus vectors to start by constructing two or more different types of vectors and determining which is best for the particular purpose. For example, if one is interested in expression of a gene in a vector with a selectable marker, either type 3 or 5 vectors could be used with the selectable marker as either gene 1 or gene 2. Since vectors made from viruses of different species are already available, it is relatively easy to insert the gene of interest into such vectors or into such vectors with one or another selectable marker. For example, to study expression of the *src* oncogene, Tarpley and Temin (1984) inserted the oncogene 5′ or 3′ of the *tk* gene into a reticuloendothelosis virus (Rev)-based type 2 vector to give type 7 vectors.

A large variety of different bacterial, viral, and animal cellular genes have been expressed in retrovirus vectors. There are no reports of genes that cannot be expressed in retrovirus vectors. However, Tarpley and Temin (1984) reported that virus derived from a type 7 Rev-based vector was produced in very low yield because of toxicity as a result of high expression of the *src* oncogene from the LTR. When expression is from the viral LTR, high levels of the gene product can be found, since retrovirus LTRs are usually highly transcribed. In a comparative study, infection with a type 5 murine leukemia virus (MLV) vector gave higher expression of gene 2 than did transfection of parallel cells with the same construct (Hwang and Gilboa, 1984). However, when there are cell-specific controls on transcription from the LTR, e.g., MLV in embryonal carcinoma cells, no gene expression from the LTR is seen (Rubenstein *et al.*, 1984; Sorge *et al.*, 1984a; Taketo *et al.*, 1985).

Two or more genes can be expressed from the retrovirus LTR in splicing vectors (types 5 and 6). However, the distal gene (gene 2) may be expressed at low levels or not at all as a result of suppression of splicing by the sequences in gene 1 (see Section 6.3.6).

Alternatively, two or more genes can be expressed in one retrovirus vector by the use of internal promoters. This type of construction (types 3, 4, and 7) allows the presence of promoters with different properties of cell specificity and inducibility. However, there can be some suppression between two promoters in one provirus that can reduce the expression of the genes (see Section 6.3.4). Since the suppression is less than an order of magnitude, it may not always be a problem.

Tissue-specific expression of genes in a retrovirus vector can be obtained. For example, Rubenstein *et al.* (1984) and Wagner *et al.* (1985) found that in mouse and embryonal carcinoma (EC) cells infected with type 1 and 2 MLV vectors, only the gene in the type 2 vector expressed from an internal promoter was expressed (see Seliger *et al.*, 1986). Advantage was taken of this lack of expression from the LTR to select rare active integrations of type 1 and 5 vectors (Sorge *et al.*, 1984a; Taketo *et al.*, 1985). In addition, Episkopou *et al.* (1984) reported cell-specific expression of an insulin promoter in a type 4 MLV vector, and Miller *et al.* (1984b) reported inducible expression of a rat growth hormone minigene in a type 3 MLV vector.

Various means have been used to increase expression of genes in retrovirus vectors. Among these methods are the use of efficient internal promoters and increasing the copy number of vector DNA. To increase copy number, two methods have been used: either a vector containing polyoma or simian virus 40 (SV40) replication origin sequences in the presence of polyoma or SV40 early proteins or a vector containing dihydrofolate reductase (DHFR) genes with subsequent stepwise amplification after methotrexate selection. With these methods, over 100 copies per cell can be obtained. Examples are MLV type 1 vector with polyoma early region outside LTRs in plasmid (Mann *et al.*, 1983); MLV type 5 vector with both SV40 and pBR322 origin regions between LTRs (Cepko *et al.*, 1984); MLV type 2 − vector with polyoma virus origin (Berger and Bernstein, 1985); and MLV type 2 vector with DHFR (Miller *et al.*, 1985).

In all the usual retrovirus vectors (types 1–7), the enhancer–promoter in the LTRs is present in the provirus. These sequences can affect the expression of internal or external promoters by suppression or enhancement. The *U3* region in the LTRs of the provirus (the region that contains the promoter and enhancer sequences) is formed from sequences at the 3′ end of viral RNA and therefore from the 3′ LTR of the parental provirus (Fig. 9). It is therefore possible to construct a retrovirus vector consisting of a provirus with a normal 5′ LTR enhancer–promoter and a deleted 3′ LTR missing the enhancer–promoter sequences, but retaining the *att*L sequences (see Fig. 7, type 8). On infection, a new provirus will be formed that lacks promoter–enhancer sequences in its LTRs (Fig. 9).

Another possibility for variation in retrovirus vectors is to substitute a heterologous promoter for the promoter in the 5' LTR of the original construction. Such a substitution would reduce the possibility of recombination with a deleted 3' LTR and could give higher or differently controlled production of the original retrovirus vector.

5.2. Studying Genes in Cell Culture

Most of the work with retrovirus vectors has involved expression of genes in cell culture, since cell cultures are easier to analyze than whole animals. In particular, promoter specificity and effects of high levels of expression of genes have been studied. As discussed in Section 5.1, both the rat insulin and growth hormone genes appear to be normally inducible in retrovirus vectors, and the MLV promoter is inactive in EC cells. Most work with oncogenes and proto-oncogenes is also related to this section. For example, expression of high levels of the c-*src* proto-oncogene in a type 6 Rous sarcoma virus (RSV) vector did not result in transformation of chicken embryo fibroblasts (Iba *et al.*, 1984). Higher expression of genes can be secured by using more active promoters or amplifying the number of copies of the vector DNA (Section 5.1).

5.3. Expressing Antisense RNA

The presence of antisense RNA in bacterial and animal cells can be used to inhibit activity of the homologous gene (Izant and Weintraub, 1985). Although

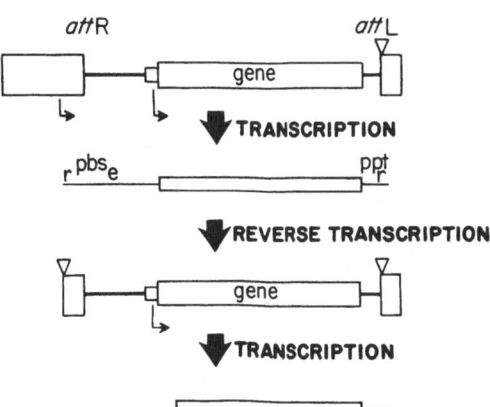

FIGURE 9. Replication of type 8 vector with deleted LTR (▽). Deletion in the 3' LTR that is present in both LTRs of progeny virus.

retrovirus vectors have been made with expression of antisense RNA resulting in reduction of gene activity (see Section 6.3.5), they have not yet been exploited to inhibit the activity of a homologous gene. As discussed in Section 4.1 (see Fig. 8), a great many different types of antisense retrovirus vectors can be designed.

5.4. Making Complementary DNA Genes and Other DNA Recombinants

Since genes with introns passaged in retrovirus vectors have their introns spliced out, such vectors provide a means to form a cDNA gene in a retrovirus vector (Shimotohno and Temin, 1982a; Sorge and Hughes, 1982). This approach has been especially useful with virus transforming genes (Donoghue *et al.*, 1984; Kriegler *et al.*, 1984; Cepko *et al.*, 1984; Iba *et al.*, 1984; Sylla and Temin, 1986). When the transforming regions of polyoma, SV40, or adenoviruses were placed in MLV vectors and the vectors were passaged, viruses expressing only a single DNA virus transforming protein were recovered. When c-*src* or c-*rel* genes with many introns were placed in RSV or Rev vectors, viruses were recovered with the introns removed and expressing the proper protein. When combined with shuttle vectors (see Section 5.5), these constructions could provide a general method to prepare cDNA genes. For example, Takeda *et al.* (1985) made a chimeric human–mouse hybrid immunoglobulin gene with a type 3 Rev vector. They put segments of a human and a mouse germ-line immunoglobulin gene in the vector and obtained a hybrid cDNA gene after passage.

In addition, type 5 MLV vectors have been used to introduce immunoglobulin genes in germ-line configurations into lymphoid cells to study the nature of gene rearrangement in development of somatic immunoglobulin genes (Lewis *et al.*, 1984, 1985). Introns were also lost in these constructions.

5.5. Retrovirus Shuttle Vectors to Tag and Clone Cellular Genes

Several MLV vectors have been constructed containing polyoma or SV40 origins, pBR322 origin, and/or a suppressor tRNA to aid in DNA cloning and to make shuttle vectors to use to recover the vector or surrounding cellular sequences or both (Cepko *et al.*, 1984; Lobel *et al.*, 1985; Reik *et al.*, 1985; Weis *et al.*, 1984; Berger and Bernstein, 1985; King *et al.*, 1985). Since retrovirus proviruses apparently integrate randomly, they can be used to tag chromosomal locations. The process is analogous to transposon tagging in bacteria or tagging with P factor in *Drosophila* (Kleckner *et al.*, 1977; Rubin and Spradling, 1982) (also see Chapter 12). For example, King *et al.* (1985) made hypoxanthine phosphoribosyl transferase (HPRT) mutations in EC cells after infection by an MLV vector. If the cells that contain the retrovirus vector integrated in a desired location can be selected, then the integrated provirus and surrounding cellular DNA sequences can be cloned using the provirus as a marker, and the cellular gene sequences can be recovered. For example, the mouse major histocompatibility complex was marked in this way by a retrovirus vector containing the *neo* gene (Weis *et al.,* 1984). Tagging of the collagen gene occurred by chance after injection of replication-competent MLV in mouse embryos when the collagen gene was mutated by MLV integration (Harbers *et al.,* 1984).

5.6. Possible Uses of Retrovirus Vectors in Animals

5.6.1. Somatic-Gene Therapy with Retrovirus Vectors

A beginning has been made in the use of retrovirus vectors in animals. The ultimate aim of this experimentation is somatic-gene therapy in humans (for reviews, see Anderson, 1984, 1985). The general idea is to infect bone-marrow or skin cells in culture and then reimplant the infected cells in an irradiated person. Some model experiments have been carried out with mice [type 2 − MLV with SV40 early promoter-*neo* (Joyner *et al.,* 1983), type 5 MLV-DHFR-*neo*-pBR322 origin (Williams *et al.,* 1984), type 1 MLV-HPRT (A. Miller *et al.,* 1984a), type 2 − MLV-*neo* (Dick *et al.,* 1985); type 1 MLV-*neo* (Eglitis *et al.,* 1985; Keller *et al.,* 1985)]. For success, there is need for efficient infection of most of the pluripotent stem cells in the bone marrow, high expression of the desired gene, no expression of other vector genes in the stem cells or the differentiated cells, stability of the vector in the infected cells, no activation of cellular proto-oncogenes or other cellular genes, and no recombination to give virus that could infect the patient's germ line or other people.

Such somatic-gene therapy could be useful in treating single-gene human diseases. There are 2000 known such single-gene diseases. Of them, 300 have been biochemically characterized (for further discussion and references, see Anonymous, 1984). It is estimated that 1–2% of newborns are affected with such diseases. The present therapies include modification of the physical environment, modification of diet, modification of metabolism, injection of proteins, transfusion, and cell transplantation. In many cases, these therapies are ineffectual. Thus, there is a desire to develop a somatic-gene therapy in which a wild-type copy of a gene could be expressed in the patient's cells even while the defective genes remain.

The diseases for which such therapies are being most discussed now are adenine deaminase deficiency (40 cases present in the world), purine nucleoside phosphorylase deficiency (9 cases present in the world), and Lesch–Nyhan disease (HPRT deficiency, 200 born per year in the United States) (See Willis *et al.*, 1984; and Friedman, 1985, for example.)

Many questions are being raised by the prospect of such therapy. These include questions about safety (will there be induction of mutations, of tumors, of germ-line infection, of infectious virus?), efficacy (what stages of development should be treated? what is the efficiency of the treatment?), cost, effects on evolution, and general social effects such as increasing the risk of human germ-line therapies. A committee of the National Institutes of Health has been considering these and other questions since 1984 (Wyngaarden, 1985).

5.6.2. Infection in Animals

Helper-free retrovirus vectors have been used to study transformation in mice (Wolff and Ruscetti, 1985). There is also discussion of the use of retrovirus vectors for infection of the germ line of farm animals. This approach is analogous to that already taken for this purpose with DNA microinjection (Hammer *et al.*, 1985). The problems of this approach are similar to those discussed above—the need for high titers of helper-free virus, efficient infection of germ-line cells, expression of the desired gene in the appropriate cells, and no activation of proto-oncogenes or other undesirable genes, or inactivation of necessary genes, or formation of replication-competent retrovirus.

In one approach, Stuhlman *et al.* (1984) injected type-1-MLV-*gpt*-infected cells into postimplantation mouse embryos along with replication-competent MLV and found vector expression in several tissues. However, expression of the helper virus was very much greater. Huszar *et al.* (1985), Jahner *et al.* (1985), and Rubenstein *et al.* (1986) have also used MLV vectors to secure germ line infection.

6. PROBLEMS OF RETROVIRUS VECTORS

6.1. General

As a consequence of the properties of retroviruses discussed above, there is strong selection in infectious retroviruses for particular types of sequence organization and for the absence of certain sequences. In addition, the level of expression of genes with internal promoters relative to the level of virus production varies with the details of the vector construction, often in an unpredictable fashion. I shall discuss each of these problems separately, taking examples from different vectors where appropriate.

6.2. Size

There appears to be a maximum limit on the size of the RNA in infectious retrovirus virions. [The size of the viral RNA is not the same as the size of the viral DNA because of the larger size of the long terminal repeat (LTR) compared to the terminal repeats in the RNA.] The possible splicing of introns present in vector DNA (see Sections 5.4 and 6.3.6) reduces the size of the virus stock during passage. However, introns that contain encapsidation sequences (see Section 2.5.1) cannot be lost because RNA without encapsidation sequences is not packaged. In addition, it is conceivable that there are differences in the size limits for different retrovirus species.

6.2.1. Minimum Size

There is some experimental difficulty in determining the minimum size of the RNA of a retrovirus vector because of the requirement for a selectable marker to mark the virus. The smallest highly oncogenic retrovirus is FBJ-murine sarcoma virus, which is 3.4 kilobases (kb) (RNA) (Curran *et al.*, 1982). However, a deleted variant of reticuloendotheliosis virus (Rev) strain T, which is 2.8 kb (RNA), can be passaged (C. K. Miller and Temin, 1986). The smallest murine leukemia virus (MLV)-based construction reported is in Perkins *et al.* (1983). Their construction is type 1 and contains a bacterial *gpt* marker. The RNA transcribed from the construct is 2.3 kb. Infectious virus could be rescued from cells transfected with this construct, but the size of the RNA in the rescued virus was not determined. Thus, it is not possible to tell whether or not recombination with the helper virus occurred. An Rev-based type 1 vector that is 1.8 kb can be passaged successfully (Dougherty and Temin, 1986), indicating that the lower limit is less than 2 kb. Therefore, it is unlikely that the lower limit will be a problem in any retrovirus vector constructions.

6.2.2. Maximum Size

Very large molecules of unintegrated viral DNA have been seen in Southern blots of extracts from cells infected with Rev-based vectors (Emerman and Temin, 1984a). However, large viruses were not isolated. The large molecules may represent aberrant by-products of viral replication that cannot be passaged.

Attempts to put additional sequences into full length replication-competent retroviruses uncovered problems in recovering infectious virus that is larger than wild-type (Shimotohno and Temin, 1981; Sorge and Hughes, 1982). A systematic attempt to make large replication-competent viruses by insertion of genes into

Rev-based vectors of types 5 and 6 demonstrates that 9–10 kb is a maximum size for such replication-competent viruses (Gelinas and Temin, 1986). However, some larger replication-defective constructions are possible. Viruses larger than this size limit are usually rapidly overgrown by smaller deleted variants.

The molecular basis for this size limit is unknown. It could be related to problems of packaging viral RNA or to problems in reverse transcription or both.

Vector constructs that contain genes with introns can be used successfully as a result of the splicing out of the intron sequences (Iba *et al.*, 1984; Sylla and Temin, 1986).

6.3. Internal Control Sequences

6.3.1. General

There are certain potential general problems with all retrovirus vectors, except possibly type 1 vectors. These potential problems are related to the presence of internal control sequences and resulting problems of instability and lack of expression. The problems of instability and lack of expression are related in that selection for expression of an internal gene can lead to selection of deleted viruses. The effects of internal control sequences depend on the relative strengths and orientations of the different control sequences.

6.3.2. 3' RNA Processing Site (+ Orientation)

The presence in a retrovirus vector of a 3' RNA processing site [poly(A) addition site] from a gene or a cDNA decreases or abolishes virus production. This problem has been reported for herpes simplex virus thymidine kinase gene in type 2 Rev and MLV vectors (Shimotohno and Temin, 1981; Joyner and Bernstein, 1983a; but see also Tabin *et al.*, 1982) human hypoxanthine phosphoribosyl transferase cDNA in a type 1 MLV vector (A. D. Miller *et al.*, 1983), simian virus 40 (SV40) early gene 3' sequences in a type 2 MLV vector (Joyner *et al.*, 1983a; Kriegler *et al.*, 1984), and c-Ha-*ras* gene in a type 3 Rev vector (Wilhelmsen *et al.*, 1984). No effect of the presence of the 3' RNA processing site of a chicken thymidine kinase gene in a type 2 Rev-based vector or the presence of the cDNA copy of the 3' end of rat growth hormone mRNA in a type 3 MLV-based vector was seen (Bandyopadhyay and Temin, 1984b; A. D. Miller *et al.*, 1984b).

It is clear that in most cases, the presence of a 3' RNA processing site prevents or reduces the formation of infectious virus. The simplest hypothesis to explain this result is that viral transcripts are processed at the internal 3' RNA processing site, and thus the proper 3' viral sequences for reverse transcription and integration are not present. Although these results also suggest that in some circumstances 3' RNA processing sites are not a problem for the formation of infectious virus, it is better to delete them from retrovirus vectors.

6.3.3. 3' RNA Processing Site (− Orientation)

The problem of a 3' RNA processing site in the opposite orientation (−) has been little studied. No effect of the 3' RNA processing site of a chicken thymidine kinase gene in the opposite orientation was seen in a type 2 − Rev vector (Bandyopadhyay and Temin, 1984b) (none had been seen in the + orientation either). Joyner and Bernstein (1983a) reported that SV40 early gene 3' RNA processing sequences in a type 2 − MLV vector were deleted in formation of a vector by recombination after transfection. However, this result could be due to the construction of the vector by recombination after transfection. No effect was seen resulting from the presence of the cDNA copy of the 3' end of rat growth hormone mRNA in a type 3 MLV-based vector (Miller *et al.*, 1984b) (none had been seen in the + orientation either).

6.3.4. Promoter (+ Orientation)

Two kinds of problems have been reported with internal promoters in type 2, 3, or 4 Rev-based vectors. Both appear to be related to suppression between nearby promoters, but differences in promoter strength may also be involved.

When selection was made for expression of gene 2 in a type 4 vector, deletion of promoter 1 was always found, and when selection was made for expression of gene 1 in type 2 vectors, deletion of promoter 1 was usually found (Emerman and Temin, 1984a; Bandyopadhyay and Temin, 1984b). Thus, if the LTR promoter is much stronger than the internal promoter(s) or if the suppression between nearby promoters is sufficiently strong, spontaneous deletions of the internal promoter are selected. (However, the LTR promoter is not deleted. If there are deletions of the LTR promoter, no infectious virus can be produced.)

When selection is made for gene 2 in type 3 vectors, activity of gene 1 is partially suppressed (Emerman and Temin, 1984b). This suppression can be large. For example, dog cells were infected by type 3 virus (Rev LTR–*tk*–tk promoter–*neo*) and were selected for activity of gene 2 (*neo*). When the cells were superinfected with helper virus, no virus could be recovered, presumably as a result of suppression of the LTR promoter (Embretson, personal communication).

Detailed study of this suppression phenomenon has shown that it is reversible, epigenetic, *cis*-acting, and at the level of transcription (Emerman and Temin, 1984b, 1986). The magnitude of the suppression is from five- to tenfold, but it depends on the relative strengths of the different promoters. Whether or not a biological effect of the suppression, e.g., cell transformation, is seen depends on the sensitivity of the biological assay.

In type 3 or 4 vectors with one selected and one nonselected marker, this suppression can be a problem—selecting for the presence of the vector by selecting for the selectable marker may result in low expression of the unselected gene (see A. D. Miller *et al.*, 1984b). However, with strong promoters, this problem can be minimized.

When the unselected gene is gene 2, its transcription may also be affected by transcription from the upstream promoter. It was reported that there was suppression of transcription from the 3' LTR by transcription from the 5' LTR during transfection of a retrovirus construct with genes expressed from both LTRs (Cullen *et al.*, 1984).

6.3.5. Promoter (− Orientation)

In all cases studied, presence of a promoter in the opposite orientation in a retrovirus vector reduces virus yield. This reduction occurred with type 2− Rev vectors with herpes simplex virus thymidine kinase or chicken thymidine kinase gene promoters (Shimotohno and Temin, 1981; Bandyopadhyay and Temin, 1984b), type 4− MLV vector with herpes simplex virus thymidine kinase promoter (Episkopou *et al.*, 1984), type 3− MLV vector with rat growth hormone promoter (A. D. Miller *et al.*, 1984b), and type 2− MLV vector with SV40 promoter (A. D. Miller *et al.*, 1985). In one case, it was shown that deletion of the promoter in the opposite orientation restored the virus yield (Bandyopadhyay and Temin, 1984b). In one case of a type 4− MLV vector, there was also low expression of gene 1− (Episkopou *et al.*, 1984).

Use of an inducible or tissue-specific promoter might alleviate this problem, allowing virus production in a cell in which the LTR is active and the internal promoter is not and then expression of the internal promoter and not the LTR in an infected cell in which the internal promoter is active and the LTR promoter is not.

6.3.6. Splice Sites (+ Orientation)

Full-length retroviral RNA is used as mRNA for *gag* and *pol* genes, is spliced to subgenomic mRNA for *env* gene expression, and is packaged into progeny virions (see Figs. 1 and 4). Thus, retroviruses normally must modulate the amount of splicing of full-length viral RNA so that proper amounts of RNA are available for these three functions.

Therefore, it is not completely surprising that insertions and deletions in the intron of the subgenomic *env* mRNA greatly influence the efficiency of splicing. These effects were first reported in Rev strain T, which normally has a deletion of *gag* and *pol* sequences relative to a related replication-competent Rev (Chen and Temin, 1982). This deletion increases the level of spliced RNA (C. K. Miller and Temin, 1986). In addition, other insertions in the intron of the replication-defective highly oncogenic retrovirus Rev strain T decrease splicing (C. K. Miller and Temin, 1986). Effects that may be similar were reported after insertion of the herpes simplex virus thymidine kinase gene in MLV spleen-focus-forming virus, in which segregation of the transforming function from the *tk* gene was found (Joyner and Bernstein, 1983b).

However, it is generally found that cellular or viral genes containing introns are normally spliced when placed in retrovirus vectors, although the efficiency is lower than for chromosomal genes (Shimotohno and Temin, 1982a; Sorge and Hughes, 1982; Cepko *et al.*, 1984; Donoghue *et al.*, 1984; Kriegler *et al.*, 1984; Iba *et al.*, 1984; Takeda *et al.*, 1985; Sylla and Temin, 1986). A chicken globin gene with an unusual 5' splice sequence was also spliced normally in a retrovirus vector (Fischer *et al.*, 1984). Thus, retrovirus vectors that contain genes with intervening sequences provide a convenient way to prepare cDNA copies of genes (see Section 5.4).

The lower efficiency of splicing of cellular genes inserted in retrovirus vectors than for the genomic copies of the genes (Shimotohno and Temin, 1982a; Sorge and Hughes, 1982; Takeda *et al.*, 1985) undoubtedly relates to the suppressing effect of viral sequences on splicing discussed above.

6.3.7. Splice Sites (– Orientation)

Cellular genes with introns inserted in retrovirus vectors in the opposite orientation do not appear to become spliced in viral genomes (Bandyopadhyay and Temin, 1984b). This failure is not surprising, since the complementary sequences are in the viral RNA. However, since such genes can be expressed from an internal promoter in molecules complementary to viral RNA, they can be spliced in such transcripts.

6.4. Instability

6.4.1. General

Retrovirus proviruses are much more stable than retroviruses replicating as infectious viruses. Proviruses undergo base-pair mutation at the same low rate as other cellular DNA and, in the mouse, undergo homologous recombination between the two LTRs to delete the internal viral sequences at about 10^{-6} per cell generation (Varmus *et al.*, 1981; Copeland *et al.*, 1983). In replicating retroviruses, there is base-pair mutation at one site at a frequency of 1 per 1000 or 10,000 (Coffin *et al.*, 1980). In addition, replicating retroviruses undergo deletion and recombination at high rates. Especially rapid is deletion of tandem duplications (Coffin *et al.*, 1980; M. Emerman, personal communication).

To prepare a stock of a retrovirus vector, there must be some virus replication to form the infecting stock of virus and then infection and formation of a provirus in the infected cells. Thus, some opportunity for variation is unavoidable in the use of retrovirus vectors. In addition, there is the possibility of inter- or intramolecular recombination of the vector DNA and recombination with the helper virus or helper cell during transfection to prepare the infecting virus stock. Therefore, it is not surprising that frequent variation has been reported in retrovirus vector stocks.

6.4.2. Recombination

6.4.2a. Recombination during Transfection. An unavoidable source of genetic variation in the use of retrovirus vectors is recombination. Any work with retrovirus vectors necessarily involves transfection or some other means of recovering virus from DNA clones. The processes of transfection and microinjection of DNA into cells are very recombinogenic. For example, after transfection, there is recombination of separated viral genomes in all cells that take up two different DNAs (C. K. Miller and Temin, 1983) and even measurable recombination when there is no homology between separated viral genomes (Bandyopadhyay *et al.*, 1984). In addition, there is recombination of transfected DNA with itself. For example, when DNA clones of a retrovirus with a deletion in one LTR are used for transfection, infectious virus is recovered as a result of recombination of the deleted LTR with the wild-type LTR (Panganiban and Temin, 1983).

Absence of homologous sequences decreases the frequency of recombination, but it still occurs. Thus, it is necessary to establish directly that recombination has not occurred in any experiments with retrovirus vectors. For example, after transfection of a helper cell with a retrovirus vector, it is necessary to demonstrate that there has not been recombination that either forms replication-competent retrovirus or modifies the vector. Both types of recombination occur.

Recombination after transfection was used in the construction of retrovirus vectors in some earlier experiments (Wei *et al.*, 1981; Joyner and Bernstein, 1983a; Joyner *et al.*, 1983). A difficulty with this approach to vector construction is the impossibility of knowing exactly the structure of the starting vector because of recombination.

6.4.2b. Recombination during Preparation of Retrovirus Vector Stock. Thus, some recombination may occur during preparation of a stock of a retrovirus vector. Lack of homology decreases recombination during transfection and during virus replication. However, the best way to reduce recombination during virus replication is to reduce the amount of viral replication as opposed to proviral replication. [A retrovirus can replicate either as a virus or as a provirus, a cellular chromosomal gene. There is much less variation when the retrovirus replicates as a cellular gene than when it replicates as a virus (see Section 6.4.1).]

The amount of viral replication as opposed to proviral replication can be reduced by the following protocol: Transfect helper cells with a retrovirus DNA clone containing a marker that can be selected in these cells; harvest helper-free virus and infect fresh helper cells, selecting for the infected helper cell; analyze the provirus in the helper-cell clone to determine whether it is altered; harvest helper-free virus from a clone of helper cells with an unaltered provirus. With this protocol, the potential for recombination is minimized. Recombination could still occur during transfection of the first helper cells, during production of virus from the first helper cells, during infection of the second helper cells, during replication of the provirus in the second helper cells, during formation of virus from the second helper cells, and during infection of recipient cells with the final vector stock. Since the structure of the provirus can be analyzed in the second helper cells, only recombination (or other variation) in the last two steps, formation of virus from the second helper cells and infection of recipient cells with the final vector stock, could escape easy detection. Unfortunately, there is no way to prevent such recombination.

6.4.3. Selection for Virus Production

By definition, an infectious retrovirus vector stock contains infectious virus. Therefore, there will be selection against nucleotide sequences that inhibit virus replication, e.g., internal 3' RNA processing sites. Since the rate of spontaneous deletion in retrovirus replication is high, variants without the inhibiting sequences arise frequently. Such variants are then strongly selected. This process of deletion and selection can change the structure of a vector population within a few cycles of virus replication. For example, Joyner and Bernstein (1983a) and Joyner et al. (1983) formed a type 2 MLV vector by recombination after transfection and found that the internal 3' RNA processing sequences were deleted from all virus recovered.

6.4.4. Selection for Expression

Retrovirus vectors are usually prepared for expression of one or more genes in infected cells. Therefore, there is selection for expression of the desired gene(s). If there are sequences in the vector that inhibit expression of the desired gene(s), they can be deleted or suppressed by spontaneous random mutations. The vectors selected for expression of the desired gene may then have deletions or other mutations with respect to the parental vector. (If the sequences that inhibit expression are needed for virus production, they cannot be deleted or mutated.) If the inhibiting sequences are part of a nonselected gene the expression of which is desired, the nonselected gene may be lost or mutated. Alternatively, the mutations may allow expression of all the genes in a vector.

Chen and Temin (1982) and Joyner and Bernstein (1983b) made insertions in the intron of transforming retroviruses and found that after selection for transformation, some of the inserted sequences were deleted. A different cause of deletion was seen by Emerman and Temin (1984a). A type 4 Rev vector was deleted of gene 1 when there was selection for expression of gene 2. A similar type 3 Rev vector was stable under the same selection, although there was epigenetic suppression in this vector (Emerman and Temin, 1984a,b) (see Section 6.3.4).

Chen and Temin (1982) and Miller and Temin (1986) also found in many cases after insertions in the intron of a type 5 transforming retrovirus and selection for transformation that there was increased expression of the transforming gene without apparent deletion. Dougherty and Temin (1986) further showed with type 5 Rev-based vectors that genes 1 and 2 could be then equally expressed in such mutant vectors.

6.4.5. Problems with Helper Cells

There have been two major problems with the use of helper cells to prepare stocks of helper-free retrovirus vectors; recombination and low titers. The frequency of recombination has been decreased by having two separate genomes in the helper cell express the viral proteins (Watanabe and Temin, 1983) or by deletion of the polypurine tract and left attachment sequences as well as the encapsidation sequences (Sorge et al., 1984b). In addition, clones of transfected helper cells have been selected to find those that have not had recombination to produce infectious virus (Cone and Mullign, 1984). Higher virus titers have been secured by selecting clones of transfected helper cells with higher virus production and by transfecting the helper cells with retrovirus vectors that contain amplifiable sequences (A. D. Miller et al., 1985). Titers of over 10^6 infectious vectors/ml have been obtained. A more usual titer is 10^4/ml.

Use of more efficient promoters to express viral proteins in the helper cells might increase the yields further. Formation of particles containing VL30 RNA has been reported from psi-2 cells (E. Keshet, personal communication).

7. DETAILED DESCRIPTION OF TYPES OF RETROVIRUS VECTORS

Each of the eight types of retrovirus vectors (see Figs. 7 and 8) will be briefly discussed in this section. A fuller discussion of uses and possible problems is presented in Sections 5 and 6.

7.1. Type 1

Vector type 1 is the simplest retrovirus vector. Transcription is only from the long terminal repeat (LTR). The efficiency of transcription depends on the strength of the LTR promoter–enhancer, and the efficiency of translation is affected by the sequences before the coding sequences of the gene (Bandyopadhyay and Temin, 1984a). Many highly oncogenic retroviruses are type 1, e.g., Fujinami sarcoma virus, myelocytomatosis virus 29 (MC29), avian erythroblastosis virus strain H, and Abelson murine sarcoma virus. A murine leukemia virus (MLV)-based type 1 vector was constructed to express human hypoxanthine phosphoribosyltransferase (HPRT) in cells obtained from humans deficient in this enzyme (A. D. Miller *et al.*, 1983). [The first retrovirus vectors to be constructed were not type 1, but type 2 (Shimotohno and Temin, 1981).]

Type 1 vectors do not have problems of promoter interactions and efficiency of splicing (see below). When the gene inserted can be selected directly, type 1 vectors are the easiest to use.

A variant of a type 1 vector with two genes expressed from the LTR promoter has been constructed (J. Dougherty, personal communication). Expression of gene 2 depends on reinitiation of translation. Thus, the vector has a bicistronic mRNA.

Examples of type 1 vectors are: MLV-*gpt* [Mann *et al.*, 1983 (also has polyoma early region in plasmid); Perkins *et al.*, 1983], MLV-human HPRT (A. D. Miller *et al.*, 1983), retriculoendotheliosis virus (REV)-*tk* (Bandyopadhyay and Temin, 1984a), MLV–simian virus 40 (SV40) early coding sequences (Kriegler *et al.*, 1984), MLV-*neo* (Rubenstein *et al.*, 1984; Sorge *et al.*, 1984a), and MLV-methotrexate-resistant dihydrofolate reductase (DHFR) gene (A. D. Miller *et al.*, 1985).

Possible problems of type 1 vectors exist. Expression in type 1 vectors is determined by transcription only from the LTR, and so transcription is limited where LTR transcription is limited, e.g., in MLV vectors in embryonal carcinoma cells (see Section 5.1). Only a single gene is present in simple type 1 vectors. Thus, it can be hard to select for the vector if that gene is not in a cell in which it can be selected. There can also be problems of lack of regulated expression and possibly, but unlikely, of too small size (see Section 6.2.1).

7.2. Type 2

Type 2 vectors have an internal promoter in addition to the LTR. The presence of an internal promoter allows expression from it, rather than from the LTR, to control expression of gene 1. However, the presence of an internal promoter also permits interactions in transcription between the internal promoter and the LTR promoter. These interactions are more easily studied with type 3 vectors (see Section 7.3). Type 2 were the first retrovirus vectors constructed, usually with the thymidine kinase gene from herpes simplex virus (Shimotohno and Temin, 1981; Tabin *et al.*, 1982). They are used primarily to study retrovirus vectors or as precursors to type 3 vectors.

Examples of type 2 vectors are: Rev-*tk* (Shimotohno and Temin, 1981) MLV-*tk* (Tabin *et al.*, 1982), MLV-SV2*neo* (Joyner *et al.*, 1983), MLV-*tk* (Joyner and Bernstein, 1983a,b), type 2 and 2 − Rev-chicken *tk* (Bandyopadhyay and Temin, 1984b); MLV-*neo* (Rubenstein *et al.*, 1984), MLV-SV-DHFR (A. D. Miller *et al.*, 1985), and MLV-tk promoter-*neo* (Wagner *et al.*, 1985).

Type 2 vectors can have two types of problems resulting from the presence of an internal promoter: (1) There can be selection of virus in which the internal promoter is deleted as a result of either (a) the LTR promoter being stronger than the internal promoter or (b) interactions between the 5′ LTR and the internal promoter resulting in suppression of the internal promoter. (2) As a result of promoter suppression, only one promoter may be fully active. Thus, there may be either expression of gene 1 or virus production, but not both (discussed in Section 6.3.4 and 6.3.5). Type 2 vectors also can conceivably be too small (see Section 6.2.1).

7.3. Type 3

Vector types 2 and 3 differ only in whether or not the sequences before the internal promoter are coding. In type 3 vectors, two genes can be separately expressed, e.g., one selectable and the other not selectable or one active when the other is not. However, the presence of an internal promoter allows interactions in transcription between the internal promoter and the LTR promoter just as for type 2 vectors. Type 3 vectors have been used to study oncogene expression after selection for infected cells using a selectable marker other than the oncogene and to prepare cDNA genes by selecting virus with the selectable gene.

Examples of type 3 vectors are: Ha-MSV-*ras-tk* [Wei *et al.*, 1981 (the structure of virus produced from the vector was not determined)], Rev-globin-*tk* (Shimotohno and Temin, 1982a), MLV-polyoma early region (Donoghue *et al.*, 1984); MLV-HPRT-rat growth hormone minigene (A. D. Miller *et al.*, 1984b), Rev-*neo-tk* and Rev-*tk*-tk promoter-*neo* (Emerman and Temin, 1984b), Rev-*ras-tk* (Tarpley and Temin, 1984), and Rev-immunoglobulin-*tk* (Takeda *et al.*, 1985).

Type 3 vectors can have two types of problems resulting from the presence of internal control sequences: (1) There can be interaction between the 5′ LTR and the internal promoter, resulting in epigenetic suppression of gene 2 leading to selection of virus in which gene 1 and the internal promoter are deleted. (2) There can be interaction between the two promoters, resulting in epigenetic suppression so only gene 1 or gene 2 is expressed at high levels, not both (discussed in Section 6.3.4).

7.4. Type 4

Vector type 4 has two internal promoters as well as the two LTRs. Two genes can be expressed from the internal promoters. However, numerous interactions can occur between the promoters leading to only one or the other gene or full-length RNA being transcribed. The efficiency of expression of the two genes depends on the relative strengths of the two promoters and, perhaps, on whether or not the LTR promoter is stronger (Emerman and Temin, 1984b, 1986). These vectors have been used to attain tissue-specific expression of the insulin promoter.

Examples of type 4 vectors are: Rev-globin-*tk* (Shimotohno and Temin, 1982a) and vector type 4 — MLV-*gpt*-insulin promoter-tk promoter-*neo* (Episkopou *et al.*, 1984).

Type 4 vectors, like types 2 and 3, can have interactions between the transcriptional control sequences causing epigenetic suppression and resulting in deletion of one or both of the internal promoters (discussed in Sections 6.3.4 and 6.3.5). The LTRs are always maintained without deletion by selection for virus production.

7.5. Type 5

Vector type 5 has a splice donor and splice acceptor. Transcription is only from the LTR, and two genes can be expressed—one from full-length viral RNA and one from spliced RNA. However, the two genes are not expressed at the same efficiency. The efficiency of expression of gene 2 depends on the efficiency of splicing of full-length viral RNA (discussed in Sections 6.3.6 and 6.3.7). Replication-competent retroviruses and some replication-defective highly oncogenic retroviruses, like avian erythroblastosis virus, Mill Hill virus strain 2, and Rev strain T, have this structure. These vectors have been used to express human adenosine deaminase in cells selected for gene 2 (Friedman, 1985; Valerio *et al.*, 1985).

Examples of type 5 vectors are: Friend F-MLV-*tk* (Joyner and Bernstein, 1983b), MLV-*neo* (Cepko *et al.*, 1984; Hwang and Gilboa, 1984), MLV-*sis* (Hannink and Donoghue, 1984), MLV-gpt-human parathyroid hormone (Hellerman *et al.*, 1984), MLV-*gpt* (Lewis *et al.*, 1984), MLV-DHFR-*neo* (Williams *et al.*, 1984), MLV-human adenosine deaminase-*neo* (Friedman, 1985; Valerio *et al.*, 1985), Rous sarcoma virus (RSV)-*lac-src* (Norton and Coffin, 1985), MLV-*tk-neo* (Taketo *et al.*, 1985) and MLV-phenylalanine hydroxylase-neo (Ledley *et al.*, 1986).

Type 5 vectors can have problems of expression of gene 2 as a result of low efficiency of splicing discussed in Section 6.3.6. If there is strong selection for expression of gene 2, deletion of gene 1 can occur, or there can be mutation to increase the level of splicing without deletion of gene 1 (see Section 6.4.4).

7.6. Type 6

Vector type 6 has a splice donor and two splice acceptors. Transcription is only from the LTR, and three genes can be expressed—one from full-length viral RNA and two from spliced RNAs. When the first two genes are the *gag–pol* genes of a replication-competent retrovirus, the vector is usually nondefective. The replication-competent highly oncogenic retrovirus RSV has this structure. In this case, no helper cell or virus is needed. As with vector type 5, transcription is only from the LTR, and the relative efficiency of expression of the genes depends on the efficiency of splicing of viral RNA. [The splicing pattern can be complex in viruses like HTLV-I and -II (Wachsman *et al.*, 1985).] Type 6 vectors have been used to study oncogene expression and splicing.

Examples of type 6 vectors are: RSV-human alpha-chorionic gonadotropin (Sorge and Hughes, 1982), RSV-*fps* (Foster and Hanafusa, 1983), RSV-globin (Fischer *et al.*, 1984), RSV-c-*src* (Iba *et al.*, 1984), and Rev-*src* (Gelinas and Temin, 1986).

Type 6 vectors can have problems of efficiency of splicing (discussed in Section 6.3.6) and of size (discussed in Section 6.2). Thus, viruses with deletions occur frequently in stocks of type 6 vectors. If a nondefective vector is used originally, a deleted replication-competent retrovirus without gene 3 can easily take over the virus population (see Estis and Temin, 1979; Coffin *et al.*, 1980).

7.7. Type 7

Vector type 7 includes any vectors with both internal promoters and splice acceptors. Thus, there are many possible structures. Such vectors, like type 3 vectors, give the possibility of one or more genes being expressed from an internal promoter and another gene or genes being expressed from the LTR promoter. However, these vectors differ from types 2, 3, and 4 in that more than one gene can also be expressed from the LTR or an internal promoter, depending on the structure. These vectors have been used to study oncogene expression.

Examples of type 7 vectors are: Rev-*src*-tk promoter-*tk* and Rev-tk promoter-*tk-src* (Tarpley and Temin, 1984) and MLV-*myc*-tk promoter-*neo* (Wagner *et al.*, 1985).

Complex interactions between promoters and effects of insertions on efficiency of splicing can occur with type 7 vectors. These problems probably would make the properties of such constructions relatively unpredictable when more than two genes are expressed.

7.8. Type 8

Vector type 8 includes any vectors with a modified right (or 3') LTR. On infection, virus made from such vector DNA will form a provirus with modified LTRs (see Fig. 9 and Section 5.6). No descriptions of such vectors were published in 1985. Possible problems with type 8 vectors include problems with 3' processing of viral RNA, recombination between the two LTRs or with helper cells, difficulty in getting high titers of virus because of only one cycle of virus replication, and suppression of virus production by internal promoters.

8. SUMMARY

Retroviruses are natural vectors for inserting genes into eukaryotic chromosomes and obtaining gene expression. Retrovirus-gene organization allows the insertion of many kinds of internal sequences including splice acceptors and promoters. Thus, many different kinds of retrovirus vectors can be made. In addition, there are a large number of possible different species of retroviruses that can be used for vectors. The existence of all these possibilities suggests that ultimately a retrovirus vector could be made to target any cell and obtain tissue-specific expression in that cell.

Methods have been developed to prepare stocks of infectious retrovirus vectors in the absence of any replication-competent retrovirus. Certain problems of suppression of splicing and suppression of nearby promoters occur in retrovirus vectors. However, the variety of different possible structures makes it feasible to design vectors without these problems. Thus, retrovirus vectors are likely to be the vectors of choice when stable integration of genes in animal cells is desired.

ACKNOWLEDGMENTS.   I thank J. Dougherty, J. Embretson, M. Emerman, C. Gelinas, T. Gilmore, C. Miller, and B. Sylla for helpful comments. The research in my laboratory is supported by Public Health Service Research Grants CA-07175 and CA-22443 from the National Cancer Institute. I am an American Cancer Society Research Professor.

REFERENCES

Anderson, W. F., 1984, Prospects for human gene therapy, *Science* **226**:401–409.
Anderson, W. F., 1985, Human gene therapy: Scientific and ethical consideration, *Recombinant DNA Tech. Bull.* **8**:55–63.
Anonymous, Human gene therapy: Background paper, U.S. Congress, Office of Technology Assessment, 1984.

Bandyopadhyay, P. K., and Temin, H. M., 1984a, Expression from an internal aug codon of herpes simplex thymidine kinase gene inserted in a retrovirus vector, *Mol. Cell. Biol.* **4:**743–748.

Bandyopadhyay, P. K., and Temin, H. M., 1984b, Expression of complete chicken thymidine kinase gene inserted in a retrovirus vector, *Mol. Cell. Biol.* **4:**749–754.

Bandyopadhyay, P. K., Watanabe, S., and Temin, H. M., 1984, Recombination of transfected DNAs in vertebrate cells in culture, *Proc. Natl. Acad. Sci. U.S.A.* **81:**3476–3480.

Battula, N., and Loeb, L., 1974, The infidelity of avian myeloblastosis virus DNA polymerase in polynucleotide replication, *J. Biol. Chem.* **249:**4086–4093.

Berger, S. A., and Berstein, A., 1985, Characterization of a retrovirus shuttle vector capable of either proviral integration or extrachromosomal replication in mouse cells, *Mol. Cell. Biol.* **5:**305–312.

Bestwick, R. K., Hankins, W. D., and Kabat, D., 1985, Roles of helper and defective retroviral genomes in murine erythroleukemia: Studies of spleen focus-forming virus in the absence of helper, *J. Virol.* **56:**660–664.

Bishop, J. M., and Varmus, H., 1985, Function and origins of retroviral transforming genes, in: *RNA Tumor Viruses,* 2nd ed. (R. Weiss, N. Teich, H. Varmus, and J. Coffin, eds.), pp. 249–356, Cold Spring Harbor Press, Cold Spring Harbor, New York.

Boeke, J. D., Garfinkel, D. J., Styles, C. A., and Fink, G. R., 1985, Ty elements transpose through an RNA intermediate, *Cell* **40:**491–500.

Cepko, C. L., Roberts, B. E., and Mulligan, R. C., 1984, Construction and applications of a highly transmissible murine retrovirus shuttle vector, *Cell* **37:**1053–1062.

Chen, I. S. Y., and Temin, H. M., 1982, Substitution of 5′ helper virus sequences into non-*rel* portion of reticuloendotheliosis virus strain T suppresses transformation of chicken spleen cells, *Cell* **31:**111–120.

Clark, S. P., Kaufhold, R., Chan, A., and Mak, T. W., 1985, Comparison of the transcriptional properties of the Friend and Moloney retrovirus long terminal repeats: Importance of tandem duplication and of the core enhancer sequence, *Virology* **144:**481–494.

Coffin, J., 1985, Genome structure, in: *Molecular Biology of Tumor Viruses,* 2nd ed., *RNA Tumor Viruses 2/Supplements and Appendixes* (R. Weiss, N. Teich, H. Varmus, and J. Coffin, eds.), pp. 1–37, Cold Spring Harbor Press, Cold Spring Harbor, New York.

Coffin, J. M., Tsichlis, P. N., Barker, C. S., Voynow, S., and Robinson, H. L., 1980, Variation in avian retrovirus genomes, *Ann. N. Y. Acad. Sci.* **354:**410–425.

Cone, R. D., and Mulligan, R. C., 1984, High-efficiency gene transfer into mammalian cells: Generation of helper-free recombinant retrovirus with broad mammalian host rnge, *Proc. Natl. Acad. Sci. U.S.A.* **81:**6349–6353.

Copeland, N., Hutchison, K., and Jenkins, H., 1983, Excision of the DBA ecotropic provirus in dilute coat-color revertants of mice occurs by homologous recombination involving the viral LTRs, *Cell* **33:**379–387.

Cullen, B. R., Lomedico, P. T., and Ju, G., 1984, Transcriptional interference in avian retroviruses— implications for the promoter insertion model of leukaemogenesis, *Nature (London)* **307:**241–245.

Cullen, B. R., Raymond, K., and Ju, G., 1985, Transcriptional activity of avian retroviral long terminal repeats directly correlates with enhancer activity, *J. Virol.* **53:**515–521.

Curran, T., Peters, G., Van Beveren, C., Teich, N. M., and Verma, I. M., 1982, FBJ murine osteosarcoma virus: Identification and molecular cloning of biologically active proviral DNA, *J. Virol.* **44:**674–682.

Dick, J. E., Magli, M. C., Huszar, D., Phillips, R. A., Bernstein, A., 1985, Introduction of a selectable gene into primitive stem cells capable of long-term reconstitution of the hemopoietic system of W/Wv mice, *Cell* **42:**71–79.

Donoghue, D. J., Anderson, C., Hunter, T., and Kaplan, P. L., 1984, Transmission of the polyoma virus middle T gene as the oncogene of a murine retrovirus, *Nature (London)* **308:**748–750.

Dougherty, J., and Temin, H. M., 1986, Analysis of spleen necrosis virus-based retrovirus splicing vector and its variation (submitted to *Mol. Cell. Biol.*).

Eglitis, M. A., Kantoff, P., Gilboa, E., Anderson, W. F., 1985, Gene expression in mice after high efficiency retroviral-mediated gene transfer, *Science* **230:**1395–1398.

Embretson, J., and Temin, H. M., 1986, Pseudotyped retroviral vectors reveal restrictions on retrovirus replication in rat cells, *J. Virol.* (in press).

Emerman, M., and Temin, H. M., 1984a, High frequency deletion in recovered retrovirus vectors containing exogenous DNA with promoters, *J. Virol.* **50:**42–49.

Emerman, M., and Temin, H. M., 1984b, Genes with promoters in retrovirus vectors can be independently suppressed by an epigenetic mechanism, *Cell* **39:**459–467.

Emerman, M., and Temin, H. M., 1986, Quantitative analysis of gene suppression in integrated retrovirus vectors, *Mol. Cell. Biol.* **6:**792–800.

Episkopou, V., Murphy, A. J. M., and Efstratiadis, A., 1984, Cell-specified expression of a selectable hybrid gene, *Proc. Natl. Acad. Sci. U.S.A.* **81:**4657–4661.

Estis, L. F., and Temin, H. M., 1979, Suppression of multiplication of avian sarcoma virus by rapid spread of transformation-defective virus of the same subgroup, *J. Virol.* **31:**389–397.

Felber, B. K., Paskalis, H., Kleinman-Ewing, C., Wong-Staal, F., and Pavlakis, G. N., 1985, The pX protein of HTLV-I is a transcriptional activator of its long terminal repeats, *Science* **229:** 675–679.

Fischer, H. D., Dodgson, J. B., Hughes, S., and Engel, J. D., 1984, An unusual 5′ splice sequence is efficiently utilized *in vivo, Proc. Natl. Acad. Sci. U.S.A.* **81:**2733–2737.

Foster, D. A., and Hanafusa, H., 1983, A *fps* gene without *gag* gene sequences transforms cells in culture and induces tumors in chickens, *J. Virol.* **48:**744–751.

Friedman, R. L., 1985, Expression of human adenosine deaminase using a transmissible murine retrovirus vector system, *Proc. Natl. Acad. Sci. U.S.A.* **82:**703–703.

Hammer, R. E., Pursel, V. G., Rexroad, C. E., Jr., Wall, R. J., Bott, D. J., Ebert, K. M., Palmiter, R. D., and Brinster, R. L., 1985, Production of transgenic rabbits, sheep and pigs by microinjection, *Nature (London)* **315:**680–683.

Hannink, M., and Donoghue, D., 1984, Requirement for a signal sequence in biological expression of the v-*sis* oncogene, *Science* **226:**1197–1199.

Harbers, K., Kuehn, M., Delius, H., and Jaenisch, R., 1984, Insertion of retrovirus into the first intron of (alpha)1(I) collagen gene leads to embryonic lethal mutation in mice, *Proc. Natl. Acad. Sci. U.S.A.* **81:**1504–1508.

Harris, J. D., Blum, H., Scott, J., Traynor, B., Ventura, P., and Haase, A., 1984, Slow virus visna: Reproduction *in vitro* of virus from extrachromosomal DNA, *Proc. Natl. Acad. Sci. U.S.A.* **81:**7212–7215.

Hawley, R. G., Shulman, M. J., Murialdo, H., Gibson, D. M., and Hozumi, N., 1982, Mutant immunoglobulin genes have repetitive DNA elements inserted into their intervening sequences, *Proc. Natl. Acad. Sci. U.S.A.* **79:**7425–7429.

Hayashida, H., Toh, H., Kikuno, R., and Miyata, T., 1985, Evolution of influenza genes, *Mol. Biol. Evol.* **2:**289–303.

Hellerman, J. G., Cone, R. C., Potts, J. T., Jr., Rich, A., Mulligan, R. C., and Kroneberg, H. M., 1984, Secretion of human parathyroid hormone from rat pituitary cells infected with a recombinant retrovirus encoding preparathyroid hormone, *Proc. Natl. Acad. Sci. U.S.A.* **81:**5340–5344.

Holland, C. A., Hartley, J. W., Rowe, W. P., and Hopkins, N., 1985, At least four viral genes contribute to the leukemogenicity of murine retrovirus MCF 247 in AKR mice, *J. Virol.* **53:**158–165.

Huszar, D., Balling, R., Kothary, R., Magli, M. C., Hozumi, N., Rossant, J., Bernstein, A., 1985, Insertion of a bacterial gene into the mouse germ line using an infectious retrovirus vector, *Proc. Natl. Acad. Sci. U.S.A.* **82:**8587–8591.

Hwang, L.-H. S., and Gilboa, E., 1984, Expression of genes introduced into cells by retroviral infection is more efficient than that of genes introduced into cells by DNA transfection, *J. Virol.* **50:**417–424.

Hwang, L.-H. S., Park, J., and Gilboa, E., 1984, Role of intron-containing sequences in formation of Moloney murine leukemia virus env mRNA, *Mol. Cell. Biol.* **4:**2289–2297.

Iba, H., Takeya, T., Cross, F. R., Hanafusa, T., and Hanafusa, H., 1984, Rous sarcoma virus variants that carry the cellular *src* gene instead of the viral *src* gene cannot transform chicken embryo fibroblasts, *Proc. Natl. Acad. Sci. U.S.A.* **81:**4424–4428.

Izant, J. G., and Weintraub, H., 1985, Constitutive and conditional suppression of exogenous and endogenous genes by anti-sense RNA, *Science* **229:**345–352.

Jahner, D., Haase, K., Mulligan, R., Jaenisch, R., 1985, Insertion of the bacterial *gpt* gene into the germ line of mice by retroviral infection, *Proc. Natl. Acad. Sci. U.S.A.* **82:**6927–6931.

Joyner, A. L., and Bernstein, A., 1983a, Retrovirus transduction: Generation of infectious retroviruses expressing dominant and selectable genes associated with *in vivo* recombination and deletion events, *Mol. Cell. Biol.* **3:**2180–2190.

Joyner, A. L., and Bernstein, A., 1983b, Retrovirus transduction: Segregation of the viral transforming function and the herpes simplex virus tk gene in infectious spleen focus-forming virus thymidine kinase vectors, *Mol. Cell. Biol.* **3:**2192–2202.

Joyner, A., Keller, F., Phillips, R. A., and Bernstein, A., 1983, Retrovirus transfer of a bacterial gene into mouse haematopoietic progenitor cells, *Nature (London)* **305:**556–558.

King, W., Patel, M. D., Lobel, L. I., Goff, S. P., and Nguyen-Huu, M. C., 1985, Insertion mutagenesis of embryonal carcinoma cells by retroviruses, *Science* **228:**554–558.

Kleckner, N., Roth, J., and Botstein, D., 1977, Genetic engineering *in vivo* using translocatable drug-resistance elements: New methods in bacterial genetics, *J. Mol. Biol.* **116:**125–159.

Kriegler, M., Perez, C. F., Hardy, C., and Botchan, M., 1984, Transformation mediated by the SV40 T antigens: Separation of the overlapping SV40 early genes with a retroviral vector, *Cell* **38:**483–491.

Ledley, F. D., Grenett, H. E., McGinnis-Shelnutt, M., and Woo, S. L. C., 1986, Retroviral-mediated gene transfer of human phenyalanine hydroxylase into NIH 3T3 and hepatoma cells, *Proc. Natl. Acad. Sci. U.S.A.* **83:**409–413.

Lewis, S., Gifford, A., and Baltimore, D., 1984, Joining of V(kappa) to J(kappa) gene segments in a retroviral vector introduced into lymphoid cells, *Nature (London)* **308:**425–428.

Lewis, S., Gifford, A., and Baltimore, D., 1985, DNA elements are asymmetrically joined during the site-specific recombination of kappa immunoglobulin genes, *Science* **228:**677–685.

Lobel, L. I., Patel, M., King, W., Nguyen-Huu, M. C., and Goff, S. P., 1985, Construction and recovery of viable retroviral genomes carrying a bacterial suppressor transfer RNA gene, *Science* **228:**329–332.

Mann, R., and Baltimore, D., 1985, Varying the position of a retrovirus packaging sequence results in the encapsidation of both unspliced and spliced RNAs, *J. Virol.* **54:**401–407.

Mann, R., Mulligan, R. C., and Baltimore, D., 1983, Construction of a retrovirus packaging mutant and its use to produce helper-free defective retrovirus, *Cell* **33:**153–159.

Miller, A. D., Jolly, D. J., Friedmann, T., and Verma, I. M., 1983, A transmissible retrovirus expressing human hypoxanthine phosphoribosyltransferase (HPRT): Gene transfer into cells obtained from humans deficient in HPRT, *Proc. Natl. Acad. Sci. U.S.A.* **80:**4709–4713.

Miller, A. D., Eckner, R. J., Jolly, D. J., Friedman, T., and Verma, I., 1984a, Expression of a retrovirus encoding human HPRT in mice, *Science* **225:**630–632.

Miller, A. D., Ong, E. S., Rosenfeld, M. G., Verma, I. M., and Evans, R. M., 1984b, Infectious and selectable retrovirus containing an inducible rat growth hormone minigene, *Science* **229:**993–998.

Miller, A. D., Law, M.-F., and Verma, I. M., 1985, Generation of helper-free amphotropic retroviruses that transduce a dominant-acting methotrexate-resistant dihydrofolate reductase gene, *Mol. Cell. Biol.* **5:**431–437.

Miller, C. K., and Temin, H. M., 1983, High efficiency ligation and recombination of DNA fragments by vertebrate cells, *Science* **220:**606–609.

Miller, C. K., and Temin, H. M., 1986, Insertion of several different DNAs in reticuloendotheliosis virus strain T suppresses transformation by reducing the amount of subgenomic mRNA, *J. Virol.* **58:**75–80.

Mizutani, S., and Temin, H. M., 1976, Incorporation of noncomplementary nucleotides at high frequences by ribodeoxyvirus DNA polymerases and *Escherichia coli* DNA polymerase I, *Biochemistry* **15:**1510–1516.

Mount, S. M., and Rubin, G. M., 1985, Complete nucleotide sequence of the *Drosophila* transposable element copia: Homology between copia and retroviral proteins, *Mol. Cell. Biol.* **5:**1630–1638.

Norton, P. A., and Coffin, J. M., 1985, Bacterial β-galactosidase as a marker of Rous sarcoma virus gene expression and replication, *Mol. Cell. Biol.* **5:**281–290.

O'Rear, J. J., and Temin, H. M., 1982, Spontaneous changes in nucleotide sequence in proviruses of spleen necrosis virus, an avian retrovirus, *Proc. Natl. Acad. Sci. U.S.A.* **79:**1230–1234.

O'Rear, J. J., Mizutani, S., Hoffman, G., Fiandt, M., and Temin, H. M., 1980, Infectious and non-infectious recombinant clones of the provirus of SNV differ in cellular DNA and are apparently the same in viral DNA, *Cell* **20:**423–430.

Palacios, R., Steinmetz, M., 1985, IL3-dependent mouse clones that express B-220 surface antigen, contain Ig genes in germ-line configuration, and generate B lymphocytes *in vivo, Cell* **41:**727–734.

Panganiban, A. T., 1985, Retroviral DNA integration, *Cell* **42:**5–6.

Panganiban, A. T., and Temin, H. M., 1983, The terminal nucleotides of retrovirus DNA are required for integration but not for virus production, *Nature (London)* **306:**155–160.

Perkins, A. S., Kirschmeier, P. J., Gatloni-Celli, S., and Weinstein, I. B., 1983, Design of retrovirus-derived vector for expression and transduction of exogenous genes in mammalian cells, *Mol. Cell. Biol.* **3:**1123–1132.

Quintrell, N., Hughes, S. H., Varmus, H. E., and Bishop, J. M., 1980, Structure of viral DNA and RNA in mammalian cells infected with avian sarcoma virus, *J. Mol. Biol.* **143:**363–393.

Reik, W., Weihar, H., and Jaenisch, R., 1985, Replication-competent Moloney murine leukemia virus carrying a bacterial suppressor tRNA gene: Selective cloning of proviral and flanking host sequences, *Proc. Natl. Acad. Sci. U.S.A.* **82:**1141–1145.

Rubenstein, J. L. R., Nicolas, J.-F., and Jacob, F., 1984, Construction of a retrovirus capable of transducing and expressing genes in multipotential embryonic cells, *Proc. Natl. Acad. Sci. U.S.A.* **81:**7137–7140.

Rubenstein, J. L. R., Nicolas, J.-F., Jacob, F., 1986, Introduction of genes ino preimplantation mouse embryos by use of a defective recombinant retrovirus, *Proc. Natl. Acad. Sci. U.S.A.* **83:**366–368.

Rubin, G. M., and Spradling, A. C., 1982, Genetic transformation of *Drosophila* with transposable element vectors, *Science* **218:**348–353.

Seliger, B., Kolleck, R., Stocking, C., Franz, T., Ostertag, W., 1986, Viral transfer, transcription, and rescue of a selective myeloproliferative sarcoma virus in embryonal cell lines: Expression of the mos oncogene, *Mol. Cell. Biol.* **6:**286–293.

Shimotohno, K., and Temin, H. M., 1981, Formation of infectious progeny virus after insertion of herpes simplex thymidine kinase gene into DNA of an avian retrovirus, *Cell* **26:**67–77.

Shimotohno, K., and Temin, H. M., 1982a, Loss of intervening sequences in genomic mouse α-globin DNA inserted in an infectious retrovirus vector, *Nature (London)* **299:**265–268.

Shimotohno, K., and Temin, H. M., 1982b, Spontaneous variation and synthesis in the U3 region of the long terminal repeat of an avian retrovirus, *J. Virol.* **41:**163–171.

Sorge, J., and Hughes, S. H., 1982, Splicing of intervening sequences introduced into an infectious retroviral vector, *J. Mol. Appl. Genet.* **1:**547–559.

Sorge, J., Cutting, A. E., Erdman, V. D., and Gautsch, J. W., 1984a, Integration-specific retrovirus expression in embryonal carcinoma cells, *Proc. Natl. Acad. Sci. U.S.A.* **81:**6627–6631.

Sorge, J., Wright, D., Erdman, V. D., and Cutting, A. E., 1984b, Amphotropic retrovirus vector system for human cell gene transfer, *Mol. Cell. Biol.* **4:**1730–1737.

Stuhlman, H., Cone, R., Mulligan, R. C., and Jaenisch, R., 1984, Introduction of a selectable gene into different animal tissue by a retrovirus recombinant vector, *Proc. Natl. Acad. Sci. U.S.A.* **81:**7151–7155.

Sylla, B., and Temin, H. M., 1986, Activation of oncogenicity of c-*rel* proto-oncogene (submitted to *Mol. Cel. Biol.*).

Tabin, C. J., Hoffman, J. W., Goff, S. P., and Weinberg, R. A., 1982, Adaptation of a retrovirus as a eucaryotic vector transmissing the herpes simplex virus thymidine kinase gene, *Mol. Cell. Biol.* **2:**426–436.

Takeda, S.-I., Naito, T., Hama, K., Noma, T., and Honjo, T., 1985, Construction of chimaeric processed immunoglobulin genes containing mouse variable and human constant region sequences, *Nature (London)* **314:**452–454.

Taketo, M., Gilboa, E., and Sherman, M. I., 1985, Isolation of embryonal carcinoma cell lines that express integrated recombinant genes flanked by the Moloney murine leukemia virus long terminal repeats, *Proc. Natl. Acad. Sci. U.S.A.* **82:**2422–2426.

Takeya, R., Hanafusa, H., Junghans, R. P., Ju, G., and Skalka, A. M., 1981, Comparison between the viral transforming gene (*src*) of recovered avian sarcoma virus and its cellular homolog, *Mol. Cell. Biol.* **1:**1024–1037.

Tarpley, W. G., and Temin, H. M., 1984, The location of v-*src* in a retrovirus vector determines whether the virus is toxic or transforming, *Mol. Cell. Biol.* **4:**2653–2660.

Temin, H. M., 1981, Structure, variation, and synthesis of retrovirus long terminal repeat, *Cell* **27:**1–3.

Temin, H. M., 1982, Function of the retrovirus long terminal repeat, *Cell* **28:**3–5.

Temin, H. M., 1985a, Developments in molecular virology: Cloning of retrovirus DNA in bacteria and cloning of other DNA in retroviruses, in: *Recombinant DNA Research and Virus* (Y. Becker, ed.), Martinus Nijhoff, Boston.

Temin, H. M., 1985b, Reverse transcription in the eukaryotic genome: Retroviruses, pararetroviruses, retrotransposons, and retrotranscripts, *Mol. Biol. Evol.* **2:**455–468.

Temin, N. M., and Miller, C. K., 1984, Insertion of oncogenes into retrovirus vectors, *Cancer Survey* **3**(2):229–246.

Valerico, D., Duyvesteyn, M. G. C., and van der Eb, A. J., 1985, Introduction of sequences encoding functional human adenosine deaminase into mouse cells using a retroviral shuttle system, *Gene* **34:**163–168.

Varmus, H. E., 1982, Form and function of retroviral proviruses, *Science* **216:**812–820.

Varmus, H. E., Quintrell, N., and Oritz, S., 1981, Retroviruses as mutagens: Insertion and excision of a nontransforming provirus alter expression of a resident transforming provirus, *Cell* **25:**23–36.

Wachsman, W., Golde, D. W., Temple, P. A., Orr, E. C., Clark, S. C., and Chen, I. S. Y., 1985, HTLV x-gene product: Requirement for the *env* methionine initiation codon, *Science* **228:**1534–1537.

Wagner, E. F., Vanek, M., and Vennstrom, B., 1985, Transfer of genes into embryonal carcinoma cells by retrovirus infection: Efficient expression from an internal promoter, *EMBO J.* **4:**663–666.

Watanabe, S., and Temin, H. M., 1982, Encapsidation sequences for spleen necrosis virus, an avian retrovirus, are between the 5′ long terminal repeat and the start of the gag gene, *Proc. Natl. Acad. Sci. U.S.A.* **79:**5986–5990.

Watanabe, S., and Temin, H. M., 1983, Construction of a helper cell line for avian reticuloendotheliosis virus cloning vectors, *Mol. Cell. Biol.* **3:**2241–2249.

Watanabe, S., Shimotohno, K., and Temin, H. M., 1984, Construction of a small retrovirus cloning vector and splicing of genomic mouse α-globin DNA inserted in this vector, in: *Progress in Cancer Research Therapy*, Vol. 30 (M. L. Pearson and N. L. Sternberg, eds.), pp. 97–103, Raven Press, New York.

Weber, F., and Schaffner, W., 1985, Enhancer activity correlates with the oncogenic potential of avian retroviruses, *EMBO J.* **4:**949–956.

Wei, C., Gibson, M., Spear, P. G., and Scolnick, E. M., 1981, Construction and isolation of transmissible retrovirus containing the *src* gene of Harvey murine sarcoma virus and the thymidine kinase gene of herpes simplex virus type 1, *J. Virol.* **39:**935–944.

Weis, J. H., Nelson, D. L., Przyborki, M. J., Chaplin, D. D., Mulligan, R. C., Housman, D. E., and Seidman, J. G., 1984, Eukaryotic chromosome transfer: Linkage of the murine major histocompatibility complex to an inserted dominant selectable marker, *Proc. Natl. Acad. Sci. U.S.A.* **81:**4879–4883.

Weiss, R., Teich, N., Varmus, H., and Coffin, J., (eds.), 1985, *Molecular Biology of Tumor Viruses, RNA Tumor Viruses,* 2nd ed., *2/Supplements and Appendixes,* Cold Spring Harbor Press, Cold Spring Harbor, New York.

Wilhelmsen, K. C., Tarpley, W. G., and Temin, H. M., 1984, Identification of some of the parameters governing transformation by oncogenes in retroviruses, in: *Cancer Cells* (G. F. Vande Woude, A. J. Levine, W. C. Topp, and J. D. Watson, eds.), pp. 303–308, Cold Spring Harbor Press, Cold Spring Harbor, New York.

Williams, D. A., Lemischka, I. R., Nathan, D. G., and Mulligan, R. C., 1984, Introduction of new genetic material into pluripotent haemopoietic stem cells of the mouse, *Nature (London)* **310:**476–483.

Willis, R. C., Jolly, D. J., Miller, A. D., Plent, M. M., Esty, A. C., Anderson, P. J., Chang, H.-C., Jones, O. W., Seegmiller, J. E., and Friedman, T., 1984, Partial phenotypic correction of human Lesch–Nyhan (hypoxanthine-guanine phosphoribosyltransferase-deficient) lymphoblasts with a transmissible retrovirus vector, *J. Biol. Chem.* **259:**7842–7849.

Wolff, L., and Ruscetti, S., 1985, Malignant transformation of erythroid cells *in vivo* by introduction of a nonreplicating retrovirus vector, *Science* **228:**1549–1552.

Wyngaarden, J., 1985, Points to consider in the design and submission of human somatic-cell gene therapy protocols, National Institutes of Health, *Fed. Reg.* **50:**2940–2944.

7

TRANSGENIC MICE: GENE TRANSFER INTO THE GERM LINE

Shirley M. Tilghman and Arnold J. Levine

1. INTRODUCTION

The discipline of biology has witnessed a virtual revolution in the past ten years, beginning with the development of recombinant DNA techniques that permitted one to isolate specific genes and to study their structure and function. The next critical stage in this revolution was the development of the technology to reintroduce these genes back into living organisms. The questions that can be posed with this combination of techniques are many, from the identification of *cis*-acting sequences required for tissue-specific developmental regulation of a gene to the consequences of inappropriate expression of a cellular oncogene. In a sense, a review of this field at this time is premature. The realization that investigators can alter both the genotype and the phenotype of an animal is just starting to be exploited, and this chapter reflects the fact that the initial attempts, while vastly encouraging, are not without problems. Nevertheless, both the advantages and limitations of the transgenic mouse system are beginning to be realized.

The experimental introduction of foreign DNA into the germ line of mice was first accomplished in 1974 by Jaenisch and Mintz (1974), who microinjected the intact viral genome of the simian virus 40 (SV40) into the blastocoel cavity of early mouse embryos. Since that time, a variety of protocols have been developed whereby the genotype of the mouse has been stably altered. These include the direct microinjection of naked DNA into the male pronucleus of the fertilized egg (Gordon *et al.*, 1980; Brinster *et al.*, 1985) and infection of early embryos with retroviruses and retrovirus vectors carrying foreign DNA (Jaenisch *et al.*, 1975, 1981; Jähner *et al.*, 1985). A third route for the introduction of

Shirley M. Tilghman ● Institute for Cancer Research, Philadelphia, Pennsylvania 19111. **Arnold J. Levine** ● Department of Molecular Biology, Princeton University, Princeton, New Jersey 08544.

genes into mice is via the formation of mouse chimeras using totipotent embryonal carcinoma cell lines into which foreign DNA has been previously transfected. This approach is being actively pursued in several laboratories, but as yet no germ-line transmission has been achieved (see Wagner and Stewart, 1986).

This review will concentrate on experiments that utilize the microinjection of naked DNA into fertilized eggs. As reviewed in detail by Brinster *et al.* (1985), this procedure normally results in the stable integration of from one to hundreds of tandem copies of a segment of DNA in a single chromosomal site, primarily by the mechanism of nonhomologous recombination. The issue as to whether the sites of integration are chosen at random has not been satisfactorily resolved. It may be that the sites are a subset of "open" or "accessible" sites in the egg chromatin, an outcome that might facilitate the subsequent expression of the introduced gene at appropriate times in ontogeny. The site of integration itself certainly plays a role in the level of expression of a given gene, in that mice carrying the same gene integrated at different sites can express it at very different levels. However, each gene appears to differ in the degree to which it is sensitive to the integration site, with the rat elastase and mouse immunoglobulin genes being remarkably insensitive (Swift *et al.*, 1984; Brinster *et al.*, 1983; Grosschedl *et al.*, 1984) and the β-globin genes being very sensitive (Chada *et al.*, 1985; Townes *et al.*, 1985).

2. REGULATION OF SPECIFIC GENE EXPRESSION IN TRANSGENIC MICE

The introduction of genes into the mouse germ line is a powerful tool for elucidating the genetic elements required for the appropriate tissue and developmental specificity of expression of genes. Prior to the use of germ-line integration, the identification of regulatory *cis*-acting DNA elements had depended almost exclusively on the introduction of genes into tissue-culture cells that were terminally differentiated (see, for example, Walker *et al.*, 1983; Queen and Baltimore, 1983; Banerji *et al.*, 1983; Gillies *et al.*, 1983; Grosschedl and Baltimore, 1985; Charnay *et al.*, 1985; Goodbourn *et al.*, 1985). While one could be certain that those elements necessary for the maintenance of expression would be identified, the procedure left answered the important question as to whether the establishment of expression during differentiation was controlled by substantially different mechanisms. With the exception of teratocarcinoma cells, which can undergo an authentic commitment choice *in vitro* (Scott *et al.*, 1984), very few culture systems exist or have been exploited by gene-transfer methods to study both the activation and maintenance of specific gene expression. Finally, few gene-transfer systems in tissue culture allow one to study the maturation of a differentiated cell type; notable exceptions being the rodent myoblast-to-myotube conversion (Seiler-Tuyns *et al.*, 1984; Nudel *et al.*, 1985; Konieczny and Emerson, 1985) and the maturation of mouse erythroleukemia cells following dimethyl sulfoxide treatment (Wright *et al.*, 1984; Charnay *et al.*, 1984).

The initial studies designed to use transgenic mice to understand tissue-specific expression of genes were largely unsuccessful in that there was either no expression (Wagner *et al.*, 1983; T. A. Stewart *et al.*, 1982) or inappropriate expression at low levels of the introduced genes (Costantini and Lacy, 1981; Lacy *et al.*, 1983). It is now generally conceded that these problems derived from multiple sources, the foremost of which was the inclusion of either plasmid or bacteriophage λ vector DNA in the microinjection (Chada *et al.*, 1985; Townes *et al.*, 1985; Krumlauf *et al.*, 1985b; Hammer *et al.*, 1985a). Additionally, it is likely that the choice of gene complicated the experiments. For example, in the first instance in which specific expression was suggested (McKnight *et al.*, 1983), preferential expression of the chicken transferrin gene in the liver, the appropriate tissue, was observed. However, the levels of messenger RNA (mRNA) were low relative to the endogenous mRNA, and mRNA could be found in other tissues as well, albeit at lower levels. Thus, confidence in the conclusions was compromised by the significant evolutionary distance between the organisms.

2.1. Rat Elastase Gene

The first clear instances in which introduced genes functioned efficiently and exclusively in the appropriate cell types were the pancreas-specific rat elastase gene (Swift *et al.*, 1984) and a B-cell-specific rearranged mouse κ light-chain immunoglobulin gene (Brinster *et al.*, 1983; Storb *et al.*, 1984). In the former case, 7 kilobases (kb) of an intact rat elastase gene, embedded in 7 kb of 5' flanking sequence and 5 kb of 3' flanking sequence as well as pBR322 plasmid vector sequence, was shown to be expressed only in the pancreas in 4 of 5 independent animals, and at levels equal to or greater than the normal mouse elastase gene. This region contains an element that is homologous with a pancreas-specific enhancer element, capable of directing specific expression of a reporter gene such as the bacterial chloramphenicol acetyltransferase gene, in cells in culture (Walker *et al.*, 1983). The region of the elastase gene responsible for this expression has subsequently been narrowed down by Ornitz *et al.* (1985) to a region 200 base pairs (bp) of DNA to the 5' side of the start of transcription of the elastase gene. This localization was accomplished by fusing the upstream region of the elastase gene to a human growth hormone structural gene, which also permitted the investigators to show by immunohistochemistry that expression of the fusion gene in the pancreas was limited to the pancreas acinar cells. However, the use of the fusion gene resulted in a higher degree of variability in the expression of introduced gene, in that 5 of 18 animals were negative for expression and several others were very low. Whether this implies that elements within the elastase gene play a role in its level of expression or whether this reflects an interaction between plasmid DNA and the rest of the construct remains to be established. Nevertheless, by showing that the upstream region was sufficient to direct expression of a heterologous reporter gene, the investigators opened up the possibility of using such dominant regulatory elements to direct expression of more biologically relevant gene products to specific cells at specific stages of development. The real importance of this is just now being realized, initially by direct expression of cellular and viral oncogenes (see Sections 3, 4, and 5).

2.2. Mouse Immunoglobulin Genes

The DNA signals to direct efficient and specific expression of the heavy- and light-chain immunoglobulin genes must, as well, be contained in a relatively restricted region surrounding the somatically rearranged genes. In the case of the functional κ light-chain gene, Brinster *et al.* (1983) and Storb *et al.* (1984) showed that when the gene was introduced with 1.5 kb of 5' flanking DNA and approximately 7 kb of 3' flanking DNA along with plasmid vector, all mice carrying DNA, irrespective of copy number, expressed the gene at high levels exclusively in B lymphocytes. Likewise, Grosschedl *et al.* (1984) injected a productively rearranged C_μ gene containing 2 kb of 5' flanking DNA and both membrane and secreted 3' exons, along with plasmid DNA. They observed that in four nonmosaic lines examined, μ transcripts from the introduced genes were detected at high levels in both B and T cells, but nowhere else. The finding of C_μ transcripts in T cells is consistent with findings by several groups (Kemp *et al.*, 1980; Alt *et al.*, 1982) of unrearranged or "sterile" μ transcripts in T-cell lines and thymus that initiate near the enhancer element in the J-C intron (Nelson *et al.*, 1983). The clear implications from this are that the cell-type specificity of the heavy-chain enhancer is less restricted than the light-chain κ enhancer and that the failure of T cells to produce C_μ is the result of a failure to rearrange the locus, rather than a failure to recognize its regulatory signals.

The regulatory domain of the heavy-chain C_μ locus is being intensively studied by gene transfer into B-cell-derived tissue-culture cells. In addition to the enhancer in the J-C intron, Grosschedl and Baltimore (1985) have identified at least two additional elements in the promoter and gene body itself. The use of transgenic mice will be invaluable in determining at what stage in B-cell ontogeny each of these elements is required.

2.3. The β-Globin Genes

The successful experiments described above were performed with intact rodent genes introduced into the germ line in the presence of vector DNA. They may, in fact, prove the exception rather the rule, in that there is a growing list of genes the activity of which in mice is significantly inhibited in the presence of the vector DNA. This phenomenon has been most clearly established for the β-globin genes of both mice and humans. These were some of the earliest genes to be introduced into mice, and success was not obtained until Chada *et al.* (1985) generated a hybrid mouse 5′–human 3′ chimeric β-globin gene in which the fusion was at the 3′ side of the second exon of the β-globin gene. This fusion gene, which contained 1.2 kb of 5′ flanking sequence and 1.9 kb of 3′ sequence, was introduced either as a fragment or as an intact plasmid into mouse eggs. Chada *et al.* (1985) detected erythroid-specific expression in 4 of 7 of the mice carrying the fragment alone, although at levels 1–2% of that of the endogenous β-globin mRNA. However, in no case (0 of 3) was the gene expressed in the presence of vector DNA. In those mice that expressed the gene at low levels, the investigators were able to demonstrate that the correct temporal developmental pattern was adhered to, in that the fusion gene was activated only in the fetal liver, and not in the embryonic erythroid cells of the yolk sac islands, in which the endogenous β-globin gene is silent (Magram *et al.*, 1985).

Townes *et al.* (1985) came to very similar conclusions about the importance of removing the vector DNA in experiments in which an intact human adult β-globin gene was introduced into mice along with 4.3 kb of 5′ and 1.7 kb of 3′ flanking sequence. In their experiments, the human β-globin gene was both tissue-specific and temporally correct in its expression. However, like Chada *et al.* (1985), they observed significant variability in the level of expression from animal to animal, from 5 of 20 animals that exhibited no human β-globin RNA to 2 animals harboring 20 copies of the gene, each of which expressed at levels approaching mouse β-globin mRNA.

Townes *et al.* (1985) also introduced deletions of the 4.8-kb 5′ flanking region of the human β-globin gene into mice and found that with as little as 48 bp 5′ of the start of transcription, the gene was appropriately expressed at approximately 10% the level of the endogenous gene in 1 of 3 mice. Expression was also observed in 5 animals bearing a construct with just 122 bp of 5′ flanking sequence at levels from 1 to 98% of the mouse β-globin mRNA. One must conclude from this surprising result that the signals for the activation of the human β-globin gene reside very near or within the gene itself. These results are consistent with those of Wright *et al.* (1984) and Charnay *et al.* (1984), who utilized transfection into mouse erythroleukemia cells to show that the sequences required for the activation of the human β-globin gene after differentiation lay both 5′ and 3′ of the position $+53$.

A question remains about the explanation for the variability in the level of expression of the human globin gene, a result that has been reproduced by Costantini *et al.* (1985), and for the striking effect of the removal of plasmid DNA. It may be simply that the globin genes are more sensitive than other genes such as elastase and immunoglobulin genes to the DNA that surrounds it. Certainly, the best efforts of a number of laboratories over the years have failed to identify a classic enhancer element in these genes. Alternatively, it may be that the globin genes have such elements, but that they are not contained within the constructs that have been introduced into mice. Clearly, this is an issue that needs resolving.

The globin-gene cluster is an especially attractive set of genes to study utilizing transgenic mice because their linkage and developmental specificity raise important issues of how coordinate expression is established and regulated. Costantini *et al.* (1985) have recently extended the studies on the human adult β-globin gene to include the human G-γ globin gene, which is expressed exclusively in the human fetal liver. The mouse has no fetal globin gene equivalent; its globin-gene complement includes only embryonic genes expressed in the yolk sac and adult genes that are activated in the fetal liver. Thus, it was not obvious which pattern the human fetal γ-globin gene would adopt in mice. In fact, it behaved as an embryonic gene, arguing that the *cis*-acting sequences responsible for its expression in human fetal liver are recognized only by embryonic *trans*-acting effectors in the mouse. This is consistent with the conclusion reached by sequence analysis that the human fetal genes arose from the duplication of embryonic and not adult β-globin genes (Efstratiadis *et al.*, 1980). However, this result suggests that the study of the switch-off of the fetal genes in humans after birth may not be amenable to detailed analysis in transgenic mice, since the phenomenon is caused by a change in the timing of expression of trans-acting factors that may not have a counterpart in the mouse.

The other attraction of the globin gene system is the availability of a significant number of naturally occurring and well-characterized mutants that are present in the DNAs of thalassemia patients. Particularly in the case of mutations that cause hereditary persistence of fetal hemoglobin, it will be of great interest to learn whether the phenotype can be studied by reproducing it with fetal globin genes in transgenic mice.

2.4. Mouse α-Fetoprotein Gene

The three gene families described above share one property, exclusive expression in one cell lineage. This is not the case for the mouse α-fetoprotein (AFP) and albumin genes, which are closely linked and coactivated in development in three separate lineages, the visceral endoderm of the yolk sac and the fetal liver and gut (Tilghman, 1985). In addition, in each cell type, the genes are transcribed at different rates; thus, the organism must accomplish diverse genetic programs with a single gene and a single promoter. The transgenic mouse technology is ideal for addressing such complexity, since it generates in one step all the biological material to answer the question.

The AFP gene was introduced into mice as an internally deleted minigene containing the first 3 and last 2 exons of the 15-exon gene, along with 14 kb of 5' flanking DNA (Krumlauf *et al.*, 1985a). This distance was specifically chosen to include the entire intergenic region between the upstream albumin and down-stream AFP genes. In initial experiments in which the plasmid vector DNA was not removed, expression of the minigene was observed exclusively in the three appropriate lineages in 50% of the mice, with levels in the yolk sac that varied between 0.1 and 25% of the endogenous level. Similar results were obtained with constructs containing only the proximal 7.6 kb of the intergenic DNA. The other 50% of the mice that carried DNA did not express it in any tissue. Despite this variability, the expression of the AFP minigene was completely suppressed in liver after birth, at the same time and the same rate as the endogenous gene, thus showing that it was able to respond to temporal signals.

The plasmid vector DNA was subsequently shown to be responsible for a high proportion of the variability in expression of the AFP gene. When constructs containing 7.6 kb of 5′ flanking DNA were introduced into mice free of vector, 100% of the mice expressed the minigene in yolk sac, at levels that were 25–100% of that of endogenous AFP mRNA (Krumlauf et al., 1985b). Much more striking, the levels in the fetal liver ranged from 100 to 1000% of the endogenous level, and in the gut, levels higher than 5000% of the endogenous mRNA in that tissue were observed. Despite this inability to appropriately modulate its level of expres-sion in a tissue-specific manner, the AFP minigene remained silent in all but these cell types (R. E. Hammer, S. Camper, R. Krumlauf, R. Brinster, and S. M. Tilghman, in preparation).

The dramatically elevated expression of the AFP minigenes in the fetal gut points to one of the limitations of the use of gene-transfer technology, whether in intact animals or tissue-culture cells, and that is the currently inescapable fact that the genes are being introduced into heterologous chromosomal sites, and often in tandem arrays at high copy number. This makes it treacherous for the investigator to infer strong, biologically relevant conclusions from the levels of expression of the gene in question, beyond whether it is on or not. The use of reporter genes, whether they be minigenes, completely heterologous genes, or even intact genes from other species, further complicates the analysis, since one must be concerned about the processing and stability of the transcript as well as its rate of transcription. This combination of position effects, copy number, and mRNA stability differences must therefore be factored into any inter-pretation designed to understand the genetic modulation of specific gene expression.

The issue of whether the three cell lineages that express the AFP gene utilize the same regulatory sequences was raised when Godbout et al. (1986) demon-strated that this gene contains at a minimum three enhancer elements distributed over 6.5 kb in the 5′ flanking region, together with a tissue-specific proximal promoter between 85 and 33 bp upstream of the start of transcription. Recent studies in which each of the three enhancer domains was introduced separately into the mouse germ line have shown that the apparent redundancy in this region, as assayed in transient expression assays, does not hold in vivo. That is, while the visceral endoderm can utilize any one of these elements to achieve high-level expression, the fetal liver can use only two of the three, and the gut is the most fastidious in that it recognizes only one element (R. E. Hammer *et al.*, in preparation). It will be important to identify the differences among these elements to finally understand the mechanism that underlies this differential usage *in vivo*.

2.5. Fusion Genes

The fact that the 5' flanking region of a gene is often sufficient to direct expression of any downstream structural gene to a specific cell type has been exploited to attain expression of structural genes at sites at which they would not normally be synthesized.

The first fusion gene to be studied in detail was the fusion between the mouse metallothionein (MT) promoter and the herpes simplex virus thymidine kinase (*tk*) gene (Brinster *et al.*, 1981; Palmiter *et al.*, 1982b). The endogenous metallothionein gene is expressed ubiquitously in the mouse, with the kidney and the liver being the predominant sites of synthesis. The fusion gene behaved similarly, implying that the choice of tissue specificity resided in the MT promoter region alone. In addition, the level of tk mRNA could be increased by the administration of heavy metal (Palmiter *et al.*, 1982b).

However, in two instances, the combination of a control region with a reporter gene has produced a totally unexpected pattern of expression in the mouse. When a fusion between the mouse MT promoter region and the human growth hormone structural gene was introduced into mice and examined for its sites of expression, Swanson *et al.* (1985) observed that the fusion gene was expressed in a very specific subset of neurons that expressed neither gene from which the fusion was generated. Likewise, a mouse MT promoter–growth-hormone-releasing factor structural gene fusion was expressed in the brain—in this instance, however, in an equally specific but different set of neurons that once again were not sites of synthesis of either MT or growth-hormone-releasing factor (Hammer *et al.*, 1985b). In both instances, the unusual pattern of expression was observed in several independently generated lines, suggesting that it derived from the fusion gene itself, and not from the heterologous site into which it had integrated. These observations, along with several examples from *Drosophila melanogaster*, in which a fusion between the SgS4 glu protein gene enhancer and the larval alcohol dehydrogenase promoter and gene was expressed in a new set of tissues in the adult fly (J. Posakony and T. Maniatis, unpublished results), argue that the signals for tissue specificity for a given gene may arise from the combination of multiple regulatory elements, each of which conveys the information for only part of the specificity. By combining two different kinds of elements, one effects a new specificity. This possibility is consistent with the studies by several groups on the multiplicity of regulatory elements in the insulin (Edlund *et al.*, 1985), immunoglobulin heavy-chain (Grosschedl and Baltimore, 1985), and AFP genes (Godbout *et al.*, 1986).

2.6. Conclusions

In conclusion, the appropriate tissue-specific expression of genes introduced into the mouse germ line is now becoming the rule rather than the exception. In the past year, investigators have shown that the rat myosin light-chain 2 (Shani, 1985), mouse crystallin (Overbeek *et al.*, 1985), mouse collagen α(I) (Westphal *et al.*, 1985), and rat insulin (Hanahan, 1985) regulatory regions behave appropriately either as intact genes or as fusions to various reporter genes. As has been emphasized above, while the *cis*-acting sequences that determine the tissue specificity clearly can be delineated by this technique, one needs to be far more cautious about deriving conclusions regarding the mechanisms that control the rate of transcription in a given tissue. This issue needs further thought.

3. PHYSIOLOGICAL CONSEQUENCES OF ALTERED GENE EXPRESSION IN TRANSGENIC MICE

The previous section described experiments that lead one to conclude that in the future it should be possible to direct expression of any gene to any tissue at any stage in development, given the right combination of control elements. The consequences of this possibility for the understanding of such fundamental processes as growth and immunity are that one can subtly perturb the system and, from the response of the animal, draw conclusions about mechanism and control. However, if the perturbations are too dramatic, it may not be possible to draw meaningful conclusions.

3.1. Growth Control

Growth control in mammals is thought to be regulated by a complex cascade of peptide hormones. As the consequence of neurotransmitters acting on the hypothalamus, growth-hormone-releasing factor (GRF) is released, which in turn acts at the pituitary to effect the release of growth hormone (GH). GH is thought to stimulate the liver to produce insulinlike growth factor I (IGF-I), which then acts on peripheral tissues to stimulate growth. Negative control is exerted by the release of somatostatin by the hypothalamus, which inhibits further GH release by the pituitary. In addition, both GH and IGF-I have been implicated in negative-feedback mechanisms on the pituitary and hypothalamus, respectively. The fine tuning of the system can be perturbed by generating mice with constitutively high levels of these hormones.

In 1982, Palmiter *et al.* (1982a) showed that several mice carrying a fusion between the mouse metallothionein promoter and the rat GH structural gene produced very high levels of GH mRNA in liver and grew as much as 87%

larger than normal littermates. Levels of circulating GH varied from line to line from 2- to 800-fold over normal levels and correlated roughly with increased body weight. Similar results were obtained with metallothionein promoter–human GH fusions (Palmiter *et al.*, 1983). In this instance, the effects of overproduction of GH were examined in some detail. Accelerated growth of these mice did not commence until 3 weeks after birth, suggesting that prenatal and neonatal growth is controlled by hormones not regulated by GH or that the levels of IGF I are limiting at this stage. One likely candidate for an alternative hormone is insulinlike growth factor II, thought to be the fetal counterpart of IGF I. Levels of IGF I elevated 2- to 3-fold were found in adult mice, consistent with the idea that GH exerts its effects in part by stimulating hepatic production of this peptide (Daughaday *et al.*, 1972). Interestingly, the IGF I levels were inversely proportional to the GH levels in four lines examined, which explained why there was not a direct correlation between GH and growth in these mice. It would appear that there is a negative-feedback inhibition of IGF I in mice that overproduce GH. In addition, there appeared to be a negative feedback by the exogenous GH on the growth of the pituitary. This tissue appeared by histological examination to be severely reduced in the number of acidophilic cells.

Unexpectedly, while transgenic males could transmit the expression of the gene to progeny, females were infertile. Whether this is the consequence of the high circulating GH levels or of production of GH at specific ectopic sites that interfered with normal female reproduction is not clear. However, the infertility cannot be attributed solely to the accelerated growth of the mice, since the introduction of metallothionein promoter–GRF structural gene fusions into mice resulted in 25–50% increased growth rates, but no infertility (Hammer *et al.*, 1985b). There was significant variability among independent lines in the circulating levels of GRF, up to 25 times the endogenous level, and indeed mouse GH itself, which ranged from the normal level of approximately 30 ng/ml to as high as 1100 ng/ml. However, there were no clear correlations between GRF and GH levels or, for that matter, between GRF levels and growth. Thus, while one can conclude that GRF can induce increased growth, the rate-limiting step is not evident from these data. Unlike the pituitaries of the GH-producing transgenic animals, those of the GRF-producing animals exhibited marked hyperplasia. That overproduction of GRF produces a less severe perturbation of the reproductive endocrine system may be of some use in applying this technology to domestic animals of commercial interest.

Somatostatin normally inhibits the production of GH, and one might expect its overproduction in mice to inhibit growth. In fact, a metallothionein promoter–somatostatin complementary DNA fusion, when introduced into mice, resulted in the elevated production of fully processed somatostatin, primarily in the anterior pituitary (Low *et al.*, 1985). However, the mice grew normally. Whether this reflects a requirement for localized somatostatin secretion by the hypothalamus or a desensitization of pituitary receptors to somatostatin inhibition of GH release is not clear.

The first instance in which a genetic defect was corrected by altering the genetic complement of a mammal was reported by Hammer *et al.* (1984). They introduced a metallothionein–rat GH gene fusion into "little" (*lit/lit*) mice homozygous for a mutation on mouse chromosome 6 (Eicher and Beamer, 1976). This mutation reduces their circulating growth hormone levels and results in reduced body weight. This defect was completely reversed in the transgenic *lit/lit* mice, which were larger than wild type. In addition, the "correction" was heritable, although the females were infertile, as had been observed with normal GH transgenic mice. Curiously, the males were somewhat more fertile than *lit/lit* mice, an effect that is not understood.

3.2. Immune System

By introducing productively rearranged and efficiently expressed immunoglobulin genes into the mouse germ line, perturbations of the mechanisms that underlie the generation of functional antibodies has been achieved. During B-cell ontogeny, immunoglobulins are generated by the juxtaposition of V_H, D_H, J_H, and C_H gene segments in the case of the heavy-chain locus and V_L, J_L, and C_L segments in either the κ or the λ light-chain loci (Tonegawa, 1983). Diversity is generated by the choice of the V, D, and J segments, at the specific sites at which they join, as well as by subsequent somatic mutation of the rearranged V-gene segment.

Normal B cells express only one heavy- and light-chain gene at any given time, and exclude from expression the other chromosome, which can exist in its germ-line or unrearranged configuration or in a nonfunctional rearrangement (Alt *et al.*, 1980; Coleclough *et al.*, 1981). Two models have been advanced to explain this allelic exclusion: that the probability of productive rearragemment is so low as to ensure that no two alleles are simultaneously producing functional mRNA (Coleclough *et al.*, 1981) or that there exists a negative feedback by the rearranged gene or its product that inhibits further rearrangement (Alt *et al.*, 1980; Bernard *et al.*, 1981; Kwan *et al.*, 1981). By generating B cells that constitutively produce one exogenous immunoglulin chain, one can ask whether the endogenous alleles are inhibited from rearranging, a direct test of the second model.

Ritchie *et al.* (1984) examined this question in three transgenic lines expressing a rearranged κ light chain. No hybridomas produced from transgenic spleen cells were identified that synthesized two κ chains. Thus, allelic exclusion was operative in those lines in which the transgenic κ chain was produced. In addition, approximately 50% of the B-cell hybridomas showed unrearranged κ and λ loci, although rearrangement of the heavy-chain locus was unaffected. More interesting, this inhibition of κ and λ gene rearrangement occurred only in those hybridomas that were producing functional antibodies, i.e., where the transgenic κ light chain was associated with a heavy chain. In hybridomas that expressed exclusively endogenous κ chains or transgenic κ chains but no heavy chain, rearrangement at the κ alleles was observed. Thus, the assembly of a functional antibody molecule is required to prevent further gene rearrangement.

A similar result was obtained in several lines bearing a transgenic heavy-chain C_μ gene. By examining pre-B cells that had been transformed by Abelson murine leukemia virus, Weaver et al. (1985) concluded that 40% of C_μ-producing lines had unrearranged J_H segments, compared to none derived from normal littermates. When more mature B cells were examined following Abelson virus transformation, 10% of the alleles remained unrearranged. Whether this inhibition of rearrangement is mediated at the DNA, RNA, or protein level has not been elucidated. However, unlike the κ case, functionally assembled antibody molecules are not required. The fact that the inhibition of rearrangement at the C_H locus is less than at the C_κ locus may be a consequence of when the transgenic C_μ gene is transcribed relative to C_H rearrangement during B-cell ontogeny.

Rusconi and Köhler (1985) also examined the question of allelic exclusion by coinjecting C_μ and κ chain genes into mice. Although they could document inhibition of heavy-chain rearrangement in hybridomas equivalent to that observed by Weaver et al. (1985), they also observed a failure of allelic exclusion in that some hybridomas were expressing more than one heavy or light chain or both. In four lines examined in detail, the endogenous κ gene was producing approximately 10 times more mRNA than the transgenic κ gene. Thus, the absence of allelic exclusion at that locus may be a concentration-dependent phenomenon, whereby the accumulation of assembled antibodies was insufficient to activate the feedback mechanism.

These transgenic lines can now be used to ask whether the localized somatic mutation that occurs at the rearranged V_H and V_L genes is dependent on their chromosomal location or on a more limited sequence domain. By introducing nonsense mutations into the introduced C_μ gene, the issue of whether functional mRNA or protein or both are required for inhibition of rearrangement may be addressed. Once that is known, the domain can be defined.

Studies on the effects of transgenic products of the major histocompatibility complex (MHC) have also been initiated, in an effort to better understand their function in areas such as self-tolerance and the maturation of the immune response. After Frels et al. (1985) introduced a porcine class I MHC gene into mice, they determined that the foreign transplantation antigen was recognized by mouse T cells in a skin-graft-rejection test. This should allow the investigators to determine the relationship between those T cells that specify xenogeneic MHC products and allogeneic murine MHC determinants and to study the acquisition of self-tolerance to the porcine antigen.

The C57BL/6 and SJL inbred mouse strains do not synthesize the Class II antigen E_α, and consequently the I region $E_{\alpha\beta}$ heterodimer does not appear on the cell surface. This glycoprotein on the surface of antigen-presenting cells activates helper T cells, an event required for the induction of most cell-mediated and B-cell antibody responses. In a series of similar experiments, Le Meur *et al.* (1985), Yamamura *et al.* (1985), and Pinkert *et al.* (1985) showed that when an E_α gene was introduced into (C57Bl/6 × SJL)$_{F2}$ or C57Bl/6 mice, the gene was expressed in a tissue-specific fashion on T cells in spleen and thymus and was inducible by γ-interferon in macrophages. In addition, the mice could then mount a B-cell-mediated immune response to synthetic antigen, which was dependent on the expression of the E complex (Yamamura *et al.*, 1985; Le Meur *et al.*, 1985) and could function in a mixed-leukocyte response assay and an antigen-presentation test (Pinkert *et al.*, 1985; Yamamura *et al.*, 1985).

Pinkert *et al.* (1985) also noted that the expression of the E_α gene rendered the transgenic males infertile. This is a curious finding, since inbred strains expressing E_α are obviously fertile. It is possible that the DNA injected, which was 43 kb in length, may contain other genes responsible for the infertility. Alternatively, it could be that some I-E alleles in conjunction with other MHC genes lead to male infertility, which would suggest that the loss of E_α expression in certain inbred strains was selected for. This observation highlights the power of the transgenic technology to uncover interesting biological phenomena.

4. INTRODUCTION OF VIRAL TRANSFORMING GENES INTO EMBRYONIC CELLS AND TRANSGENIC MICE

The exposure of embryos or embryonic stem cells to infectious viral agents has been carried out in the past for several different reasons. First, the study of how viral genetic information is regulated or expressed in diverse developmental backgrounds has been used as a model system to understand the regulation of cellular genes (F. Kelly and Condamine, 1982; Maltzman and Levine, 1981; Levine, 1982). Second, the impact of viral-induced tissue pathology in the fetus has been explored to examine the role of these agents in specific types of birth defects (Notkins and Oldstone, 1984). Third, a number of viruses are capable of producing tumors in animal hosts after inoculation of a fetus or newborn with an infectious agent. While these viruses may infect and replicate in a large number of cell types, only a subset of characteristic tumors of specific tissue types are usually observed (polyoma virus is a notable and interesting exception) (Tooze, 1981; Weiss *et al.*, 1982). The study of viral-induced tumorigenesis, after exposure of the embryo to these viruses or even with germ-line transmission (vertical transmission) of viruses, has been an informative approach to study the problem of tissue-specific tumorigenesis (Brinster *et al.*, 1984; Small *et al.*, 1985). Similarly, the introduction of the normal cellular forms of oncogenes (protooncogenes, c-onc) or altered forms of oncogenes into fertilized eggs and the subsequent analysis of transgenic animals and their progeny has been and will continue to be a fruitful experimental approach to answer questions of gene regulation and tumorigenesis (T. A. Stewart *et al.*, 1985). Indeed, these systems have yielded new experimental animal models for testing or refining diagnostic techniques whereby, for example, tumors appear a reproducible number of times in an animal's life span (van Dyke *et al.*, 1985). In addition, these experimental approaches provide, for the first time, animal model systems for diseases the etiology of which has been difficult to study, such as peripheral neuropathies associated with particular transgenic animal lines (Messing *et al.*, 1985). Clearly, understanding the many consequences of viral infections or germ-line transmission of viruses in zygotes, embryos, or adult animals depends on a better knowledge of gene regulation of viral chromosomes in diverse cellular developmental backgrounds. For this reason, it will be worthwhile to first review the available information concerning the expression of viral genomes in embryonic cells or related embryonic stem cells such as embryonal carcinoma cells. This will then be followed by a discussion of viral genes in transgenic mice. The viruses most commonly studied in this context have been simian virus 40 (SV40), polyoma virus, and the Moloney murine leukemia virus (M-MuLV). The focus of this review is the mouse embryo and derived transgenic mice, and for that reason, only the interactions of these three viruses with murine cells will be discussed.

4.1. Expression of Viruses in Embryonic Cells and Stem Cells

4.1.1. Simian Virus 40

SV40 undergoes a complete productive infection in African green monkey kidney cells in culture (Tooze, 1981). The virus was first isolated as a contaminant in rhesus monkey kidney cells brought into cell culture. SV40 infects a large proportion of the monkey population and persists (in the kidney at least) throughout the lifetime of the animal, producing little or no known disease in its natural host (Tooze, 1981). The virus may be secreted in the urine and replicated in kidney cells, suggesting that an equilibrium is attained between the parasite and its host that results in long-term persistence. When SV40 is employed to infect mouse cells in culture, the virus enters the cell and produces two proteins, the large and small tumor antigens (T and t antigens), the genetic information of which represents about one half the viral genome (2.6 kb, the early viral genes). Viral DNA fails to replicate in mouse cells in culture, and the late or structural proteins of the virus are not synthesized (Tooze, 1981). In a small percentage of cells infected with SV40, the viral DNA integrates and expresses the viral tumor antigens, and the growth properties of these cells are altered (i.e., transformation). The expression and function of the SV40 large-T antigen (and in some cases the small-t antigen) are required for the establishment and maintenance of this transformed state (Tooze, 1981).

The exposure of preimplantation mouse embryos to SV40 has been reported to kill the embryos. The early large-T antigen gene product and late structural protein antigens could be detected in the trophectoderm cells, but not in the inner mass cells (Abramczuk et al., 1978) that will eventually make up the embryo. Jaenisch (1974) infected preimplantation mouse embryos with SV40 and failed to detect viral DNA replication in these embryos. This would appear to contradict the results of Abramczuk et al. (1978), because in monkey cells, SV40 DNA replication is normally required for the expression of the late viral structural proteins (Tooze, 1981). It remains possible that the trophectoderm cell type permits expression of the SV40 late viral genes even in the absence of DNA replication. On the other hand, the approaches employed to detect viral DNA replication or persistence in morulae may not have been sensitive enough to observe this. Willison et al. (1983) infected morulae with SV40, and both integrated SV40 DNA and episomal SV40 DNA were detected in a few individuals subsequently analyzed by much more sensitive techniques than first employed by Jaenisch (1974). It appears that the inner cell mass that will eventually make up the embryo is composed of cells that are not permissive for viral gene expression at the preimplantation stage, but that trophectoderm cells can support some SV40 gene expression, as can cells derived from the inner cell mass at later stages in differentiation, i.e., 9- to 10-day postimplantation embryos (see F. Kelly and Condamine, 1982).

This conclusion is consistent with more detailed studies of SV40 gene expression in embryonal carcinoma (EC) cells (Levine, 1982). These stem cells are thought to be most analogous to inner cell mass cell types. As anticipated, then, the SV40 genome in EC cells failed to produce detectable levels of the viral tumor antigens, and viral DNA was not replicated (Swartzendruber and Lehman, 1975) in these cells. When EC cells were induced to differentiate in cell culture, SV40 virus could then synthesize its viral tumor antigens (Levine, 1982). These EC-derived and differentiated cells were then able to be transformed by SV40 (Maltzman et al., 1979). The nature of the block to SV40 early gene expression in EC cells is at present controversial. What is clear is that the steady-state levels of SV40 early mRNA in virus-infected EC cells is lower than that found in virus-infected differentiated mouse cells or SV40-infected murine cells (3T3 cells) in culture (Levine, 1982). Whether this reduced level of early mRNA is due to a reduced rate of transcription (Segal and Khoury, 1979), an altered or incorrect splicing mechanism (Segal et al., 1980), or a translational block (Linnenbach et al., 1980, 1981; Huebner et al., 1983) is unclear. Whatever the mechanism, the SV40 genome fails to direct the synthesis of viral proteins in early embryonic cells (inner cell mass or EC cells), while cells derived from these stem cells contain the cellular component to support the expression of viral proteins such as large-T antigen.

4.1.2. Polyoma Virus

Polyoma virus was isolated from mouse tissue and has the interesting property of producing tumors and expressing its viral gene products in a wide variety of murine cell and tissue types (Tooze, 1981). The virus undergoes a productive infection in mouse cells, most often resulting in cell death. Rare survivors of the virus infection may be transformed and then contain a defective viral genome integrated into the transformed cell line. The viral DNA expresses a portion of its genome, the tumor antigens, resulting in the transformed phenotype. These cells no longer produce infectious virus. Infection of mouse preimplantation embryos with polyoma virus does not kill these cells, and the embryos fail to support viral gene expression or viral DNA replication (see Jaenisch and Berns, 1977; F. Kelly and Condamine, 1982). In a manner similar to SV40, postimplantation mouse embryos on day 9 of gestation become permissive for both early and late viral gene expression (F. Kelly and Condamine, 1982). In addition, it has been reported that polyoma can express its late proteins in trophectoderm cells of preimplantation embryos (Abramczuk et al., 1978), but these cells do not die. It is clear that polyoma gene expression, like SV40, is restricted in early embryonic cell types (inner mass cells).

Polyoma virus can infect EC cells in culture and produces very low steadystate levels of viral mRNA. The mRNA that is produced is correctly spliced and processed (Dandolo et al., 1983), but viral tumor antigens are not detectable. Mutants of polyoma virus that can now direct the synthesis of viral early tumor antigens and infectious virus have been isolated by three different laboratories. In all cases, these mutations could be mapped to the enhancer region of the viral chromosome, 5' to the viral gene promotor and start of early mRNA synthesis (Vasseur et al., 1980; Fujimura et al., 1981; Sekikawa and Levine, 1981; Levine, 1982). The polyoma enhancer region is required for the transcription of the viral early genes and for the replication of viral DNA even when the early gene products are supplied in trans. Thus, these enhancer mutations do not discriminate between a requirement of the mutant enhancer in EC cells for increased transcription or for enhanced ability to replicate viral DNA, or for both (Fujimura and Linney, 1982). Polyoma mutants that were selected for virus replication in one EC cell line (F9) failed to replicate in another EC cell line (PCC-4) (Sekikawa and Levine, 1981) and vice versa (Vasseur et al., 1980). These two classes of polyoma EC cell mutants mapped to different parts of the enhancer region on the viral chromosome. These observations suggest that different EC cell lines are expressing different regulatory elements that act on distinguishable regions or signals in the viral enhancer region of the chromosome. The viral mutants might then discriminate among different stages of development reflected in different EC cell lines. This in turn could identify different stages of embryonic development when studying inner cell mass cells of the embryo.

4.1.3. Moloney Murine Leukemia Virus

M-MuLV replicates productively in cultured mouse fibroblasts. The virus is not permissive for replication in preimplantation mouse embryos (Jaenisch *et al.*, 1975), although a productive infection of embryonic cells can be achieved in 8- to 9-day-old postimplantation embryos. In this case, many different cell or tissue types can participate in the productive infection of 8- to 9-day-old embryos (Jähner *et al.*, 1982). In preimplantation embryos, M-MuLV enters these cells and produces a DNA copy that integrates into the host chromosome, but this DNA fails to be expressed and no viral proteins are produced (Jähner *et al.*, 1982). In this instance, the proviral genome may be transmitted vertically, in a Mendelian fashion, over many generations (Janenisch *et al.*, 1983). In some cases, the offspring of these mice will routine activate proviral transcription and virus production, and leukemia results (Jaenisch *et al.*, 1981). The time at which activation and virus production occur in an animal's life varies among substrains of mice with different integration sites of the provirus. This observation has been used to suggest that a temporally regulated event could be involved in initiating the expression of the proviral genome and subsequently the disease state (Jaenisch *et al.*, 1981). Since the M-MuLV agent can replicate in adult mouse tissues, rare events resulting in activation of the provirus in one cell are propagated and amplified in the whole animal.

Similarly, M-MuLV can infect EC cells and integrate its proviral DNA into the chromosomes of this host cell (C. L. Stewart *et al.*, 1982; Jähner *et al.*, 1982). The proviral DNA produces much reduced (1/100) steady-state levels of mRNA, and no infectious virus is detectable (Teich *et al.*, 1977; C. L. Stewart *et al.*, 1982; Gautsch and Wilson, 1983). When these EC cells were induced to differentiate, proviral transcripts were not detected and infectious virus was not produced (C. L. Stewart *et al.*, 1982; Jähner *et al.*, 1982). This is in contrast to the M-MuLV productive infection of cells differentiated from EC cells in culture or postimplantation embryos (day 9 embryos), in which viral mRNA and infectious virus are produced. Therefore, a retrovirus infection of EC cells or preimplantation embryonic cells results in a provirus that cannot be efficiently activated on differentiation of those stem cells, even though the differentiated cells themselves can support M-MuLV proviral gene expression if infected from without.

In some studies, the lack of provirus expression in EC cells could be correlated with methylation of cytosine residues in the proviral DNA and adjacent regions (C. L. Stewart *et al.*, 1982). In agreement with this observation, M-MuLV proviral DNA copies in mice derived from virus-infected preimplantation embryos were also highly methylated at the cytosine residues (Jähner *et al.*, 1982, Jähner and Jaenisch, 1985). The insertion of such a provirus can eliminate gene expression in cellular DNA sequences adjacent to the provirus integration site and result in hypermethylation of residues in this cellular DNA (Jähner and Jaenisch, 1985). Hypermethylation of cytosine residues has been correlated with lower levels of steady-state mRNA produced from that region of the chromosome (Doerfler, 1983). Experimental results consistent with this interpretation were obtained by transfecting integrated proviral DNA sequences derived from EC cells into NIH-3T3 cells permissive for M-MuLV. When the recipient cells were treated with 5-azacytidine to inhibit methylation of newly replicated C-residues prior to or after transfection of the proviral DNA, infectious virus was produced (C. L. Stewart *et al.*, 1982; Gautsch and Wilson, 1983). In the absence of 5-azacytidine treatment, infectious M-MuLV was not detectable in the transfected cells. These results are consistent with the possibility that *de novo* methylation of proviral DNA sequences (and surrounding cellular DNA sequences) in EC cells or preimplantation embryo cells shuts off the expression of the provirus, an event that is not efficiently reversed by differentiation. However, additional experimental tests of this idea have failed to produce a good correlation between the timing of methylation of the proviral DNA and provirus inactivation (Niwa *et al.*, 1983; Gautsch and Wilson, 1983). This interpretation therefore requires further testing and remains to be proven correct or not.

The M-MuLV enhancer and promotor DNA sequences localized in the long terminal repeat (LTR) of the viral DNA function poorly in short-term DNA transfection assays in EC cells (Linney *et al.*, 1984). Substitution of the polyoma enhancer mutants that permit polyoma virus to replicate in EC cells for the M-MuLV enhancer restored high-level expression of a test gene in EC cells (Linney *et al.*, 1984). When an M-MuLV LTR is used to express a selectable gene, such as the neomycin-resistance gene, which provides for cellular resistance to the drug G418, then the LTR can function, but at a frequency that is reduced 10,000 to 100,000-fold in EC cells as compared to fibroblasts (Sorge *et al.*, 1984). Thus, the poor expression of M-MuLV-LTR-promoted genes in EC cells can be overcome at a low frequency. Whether that results from the integration site of the provirus, a varient cell line selected in the EC cell culture, transient expression during one stage in the cell cycle, or other possibilities remains to be explored.

4.1.4. Conclusions

Viruses as diverse as SV40,. polyoma (extrachromosomal), and M-MuLV (integrated) fail to express their genetic information in early developmental cell types such as the inner cell mass of preimplantation embryos or EC cells. Mutations that permit the expression of viral gene products in such cells map to the enhancer regions of the viral chromosomes, and the steady-state levels of viral gene products and mRNAs are much reduced in these early embryonic cell types. Different portions of the enhancer sequences may be recognized by *trans*-acting factors that regulate gene expression at different stages of early embryonic development. As cell and tissue types differentiate in the postimplantation embryo or EC cells are induced to differentiate in culture, these tissues become permissive for the replication or expression of these viral genes. There is a dramatic change in the ability of differentiated cell types to recognize the cis-acting signals carried by these viruses. There is some indication that these same viral cis-acting signals are also important for the replication of the viral DNA itself.

The idea that differentiated cells acquire a set of new trans-acting gene functions that positively regulate cis-acting signals adjacent to a structural gene (enhancer sequences) and that the viruses discussed here are good measures of that new cellular function cannot by itself be the entire explanation. The de novo methylation of M-MuLV proviral sequences (or some other event) is not readily reversed on differentiation, so at the very least, the trans-acting activities must respect methylation patterns or other signals. Because infection of differentiated cells with M-MuLV results in viral gene expression, an activity, such as de novo methylation, also needs to be altered during these developmental processes. It is not surprising that two distinct levels of regulatory events (e.g., hypomethylation and protein recognition of enhancer signals) might be involved in the changing pattern of gene expression during development. Furthermore, that they might act sequentially and at specific regulatory sites in the DNA should not come as a surprise and additionally has the virtue of being a testable set of ideas.

4.2. Production of Transgenic Mice: Simian Virus 40

The first injections of viral DNA into mouse embryos were reported by Jaenisch and Mintz (1974). SV40 DNA was microinjected into the blastocoel cavity of embryos, and the embryos were implanted into pseudopregnant females and allowed to develop to term. Adult mice derived from these experiments were found to contain varying amounts of SV40 DNA sequences in different organs ranging from about 0.5 to 13 genome equivalents of SV40 DNA per cell (Jaenisch and Mintz, 1974). While some mice died prematurely or unexpectedly, no pathology or abnormal reactions were reported in this study. Attempts to repeat this study have been reported (Willison et al., 1983; Abramczuk et al., 1983), and both integrated and extrachromosomal copies of SV40 circular molecules could be shown to persist in the mice derived from either infection or microinjection of viral DNA into the blastocoel cavity (Willison et al., 1983). Little or no information is available in these studies on the possible expression of the SV40 genome in these mice.

More recently, Brinster et al. (1984) injected two different plasmids containing the SV40 early region genes for the large and small tumor antigens into the pronucleus of fertilized mouse eggs. The plasmids contained the SV40 early region from KpnI (nucleotide 294) to BamHI (nucleotide 2533) in a pBXΔ bacterial plasmid background. In one case, the plasmid contained a metallothionein–thymidine kinase fusion gene (called *SV-MK*), and in a second experiment, a metallothionein–human growth hormone gene (called *SV-MGH*) fusion was in the plasmid construction. Between 240 and 830 plasmid DNA molecules were microinjected into the zygote pronucleus in three separate experiments. A total of 925 eggs were injected and transferred to foster mothers in this study. Of this total, 95 pups were born (about 10% efficiency). The frequency of pups born was similar whether or not the SV40 early gene region was present in these plasmids. Of the 95 mice born in this study, 25 mice (26%) were shown to be transgenic and to contain SV40 DNA sequences in DNA obtained from the tails. The 26% transgenic mice produced is a very high frequency and indicates no evidence for lethality of the SV40 early region DNA (or the gene products that might have been expressed) by these plasmid constructions carried during embryogenesis.

Of the 25 mice that carried SV40 DNA in all their tissues, 16 (64%) died prematurely (by 6 months of age), with histologically confirmed tumors of a specific organ, the choroid plexus. The choroid plexus is the organ in the brain that filters the serum from the blood and produces the cerebrospinal fluid that bathes the brain and spinal chord. The organ is the blood–brain barrier, composed of a semipermeable filter permitting a subset of the serum proteins into the central nervous system. In all cases, the tumors were classified papillomas of the choroid plexus. Founder mice were bred, and through at least six generations, the SV40 DNA segregated in a Mendelian fashion (Brinster *et al.*, 1984; van Dyke *et al.*, 1985). Mice that carried SV40 DNA developed tumors of the choroid plexus showing a cosegregation of the SV40 DNA and the tumorigenic phenotype over several generations. Much like the mice carrying the M-MuLV provirus (Jaenisch *et al.*, 1981), independently derived mouse lines carrying SV40 DNA at different integration sites developed tumors of the choroid plexus at different times after birth (van Dyke *et al.*, 1985). For example, the 427 line of mice developed choroid-plexus tumors and died (24 mice over 4 generations) at 156 ± 10 days, while SV40-containing mice of the pSV11 lineage (26 mice over 6 generations) developed tumors and died at 110 ± 15 days. These nonoverlapping distributions suggest that additional events play a role in the complex phenotype under study here.

There were two other phenotypes noted in some of the original 16 mice that died prematurely in this study. Some mice also had hyperplasia of the thymus and some developed cysts in the cortex of the kidney (Brinster *et al.*, 1984). The expression of SV40 large-T antigen was restricted to the tissues that showed these pathologies or phenotypes. Papilloma tumor tissue contained high levels of the SV40 large-T antigen, while hyperplastic thymic tissue or kidney, with cortical cyst pathology, often contained lower levels of this viral gene product (Brinster *et al.*, 1984; van Dyke *et al.*, 1985). Analysis of the steady-state levels of SV40 large-T antigen mRNA confirmed that tumor tissue and hyperplastic thymus and kidney were the only tissues containing RNA that hybridized with

SV40 probes (Brinster *et al.*, 1984). Thus, the differential tissue expression of SV40 mRNA and protein correlated well with tissue pathology detected in these mice. When normal and tumor tissue were compared for SV40-related DNA sequences, there were some examples (but not in every case) of amplified copies of SV40 DNA in the brain tumors. As cell lines were established from these brain tumors or hyperplastic thymic cells, additional amplifications of SV40 DNA copy numbers could be detected (up to 2- to 10-fold). Apparently, the amplification of SV40 DNA copies in tumor tissue or cell lines derived from tumor tissue can occur and be selected for as lines are cloned or subcloned (Brinster *et al.*, 1984). DNA rearrangements at the SV40 integrated locus were rare or not detected.

The tissue-specific expression of the SV40 mRNA and large-T antigen in transgenic mice and the resultant pathology could be shown to reside, at least in part, in the enhancer of the SV40 chromosome (Palmiter *et al.*, 1985; Hanahan, 1985). Deletion of the SV40 enhancer sequences (the 72-bp repeat) reduced the incidence of papillomas of the choroid plexus and led to the new pathologies such as liver and pancreatic tumors (Palmiter *et al.*, 1985). This new spectrum of tissue pathology could have resulted from the metallothionein promoter in the plasmid constructs employed in that study. In another approach, Hanahan (1985) replaced the SV40 enhancer with an upstream region of the insulin gene. The SV40 large-T antigen was then detected in adult transgenic mice carrying this construction exclusively in the beta cells of the pancreas. Beta-cell tumors (insulinomas) were detected in these mice, and progeny mice inherited the tumorigenic phenotype and the hybrid gene (Hanahan, 1985). While every mouse carrying this hybrid gene eventually developed a tumor of the beta cells and many independent islets of beta cells were large-T-antigen positive, only a small fraction of the islets became tumorigenic (Hanahan, 1985). These observations suggest that a second event is required to produce such tumors. These studies also provide good evidence that the tissue-specific expression of the SV40 large-T antigen is regulated, at least in part, by the viral enhancer region of the chromosome.

That this type of regulation is not the entire explanation for tissue-specific gene expression and pathology can be seen by comparing the results of experiments in which the SV40 enhancer was deleted from the large-T antigen construct but the human growth hormone gene was present or not in the plasmid. When the human growth hormone gene was present in an SV40-enhancer-negative large-T-antigen gene plasmid, the mice were observed to have peripheral neuropathies (13 of 16 mice), liver tumors (11 of 16 mice), pancreatic tumors (8 of 16 mice), and adrenal tumors (3 of 16 mice) and 1 mouse had a tumor of the choroid plexus. In the absence of the SV40 enhancer and no human growth hormone gene, there was a reduction in detectable pathology (1 of 7 mice with a pancreatic tumor). It appears likely that the nucleotide sequences within the metallothionein promoter human growth hormone gene itself (or the gene product) could play an important role in directing tissue-specific gene expression when the SV40 control elements are deleted (Palmiter *et al.*, 1985).

That the SV40 large-T-antigen gene product is itself essential for tumorigenesis was shown by employing viral deletion mutations. Transgenic mice were constructed that contained only the large-T antigen gene intact and deleted the small-t antigen gene. Of 7 mice with this construction, 5 died with papillomas of the choroid plexus (Palmiter *et al.*, 1985). Deletion mutations that altered the structure of the SV40 large-T antigen have failed (in 9 transgenic mice and the progeny of two transgenic mouse lines) to produce tumors of the choroid plexus or any other detectable pathologies associated with transgenic mice expressing the SV40 large-T antigen.

SV40 DNA can be transmitted through the germ line of transgenic mice in a Mendelian fashion, and with some lines of mice (pSV-11), every mouse containing SV40 DNA develops a tumor of the choroid plexus at a characteristic time (110 ± 15 days). With other lines of mice (419), the SV40 DNA is inherited in a Mendelian fashion, but mice carrying this DNA fail to develop any pathology (Brinster *et al.*, 1984; van Dyke *et al.*, 1985). This expression of SV40 large-T antigen in the pSV-11 mice has been followed by several techniques. Employing Western blotting, the large-T antigen was detected in extracts made from the brain tissue of pSV-11 mice as early as 7–14 days after birth (van Dyke *et al.*, 1985; van Dyke and Levine, unpublished results). In agreement with that observation is the reproducible ability to develop cell lines expressing SV40 large-T antigen from brain tissue obtained from 1- to 2-week-old pSV-11 animals (van Dyke *et al.*, 1985). Brain tissue from this same line of mice failed to establish cell lines in culture when the cells were derived from animals less than 1 week old.

The first indication of a precancerous pathology in pSV-11 mice detected histologically is at about 8–12 weeks, and frank tumors appear by 14–16 weeks coincident with a large increase in large-T antigen levels and rapid tumor-cell growth (van Dyke *et al.*, 1985; van Dyke and Levine, unpublished results). Thus, SV40 large-T antigen levels are detectable in the brain tissue of mice prior to any detectable pathology. This is consistent with a possible second event being required for tumorigenesis.

In contrast, during this entire time period, 419 mice, which fail to develop tumors but carry SV40 DNA in the germ line, do not produce detectable levels of SV40 large-T antigen in the brain. In addition, brain tissue fails to rapidly produce cell lines in culture (van Dyke *et al.*, 1985). Very low levels of SV40 large-T antigen have been detected in some 419 mice at older ages (32 and 47 weeks of age), but only in brain tissue (van Dyke *et al.*, 1985). More experimentation is required to determine whether this is a general feature of 419 mice or specific to a few examples of 419 mice, but this observation is also consistent with a second event being required for tumorigenesis.

In over five generations of breeding 419 mice, 52 mice have been born that carry SV40 DNA in the germ line. One of these mice (number 419-2), at 1.5 years of age, developed a papilloma of the choroid plexus. Western-blot analysis showed large-T antigen in tumor tissue, but no other tissue tested expressed detectable levels of large-T antigen. The brother of this mouse is healthy at 1.75 years of age, suggesting that this single occurrence is the result of a rare event rather than an example of tumors arising at 1.5 years of age in this family. The availability of the mouse that developed the papilloma permits an experimental test of the question as to why 419 mice, which contain SV40 DNA, fail to express large-T antigen and have no detectable pathology. One possibility is that the position of integration of this SV40 DNA has kept it transcriptionally silent— a position effect. If this is correct, then tumor-tissue DNA from the 419-2 mouse contain a transposition (or translocation) of SV40 DNA, placing it in a new position (as a rare event with age). Southern-blot analysis of tumor DNA or kidney DNA from 419-2 or genomic DNA from the related 419 family mice has failed to detect any transpositions or gross alterations in the integration site of SV40 DNA in tumor tissue (van Dyke *et al.*, 1985). Restriction-enzyme analysis of this DNA did reveal that tumor DNA, when compared to kidney DNA from the same animal or DNA from related 419 mice (kidney or brain DNA), was undermethylated at the cytosine residues in or around the integrated SV40 DNA sequences (van Dyke *et al.*, 1985). The decreased methylation with age correlated well with SV40 large-T antigen expression in tumor tissue, similar to the events found with the insertion of M-MuLV DNA in embryonic tissue (reviewed in the previous section). In both cases, genetically silent copies of viral DNA could be activated by mistakes in regulation, such as the failure to methylate cytosine residues during DNA replication. The expression of M-MuLV virus or SV40 large-T antigen results, even as a rare event, in an amplified signal (more virus or tumor cells) that can then be detected by the pathological consequences of the rare event. That this might happen as a function of age is interesting, and what is useful is that these experimental systems provide an amplified signal or selection for examining even rare events in gene regulation in these animals.

5. CELLULAR ONCOGENES IN TRANSGENIC MICE

5.1. *myc* Oncogene

The *myc* oncogene was first recognized as a cellular-related nucleotide sequence in the genome of the avian retrovirus MC29 or myelocytomatosis virus. This oncogene has been isolated as part of a retrovirus at least three additional times (CM-II, OK-10, MH-2). In each case, the altered *myc* gene carried in the retrovirus genome produced a *gag–myc* or *gag–pol–myc* (OK-10) fusion protein with molecular weights ranging between 90,000 and 200,000 daltons (Weiss *et al.*, 1982). All four viruses have similar pathology in birds and transform the same cell types *in vitro*. The available evidence strongly suggests that the *myc* protein is essential for the tumorigenic events associated with these viruses (Weiss *et al.*, 1982). The homologous protooncogene or *c-myc* gene was first implicated in lymphomagenesis in a study of the nonacute or slow retrovirus-induced pathologies in the chicken. The development of bursal lymphomas (B-cell lymphomas) in chickens, after long or chronic infections with avian leukosis virus, results from the insertion of a viral long terminal repeat (LTR) (promoter–enhancer) near the *c-myc* locus. The LTR is able to enhance *c-myc* expression by virtue of a promoter-insertion event (Neel *et al.*, 1981; Payne *et al.*, 1982; Hayward *et al.*, 1981). Some evidence for an additional activated oncogene in these tumors has been documented (Cooper and Neiman, 1980), so a second event in the development of these lymphomas remains possible. A similar activation of the *c-myc* gene and gene products results from a translocation between the chromosomal *c-myc* locus and the immunoglobulin heavy-chain or light-chain loci in human Burkitt lymphomas (Taub *et al.*, 1982) and murine plasmocytomas (Shen-Oneg *et al.*, 1982; Stanton *et al.*, 1983). In all these cases, the activation of the *c-myc* protooncogene in B cells apparently involves the increased levels of expression of the normal gene product leading to lymphomagenesis.

In some experimental models of cellular transformation, i.e., primary rat embryo or newborn rat fibroblasts, a *c-myc* gene or gene product is not by itself sufficient for a complete transformation event. The *c-myc* gene along with a second oncogene (*EJ-ras*), however, provides the genetic information that is both necessary and sufficient to obtain a completely transformed clone of cells (Land *et al.*, 1983). It has been suggested that the role of *c-myc* in this model system is to "immortalize" the primary rat cell or to otherwise prepare it for the transformation even by the *ras* oncogene (Land *et al.*, 1983). The *EJ-ras* oncogene, in the absence of additional regulatory signals, fails to transform primary rat cells in culture (Land *et al.*, 1983). The addition of a Moloney murine sarcoma virus LTR (enhancer–promoter) to the *Ha-ras-1* gene results in a *ras* oncogene capable of a complete transformation event with primary rat cells in culture without the help of the *c-myc* gene (Spandidos and Wilkie, 1984). These experiments indicate the role that both quantitative and qualitative elements may play in affecting oncogenic events.

The normal function of the *c-myc* gene product (a 68,000-dalton nuclear protein) is not known. The protein appears to have a short half-life (Hann *et al.*, 1984) and is regulated by or with events in the cell cycle. The levels of *c-myc* have been shown to increase as cells are stimulated from a G_0 resting state into the cell cycle (G1) under the influence of growth factors or mitogens (K. Kelly *et al.*, 1983). It would appear that *c-myc* protein might play a role in regulating cell division or growth.

5.2. *myc* Transgenic Mice

To continue the study of the role of the *myc* oncogene in tumorigenesis, T. A. Stewart *et al.* (1984) constructed a regulatable enhancer–promoter fusion with three different regions of the mouse *c-myc* gene. The murine mammary tumor virus (MMTV) LTR was employed to supply the inducible promoter–enhancer functions. This is composed of a region of DNA required for glucocorticoid control (positive control) of transcription, the MMTV promoter, and the MMTV cap site for the start of mRNA synthesis. These three elements were fused to either a normal murine *c-myc* gene, a *c-myc* gene with the normal chromosomal promoter element deleted, a *c-myc* gene with the normal promoter and cap site deleted, or a *c-myc* gene with the first exon of the gene deleted. The first exon encodes information for mRNA sequences but does not contain any protein-coding sequences, so all these fusion-gene constructions could produce a normal *c-myc* protein (T. A. Stewart *et al.*, 1984). These four different MMTV-LTR–*c-myc* gene constructions were employed to produce transgenic mice (CD-1 × C57 B1/6J). Thirteen lines of mice were produced that contained 1–5 copies of one of the four MMTV-LTR–*c-myc* fusion genes. The DNA was integrated into chromosomes as tandem repeated sequences and inherited in a Mendelian fashion.

Transcripts derived from the *myc* fusion genes could be detected in the presence of the normal *c-myc* mRNA by a differential S1 assay. One transgenic mouse line expressed the fusion gene in all tissues tested, while other mice failed to contain detectable levels of fusion gene mRNA in any tissue. Of 12 independently derived transgenic mice, 11 expressed the fusion-gene mRNA in salivary-gland tissue. Other tissues contained either no fusion-gene mRNA (in most mice) or occasional mRNA expression in selected tissues, usually the testes. No correlation between fusion-gene regulation and the *c-myc* gene construction utilized was apparent. The only consistent expression of the *c-myc* fusion gene was in the salivary gland.

Within the first year after the production of these transgenic mice, two multiparous pregnant females (one 206 days old, the second 228 days old) developed breast tumors that were shown to be mammary adenocarcinomas. One female carried the MMTV-LTR–*c-myc* gene construction with no *c-myc* promoter, and the other carried the deletion of the first *c-myc* exon. In the F_1 progeny from these mice, three multiparous (two litters each) pregnant females also developed mammary adenocarcinomas. The tumor tissue could grow and be passed in nude mice. It appears, therefore, that multiple pregnancies and elevated hormones during pregnancy are important variables in the production of these carcinomas. This is the same type of control and pattern of pathology as observed with MMTV, suggesting that the LTR can direct the tissue specificity and development of *myc*-induced tumors. Not every transgenic female with multiple pregnancies develops this adenocarcinoma, suggesting that the MMTV-LTR–*c-myc* fusion gene is not by itself sufficient to produce these tumors. Indeed, not every mammary gland in a multiparous pregnant transgenic mouse with an adenocarcinoma contains tumor tissue (but every cell contains the transgene). This observation is similar to the one made with the SV40 large-T-antigen-induced papilloma of the choroid plexus, in which a portion of the choroid-plexus tissue is normal and another part has a papilloma even though all cells have the transgene. These observations suggest that second events may be required to induce tumors in these systems.

The tumor tissue obtained from the adenocarcinomas expressed high levels of the *c-myc* fusion-gene mRNA. The fusion gene produced about 10 times higher levels of *c-myc* mRNA in the tumors that the normal *c-myc* gene synthesized in this tissue. The transgenic DNA from tumor tissues was not amplified and showed no gross rearrangements when analyzed by Southern-blotting procedures. Presumably, the hormone-induced enhanced transcription of the *c-myc* fusion gene results in an adenocarcinoma of mammary tissue.

5.3. Conclusions

The studies of T. A. Stewart *et al.* (1984) with the MMTV-LTR–*c-myc* fusion gene and of Brinster *et al.* (1984) with the SV40 early gene regions introduced into transgenic mice begin to outline some clear principles. First, the role of the enhancer–promoter in guiding and directing tissue-specific gene expression and therefore pathology is well illustrated by these studies. The *myc*-induced adenocarcinomas occur only in multiparous pregnant females. The promoter–enhancer elements of the MMTV-LTR are responsive to prolactogenic stimuli and glucocorticoids. The *c-myc*-mediated adenocarcinoma pathology observed with transgenic mice has the same requirements and tissue specificity as MMTV-induced pathology, even though the oncogene mediating the carcinoma may be different. Similarly, the enhancer (promoter) elements of SV40 appears to direct SV40 large-T antigen synthesis in the choroid-plexus tissue. Deletion of the enhancer alters the tissue specificity. Substitution of the SV40 enhancer–promoter with an element of a different specificity (beta cells of the pancreas) results in a new type of tumor pattern (Hanahan, 1985). The SV40 viral transforming gene can produce tumors in several tissue types, but its expression is limited or directed by nucleotide signal sequences 5' to the structural gene. In a similar manner, polyoma virus early gene expression is regulated in cells of diverse developmental backgrounds by base sequences in the polyoma virus genome enhancer elements (Sekikawa and Levine, 1981). It is clear that for many oncogenes, or viral-encoded transforming genes, tissue specificity, pathology, and disease patterns can be influenced by the regulatory elements that control these genes.

Second, the MMTV-LTR–c-myc fusion gene and SV40 early region transgenic mice express the transgene (mRNA or protein) in tissues that do not subsequently produce tumors. The expression of these gene products may result in pathology (cortical cysts of the kidney, peripherial neuropathies), or it may have no apparent impact. The absence of tumors in these tissues could result from a quantitative (too little) or qualitative (need for a second oncogene to be expressed) difference when compared with tumor tissue from these mice. In either case, there is a good deal of evidence suggesting the need for additional factors or events in the formation of a tumor in these transgenic mice. In the case of the transgenic mice containing the SV40 large-T antigen gene, histological sections of the choroid-plexus tumor tissue show both normal and papillomatous tissue. Either expression of the transgene is not uniform, the level of expression is not uniform (quantitative differences), or qualitative changes (second gene) are required for tumorigenesis. SV40 large-T antigen is expressed in normal pancreatic tissue in fetal and newborn mice containing the fusion transgene composed of the insulin promotor and the large-T antigen structural gene (Hanahan, 1985). It would thus appear that the expression of SV40 large-T antigen (or myc) in a cell is not sufficient for the production of a tumor from that cell in an animal. Perhaps it is not surprising to recall that in 1974, an identical conclusion resulted from a study of SV40 transformation of cells in culture employing a nonselective analysis of cell lines produced on SV40 virus infection (Risser and Pollack, 1974).

Finally, it is worth mentioning that a number of oncogenes and their products may be involved in promoting developmental events in a stem cell or cell lineage (see Levine, 1984; and Vande Woude et al., 1984). It may therefore be fruitful, when introducing oncogenes into transgenic mice, to look for more than the obvious phenotypes.

REFERENCES

Riser, R., and Pollack, R., 1974, A nonselective analysis of SV40 transformation of mouse 3T3 cells, Virology 59:477–489.

Abramczuk, J., Vorbrodt, A., Solter, D., and Koproswki, H., 1978, Infection of mouse preimplantation embryos with simian virus 40 and polyoma virus, Proc. Natl. Acad. Sci. U.S.A. 75:999–1003.

Alt, F. W., Enea, V., Bothwell, A. L. M., and Baltimore, D., 1980, Activity of multiple light chain genes in murine myeloma cells producing a single functional light chain, Cell 21:1–12.

Alt, F. W., Rosenberg, N., Enea, V., Siden, E., and Baltimore, D., 1982, Multiple immunoglobulin heavy-chain gene transcripts in Abelson murine leukemia virus-transformed lymphoid cell lines, Mol. Cell. Biol. 2:286–400.

Banerji, J., Olson, L., and Schaffner, W., 1983, A lymphocyte-specific cellular enhancer is located downstream of the joining region in immunoglobulin heavy chain genes, Cell 33:729–740.

Bernard, O., Gough, N., and Adams, J., 1981, Plasmacytomas with more than one immunoglobulin κ mRNA: Implications for allelic exclusion, *Proc. Natl. Acad. Sci. U.S.A.* **78:**5812–5816.

Brinster, R. L., Chen, H. Y., Trumbauer, M., Denear, A. W., Warren, R., and Palmiter, R. D., 1981, Somatic expression of herpes thymidine kinase in mice following injection of a fusion gene into eggs, *Cell* **27:**223–231.

Brinster, R. L., Ritchie, K. A., Hammer, R. E., O'Brien, R. L., Arp, B., and Storb, U., 1983, Expression of a microinjected immunoglobulin gene in the spleen of transgenic mice, *Nature (London)* **306:**332–336.

Brinster, R. L., Chen, H. Y., Messing, A., van Dyke, T., Levine, A. J., and Palmiter, R. D., 1984, Transgenic mice harboring SV40 T-antigen genes develop characteristic brain tumors, *Cell* **37:**367–379.

Brinster, R. L., Chen, H. Y., Trumbauer, M. E., Yagle, M. K., and Palmiter, R. D., 1985, Factors affecting the efficiency of introducing foreign DNA into mice by microinjecting eggs, *Proc. Natl. Acad. Sci. U.S.A.* **82:**4438–4442.

Chada, K., Magram, J., Raphael, K., Radice, G., Lacy, E., and Constantini, F., 1985, Specific expression of a foreign β-globin gene in erythroid cells of transgenic mice, *Nature (London)* **314:**377–380.

Charnay, P., Treisman, R., Mellon, P., Chao, M., Axel, R., and Maniatis, T., 1984, Differences in human α- and β-globin gene expression in mouse erythroleukemia cells: The role of intragenic sequences, *Cell* **38:**251–263.

Coleclough, C., Perry, R., Karjalainen, K., and Weigert, M., 1981, Aberrant rearrangements contribute significantly to the allelic exclusion of immunoglobulin gene expression, *Nature (London)* **290:**372–378.

Cooper, G. M., and Neiman, P. E., 1980, Transforming gene of neoplasms induced by avian leukosis viruses, *Nature (London)* **287:**656–659.

Costantini, F., and Lacy, E., 1981, Introduction of a rabbit β-globin gene into the mouse germ line, *Nature (London)* **294:**92–94.

Costantini, F., Radice, G., Magram, J., Stamatoyannopoulos, G., Papayannopoulou, T., and Chada, K., 1985, Developmental regulation of human globin genes in transgenic mice, *Cold Spring Harbor Symp. Quant. Biol.* **50:**361–370.

Dandolo, L., Blangy, D., and Kamen, R., 1983, Regulation of polyoma virus transcription in murine embryonal carcinoma cells, *J. Virol.* **47:**55–64.

Daughaday, W. H., Hall, K., Rahen, M. S., Salmon, W. R., Jr., Van den Brande, J. L., and Van Wyke, J. J., 1972, Somatomedin: Proposed designation for sulfation factor, *Nature (London)* **235:**107–109.

Doerfler, W., 1983, DNA methylation and gene activity, *Annu. Rev. Biochem.* **52:**93–124.

Eicher, E. M., and Bearner, W. G., 1976, Inherited ateliotic dwarfism in mice, *J. Hered.* **67:**87–91.

Edlund, T., Walker, M. D., Barr, P. J., and Rutter, W. J., 1985, Cell specific expression of the rat insulin gene: Evidence for role of two distinct 5' flanking elements, *Science* **230:**912–916.

Efstratiadis, A., Posakony, J. W., Maniatis, T., Lawn, R. M., O'Connell, C., Spritz, R. A., DeRiel, J. K., Forget, B. G., Slighton, L., Blechl, A. E., Smithies, O., Baralle, F. E., Shoulders, C. C., and Proudfoot, N. J., 1980, The structure and evolution of the β-globin gene family, *Cell* **21:**653–668.

Frels, W. I., Bluestone, J. A., Hodes, R. J., Capecchi, M. R., and Singer, D. S., 1985, Expression of a microinjected porcine class I major histocompatibility complex gene in transgenic mice, *Science* **228:**577–580.

Fujimura, F., and Linney, E., 1982, Polyoma mutants that productively infect F9 embryonal carcinoma cells do not rescue wild-type polyoma in F9 cells, *Proc. Natl. Acad. Sci. U.S.A.* **79:**1479–1483.

Fujimura, F. K., Deininger, P. L., Friedmann, T., and Linney, E., 1981, Mutation near the polyoma DNA replication origin permits productive infection of F9 embryonal carcinoma cells, *Cell* **23:**809–814.

Gautsch, J. W., and Wilson, M. C., 1983, Delayed *de novo* methylation in teratocarcinoma cells suggests additional tissue specific mechanisms for controlling gene expression, *Nature (London)* **301**:32–35.

Gillies, S. D., Morrison, S. L., Oi, V. T., and Tonegawa, S., 1983, A tissue-specific transcription enhancer element is located in the major intron of a rearranged immunoglobulin heavy chain gene, *Cell* **33**:717–728.

Godbout, R., Ingram, R., and Tilghman, S. M., 1986, Multiple regulatory elements in the intergenic region between the α-fetoprotein and albumin genes, *Mol. Cell. Biol.* **6**:477–487.

Goodbourn, S., Zinn, K., and Maniatis, T., 1985, Human β-interferon gene expression is regulated by an inducible enhancer element, *Cell* **41**:509–520.

Gordon, J. W., Scangos, G. A., Plotkin, D. J., Barbosa, J. A., and Ruddle, F. H., 1982, Genetic transformation of mouse embryos by microinjection of purified DNA, *Proc. Natl. Acad. Sci. U.S.A.* **77**:7380–7384.

Grosschedl, R., and Baltimore, D., 1985, Cell-type specificity of immunoglobulin gene expression is regulated by at least three DNA sequence elements, *Cell* **41**:885–897.

Grosschedl, R., Weaver, D., Baltimore, D., and Costantini, F., 1984, Introduction of a μ immunoglobulin gene into the mouse germ line: Specific expression in lymphoid cells and synthesis of functional antibody, *Cell* **38**:647–658.

Hammer, R. E., Palmiter, R. D., and Brinster, R. L., 1984, Partial correction of murine hereditary growth disorder by germ line incorporation of a new gene, *Nature (London)* **311**:65–67.

Hammer, R. E., Brinster, R. L., and Palmiter, R. D., 1985a, Use of gene transfer to increase animal growth, *Cold Spring Harbor Symp. Quant. Biol.* **50**:379–388.

Hammer, R. E., Brinster, R. L., Rosenfeld, M. G., Evans, R. M., and Mayo, K. E., 1985b, Expression of human growth hormone releasing factor in transgenic mice results in increased somatic growth, *Nature (London)* **315**:413–416.

Hanahan, D., 1985, Heritable formation of pancreatic β-cell tumors in transgenic mice expressing recombinant insulin/SV40 oncogene, *Nature (London)* **315**:115–122.

Hann, S. R., and Eisenman, R. N., 1984, Proteins encoded by the human c-myc oncogene: Differential expression in neoplastic cells, *Mol. Cell. Biol.* **4**:2486–2497.

Hayward, W. S., Neel, B. G., and Astrin, S. M., 1981, Activation of a cellular onc gene by promoter insertion in ALV-induced lymphoid leukosis, *Nature (London)* **290**:475–480.

Huebner, K., Linnenbach, A., Ghosh, P. K., Rushdi, A., Romanczuk, H., Tsuchida, N., and Croce, C. M., 1983, Tumor virus genomes in DNA-transformed F9 cells, in: *Cold Spring Harbor Conferences on Cell Proliferation 10* (L. M. Silver, G. R. Martin, and S. Strickland, eds.), pp. 343–361, Cold Spring Harbor Press, Cold Spring Harbor, New York.

Jaenisch, R., 1974, Infection of mouse blastocysts with SV40 DNA: Normal development of the infected embryos and persistence of SV40 specific DNA sequences in the adult animals, *Cold Spring Harbor Symp. Quant. Biol.* **39**:375–380.

Jaenisch, R., and Berns, A., 1977, Tumour virus expression during mammalian embryogenesis, in: *Concepts in Mammalian Embryogenesis* (M. J. Sherman, ed.), pp. 267–314, MIT Press, Cambridge, Massachusetts.

Jaenisch, R., and Mintz, B., 1974, Simian virus 40 DNA sequences in DNA of healthy adult mice derived from preimplantation blastocysts injected with viral DNA, *Proc. Natl. Acad. Sci. U.S.A.* **71**:1250–1254.

Jaenisch, R., Fan, H., and Croker, B., 1975, Infection of preimplantation mouse embryos and of newborn mice with leukaemia virus: Tissue distribution of viral DNA and RNA leukemogenesis in the adult animal, *Proc. Natl. Acad. Sci. U.S.A.* **72**:4008–4012.

Jaenisch, R., Jähner, D., Nobis, P., Simon, I., Löhler, J., Harbers, K., and Grotkopp, D., 1981, Chromosomal position and activation of retroviral genomes inserted into the germ line of mice, *Cell* **24**:519–529.

Jaenisch, R., Harbers, K., Schnieke, A., Löhler, J., Chumakov, I., Jähner, D., Grotkopp, D., and Hoffman, E., 1983, Germline integration of Moloney murine leukaemia virus at the *Mov13* locus leads to recessive lethal mutation and early embryonic death, *Cell* **32**:209–216.

Jähner, D., and Jaenisch, R., 1985, Chromosomal position and specific demethylation in enhancer sequences of germ line-transmitted retroviral genomes during mouse development, *Mol. Cell. Biol.* **5:**2212–2220.

Jähner, D., Stuhlmann, H., Stewart, C. L., Harbers, K., Löhler, J., Simon, I., and Jaenisch, R., 1982, De novo methylation and expression of retroviral genomes during mouse embryogenesis, *Nature (London)* **298:**623–628.

Jähner, D., Haase, K., Mulligan, R., and Jaenisch, R., 1985, Insertion of the bacterial gpt gene into the germ line of mice by retroviral infection, *Proc. Natl. Acad. Sci. U.S.A.* **82:**6927–6931.

Kelly, F., and Condamine, H., 1982, Tumor viruses and early mouse embryos, *Biochim. Biophys. Acta* **651:**105–141.

Kelly, K., Cochran, B. H., Stiles, C. D., and Leder, P., 1983, Cell-specific regulation of the c-myc gene by lymphocyte mitogens and platelet-derived growth factor, *Cell* **35:**603–610.

Kemp, D. J., Harris, A. W., Corey, S., and Adams, T. M., 1980, Expression of the immunoglobulin Cμ gene in T and B lymphoid and myeloid cell lines, *Proc. Natl. Acad. Sci. U.S.A.* **77:**2876–2880.

Konieczny, S. F., and Emerson, C. P., Jr., 1985, Differentiation, not determination, regulates muscle gene activation: Transfection of troponin I genes in multipotential and muscle lineages of 10T1/2 cells, *Mol. Cell. Biol.* **5:**2423–2432.

Krumlauf, R., Hammer, R., Tilghman, S., and Brinster, R. L., 1985a, Developmental regulation of a α-fetoprotein genes in transgenic mice, *Mol. Cell. Biol.* **5:**1639–1648.

Krumlauf, R., Hammer, R. E., Brinster, R., Chapman, V. M., and Tilghman, S. M., 1985b, Regulated expression of α-fetoprotein genes in transgenic mice, *Cold Spring Harbor Symp. Quant. Biol.* **50:**371–378.

Kwan, S.-P., Max, E. E., Seidman, J. G., Leder, P., and Scharff, M. D., 1981, Two kappa immunoglobulin genes are expressed in the myeloma S107, *Cell* **26:**57–66.

Lacy, E., Roberts, S., Evans, E. P., Burtenshaw, M. D., and Costantini, F., 1983, A foreign β-globin gene in transgenic mice: Integration at abnormal chromosomal positions and expression in inappropriate tissues, *Cell* **34:**343–358.

Land, H., Parada, L. F., and Weinberg, R. A., 1983, Tumorigenic conversion of primary embryo fibroblasts requires at least two cooperating oncogenes, *Nature (London)* **304:**596–602.

Le Meur, M., Gerlinger, P., Benoist, C., and Mathis, D., 1985, Correcting an immune response deficiency by creating Eα gene transgenic mice, *Nature (London)* **316:**38–42.

Levine, A. J., 1979, Permanent teratocarcinoma-derived cell lines stabilized by transformation with SV40 and SV40 tsA mutant viruses, *J. Int. Cytol. Suppl.* **10:**173–189.

Levine, A. J., 1982, The nature of the host range restriction of SV40 and polyoma viruses in embryonal carcinoma cells, *Curr. Top. Microbiol. Immunol.* **101:**1–30.

Levine, A. J. (ed.), 1984, *The Transformed Phenotype,* Cold Spring Harbor Press, Cold Spring Harbor, New York.

Linnebach, A., Huebner, K., and Croce, C. M., 1980, DNA transformed murine teratocarcinoma cells: Regulation of expression of simian virus 40 tumor antigen in stem versus differentiated cells, *Proc. Natl. Acad. Sci. U.S.A.* **77:**4875–4879.

Linnebach, A., Huebner, K., and Croce, C. M., 1981, Transcription of the simian virus 40 genome in DNA-transformed murine teratocarcinoma stem cells, *Proc. Natl. Acad. Sci. U.S.A.* **78:**6386–6390.

Linney, E., David, B., Overhauser, J., Chao, E., and Fan, H., 1984, Non-function of a Moloney murine leukaemia virus regulatory sequence in F9 embryonal carcinoma cells, *Nature (London)* **380:**470–472.

Low, M. J., Hammer, R. E., Goodman, R. H., Habener, J. F., Palmiter, R. D., and Brinster, R. L., 1985, Tissue-specific posttranslational processing of pre-prosomatostatin encoded by a metallothionein–somatostatin fusion gene in transgenic mice, *Cell* **41:**211–219.

Magram, J., Chada, K., and Costantini, F., 1985, Developmental regulation of a cloned adult β-globin gene in transgenic mice, *Nature (London)* **315:**338–340.

Maltzman, W., Linzer, D. I., Brown, F., Teresky, A. K., Rosenstraus, M., and Levine, A. J., 1979, Permanent teratocarcinoma-derived cell lines stabilized by transformation with SV40 and SV40tsA mutant viruses, *Int. Rev. Cytol.* **10:**173–89.

Maltzman, W., and Levine, A. J., 1981, Viruses as probes for development and differentiation, *Adv. Virus. Res.* **26:**65–117.

McKnight, G. S., Hammer, R. E., Kuenzel, E. A., and Brinster, R. L., 1983, Expression of the chicken transferrin gene in transgenic mice, *Cell* **34:**335–341.

Messing, A., Chen, H. Y., Palmiter, R. D., and Brinster, R. L., 1985, Peripheral neuropathies, hepatocellular carcinomas, and islet cell adenomas in transgenic mice, *Nature (London)* **316:**461–463.

Neel, B. G., Hayward, W. S., Robinson, H. L., Fang, J., and Astrin, S. M., 1981, Avian leukosis virus-induced tumors have common proviral integration sites and synthesize discrete new RNAs: Oncogenesis by promoter insertion, *Cell* **23:**323–334.

Nelson, K. J., Haimovich, J., and Perry, R., 1983, Characterization of productive and sterile transcripts from the immunoglobulin heavy chain focus: Processing of $C_{\mu m}$ and $C_{\mu s}$ mRNA, *Mol. Cell. Biol.* **3:**1317–1332.

Niwa, O., Yokota, Y., Ishida, H., and Sugahara, T., 1983, Independent mechanisms involved in suppression of the Moloney leukaemia virus genome during differentiation of murine teratocarcinoma cells, *Cell* **22:**1105–1113.

Notkins, A. L. and Oldstone, M. B. A. (eds.), 1984, *Concepts in Viral Pathogenesis,* Springer-Verlag, New York.

Nudel, U., Greenberg, D., Ordahl, C. P., Saxel, O., Neuman, S., and Yaffe, D., 1985, Developmentally regulated expression of a chicken muscle-specific gene in stably transfected rat myogenic cells, *Proc. Natl. Acad. Sci. U.S.A.* **82:**3106–3109.

Ornitz, D. M., Palmiter, R. D., Hammer, R. E., Brinster, R. L., Swift, G. H., and McDonald, J. R., 1985, Specific expression of an elastase–human growth hormone fusion gene in pancreatic acinar cells of transgenic mice, *Nature (London)* **313:**600–602.

Overbeek, P. A., Chepelinsky, A., Khillan, J. S., Piatigorsky, J., and Westphal, H., 1985, Lens-specific expression and developmental regulation of the bacterial chloramphenicol acetyltransferase gene driven by the murine αA-crystallin promoter in transgenic mice, *Proc. Natl. Acad. Sci. U.S.A.* **82:**7815–7819.

Palmiter, R. D., Brinster, R. L., Hammer, R. E., Trumbauer, M. E., Rosenfeld, M. G., Birnberg, N. C., and Evans, R. M., 1982a, Dramatic growth of mice that develop from eggs microinjected with metallothionein–growth hormone fusion genes, *Nature (London)* **300:**611–615.

Palmiter, R. D., Chen, H. Y., and Brinster, R. L., 1982b, Differential regulation of metallothionein–thymidine kinase fusion gene in transgenic mice and their offspring, *Cell* **29:** 701–710.

Palmiter, R. D., Norstedt, G., Gelinas, R. E., Hammer, R. E., and Brinster, R. L., 1983, Metallothionein–human GH fusion genes stimulate growth of mice, *Science* **222:**809–814.

Palmiter, R. D., Chen, H. Y., Messing, A., and Brinster, R. L., 1985, SV40 enhancer and large-T antigen are instrumental in development of choroid plexus tumours in transgenic mice, *Nature (London)* **316:**457–460.

Payne, G. S., Bishop, J. M., and Varmus, H. E., 1982, Multiple arrangements of viral DNA and an activated host oncogene in bursal lymphomas, *Nature (London)* **295:**209–214.

Pinkert, C. A., Widera, G., Cowing, C., Heber-Katz, E., Palmiter, R. D., Flavel, R. A., and Brinster, R. L., 1985, Tissue-specific, inducible and functional expression of the E_α^d MHC class II gene in transgenic mice, *EMBO J.* **4:**2225–2230.

Queen, C., and Baltimore, D., 1983, Immunoglobulin gene transcription is activated by downstream sequence elements, *Cell* **33:**741–748.

Riser, R., and Pollack, R., 1974, A nonselective analysis of SV40 transformation of mouse 3T3 cells, *Virology* **59:**477–489.

Ritchie, K. A., Brinster, R. L., and Storb, U., 1984, Allelic exclusion and control of endogenous immunoglobulin gene rearrangement in κ transgenic mice, Nature (London) 312:517–520.

Rusconi, S., and Köhler, G., 1985, Transmission and expression of a specific pair of rearranged immunoglobulin μ and κ genes in a transgenic mouse line, Nature (London) 314:330–334.

Scott, R. W., Vogt, T. F., Croke, M. E., and Tilghman, S. M., 1984, Tissue-specific activation of a cloned α-fetoprotein gene during differentiation of a transfected embryonal carcinoma cell line, Nature (London) 310:562–567.

Segal, S., and Khoury, G., 1979, Differentiation as a requirement for SV40 gene expression in F9 embryonal carcinoma cells, Proc. Natl. Acad. Sci. U.S.A. 76:5611–5615.

Segal, S., Levine, A. J., and Khoury, G., 1980, Evidence for non-spliced SV40 RNA in undifferentiated murine teratocarcinoma stem cells, Nature (London) 20:335–337.

Seiler-Tuyns, A., Eldridge, J. D., and Paterson, B. M., 1984, Expression and regulation of chicken actin genes introduced into mouse myogenic and nonmyogenic cells, Proc. Natl. Acad. Sci. U.S.A. 81:2980–2984.

Sekikawa, K., and Levine, A. J., 1981, Isolation and characterization of polyoma host mouse mutants that replicate in multipotential embryonal carcinoma cells, Proc. Natl. Acad. Sci. U.S.A. 78:110–1104.

Shani, M., 1985, Tissue-specific expression of rat myosin light-chain 2 gene in transgenic mice, Nature (London) 314:283–286.

Shen-Ong, G. L. C., Keath, E. J., Piccoli, S. P., and Cole, M. D., 1982, Novel myc oncogene RNA from abortive immunoglobulin-gene recombination in mouse plasmacytomas, Cell 31:443–452.

Small, J. A., Blair, D. G., Showalter, S. D., and Scangos, G. A., 1985, Analysis of a transgenic mouse containing SV40 and v-myc sequences, Mol. Cell. Biol. 5:642–648.

Sorge, J., Cutting, A. E., Erdman, V. D., and Gautsch, J. W., 1984, Integration-specific retrovirus expression in embryonal carcinoma cells, Proc. Natl. Acad. Sci. U.S.A. 81:6627–6632.

Spandidos, D. A., and Wilkie, N. M., 1984, Malignant transformation of early passage rodent cells by a single mutated human oncogene, Nature (London) 310:469–475.

Stanton, L. W., Watt, R., and Marcu, K. B., 1983, Translocation, breakage and truncated transcripts of c-myc oncogene in murine plasmacytomas, Nature (London) 303:401–406.

Stewart, C. L., Stuhlman, H., Jähner, D., and Jaenisch, R., 1982, De novo methylation, expression and infectivity of retroviral genomes introduced into embryonal carcinoma cells, Proc. Natl. Acad. Sci. U.S.A. 79:4098–4102.

Stewart, T. A., Wagner, E. F., and Mintz, B., 1982, Human β-globin gene sequences injected into mouse eggs, retained in adults, and transmitted to progeny, Science 217:1046–1048.

Stewart, T. A., Pattengale, P. K., and Leder, P., 1984, Spontaneous mammary adenocarcinomas in transgenic mice that carry and express MTV/myc fusion genes, Cell 38:627–637.

Storb, U., O'Brien, R. L., McMullen, M. D., Gollahon, K. A., and Brinster, R. L., 1984, High expression of cloned immunoglobulin K gene in transgenic mice is restricted to B lymphocytes, Nature (London) 310:238–241.

Swanson, L. W., Simmons, D. M., Arriza, J., Hammer, R., Brinster, R., Rosenfeld, M. G., and Evans, R. M., 1985, Novel developmental specificity in the nervous system of transgenic animals expressing growth hormone fusion genes, Nature (London) 317:363–366.

Swartzendruber, E. C., and Lehmann, J. M., 1975, Neoplastic differentiation: Interaction of SV40 and polyoma virus with murine teratocarcinoma cells in vitro, J. Cell Physiol. 85:179–180.

Swift, G. H., Hammer, R. E., McDonald, R. J., and Brinster, R. L., 1984, Tissue-specific expression of the rat pancreatic elastase I gene in transgenic mice, Cell 38:639–646.

Taub, R., Kirsch, I., Morton, C., Lenoir, G., Swan, D., Tronick, S., Aronson, S., and Leder, P., 1982, Translocation of the c-myc gene into the immunoglobulin heavy chain locus in human Burkitt lymphoma and murine plasmacytoma cells, Proc. Natl. Acad. Sci. U S A 79:7837–7841.

Teich, N. M., Weiss, R. A., Martin, G. R., and Lowy, D. R., 1977, Virus infection of murine teratocarcinoma stem cell lines, Cell 12:973–982.

Tilghman, S. M., 1985, The structure and regulation of the α-fetoprotein and albumin genes, in: *Oxford Surveys in Eukaryotic Genes,* Vol. 2. (N. Maclean, ed.), Oxford University Press, Oxford, p. 160.

Tonegawa, S., 1983, Somatic generation of antibody diversity, *Nature* **302:**575–581.

Tooze, J. (ed.), 1981, *Molecular Biology of Tumor Viruses,* 2nd ed., Part 2, Cold Spring Harbor Press, Cold Spring Harbor, New York.

Townes, T. M., Lingrel, J. B., Chen, H. Y., Brinster, R. L., and Palmiter, R. D., 1985, Erythroid specific expression of human β-globin genes in transgenic mice, *EMBO J.* **4:**1715–1723.

Van Dyke, T., Finlay, C., and Levine, A. J., 1985, A comparison of several lines of transgenic mice containing the SV40 early genes, *Cold Spring Harbor Symp. Quant. Biol.* **50:**671–678.

Vande Woude, G., Levine, A. J., Topp, W. C., and Watson, J. D. (eds.), 1984, *Oncogenes and Viral Genes,* Cold Spring Harbor Press, Cold Spring Harbor, New York.

Vasseur, M., Kress, C., Montreau, N., and Blangy, D., 1980, Isolation and characterization of polyoma virus mutants able to develop in multipotent murine embryonal carcinoma cells, *Proc. Natl. Acad. Sci. U.S.A.* **77:**1068–1072.

Wagner, E. F., and Stewart, C. L., 1986, Integration and expression of genes introduced into mouse embryos, in: *Experimental Approaches to Mammalian Embryonic Development* (J. Rossant and R. Pederson, eds.) Cambridge University Press, Cambridge, England (in press).

Wagner, E. F., Covarrubias, L., Stewart, T. A., and Mintz, B., 1983, Prenatal lethalities in mice homozygous for human growth hormone gene sequences integrated in the germ line, *Cell* **35:**647–655.

Walker, M. C., Edlund, T., Boulet, A. M., and Rutter, W. J., 1985, Cell-specific expression controlled by the 5'-flanking region of insulin and chymotrypsin genes, *Nature (London)* **306:**557–561.

Weaver, D., Costantini, F., Imanishi-Kari, T., and Baltimore, D., 1985 A transgenic immunoglobulin Mu gene prevents rearrangement of endogenous genes, *Cell* **42:**117–127.

Weiss, R., Teich, N., Varmus, H., and Coffin, J. (eds.), 1982, *RNA Tumor Viruses,* 2nd ed., Cold Spring Harbor Press, Cold Spring Harbor, New York.

Westphal, H., Overbeek, P. A., Khillan, J. S., Chepelinsky, A. B., Schmidt, A., Mahon, K. A., Bernstein, K. E., Piatigorsky, J., and de Crombrugghe, B., 1985, Promoter sequences of murine αA crystallin, murine α2(1) collagen or avian sarcoma virus genes linked to the bacterial CAT gene direct tissue specific patterns of CAT expression in transgenic mice, in: *Cold Spring Harbor Symp. Quant. Biol.* **50:**411–416.

Willison, K., Babinet, C., Boccara, M., and Kelly, F., 1983, Infection of preimplantation mouse embryos with simian virus 40, in: *Cold Spring Harbor Monographs on Cell Proliferation,* Vol. 10 (L. M. Silver, G. F. Martin, and S. Strickland, eds.), pp. 307–317, Cold Spring Harbor Press, Cold Spring Harbor, New York.

Wright, S., Rosenthal, A., Flavell, R., and Grosveld, F., 1984, DNA sequences required for regulated expression of β-globin genes in murine erythroleukemia cells, *Cell* **38:**265–273.

Yamamura, K., Kikutani, H., Folson, V., Clayton, L. K., Kimoto, M., Akira, S., Kashiwamura, S., Tonegawa, S., and Kishimoto, T., 1985, Functional expression of a microinjected E_α^d gene in C57BL/6 transgenic mice, *Nature (London)* **316:**67–69.

8

EXPRESSION OF TRANSFECTED GENES

Moses V. Chao

1. INTRODUCTION

The stable introduction of cloned genes into mammalian cells has proven to be a widely used method to study the DNA sequences responsible for the regulation of eukaryotic gene expression. An implicit assumption of this approach is that transfected genes are capable of expression in the foreign cell and that transcription occurs at levels high enough to detect. There are now abundant examples of tissue-specific genes introduced into mammalian cells that are transcribed at measurable levels. This chapter will review examples of cell-specific genes and developmentally regulated genes that are expressed in foreign cellular environments and examine the parameters that govern the expression of these genes in mammalian cells. A generality that emerges from these examples is that cloned genes that are introduced into cells are frequently expressed even though the endogenous counterpart genes are not. Furthermore, several genes expressed normally in a tissue-specific manner can be regulated appropriately even when they are present in a foreign cellular environment.

2. METHODOLOGY

When cultured cells are exposed to a calcium phosphate precipitate of DNA (Graham and van der Eb, 1973), transformants result that integrate DNA randomly at one or more locations in the host chromosome (Robins *et al.*, 1981). The uptake of DNA in mammalian cells occurs through internalization of exogenous DNA due to endocytosis. Transformed cells are identified by the ability

Moses V. Chao • Department of Cell Biology and Anatomy and Hematology/Oncology Division, Cornell University Medical College, New York, New York 10021.

to survive under selective conditions. The herpes simplex virus thymidine kinase gene (*tk*) as well as dominant selectable genes such as *Escherichia coli* aminoglycoside transferase (*neo*) and xanthine-guanine phosphoribosyltransferase (*gpt*) have been used extensively as selectable markers to transfect mammalian cell lines. The selectable gene is either linked on the same plasmid with the gene of interest or cotransformed in separate plasmids. Multiple copies of the exogenous DNA are usually introduced into the host cell as part of a concatenated molecular complex (Perucho *et al.*, 1980).

Other methods that are employed to introduce DNA include microinjection (Capecchi, 1980), protoplast fusion (Schaffner, 1980), and electroporation (Neumann *et al.*, 1982). For this discussion, emphasis will be placed on the expression of genes transferred into cells in which there is stable integration of the gene. Expression experiments utilizing transient expression assays will not be emphasized here.

In the majority of experiments discussed herein, stable transformants were isolated and grown in mass culture. The presence and gene copy number can be estimated by Southern-blot analysis. Expression of the transferred gene in these stable transformants is assayed by an S1 nuclease analysis or gel-transfer analysis of isolated RNA with radioactive DNA probes. Protein expression is most usually measured by metabolic labeling and immunoprecipitation, isoelectric focusing, radioimmunoassay, or cell-surface tagging with immunofluorescence or rosetting.

Transfection using total cellular DNA to detect single-copy gene transfer indicates that the efficiency of gene transfer is a function of the size of the gene. Transfer of dihydrofolate reductase and hypoxanthine-guanine phosphoribosyltransferase genes, which span over 30 kilobases (kb) of information, can be transferred and expressed in mammalian cells. However, the efficiency of transfer is somewhat lower than for cellular transfer of smaller genes.

Bacteriophage λ clones containing genomic fragments of DNA have been used extensively in gene-transfer experiments. In fact, Ishiura *et al.* (1982) have shown that the herpes virus *tk* gene can be transferred in phage particles to mouse Ltk⁻cells. The phage particles, which were precipitated with calcium phosphate without carrier DNA, gave efficiencies of *tk* transfer at least an order of magnitude higher than DNA-mediated gene transfer. The inclusion of phage or plasmid sequences or carrier DNA sequences in the transfection does not inhibit the expression of cotransformed genes.

3. EXPRESSION IN L CELLS

Mouse fibroblasts have been used extensively for studying the expression of cloned genes due to the relative ease of introducing foreign genes into fibroblasts by cotransformation, the availability of a selectable marker (*tk* gene), and the absence of expression of the endogenous counterpart gene in a wide number of cases. The rabbit β-globin gene was the first gene to be systematically examined in transformed L cells (Wold *et al.*, 1979; Mantei *et al.*, 1979; Dierks *et al.*, 1981). Levels of globin messenger RNA (mRNA) expressed were generally low; on the average, 1–10 copies were detected per cell. Nevertheless, several lines exhibited very high levels of globin transcription, up to 2000 globin mRNAs per cell (Mantei *et al.*, 1979). Processing of the mRNA transcript appeared to be appropriate, since polyadenylation and splicing of the globin message took place in the fibroblast to give a mature 9 S transcript. In a similar experiment, globin mRNA transcripts from an L-cell transfectant were examined in a denaturing agarose gel. Figure 1 is a blot-transfer analysis of transcripts from an L-cell transformant containing a mouse β-globin gene fused at the 3′ end to a human β-globin gene (Chao *et al.*, 1983). A specific 9 S transcript from the exogenous globin gene is produced by transformed fibroblasts.

Similar cotransformation experiments with the chicken ovalbumin gene (Breathnach *et al.*, 1980; Lai *et al.*, 1980) indicates that accurate splicing of the ovalbumin message is carried out in fibroblast cells. Hence, the mouse fibroblast cell is capable of transcribing genes that are not normally expressed in the fibroblast cell. Correct processing of the globin and ovalbumin genes occurs, indicating that fibroblasts contain the necessary enzymatic activities to splice and polyadenylate transcripts transcribed from foreign genes. Genes from different species are not refractory to expression. Table I lists a variety of genes that have been transferred to mouse L cells. A wide variation in the levels of mRNA has been observed in each independent transformed line. In general, there is no strict correlation between the number of copies introduced into the

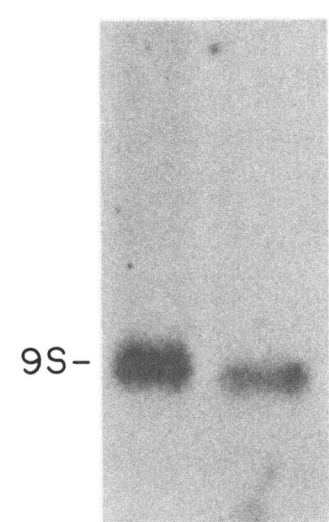

FIGURE 1. RNA analysis of L cells cotransformed with mouse 5′–human 3′ fusion globin genes. L-cell transformants containing the globin gene were grown in the absence (−) or presence (+) of dimethyl sulfoxide (DMSO). Total cellular RNA was isolated and fractionated on a 1% agarose–2.2 M formaldehyde gel, transferred to nitrocellulose, and probed with a nick-translated human β-globin fragment. From Chao *et al.* (1983).

TABLE I. Heterologous Genes Stably
Introduced and Expressed in Mouse L
Cells

Gene	Species
β-Globin	Rabbit
γ-Globin	Human
α-Globin	Human
Growth hormone	Human
$\alpha_{2\mu}$-Globulin	Rat
β-Interferon	Human
α-Interferon	Human
Heat shock *hsp70*	*Drosophila*
Ovalbumin	Chicken
γ_{2b} Heavy chain	Mouse
κ Light chain	Mouse
λ Light chain	Mouse
H-2L antigen	Mouse
HLA-DR	Human
HLA-A2	Human
HLA B7	Human
T8 differentiation antigen	Human
T4 differentiation antigen	Human
Dihydrofolate reductase	Hamster
Mammary tumor virus	Mouse
β-Tubulin	Chicken
β-Actin	Chicken
Metallothionein	Mouse
Transferrin receptor	Human
NGF receptor	Human
EGF receptor	Human

cell and the level of expression for a large number of genes. Not listed in Table
I are a number of transformed cell lines that do not express the transfected gene.
The lack of expression is likely due to the cellular environment into which the
gene integrates (heterochromatin) or to chromosome-position effects. In general,
at least half the transformed lines examined have been found to express appreciable levels of mRNA in L cells.

Although not all transcripts of transfected genes have been carefully examined in transformed mouse L cells, in several cases, incorrect 5' termini have
been observed. The rabbit β-globin gene lacks about 50 bases of 5' information
in transcripts from L cells (Wold *et al.*, 1979; Dierks *et al.*, 1981), and the
ovalbumin-gene transcript in a transformed L cell begins approximately 600
bases upstream of the actual start point of transcription. It is not clear why
transcription begins aberrantly for ovalbumin and globin in transfected fibroblast
cells. One explanation for the aberrant transcription in L cells is the absence of

regulatory factors in L cells that are normally necessary for the correct transcription of these two tissue-specific genes.

In many of the cases cited in Table I, the endogenous native gene is not expressed in the fibroblast cell. Fundamental differences must exist between the genes introduced into cells by transformation and their endogenous counterparts. The transfected gene appears to bypass the negative regulatory elements that restrict the expression of the endogenous gene. The introduced gene may not have the same chromatin conformation or the same methylation status as the endogenous gene. Regulatory proteins may be localized in actively transcribed regions and not be accessible to the native gene. Due to the random nature of integration, the introduced gene is likely to reside in a chromosome position quite different from that of the endogenous counterpart. In addition, the transfected gene has not undergone the developmental modifications that have occurred during normal development to silence the host gene.

Methylation has been shown to inhibit the expression of genes such as the hamster adenine phosphoribosyltransferase (Stein *et al.*, 1982) and the γ-globin gene (Busslinger *et al.*, 1983) in transfection experiments. When these genes were methylated *in vitro* and cotransformed into the genome of mouse tissue-culture cells, no expression was detected from the methylated gene. However, the undermethylated gene is expressed after gene transfer. The correlation that has been made between methylation and gene inactivity appears to apply to methylated genes introduced into cultured cells.

4. INDUCIBLE GENES IN FIBROBLASTS

A number of tissue-specific genes have been introduced into mammalian cells and shown to be expressed in a majority of stable transformants. In many cases, the cell line never expresses the endogenous counterpart. Nevertheless, several tissue-specific genes respond frequently to inducing agents in cultured cells. Genes that are responsive to glucocorticord hormones were among the first to be shown to be inducible in mouse L cells. These genes include the $\alpha_{2\mu}$-globulin gene (Kurtz, 1981), the mouse mammary tumor virus (Buetti and Diggelmann, 1981; Hynes *et al.*, 1981), and the human growth hormone gene (Robins *et al.*, 1982). In the presence of dexamethasone, Ltk$^+$ cells with mouse mammary tumor virus sequences, the rat $\alpha_{2\mu}$-globulin gene, or the human growth hormone gene display increased levels of mRNA for these sequences. The level of induction was approximately 10- to 15-fold higher in the amounts of steady-state RNA. Immunoprecipitable protein products of these hormonally responsive genes could also be detected. The hormone inducibility of these genes in transfected cells implies that the induction is an inherent property of DNA sequences about the genes. Steroid hormones function by interacting with a soluble glucocorticord receptor. Fibroblasts express functional glucocorticord receptors, hence hormonally responsive genes in these cells have the potential to be activated. A striking result from these experiments is that the native counterpart genes in L cells are transcriptionally inactive. It follows from these observations that introduced genes must be in a more permissive environment for induction than the corresponding endogenous gene. Since the same hormonally responsive sequences are present in both introduced and endogenous genes, the difference must lie either in the packaging of the host gene in inactive chromatin or modifications that mark the endogenous gene as unresponsive.

Induction of interferon genes by viral infection or by treatment with double-stranded RNA such as poly(rI : rC) has been reproduced in mouse fibroblast cells cotransformed with the human interferon genes (Hauser *et al.*, 1982; Mantei and Weissmann, 1982; Canaani and Berg, 1982; Ohno and Taniguchi, 1982). Not only have genomic interferon genes been introduced in mouse L cells, but also interferon complementary DNAs (cDNAs) without 5' and 3' regulatory flanking sequences have been expressed in mouse L cells and found to be inducible with poly(rI : rC) (Pitha *et al.*, 1982). Introduced interferon genes in mouse fibroblast cells express roughly the same amounts of RNA as the native gene in response to inducers. In addition, bovine papilloma virus vectors carrying the human β-interferon gene have been introduced into mouse cells (Zinn *et al.*, 1982). The interferon gene in this case exists extrachromosomally and still is appropriately induced.

The increase in mRNA for interferon has been found to be due to an activation of transcription (Ohno and Taniguchi, 1982; Weidle and Weissmann, 1983). Transfection of deletion mutants of the interferon gene has led to the identification of sequences in the promoter that are required for induction. The 5' sequences responsible for the induced transcription appear to be located approximately between − 40 from the start point of transcription and upstream to sequences near − 110 (Fujita *et al.*, 1985; Goodbourn *et al.*, 1985). This region has many features expected for an enhancer element.

Upstream sequences in the same region have been found in the metallothionein promoter that can confer heavy-metal responsiveness in transfected cell lines (Mayo *et al.*, 1982). Metallothionein proteins function by protecting cells from high concentrations of heavy metals such as cadmium. The genes that encode these proteins are in turn induced by heavy metals and also glucocorticords. When the metallothionein promoter was fused to the *tk* gene and introduced in mouse L cells, *tk*-specific mRNA increased in response to cadmium. Metallothionein sequences responsive to heavy metals function when placed in the fibroblast cell.

Perhaps the most unusual induction of a foreign gene is the heat-shock responsiveness of the *Drosophila hs70* gene introduced into mouse L cells (Corces *et al.*, 1981). Heat-shock regulatory elements appear to be extraordinarily conserved, since mammalian cells are capable of regulating the *Drosophila* heat shock 70 gene in response to heat by an increase in *hs70* transcription. Similar results were found in a transient assay using a simian virus 40 vector (Pelham, 1982). Since heat-shock genes have been found in bacteria as well as vertebrates, the sequences that are required for induction are recognized across species.

Several structural-protein genes such as tubulin and actin have been transfected into mouse L cells. In response to microtubule inhibitors, tubulin mRNA synthesis is decreased and actin mRNA synthesis is unaffected. The same responses appear to occur with transfected tubulin and actin genes. Down-regulation of tubulin mRNA synthesis in response to microtubule inhibitors occurred for the transfected β-tubulin gene in L cells in the same manner as the endogenous tubulin genes, whereas no increase was detected for the transfected actin gene. In addition, the transfected tubulin gene underwent processing in L cells to produce a correct transcript (Lau *et al.*, 1985).

The examples of inducible genes discussed thus far demonstrate that mouse fibroblasts can recapitulate the pattern of regulation for these genes. Transfection has allowed for the identification of DNA sequences that are important for the regulated expression of these genes. Two tissue-specific genes that do not exhibit regulated expression in mouse L cells are the globin gene and immunoglobulin genes. They will be elaborated on in Sections 5 and 6.

5. EXPRESSION OF GLOBIN GENES

As noted above, globin genes introduced into mouse L cells are capable of expression, in a few cases at relatively high levels (Mantei *et al.*, 1979; Hsiung *et al.*, 1982). Regulation of the transcription of foreign globin genes in stable L-cell transfectants similar to what has been observed in erythroid cells with chemical inducers has not been successful. For example, when L cells transfected with a mouse human β-fusion globin gene are treated with dimethyl sulfoxide (DMSO), globin-gene transcription remains at a constitutive level. Figure 1 indicates that L cells can transcribe and process the fusion globin gene to a 9 S message. Polyadenylated [poly(A)]-enriched RNA was analyzed by Northern-blot hybridization to nick-translated DNA containing 3′ human globin sequences. No effect of DMSO on the transcription of the fusion globin gene was observed. Other cotransformed lines that have been examined have given similar results. In contrast, other inducible genes are capable of appropriate regulation in L cells. Since the fibroblast cell is an end-stage cell that is not capable of differentiation in culture, the globin gene would not expected to be responsive in a fibroblast. Presumably, mouse fibroblast cells lack erythroid-specific factors that are necessary for regulated expression.

Mouse erythroleukemia (MEL) cells are arrested at the proerythroblast stage, and treatment with a wide number of inducers triggers a pattern of differentiation that mimics erythroid maturation. Globin genes have been introduced via the calcium phosphate method and protoplast fusion into MEL cells in the hope that the transfected globin gene would become responsive to chemical inducers of erythroid differentiation (Spandidos and Paul, 1982; Chao *et al.*, 1983; Wright *et al.*, 1983, 1984; Charney *et al.*, 1984). Several generalizations can be made from these experiments. First, the transfected globin gene appears to be regulated by chemical inducers such as DMSO and hexamethylenebisacetylimide (HMBA) at the same levels of induction that the endogenous globin gene undergoes. Second, the induction observed is specific for the globin gene. Other genes introduced into MEL cells are not regulated by these inducers. Third, the transcription of the β-globin gene in MEL cells begins at the appropriate 5' end, in contrast to the results with L cells. Fourth, the accumulation in the levels of mRNA in response to inducer is due to an increase in the rate of transcription of the gene. These studies indicate that a tissue-specific gene such as β-globin can be regulated properly when transfected in an erythroid cell and that induction is a property of the globin DNA sequence and a property of the recipient cell.

The cell specificity of induction for the globin gene is strikingly illustrated in experiments in which the globin gene is introduced into a cell type that is capable of differentiation but along a different pathway. Myoblast cells can be induced to differentiate by growing to confluence. During differentiation, these cells fuse and begin to express muscle-specific genes at much higher levels. Skeletal actin genes have been introduced into rat muscle cells, which can be induced to undergo this differentiation. Induction of these transfected cells leads to an increase in the actin RNA transcribed from the introduced gene (Nudel *et al.*, 1985). Transfection of the fusion mouse–human globin gene into the same muscle cells does not lead to induction of the globin gene during muscle differentiation. Hence, actin-gene regulation occurs in differentiating muscle cells; β-globin regulation requires an erythroid cell for appropriate regulation. When the globin gene is placed into a differentiating muscle cell, no developmental regulation is observed.

Several differences between the globin genes introduced into MEL cells and endogenous globin genes deserve mention. When the number of transcripts in an induced MEL cell was normalized on a per-gene basis, the absolute level of globin mRNA synthesis was lower for the transferred globin gene than for the endogenous gene. The lower relative level of transcription of the foreign globin gene may be due to mRNA stability differences between the host globin message and the exogenous globin message. Alternatively, differences in chromosome structure or location or in the pattern of DNA modification may influence the level of expression. It should be noted that in many of the transformed lines, multiple copies (>5 copies per cell) are found. Whether each introduced copy is inducible or only a few genes that are introduced become inducible is not certain.

Interestingly, the α-globin gene is not appropriately regulated in MEL cells (Charney *et al.*, 1984), though it is expressed at high constitutive levels relative to the β-globin gene. The inference from this finding is that α-globin-gene expression differs significantly from β-globin-gene expression in erthyroid cells. Earlier work showed that hybrids made between MEL cells and human fibroblasts that retained human chromosome 16 displayed induction of the α-globin gene (Deisseroth and Hendrick, 1978). The difference in behavior of the α- and β-globin genes is reflected in the temporal difference of expression during development. The human α-globin gene is transcribed earlier in fetal life and the β-globin gene expression begins after birth. Hence, different mechanisms must govern these adult globin genes during differentiation.

The mouse adult globin gene is induced in MEL cells significantly in response to inducers. However, the mouse embryonic gene does not respond to chemical inducers. Most likely, differences in the chromatin conformation between the adult and embryonic β-globin genes account for the lack of expression in MEL cells. Differences between the promoter or coding sequences between the adult and embryonic globin genes could also preclude induction. To test this possibility, a fusion gene consisting of the 5' portion of the mouse embryonic *ey3* gene was joined in the second exon to the 3' portion of the adult human β-globin gene (Fig. 2A). This embryonic fusion gene (*MHε*) was introduced into MELaprt⁻ cells by cotransformation with the hamster *aprt* gene (Chao *et al.*, 1983). Two of three stable tranformants displayed a low level of induction in an S1 nuclease analysis of poly(A) RNA from the transformants (Fig. 2B). In contrast, no mRNA was produced from the endogenous mouse *ey3* gene in the transformed MEL cells even under conditions of induction (Fig. 3). Isolated total RNA from 14-day-old mouse embryos was used as a positive control. Hence, embryonic gene expression and low levels of induction can take place in an erythroid cell in which the endogenous gene is totally silent. The embryonic globin gene certainly has the potential of being expressed. The DNA sequence about the embryonic gene does not restrict expression of this gene in MEL cells. The fusion embryonic globin gene has been able to circumvent the restraints that do not allow expression of the endogenous gene.

Experiments of a similar nature with fusion genes constructed with either the adult β-globin promoter or globin-coding regions indicate that there are sequences 3' to the cap site as well as sequences in the globin promoter that are necessary for induction in MEL cells (Charney *et al.*, 1984; Wright *et al.*, 1984). Human 3' globin sequences may in fact render the embryonic ε-fusion gene responsive to induction. However, the embryonic human globin gene is capable of undergoing induction in the K562 erythroleukemic cell line, indicating that this gene is not transcriptionally silent in the appropriate erythroid cell (Kioussis *et al.*, 1985).

Further evidence that developmentally regulated genes can be expressed and regulated after gene transfer comes from transfection experiments with the human fetal γ^G-globin gene. The switch in expression from fetal globin to adult globin occurs about the time of birth. No fetal globin synthesis occurs later in development. MEL cells containing the human γ^G gene are also capable of

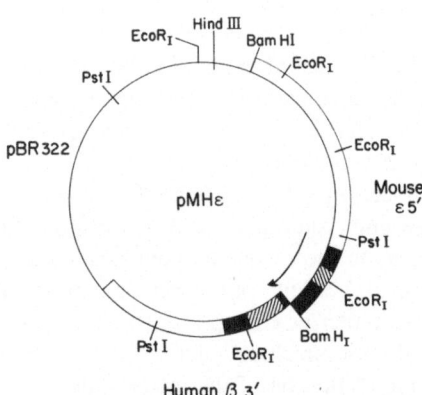

FIGURE 2. Hybrid embryonic globin gene expression. (Left) A 5′ segment of the mouse *ey3* gene was ligated to a 3′ fragment of human β-globin DNA extending from the conserved *Bam*HI site in the second exon. This plasmid was introduced into MEL cells, and cotransformants were identified by blot hybridization. (Below) Three cotransformant lines were analyzed by 3′ S1 nuclease analysis. Poly(A)-RNA was isolated from these in the absence (A, C, and E) or presence (B, D, and F) of 2% DMSO. The probe was a human 3′ *Bam–Eco*RI fragment end-labeled at the *Bam* site. After hybridization and S1 nuclease digestion, fragments were analyzed by electrophoresis on an 8% polyacrylamide sequencing gel. A protected fragment of 212 bases appears with human cord blood RNA (G).

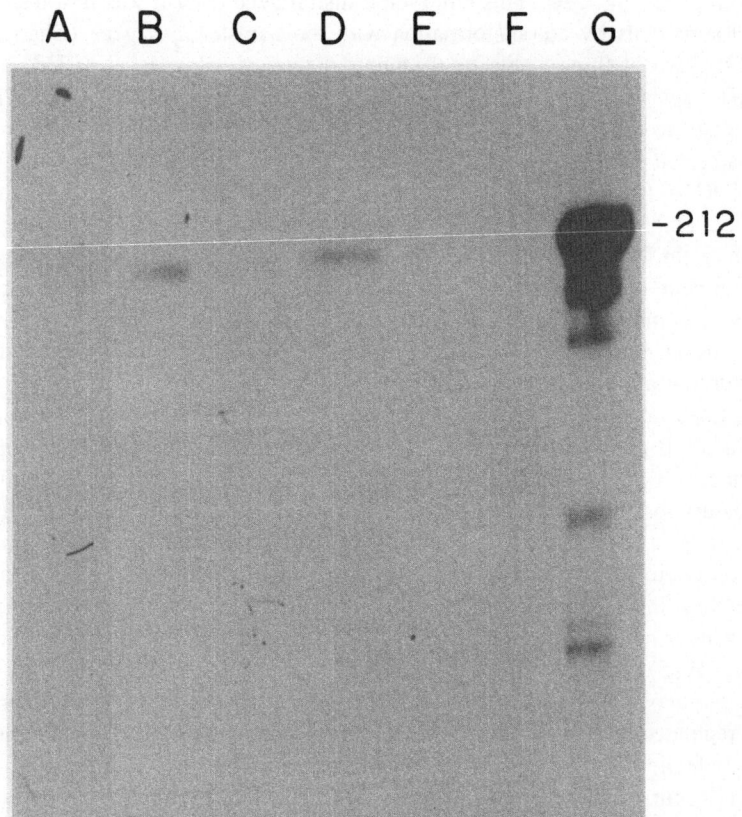

proper regulation of the fetal globin gene. In Fig. 4, two of three transformants containing the γ^G gene are responsive to HMBA induction. Approximately 5-fold induction of the γ^G gene can be seen. The mouse globin locus does not carry a fetal globin gene and therefore does not undergo the fetal-to-adult globin switch during development. Nevertheless, MEL cells can induce transcription of a heterologous globin gene that is expressed only at a very restricted period during development. The mechanism of switching in hemoglobin biosynthesis must involve restraints on expression at specific points during development. These developmental restrictions do not apply for transformed genes. Trans-

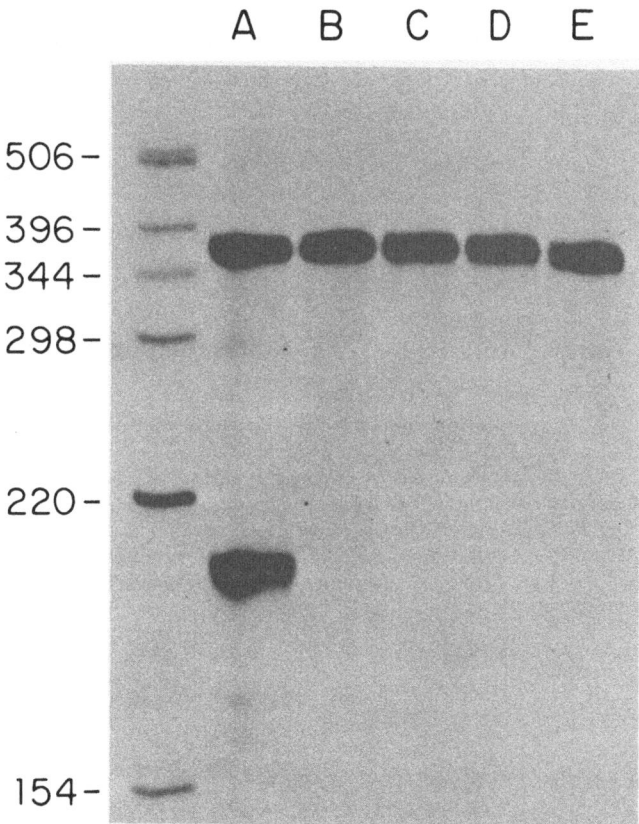

FIGURE 3. Lack of expression of the mouse embryonic gene in MEL cells. Poly(A) RNA was isolated from two independent cotransformants treated with HMBA, hybridized with a labeled 360-base-pair *Eco*RI restriction fragment from the 3' end of the mouse embryonic globin gene (Hansen *et al.*, 1982), and digested with S1 nuclease. A protected fragment 190 bases long from 14-day-old mouse embryo RNA is seen (A). (B, D) Uninduced RNA; (C, E) induced RNA from the two cotransformant MEL lines.

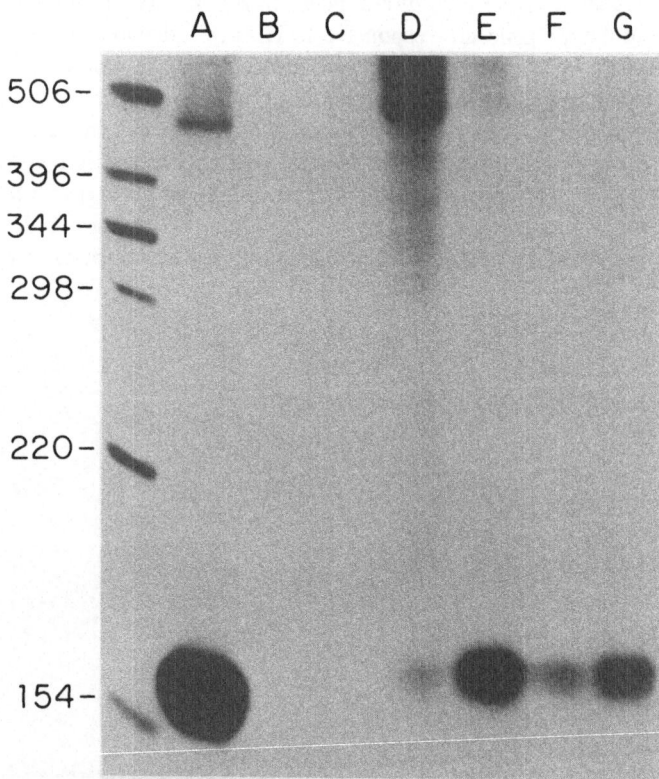

FIGURE 4. Induction of the human fetal γ^G-globin genes in MEL cells. The human γ^G-globin gene was cotransformed into MEL cells with the hamster *aprt* gene and verified by Southern-blot hybridization. RNA was isolated from three independent lines treated with (C, E, and G) and without inducer (B, D, and F) and analyzed with a 3'-labeled *Eco*RI restriction fragment from the γ^G gene by S1 nuclease analysis. A protected fragment of 160 bases is observed with human cord blood RNA (A).

formed genes appear to escape the developmental history of the cell and may be regulated if the appropriate factors exist in the cell.

6. EXPRESSION OF IMMUNOGLOBULIN GENES

When rearranged heavy- and light-chain immunoglobulin genes are introduced into mouse Ltk⁻ cells, only very low-level transcription is observed for these genes (Stafford and Queen, 1983; Gillies and Tonegawa, 1983). Accurate transcription has been observed from a heavy-chain γ_{2b} gene, but transcripts originating upstream of the authentic start site of transcription were detected for the light-chain genes. These aberrant transcripts are reminiscent of the globin and ovalbumin gene transcripts in L cells. The overwhelming evidence indicates that little or no transcription occurs for the immunoglobulin genes transfected into L cells. In fact, the level of γ_{2b} was at least two orders of magnitude lower in L cells than in myeloma cells (Gillies and Tonegawa, 1983).

The cellular *cis*-acting enhancer in the heavy- and light-chain immunoglobulin genes promotes the transcription of these genes in lymphoid cells (Gillies *et al.*, 1983; Banerji *et al.*, 1983; Picard and Schaffner, 1984). The characterization of these sequences was accomplished by transfection into lymphoid cells. A wide variety of techniques including electroporation (Potter *et al.*, 1984), protoplast fusion (Gillies *et al.*, 1983), and transient expression assays (Picard and Schaffner, 1984) have been employed to identify the internal enhancer sequence. When these sequences are deleted from the immunoglobulin gene, transcription is severely reduced in B cells. The immunoglobulin enhancer is similar to viral enhancers, acting without regard to orientation and distance from the gene. The position of the enhancer within the major intron means that the enhancer can operate on the promoter of the rearranged μ-immunoglobulin gene as well as other heavy-chain genes later in B-cell development. The tissue specificity of the immunoglobulin enhancer sequence is quite striking, being active in mouse B cells but not in mouse fibroblasts. The implication of these studies is that tissue-specific factors present in lymphoid cells and not present in fibroblasts are required for high-level expression of immunoglobulin genes.

The effect of the enhancer sequence on immunoglobulin-gene expression is but one aspect that is responsible for tissue-specific expression. Other elements have also been discovered by transfection of immunoglobulin genes in different lymphoid-cell types. Not only is the immunoglobulin enhancer necessary, but also the promoter region confers cell-type specificity of immunoglobulin transcription (Mason *et al.*, 1985; Gopal *et al.*, 1985). The promoter sequence alone can give cell-type specificity to transcription of the immunoglobulin gene. Gene fusions made between the heavy-chain V_H promoter and the coding region of the β-globin gene were introduced by cotransformation into J558L mouse plasmacytoma cells, mouse L cells, and HeLa cells. Little or no expression was observed in L cells and HeLa cells. Correct initiation at the V_H cap sites was seen in the plasmacytoma cell, indicating that the heavy-chain promoter confers cell-type specificity to a gene not normally expressed in B cells.

7. EXPRESSION OF CELL-SURFACE ANTIGENS

Transfer of genes that encode cell-surface molecules into mammalian cells has been accomplished not only by cotransformation with cloned genes but also by using total cellular DNA. Identification of transformants that express the membrane molecules has utilized fluorescence-activated cell sorting (Kavathas and Herzenberg, 1983; Kuhn *et al.*, 1984), radioimmunoassay (Herman *et al.*, 1983), and immunological rosetting using monoclonal antibodies directed against the molecule (Albino *et al.*, 1985; Littman *et al.*, 1985).

Using cotransformation with the herpes simplex *tk* gene, the human histocompatibility Class I antigens HLA-A2 and HLA-B7 (Barbosa *et al.*, 1982; Herman *et al.*, 1983) and Class II HLA-DR antigen (Rabourdin-Combe and Mach, 1983) have been introduced into mouse L cells. In most cell lines, the level of expression approximated the levels found for the molecules in lymphoid cells. By immunofluorescence and antibody-binding assays, the histocompatibility molecules were found to be present on the cell surface. The Class I major histocompatibility complex molecules are normally associated with a second molecule, β_2-microglobulin. Mouse fibroblasts normally express β_2-microglobulin. This molecule has been found in association with the transfected HLA molecules. Mouse L cells bearing the HLA antigens can function appropriately as targets for lysis by cytotoxic T cells. These antigens are assembled intracellularly in a correct manner, processed appropriately, and appear on the cell surface of the fibroblast with the correct antigenic specificity.

Total cellular DNA has been used for gene transfer of single-copy genes encoding cell-surface antigens. Since one positive Ltk$^+$ colony can be generated from 10^6 cells with 20 pg of the cloned *tk* gene, a single-copy gene can be transferred using 20 μg of total cellular DNA, assuming the gene is present at one in a million DNA sequences. Since a considerable amount of carrier DNA is introduced during the calcium phosphate procedure, the only limitation in this scheme is whether the cell-surface antigen is capable of expression in a heterologous cell.

The T-cell differentiation antigens T4 and T8 have been transferred to L cells using total DNA (Kavathas and Herzenberg, 1983; Littman *et al.*, 1985; Madden *et al.*, 1985) as well as an antigen from a chronic lymphocytic leukemia cell (Stanners *et al.*, 1981). The efficiencies of transfer vary between 1 in 1000 tk$^+$ colonies and 1 in approximately 10,000 tk$^+$ colonies. In addition, platelet glycoproteins IIb/IIIa (Cosgrove *et al.*, 1986), a melanoma antigen (Albino *et al.*, 1985) and growth-related receptors such as the transferrin receptor (Kuhn *et al.*, 1984) and the nerve-growth-factor receptor (Chao *et al.*, 1986) have been transferred to fibroblast cells. In the great majority of these experiments, the foreign antigen is expressed in varying levels from 10,000 copies per cell to well over 100,000 molecules per cell. The molecules are processed to a size identical to the size seen on cells that normally express the molecule. Fibroblasts can also glycosylate the transferred gene product appropriately. The ability of these gene products to be identified in heterologous cells provides the opportunity to isolate the genes either by subtractive complementary DNA (cDNA) hybridization or by rescue of the transforming sequence.

The source of DNA for genomic DNA transfections appears not to be a crucial determinant for expression. Isolated DNAs from a number of different sources have been used to transfer the genes for cell-surface antigens (Hsu *et al.*, 1984). Even DNA isolated from cell lines that do not express the gene is nevertheless fully competent in transferring the gene. Heavily methylated DNA does not serve as appropriate donor for gene transfer.

An additional requirement for gene transfer using total DNA is the requirement for a single gene product. The likelihood of transferring two unlinked genes simultaneously from total genomic DNA is very low. However, expression of one chain of a cell-surface protein can complement the expression of associated antigens, as in the case of the β_2-microglobulin with histocompatibility proteins. In addition, a cDNA for the T-cell-receptor chain has been expressed in receptor-minus mutants of the Jurkat line after transfection (Ohashi *et al.*, 1985). A functional T-cell receptor was reconstituted with the β-chain as well as with an associated T-cell antigens of T3.

8. SUMMARY

A large number of eukaryotic genes have been shown to be accurately and efficiently expressed in tissue-culture cells. What governs the ability of a gene to be expressed in a foreign environment? From the majority of the cases observed, the mere introduction of the gene is sufficient to assure expression. A number of genes will respond to environmental changes in a heterologous host. Specific DNA sequences responsible for induction have been identified using transfection of modified genes. On one level, these sequences dictate the level of expression of inducible genes. However, the mere presence of a particular DNA sequence does not guarantee expression. Superimposed on the primary DNA sequence are developmental specificities and higher orders of structure.

The failure to observe expression of transfected genes may be due to an enhanced methylation state of the gene in the donor DNA, the chromosomal location of the integrated gene, or a chromatin structure around the foreign DNA that makes the gene inaccessible for transcription.

The chromatin environment undoubtedly plays a role in the examples in which the host-cell gene is silent and yet the analogous gene that is transfected is active. Transfected genes appear to be primed for expression and are probably readily available to the transcription machinery of the cell. Since cotransformed genes are found with the selectable marker and carrier DNA in a large concatenated structure after gene transfer, transferred sequences are closely linked on the same chromosome segment (Perucho *et al.*, 1980). How many sequences within the concatenate are actually transcribed is not known, but it is likely many sequences have the potential of expression. The ability to detect mRNA coding for T4 and T8 in a primary fibroblast transformant using subtractive hybridization

(Littman *et al.*, 1985) implies that many transferred sequences are frequently expressed. Studies with linked genes indicate that at least 20 kb of transferred information is regulated coordinately in mouse cells (Roginski *et al.*, 1983).

Examination of the chromatin structure of transfected genes indicates that the gene retains an active structure after transfection, yet remains in a nucleosomal structure (Camerini-Otero and Zasloff, 1980). Weintraub (1983) has been able to detect specific hypersensitive sites in DNA transfected into mouse fibroblasts. It is probable that most transferred genes are in an active configuration after gene transfer. Sweet *et al.* (1982) found that the herpes *tk* gene in mouse fibroblasts could undergo changes in chromatin structure that correlated with *tk* expression. Specifically, hypersensitive sites were detected in the 5' end of the *tk* gene in cell lines that were tk$^+$. This structure presumably reflects a more open region available for transcription by RNA polymerase II or an increased accessibility of the transfected gene to transcription factors. Selection for tk$^-$ cells yielded a loss of the 5' hypersensitive sites. Different chromatin structures can exist for the same gene in the same cell and direct the expression of the gene. Regulated expression of introduced genes requires more than just a permissive environment for expression. Specific factors in appropriate cell types are necessary for proper regulation of tissue-specific genes such as globin and immunoglobulin. The absence of these putative tissue-specific factors, however, does not preclude expression of cloned genes in mammalian cells, since nearly every gene is capable of expression in heterologous cells.

REFERENCES

Albino, A. P., Graf, L. H., Kantor R. R. S., McLean, W., Silagi, S., and Old, L., 1985, DNA-mediated transfer of human melanoma cell surface glycoprotein gp130: Identification of transfectants by erythrocyte rosetting, *Mol. Cell. Biol.* **5**:692–697.

Banerji, J., Olson, L., and Schaffner, W., 1983, A lymphocyte-specific cellular enhancer is located downstream of the joining region in immunoglobulin heavy chain genes, *Cell* **33**:729–740.

Barbosa, J. A., Kamarck, M. E., Biro, P. A., Weissman, S. M., and Ruddle, F. H., 1982, Identification of human genomic clones coding the major histocompatibility antigens HLA-A2 and HLA-B7 by DNA mediated gene transfer, *Proc. Natl. Acad. Sci. U.S.A.* **79**:6327–6331.

Breathnach, R., Mantei, N., and Chambon, P., 1980, Correct splicing of a chicken ovalbumin gene transcript in mouse L cells, *Proc. Natl. Acad. Sci. U.S.A.* **77**:740–744.

Buetti, E., and Diggelmann, H., 1981, Cloned mouse mammary tumor virus DNA is biologically active in transfected mouse cells and its expression is stimulated by glucocorticoid hormones, *Cell* **23**:335–345.

Busslinger, M., Hurst, J., and Flavell, R., 1983, DNA methylation and the regulation of globin gene expression, *Cell* **34**:197–206.

Camerini-Otero, R. D., and Zasloff, M. A., 1980, Nucleosome packaging of the thymidine kinase gene of herpes simplex virus transferred into mouse cells: An actively expressed single-copy gene, *Proc. Natl. Acad. Sci. U.S.A.* **77**:5079–5083.

Canaani, D., and Berg, P., 1982, Regulated expression of human interferon β1 gene after transduction into cultured mouse and rabbit cells, *Proc. Natl. Acad. Sci. U.S.A.* **79**:5166–5170.

Capecchi, M. R., 1980, High efficiency transformation by direct microinjection of DNA into cultured mammalian cells, *Cell* **22:**479–488.

Chao, M. V., Mellon, P., Charney, P., Maniatis, T., and Axel, R., 1983, The regulated expression of β-globin genes introduced into mouse erythroleukemia cells, *Cell* **32:**483–493.

Chao, M. V., Bothwell, M. A., Ross, A., Koprowski, H., Lanahan, A., Buck, C. R., and Sehgal, A., 1986, Gene transfer and molecular cloning of the human NGF receptor, *Science* **232:**518–521.

Charney, P., Treisman, R., Mellon, P., Chao, M., Axel, R., and Maniatis, T., 1984, Differences in human α- and β-globin gene expression in mouse erythroleukemia cells: The role of intragenic sequences, *Cell* **38:**251–263.

Corces, V., Pellicer, A., Axel, R., and Meselson, M., 1981, Integration, transcription, and control of a *Drosophila* heat shock gene in mouse cells, *Proc. Natl. Acad. Sci. U.S.A.* **78:**7038–7042.

Cosgrove, L. H., Sandrin, M. S., Rajaskariah, P., and Makausie, I. F. C., 1986, A genomic clone encoding the α chain of the OKM1 LFA-1, and platelet glycoprotein IIb-IIIa molecules, *Proc. Natl. Acad. Sci. U.S.A.* **83:**752–756.

Deisseroth, A., and Hendrick, D., 1978, Human α-globin gene expression following chromosomal dependent gene transfer into mouse erythroleukemia cells, *Cell* **15:**55–63.

Dierks, P., van Ooyen, A., Mantei, N., and Weissmann, C., 1981, DNA sequences preceding the rabbit β-globin gene are required for formation in mouse L cells of β-globin RNA with the correct 5' terminus, *Proc. Natl. Acad. Sci. U.S.A.* **78:**1411–1415.

Fujita, T., Ohno, S., Yasumitsu, H., and Toniguchi, T., 1985, Delimitation and properties of DNA sequences for the regulated expression of human interferon-β gene, *Cell* **41:**489–496.

Gillies, S. D., and Tonegawa, S., 1983, Expression of cloned immunoglobulin genes introduced into mouse L cells, *Nucleic Acids Res.* **11:**7981–7997.

Gillies, S. D., Morrison, S. L., Oi, V. T., and Tonegawa S., 1983, A tissue-specific transcription enhancer element is located in the major intron of a rearranged immunoglobulin heavy chain gene, *Cell* **33:**717–728.

Goodbourn, S., Zinn, K., and Maniatis, T., 1985, Human B-interferon gene expression is regulated by an inducible enhancer element, *Cell* **41:**509–520.

Gopal, T. V., Shimada, T., Baur, A. W., and Nienhuis, A. W., 1985, Contribution of promoter to tissue-specific expression of the mouse immunoglobulin kappa genes, *Science* **229:**1102–1104.

Graham, F. L., and van der Eb, A. J., 1973, A new technique for the assay of infectivity of human adenovirus 5 DNA, *Virology* **52:**456–467.

Hansen, J. N., Konkel, D. A., and Leder, P., 1982, The sequence of a mouse embryonic β-globin gene, *J. Biol. Chem.* **257:**1048–1052.

Hauser, H., Gross, G., Bruns, W., Hochkeppel, H.-K., Mayr, U., and Collins, J., 1982, Inducibility of human β-interferon gene in mouse L-cell clones, *Nature (London)* **297:**650–654.

Herman, A., Parham, P., Weissman, S. M., and Engelhard, V. H., 1983, Recognition by xenogeneic cytotoxic T lymphocytes of cells expressing HLA-A2 or HLA-B7 after DNA-mediated gene transfer, *Proc. Natl. Acad. Sci. U.S.A.* **80:**5056–5060.

Hsiung, N., Roginski, R. S., Henthorn, P., Smithies, O., Kucherlapati, R., and Skoultchi, A. I., 1982, Introduction and expression of a fetal human globin gene in mouse fibroblasts, *Mol. Cell. Biol.* **2:**401–411.

Hsu, C., Kavathas, P., and Herzenberg, L. A., 1984, Cell-surface antigens expressed on L cells transfected with whole DNA from non-expressing and expressing cells, *Nature (London)* **312:**68–69.

Hynes, N. E., Kennedy, N., Rahmsdorf, U., and Groner, B., 1981, Hormone-responsive expression of an endogenous provirus gene of mouse mammary tumor virus after molecular cloning and gene transfer into cultured cells, *Proc. Natl. Acad. Sci. U.S.A.* **78:**2038–2042.

Ishiura, M., Hirose, S., Uchida, T., Hamada, Y., Suzuki, Y., and Okada, Y., 1982, Phage particle-mediated gene tranfer to cultured mammalian cells, *Mol. Cell. Biol.* **2:**607–616.

Kavathas, P., and Herzenberg, L., 1983, Stable transformation of mouse L cells for human membrane T-cell differentiation antigens, HLA, and β$_2$-microglobulin: Selection by fluorescence-activated cell sorting, *Proc. Natl. Acad. Sci. U.S.A.* **80:**524–528.

Kioussis, D., Wilson, F., Khazaie, K., and Grosveld, F., 1985, Differential expression of human globin genes introduced into K562 cells, *EMBO J.* **4**:927–931.

Kuhn, L. C., McClelland, A., and Ruddle, F. H., 1984, Gene transfer, expression, and molecular cloning of the human transferrin receptor gene, *Cell* **37**:95–103.

Kurtz, D. T., 1981, Hormonal inducibility of rat $\alpha_{2\mu}$-globulin genes in transfected mouse cells, *Nature (London)* **291**:629–631.

Lai, E. C., Woo, S. L. C., Bordelon-Riser, M. E., Fraser, T. H., and O'Malley, B. A., 1980, Ovalbumin is synthesized in mouse cells transformed with the natural chicken ovalbumin gene, *Proc. Natl. Acad. Sci. U.S.A.* **77**:244–248.

Lau, J. T. Y., Pittenger, M. F., and Cleveland, D. W., 1985, Reconstruction of appropriate tubulin and actin gene regulation after transient transfection of cloned β-tubulin and β-actin genes, *Mol. Cell. Biol.* **5**:1611–1620.

Littman, D. R., Thomas, Y., Madden, P. J., Chess, L., and Axel, R., 1985, The isolation and sequence of the gene encoding T8: A molecule defining functional classes of T lymphocytes, *Cell* **40**:237–246.

Madden, P. J., Littman, D. R., Godfrey, M., Maddon, D. E., Chess, L., and Axel, R., 1985, The isolation and nucleotide sequence of a cDNA encoding the T cell surface protein T4: A new member of the immunoglobulin gene family, *Cell* **42**:93–104.

Mantei, N., and Weissmann, C., 1982, Controlled transcription of a human A-interferon gene introduced into mouse L cells, *Nature (London)* **297**:128–132.

Mantei, N., Boll, W., and Weissmann, C., 1979, Rabbit β-globin mRNA production in mouse L cells transformed with cloned rabbit β-globin chromosomal DNA, *Nature (London)* **281**:40–46.

Mason, J. O., Williams, G. T., and Neuberger, M. S., 1985, Transcriptional cell type specificity is conferred by an immunoglobulin V_H gene promoter that includes a functional consensus sequence, *Cell* **41**:479–487.

Mayo, K. E., Warren, R., and Palmiter, R. D., 1982, The mouse metallothionein-I gene is transcriptionally regulated by cadmium following transfection into human or mouse cells, *Cell* **29**:99–108.

Neumann, E., Shaefer-Ridder, M., Wang, Y., and Hofschneider, F., 1982, Gene transfer into mouse lyoma cells by electroporation in high electric fields, *EMBO J* **1**:841–845.

Nudel, U., Greenberg, D., Ordahl, C. P., Saxel, O., Neuman, S., and Yaffe, D., 1985, Developmentally regulated expression of a chicken muscle-specific gene in stably transfected rat myogenic cells, *Proc. Natl. Acad. Sci. U.S.A.* **82**:3106–3109.

Ohashi, P. S., Mak, T. W., Van den Elsen, P., Yanagi, Y., Yoshikai, Y., Calman, A. F., Terhorst, C., Stobo, J. D., and Weiss, A., 1985, Reconstitution of an active surface T3/T-cell antigen receptor by DNA transfer, *Nature (London)* **316**:606–609.

Ohno, S., and Taniguchi, T., 1982, Inducer-responsive expression of the cloned human interferon β₁ gene introduced into cultured mouse cells, *Nucleic Acids Res.* **10**:967–977.

Pelham, H. R. B., 1982, A regulatory upstream promoter element in the *Drosophila* hsp 70 heatshock gene, *Cell* **30**:517–527.

Perucho, M., Hanahan, D., Lipsich, L., and Wigler, M., 1980, Genetic and physical linkage of exogenous sequences in transformed cells, *Cell* **22**:309–317.

Picard, D., and Schaffner, W., 1984, A lymphocyte specific enhancer in the mouse immunoglobulin κ gene, *Nature (London)* **307**:80–82.

Pitha, P. M., Ciufo, D. M., Kellum, M., Raj, N. B. K., Reyes, G. R., and Hayward, G. S., 1982, Induction of human β-interferon synthesis with poly(rI rC) in mouse cells transfected with cloned cDNA plasmids, *Proc. Natl. Acad. Sci. U.S.A.* **79**:4337–4341.

Potter, H., Weir, L., and Leder, P., 1984, Enhancer-dependent expression of human μ immunoglobulin genes introduced into mouse pre-B lymphocytes by electroporation, *Proc. Natl. Acad. Sci. U.S.A.* **81**:7161–7165.

Rabourdin-Combe, C., and Mach, B., 1983, Expression of HLA-DR antigens at the surface of mouse L cells co-transfected with cloned human genes, *Nature (London)* **303**:670–674.

Robins, D., Ripley, S., Henderson, A. S., and Axel, R., 1981, Transforming DNA integrates into the host chromosome, *Cell* **23**:29–39.

Robins, D., Paek, I., Seeburg, P. H., and Axel, R., 1982, Regulated expression of human growth hormone genes in mouse cells, *Cell* **29**:623–631.

Roginski, R. S., Skoultchi, A. I., Henthron, P., Smithies, O., Hsiung, N., and Kucherlapati, R., 1983, Coordinate modulation of transfected HSV thymidine kinase and human globin genes, *Cell* **35**:149–155.

Schaffner, W., 1980, Direct transfer of cloned genes from bacteria to mammalian cells, *Proc. Natl. Acad. Sci. U.S.A.* **77**:2163–2167.

Spandidos, D. A., and Paul, J., 1982, Transfer of human globin genes to erythroleukemic mouse cells, *EMBO J.* **1**:15–20.

Stafford, J., and Queen, C., 1983, Cell-type specific expression of a transfected immunoglobulin gene, *Nature (London)* **306**:77–79.

Stanners, C. P., Lam, T., Chamberlain, J. W., Stewart, S. S., and Price, G. B., 1981, Cloning of a functional gene responsible for the expression of a cell surface antigen correlated with human chronic lymphocytic leukemia, *Cell* **27**:211–221.

Stein, R., Razin, A., and Cedar, H., 1982, *In vitro* methylation of the hamster adenine phosphoribosyltransferase gene inhibits its expression in mouse L cells, *Proc. Natl. Acad. Sci. U.S.A.* **79**:3418–3422.

Sweet, R. W., Chao, M. V., and Axel, R., 1982, The structure of the thymidine kinase gene promoter: Nuclease hypersensitivity correlates with expression, *Cell* **31**:347–353.

Weidle, U., and Weissmann, C., 1983, The 5′-flanking region of a human IFN-α gene mediates viral induction of transcription, *Nature (London)* **303**:442–446.

Weintraub, H., 1983, A dominant role for DNA secondary structure in forming hypersensitive structures in chromatin, *Cell* **32**:1191–1203.

Wold, B., Wigler, M., Lacy, E., Maniatis, T., Silverstein, S., and Axel, R., 1979, Introduction and expression of a rabbit β-globin gene in mouse fibroblasts, *Proc. Natl. Acad. Sci. U.S.A.* **76**:5684–5688.

Wright, S., de Boer, E., Grosveld, F. G., and Flavell, R. A., 1983, Regulated expression of the human B-globin family in erythroleukemia cells, *Nature* **305**:333–336.

Wright, S., Rosenthal, A., Flavell, R., and Grosveld, F., 1984, DNA sequences required for regulated expression of B-globin in murine erythroleukemia cells, *Cell* **38**:265–273.

Zinn, K., Mellon, P., Ptashne, M., and Maniatis, T., 1982, Regulated expression of an extrachromosomal human β-interferon gene in mouse cells, *Proc. Natl. Acad. Sci. U.S.A.* **79**:4897–4901.

9

MUTATION OF AUTONOMOUSLY REPLICATING PLASMIDS

Michele P. Calos

1. INTRODUCTION

Transfection of mammalian cells with DNA vectors that can replicate autonomously as plasmids is advantageous for several types of experiments. Plasmids that replicate to high copy number provide amplification of signals for DNA, RNA, and protein synthesis. Also, plasmid DNA can be easily retrieved from mammalian cells, separate from the mass of chromosomal DNA. Thus, the use of autonomously replicating recombinant DNA vectors, in conjunction with transfection techniques for transfer of the vectors to mammalian cells, has produced rapid and powerful experiments that avoid many of the complexities associated with the full-sized mammalian genome.

Experiments designed to study mutation using autonomously replicating plasmids have revealed, however, that transfection of these vectors is associated with a high mutation frequency. Moreover, it is clear that many of these results apply to all transfected DNA, whether or not it can replicate. The mutations detected run the gamut from deletions, insertions, and other rearrangements to point mutations such as single base-pair substitutions. Furthermore, introduction of DNA by all methods described to date appears to have the potential to introduce mutations. Therefore, experimenters engaged in introduction of DNA into mammalian cells need to be aware of the mutagenesis associated with transfection and to take it into consideration, if necessary, in the interpretation of data.

Fortunately, a great deal has been learned in recent years about the nature and causes of mutation in transfected DNA. This work has taken advantage of

Michele P. Calos • Department of Genetics, Stanford University School of Medicine, Stanford, California 94305.

shuttle vectors that can replicate autonomously in both mammalian and bacterial cells. Bacterial genes placed on the shuttle vectors have been used as targets for mutation. Thus, the shuttle vectors can be transfected into mammalian cells where mutagenesis takes place. Vector DNA can then be purified and reintroduced into *Escherichia coli,* in which mutations can be rapidly scored and analyzed. This methodology has permitted a thorough study of the mutations incurred during transfection, and these data are reviewed in this chapter.

Many different systems have been used to drive autonomous replication of transfected DNA. Because no bona fide origin of replication from a mammalian chromosome has yet been isolated, the origins of replication used have been derived from animal viruses. Depending on the nature of the origin of replication, vectors derived from these viruses have been divided into the categories of transiently replicating or stably replicating plasmids. Each of these replication systems requires a *cis*-acting sequence at which replication initiates and *trans*-acting sequences that encode one or more products required to activate the replication origin. The term "transiently replicating" refers to vectors that, once inside the nucleus of a permissive cell, replicate continuously in the presence of the *trans*-acting products. This type of replication is called transient because the high vector copy numbers attained are not generally compatible with long-term survival of the transfected cells. Examples of transiently replicating plasmids are those based on the papovaviruses simian virus 40, polyoma virus, and BK virus.

In contrast, stably replicating plasmids replicate in the nuclei of permissive cells in synchrony with the DNA-synthesis period of the cell cycle of the transfected cells. They generally attain a fairly stable copy number per cell that is lower than that achieved by transiently replicating plasmids. Immortalized cell lines transfected with stably replicating vectors are generally able to carry the vectors as plasmids indefinitely. Examples of stably replicating plasmids are those derived from the papovavirus bovine papillomavirus and the herpesvirus Epstein–Barr virus.

It is apparent that autonomously replicating plasmids encompass a variety of vector types with different replication properties. In turn, cells from a wide range of mammalian species and tissue types can serve as hosts for one or more of the vector systems. A given combination of vector and host cell leads to a characteristic mutational behavior due to contributions from both partners. Data on all these systems are presented below. Furthermore, it has been possible, with certain combinations of vectors and host cells, to arrive at spontaneous mutation frequencies in the transfected DNA that are relatively low. In these situations, mutations induced by mutagens and carcinogens can be readily studied. Thus, the mutagenic specificity of carcinogens is being defined at the DNA-sequence level. These autonomously replicating plasmid systems and the information they have produced to date are also described.

2. SPONTANEOUS MUTATION OF AUTONOMOUSLY REPLICATING PLASMIDS

2.1. Simian-Virus-40-Based Vectors

Evidence existed, dating from some of the first transfection experiments, that transfected DNA could suffer damage and rearrangement. These data took the form of sporadic observations of discrepancies between the restriction patterns of DNA integrated into mammalian chromosomes following transfection and those of the starting DNA (e.g., Wigler *et al.*, 1978). However, the degree and variety of mutation sustained by DNA transfected by commonly used techniques did not become fully apparent until experiments designed to measure mutation frequency were performed.

These experiments used vectors based on simian virus 40 (SV40). SV40 replicates efficiently in monkey kidney cells such as CV-1, BS-C-1, and the SV40-transformed CV-1 lines COS-1 and COS-7 (Gluzman, 1981). Replication of SV40 vectors in CV-1 or BS-C-1 requires the SV40 large tumor (large-T) antigen and the SV40 origin of replication. The vectors therefore carry these regions of the SV40 genome. Since COS cells synthesize large-T antigen constitutively, vector replication in COS cells requires only that the SV40 origin be present on the vector. Use has been made of shuttle vectors containing one or both of these SV40 sequence elements, plus portions of the bacterial plasmid pBR322 to provide for replication and selection (ampicillin resistance) in bacterial cells. To these vectors were added bacterial genes to be used as targets for mutation. *LacI* (Calos *et al.*, 1983) and *galK* (Razzaque *et al.*, 1983) have been used for this purpose. (Figure 1 depicts some of the vectors used.) The exper-

FIGURE 1. SV40-based plasmids. (■) SV40 DNA; (□) *lac* sequences; (——) pML (a derivative of pBR322) sequences. The 1.7-kilobase (kb) *lac* fragment contains the wild-type *lacI* gene, as well as the beginning of *lacZ*. The *lacZ* portion is necessary to give a blue colony in our assay. pSVi2 and pSVi4 contain the 1.7-kb *lacI* fragment inserted on opposite sides of a 312-base-pair fragment containing the SV40 origin of replication, as well as the chicken thymidine kinase gene (ChTK). pJYMib contains the *lacI* fragment, plus all the SV40 inserted in the *Bam*H1 site of pML. (B) *Bam*H1; (H) *Hind*III; (R) *Eco*R1.

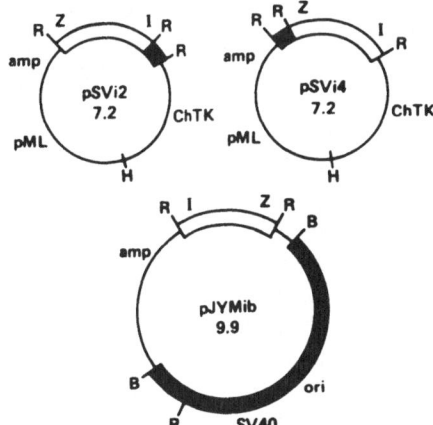

imental scheme thus includes transfection of vector DNA into mammalian cells, passage time of approximately 48 hr to allow for maximal vector replication, and purification of vector DNA from the mammalian cells using the Hirt (1967) procedure. The DNA is then transformed into an appropriate *E. coli* strain such that mutations in the target gene can be easily selected or screened. For *lacI*, a strain (MC1061 F'150kan) is used in which I$^+$ colonies appear white and I$^-$ colonies are blue when plated on medium containing the indicator dye X-gal. The ratio of blue to white colonies thus gives a direct measure of the I$^-$ mutant frequency. This simple screening procedure in *E. coli* allows for much faster and easier detection of mutations than if mutants were selected in the mammalian cells.

Furthermore, the mutant plasmids isolated in *E. coli* can be easily analyzed by physical and genetic techniques. Restriction digests of small-scale plasmid preparations from the mutant colonies reveal whether any gross rearrangement of the target gene such as a large deletion or insertion has taken place. In addition, for *lacI*, a sophisticated genetic system exists (J. H. Miller, 1978) that can provide detailed information about the location of the mutation within *lacI*. For base substitutions that lead to nonsense mutations, the DNA-sequence change involved can be determined using simple genetic techniques (Coulondre and Miller, 1977a). Rapid DNA-sequencing techniques are also available and are facilitated in *lacI* by the ability to localize mutations precisely by genetic methods.

Thus, shuttle-vector systems have the capability of rapidly providing detailed information about mutation in mammalian cells. One intention was to use the systems for detection of mutations in mammalian cells caused by carcinogens. Instead, the shuttle systems immediately revealed a high mutation frequency in the transfected DNA. For example, when pSVi2 DNA (Fig. 1) was transfected into COS-7 cells using the DEAE–dextran technique, allowed to remain in the cells and replicate for 48 hr, then purified and transformed into *E. coli*, a mutation frequency of 1% in *lacI* resulted (Table I) (Calos *et al.*, 1983). This mutation frequency of 10^{-2} in the 1000-base-pair (bp) *lacI* target is several orders of magnitude higher than the spontaneous mutation rate of 2.5×10^{-6} for the gene in *E. coli* (J. H. Miller *et al.*, 1977). It is also much higher than the spontaneous mutation rate for genes in mammalian cells, which lies in the range of 10^{-6}–10^{-8} (Lewin, 1980). Plasmid DNA was examined from a large number of I$^-$ colonies containing pSVi2. Over half the I$^-$ plasmids (57%) did not differ from wild-type in restriction pattern and were classified as putative point mutations, while most of the remaining mutations were deletions (38%). A small fraction (4%) of the mutant plasmids were larger than pSVi2 and were shown to contain duplications or insertions from the simian genome (Lebkowski *et al.*, 1984). Comparable results were obtained after passage of an SV40-*galK* vector in simian cells (Razzaque *et al.*, 1983). Experiments with an SV40-*gpt* vector in COS cells produced similar data (Ashman and Davidson, 1984).

Controls were done to show that the mutations were not due to the

TABLE I. Spontaneous Mutation Frequencies of Transfected Autonomously Replicating Plasmids[a]

Organism	Cell line	Vector	I mutant frequency (%)	Point mutations (%)	Rearrangements (%)
Monkey	COS-7	pSVi2	1.0	57	43
Monkey	COS-7	pSVi4	2.5	13	87
Monkey	COS-7	pJYMib	0.5	49	51
Monkey	CV-1	pJYMib	0.4	40	60
Human	293	pJYMib	0.035	70	30
Human	293	pBKib	0.05	85	15
Mouse	L	pPyib	1.0	8	92
Mouse	3T3	pPyib	0.9	44	55
Mouse	C127	pPyib	0.9	12	88
Mouse	3T6	pPyib	0.8	17	83
Syrian hamster	BHK-21	pPyib	1.4	0	100

[a] Data are derived from Lebkowski et al. (1984, 1986).

DEAE–dextran protocol or Hirt extraction procedures themselves, and it was shown that introduction of the DNA by DEAE–dextran, calcium phosphate coprecipitation, or protoplast fusion all led to elevated mutation frequencies (Calos et al., 1983; Razzaque et al., 1983; Lebkowski et al., 1984). That the target gene itself was not causing the high mutation frequency was shown by experiments using vectors from which lacI was removed and the number of rearrangements scored by restriction analysis (Calos et al., 1983). The possibility that a particular sequence on the vector was triggering mutation was decreased by testing a variety of vector constructs, all of which showed high mutation frequencies (Lebkowski et al., 1984). The conclusion from these experiments and from localization of the mutations in lacI and elsewhere in the plasmid was that the mutations were random in position and detectable in any portion of the vector not forbidden by the selection and screening assays.

The mutation frequency of pSVi4 (Fig. 1) was 2.5 times higher than that for pSVi2, and a higher fraction of the mutations were deletions (Table I) (Calos et al., 1983). The reason for this observation is apparent from examination of the position of lacI in the two vectors. In pSVi2, lacI is closely flanked by sequences selected or screened in the experiments. On the left of lacI is the beginning of lacZ, which is required to produce a blue colony in E. coli. The gene that confers ampicillin resistance, essential for plasmid survival in E. coli, is adjacent to Z. On the right of lacI is the SV40 origin of replication, which is required for vector replication in the mammalian cells. Thus, in pSVi2, I⁻ deletion mutations must be confined to the lacI coding sequence. Deletions extending into neighboring regions on either side of the gene would not produce plasmids that would be detected in this system.

On the other hand, in pSVi4, lacI is flanked by important sequences only on the left side. A deletion can extend from *lacI* into the neighboring chicken thymidine kinase (TK) gene, since that gene is not selected in these experiments. Therefore, large deletions involving *lacI* on pSVi4 are detected with the system and cause a net increase in the percentage of I⁻ molecules recovered and a proportionate increase in the fraction of I⁻ mutations that are deletions (Table I). This example demonstrates that flanking the target gene with selected or screened sequences will decrease the level of mutations detected whenever there is an appreciable fraction of large deletions. It also proves that the deletions formed in the mammalian cells, since the SV40 origin, which has no function in bacteria, acts as a selected sequence to restrict deletion size in pSVi2. The principle of flanking the target gene with selected or screened sequences to reduce the number of deletions recovered was further studied and borne out by testing a variety of vectors containing flanked and nonflanked arrangements of *lacI* in numerous mammalian cell lines (Lebkowski *et al.*, 1984).

It was of interest to determine whether these large deletions conferred any replicative advantage on the mutant plasmids. Results demonstrating an inverse relationship between size and SV40 plasmid copy number had been reported by Santangelo and Cole (1983). By using Southern-blot analysis, these investigators noted that SV40 vectors with large inserts of African green monkey DNA replicated less efficiently in COS-7 cells than did the starting vectors with small inserts. Lebkowski (unpublished results) performed experiments in my laboratory to examine whether plasmid size influences replication of the SV40-*lacI* vectors. Six different I⁻ pSVi4 plasmids that had been generated as a result of passage through COS-7 cells were collected. The lesions in these mutants ranged from a point mutation in *lacI* to a large deletion spanning approximately 4 kilobases (kb) of the *lacI* and the chicken TK genes. Wild-type plasmid was mixed with each I⁻ plasmid, and the exact proportion of each plasmid in these starting mixtures was determined by transforming into *E. coli* and scoring I⁺ (white) and I⁻ (blue) colonies. In each case, the ratio was approximately 1 : 1 (range: 40–56% I⁻).

Each plasmid mixture was transfected into COS-7 cells, and after 72 hr, the replicated vectors were extracted. Plasmid DNA was purified, digested with the restriction enzyme *Dpn*I to remove unreplicated DNA (see Lebkowski *et al.*, 1984), and transformed into *E. coli*. The numbers of I⁺ and I⁻ colonies were used to determine the final relative amounts of wild-type and mutant molecules in the monkey cells at the time of extraction. Any change in the ratio of mutant to wild-type plasmids after passage through the monkey cells indicated that one of the plasmids had a replicative advantage in the simian cells. The results of these experiments are shown in Fig. 2a.

When pSVi4 was cotransfected with an equal-sized I⁻ pSVi4 vector containing a point mutation of *lacI*, there was no significant change in the percentage of mutants observed in the final pool of molecules after replication in the monkey cells; neither plasmid had any replicative advantage. However, all pSVi4 deriv-

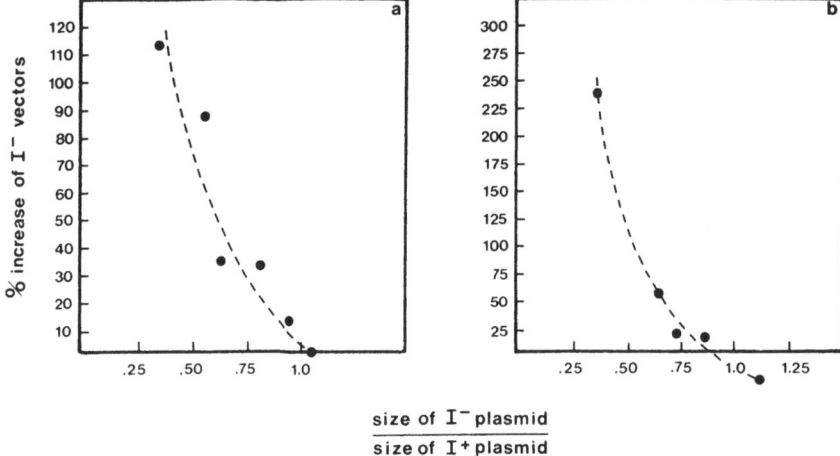

$$\frac{\text{size of } I^- \text{plasmid}}{\text{size of } I^+ \text{plasmid}}$$

FIGURE 2. Replication efficiency vs. vector size. Mixtures of I^+ and I^- vectors were made, and the proportion of each plasmid in this initial mixture was determined by the number of I^+ and I^- colonies that resulted after transformation of MC1061 F'150kan *recA51*. Next, 4 μg of a mixture of two homologous SV40 vectors was transfected into approximately 6×10^5 COS-7 cells using the calcium phosphate coprecipitation method. After 24 hr, the precipitate was removed and fresh medium was added. At 72 hr after transfection, the plasmids were extracted from the monkey cells using the procedure of Hirt (1967). The plasmids were purified, digested with *Dpn*I to eliminate unreplicated DNA, and returned to *E. coli*. I^- vectors gave rise to blue colonies on plates containing 5-bromo-4-chloro-3-indolyl-β-D-galactoside (X-gal), whereas colonies containing I^+ vectors were white. Colony numbers were used to determine the relative amount of each plasmid in the monkey cells at the time of extraction. Transfection of COS-7 cells with these plasmids typically led to the analysis of 1000–10,000 *E. coli* colonies. Each plasmid was also transfected alone into COS-7 cells to estimate the I^- (for wild-type vectors) or Z^- (for I^- vectors) spontaneous mutation frequencies arising due to transfection. These frequencies were used to correct the final percentages of each plasmid in the cotransfection experiments, though they had a very small impact on the outcome. The percentage increase in the I^- vector was calculated as follows: % mutant in final mixture − % mutant in starting mixture/% mutant in starting mixture × 100%. (a) Experiments using I^+ and I^- pSVi4 plasmids; (b) experiments using I^+ and I^- pSV2gptia plasmids. The ordinate values for the pSV2gptia mutants are greater than those for the pSVi4 plasmids because the proportions of the pSV2gptia I^- mutants in the starting mixture were lower than in the experiments with pSVi4 plasmids.

atives containing deletions replicated faster than their wild-type counterparts. The greater the deletion size, the greater the proportion of the mutant vector in the pool after passage through the monkey cells. In the most extreme case, in which the wild-type vector of 7.2 kb was cotransfected with an I^- plasmid of approximately 3.2 kb (size ratio = 0.44), the fraction of mutants in the plasmid population rose from 40.0 to 85.5% after replication in the simian cells. This increase represents a 114% increase in the proportion of mutant molecules during the replication period in COS-7 cells. In other words, the smaller the mutant molecule, the higher the copy number it attained in monkey cells.

This increase in the replication efficiency of smaller plasmids is also observed with other SV40-based shuttle vectors. The plasmid pSV2gptia (Lebkowski *et al.*, 1984) is a derivative of the vector pSV2gpt (Mulligan and Berg, 1980). Five I⁻ mutants of this vector were generated by passage through COS-7 cells and cotransfected with wild-type pSV2gptia. After 72 hr, the proportions of mutant and wild-type plasmids in the replicated populations were determined by transformation into *E. coli*. Again, the smaller the mutant plasmid, the better it replicated in the monkey cells (Fig. 2b). In one case, the mutant plasmid was larger than the wild-type molecule due to an insertion or a duplication, or both, of 500 bp of DNA (size ratio = 1.1). This mutant replicated less effectively than the wild-type plasmid, supporting the theory that size influences the replication efficiency of SV40.

In every case in which a small plasmid replicated more efficiently than the wild-type vector, the small plasmid contained a substantial deletion of *lacI*. To rule out the possibility that these small vectors replicated more efficiently because they had lost some sequence deleterious to replication, pMLRIIia was constructed. This vector is the same as pSVi4 except that it lacks the chicken TK gene and 346 bp of pML. The vector has the entire *lacI* gene, yet is small (4.6 kb). It was cotransfected into COS-7 cells with a pSVi4 I⁻ point mutant of 7.2 kb. The starting plasmid mixture was 41.4% pMLRIIGia, yet after 72 hr in the monkey cells, 89.4% of the final pool of molecules were pMLRIIGia. Again the smaller molecule, although I⁺, replicated more efficiently than the larger plasmid.

Other experiments showed that sequence differences among plasmids of similar size can also influence their ability to replicate. Two plasmids of equal size but different composition were cotransfected into COS-7 cells. The plasmid pSVgptia is 6.9 kb; it contains the *Ecogpt* gene of *E. coli*, the entire pBR322 molecule, the SV40 enhancer, and other SV40 sequences. The plasmid pSVi4-31 is a 6.9-kb deletion derivative of pSVi4. It lacks both the SV40 enhancer and 300 bp of *lacI* and contains the chicken TK gene and pML. Although both plasmids are of equal size, pSV2gptia replicated faster than pSVi4-31 in each of the three separate experiments. Therefore, sequence composition also influences how well an SV40-based vector replicates.

The mechanisms by which size and sequence affect replication are unknown. However, a smaller molecule would complete a round of replication before a larger plasmid. Since the smaller plasmid would complete replication first, its effective copy number would be greater at any given time, possibly giving it an advantage for reinitiation. It is clear from these studies that there is an inverse correlation between the replication efficiency of an SV40 vector and its size. This phenomenon would tend to cause an overrepresentation of deletion mutants with increasing replication time in the mammalian cell.

All our SV40 vectors lead to an elevated mutation frequency in COS-7 simian cells. Vectors such as pJYMib that encode their own large-T antigen also

have the potential for replication in non-SV40-transformed monkey cells. Therefore, pJYMib (see Fig. 1) was transfected into CV-1 cells, where it also showed
a high mutation frequency. Human cells are considered semipermissive for SV40
replication, so pJYMib was transfected into a variety of human cell lines. Vig-
orous SV40 replication at levels similar to those obtained in COS-7 cells was
obtained only in human 293 cells (Lebkowski *et al.*, 1984, 1985a). This line is
derived from human embryonic kidney cells transformed by adenovirus 5 (Graham *et al.*, 1977). The mutation frequency for pJYMib in 293 cells was 0.035%,
approximately 20-fold below that obtained in COS or CV-1 cells, and a higher
proportion of the mutations were of the point-mutation class (Table I).

It should be borne in mind that a mutation frequency of 3.5×10^{-4} is still
well above that expected for a mammalian gene. Yet this background is low
enough to permit experiments with induced mutation in human cells with this
system (see Section 3). The reason for the lower transfection-associated background in 293 cells is unknown, but may be a general property of human cells,
since similar frequencies were obtained in two other human lines, HeLa and
JW-2, in which SV40 replication was obtained (Lebkowski *et al.*, 1984, 1985a).

Given that a high mutation frequency occurred in DNA transfected into
mammalian cells, what was the cause of this phenomenon and the mechanism
of formation of the different types of lesions encountered? Because the vectors
were autonomously replicating, a role for replication errors in causing the mutations had to be considered. Alternatively, the mutations could be due to damage
inflicted on the transfected DNA by the mammalian-cell environment. A role
for replication in creating mutations was tested by performing time–course experiments and by examining the mutational behavior of vectors that could not
replicate in mammalian cells.

If replication were responsible for generation of the mutations, the mutation
frequency should increase with time over the replication period. When the mutation frequency of an SV40-*galK* shuttle vector replicating in BS-C-1 cells was
sampled 24, 48, 72, and 96 hr after transfection, no increase in mutation frequency was detected (Razzaque *et al.*, 1983). This result argues against a major
role for replication in formation of the mutations.

Time–course experiments were also undertaken with SV40-*lacI* vectors to
determine when mutagenesis occurs and whether the mutation frequency was
established immediately or increased throughout the replication period (Lebkowski *et al.*, 1984). These experiments showed that replication of the transfected
vector DNA was first detectable about 24 hr posttransfection and that I⁻ molecules
were also first detectable at this time; i.e., there was a coincidence in time
between the occurrence of mutations and the commencement of replication. It
was thus inferred that the molecules must have been in the nucleus when the
mutations were completed, since replication of the vectors was presumably confined to the nucleus. However, the mutation frequency did not increase with
time. Instead, the mutation frequency was established at the time replication

began and was maintained at that level. These data could be interpreted to mean that the mutations formed as an early event and required the nucleus for completion.

That the nuclear environment alone, without replication, was sufficient to produce mutations was shown by removing the SV40 origin of replication from pSVi4 and transfecting the resultant, nonreplicating plasmid into COS-7 cells (Lebkowski *et al.*, 1984). Time–course experiments were done with the nonreplicating vector, and, as before, mutations appeared after about 24 hr. Both point mutations and rearrangements were detected. These observations suggested that mutation again occurred in the nucleus, but that overall replication of the vector was not required.

All these findings fortify the conclusion that most of the mutations are introduced early, rather than continually during replication. It thus seems likely that intracellular DNA damage is the driving force in the formation of mutations in transfected DNA, and the nucleus appears to be required to complete a mutation. For the deletions, the most obvious route of formation would involve a double-stranded break, followed by degradation and re-ligation. The required endonuclease, exonuclease, and ligase activities are present in the mammalian nucleus. If newly transfected DNA were subject to attack by these enzymes, deletions could readily be generated.

As a model for what may happen to transfected DNA, DNA with specific preexisting damage was introduced into mammalian cells. For example, Razzaque *et al.* (1984) observed a stimulation of deletions when DNA containing a double-stranded break was transfected. This finding was used to argue that a double-stranded break may be an intermediate in deletion formation and that the bulk of deletion formation must precede replication. Introduction of a double-stranded break into transfected SV40-*lacI* vectors led to a sharp increase in the incidence of deletions recovered (J. H. Miller *et al.*, 1984). A double-stranded break may be an intermediate in deletion formation by providing a substrate for exonucleolytic digestion. If a double-stranded break formed intracellularly, presumably a similar course of events would ensue. Nuclear ligase activity would restore a circular conformation to a degraded molecule, making it a substrate for replication. Abundant evidence for plentiful ligase activity in the nuclei of mammalian cells has been presented (Wigler *et al.*, 1979; Winocour and Keshet, 1980; Folger *et al.*, 1982; Wilson *et al.*, 1982; C. K. Miller and Temin, 1983). The studies of Wake *et al.* (1984) indicate that fragmentation of transfected DNA is common and that the subpopulation that reaches the nucleus suffers about one double-stranded break per 5–15 kb. It is noteworthy that SV40 viral DNA suffers rearrangements similar to those observed in transfected DNA, and these rearrangements may arise in the same way (Kelly and Nathans, 1977). The duplications observed in transfected SV40 vectors could result from deletions in molecules joined into dimers by recombination. The insertions could occur by ligation of host sequences to vectors containing double-stranded breaks.

A stimulation of deletions and base substitutions was also observed when nicked or gapped molecules were used as the substrates for transfection (J. H. Miller et al., 1984). Gaps may increase the susceptibility of a single-stranded region to breakage compared with a double-stranded region. Single-stranded DNA is also a better substrate for base damage, possibly explaining the stimulation of base substitutions. The similar behavior of nicked molecules suggests that they are predecessors to gapped molecules (J. H. Miller et al., 1984).

As for the point mutations, it appears that most of them are base substitutions (J. H. Miller et al., 1984). A study of the sequence of 93 independent base substitutions induced in lacI during passage through COS-7 cells shows that the mutations have a distinct specificity. All 93 mutations occur at G : C base pairs and are approximately equally divided between G : C to A : T transitions and G : C to T : A transversions (J. H. Miller et al., 1984). This specificity led us to propose that preferential damage to G : C base pairs may occur in transfected DNA. The two changes seen could be accounted for by deamination of cytosine and depurination of guanine. Both reactions are acid-catalyzed and occur more readily on single-stranded DNA. Since the bulk of transfected DNA enters the cells by endocytosis and is delivered to lysosomes that maintain a pH of 5, transfected DNA could potentially acquire base damage under these conditions. A study of 50 spontaneous mutations generated in lacI after passage in 293 cells led to results similar to those obtained with COS-7 cells. Again, the majority of the mutations (47 of 50) were at G : C base pairs, and the mutations were evenly divided between transitions and transversions (Lebkowski et al., 1985b, 1986).

2.2. Polyoma-Virus- and BK-Virus-Based Vectors

The foregoing studies demonstrated a high mutation frequency of SV40-based vectors in monkey cells. Similar types of mutations occurred in human cells, but at lower frequencies. Could these results be generalized to other species and different replicons?

To extend the studies into rodent cells, lacI was added to pSV5gpt (Mulligan and Berg, 1980), which contains the origin of replication from polyoma virus. This papovavirus is a relative of SV40 with a similar mode of replication. However, it can replicate in mouse cells (Tooze, 1981). When these vectors were tested in a variety of mouse cell lines and in Syrian hamster cells, it was apparent that a mutation frequency of about 1% was again occurring (Table I) (Lebkowski et al., 1984). However, unlike the mutations obtained in primate cells, the mutations from rodent cells were dominated by rearrangements rather than point mutations (Table I).

In an attempt to gain access to replication in a wide range of human cells, vectors based on the human papovavirus BK virus (BKV) were constructed

(Lebkowski *et al.*, 1984). This close relative of SV40 has humans as its natural hosts (Tooze, 1981). However, as with SV40, BKV vectors failed to replicate well in most human lines tested. Efficient replication was obtained in 293 cells, and the mutations obtained with the BKV-based vectors closely resembled in frequency and nature those detected when vectors based on SV40 were used (Table I) (Lebkowski *et al.*, 1984). It can be concluded that an elevated mutation frequency in transfected DNA is a general feature of mammalian cells.

2.3. Bovine-Papillomavirus- and Epstein–Barr-Virus-Based Vectors

If most of the mutations incurred by transiently replicating vectors are early events targeted to intracellular damage inflicted during transfection, then stably replicating vectors could offer some immunity from the problem. The initial establishment of stable vectors via transfection would be expected to lead to some mutation of the incoming DNA. However, since this mutation frequency is about 1% per 1000 bp, it should be quite easy to obtain cell lines that carry only vector DNA unmutated in the target gene. If no appreciable mutagenesis occurs during vector replication, the mutation frequency of vector DNA purified from such cell lines should remain very low. In other words, stable vector systems potentially allow one to, in effect, clone molecules that have safely passed the transfection–mutation barrier. This procedure is not possible in cells transfected with transiently replicating plasmids because of the short life of these cells.

Plasmids based on bovine papillomavirus (BPV) were the first stably replicating vector systems to become available. BPV is an 8-kb plasmid that replicates autonomously in the nuclei of various rodent tissue-culture lines to the extent of approximately 50 copies/cell (Law *et al.*, 1981). The vector copy number seems to be tightly controlled and is stably maintained (Lusky and Botchan, 1984, 1985). Amplification systems exist to transiently boost this copy number (DuBridge *et al.*, 1985). Thus, for studying mutation in mammalian cells, BPV would appear to offer promise as a vector system that would be free of the high mutational background associated with transfection.

However, problems with BPV vectors were apparent from the early studies. Vectors based on BPV seemed to be very sensitive to the nature of the specific DNA that was cloned into them. For example, the first BPV vectors replicated autonomously in mouse cells only if their prokaryotic portions were removed before transfection (Law *et al.*, 1981). Only vectors with certain combinations of BPV, prokaryotic, and other DNA were found to replicate autonomously (DiMaio *et al.*, 1982; Sarver *et al.*, 1982). When extraneous DNA was added to these vectors, they often rearranged or failed to replicate autonomously (e.g., Schenborn *et al.*, 1985; Mangin *et al.*, 1985), and there were ominous reports that variant mutant forms could arise during passage of the cell lines (e.g., DiMaio *et al.*, 1982).

Our experience with BPV-*lacI* vectors has borne out these observations (R. B. DuBridge, unpublished observations). Some constructs were very prone to deletions, with the deleted forms apparently having a strong selective advantage. Other constructs started with a low mutational background, but this mutation frequency rapidly increased over time. We concluded that BPV-based vectors were unsuitable for mutagenesis experiments because of their instability. The weight of the evidence indicates that the BPV replicon is very carefully adjusted and that perturbations are often triggered by the insertion of foreign DNA. Rearrangements of the vector may partially compensate for these imbalances and are therefore strongly selected. The net result is a vector system that rapidly evolves. This phenomenon is manifested by a high and increasing I⁻ mutation frequency in BPV-*lacI* vectors.

A contrasting picture is presented by vectors based on the human herpesvirus Epstein–Barr virus (EBV). Recently, autonomously replicating minireplicons have been derived from EBV by Sugden and colleagues (Yates *et al.*, 1984, 1985; Sugden *et al.*, 1985) (see Chapter 5). These vectors contain the origin for EBV latent replication, the *trans*-acting gene that codes for EBNA-1 that is required for activation of the origin, a gene that confers resistance to hygromycin for selection in mammalian cells, and portions of pBR322 for selection and replication in *E. coli*. These vectors are carried stably at copy numbers of approximately 10–100 per cell in a wide range of human and other primate cells as long as selection pressure for hygromycin is maintained.

Our experiments with *lacI* derivatives of EBV vectors in human cells indicate that apart from a low percentage of lines that harbored *lacI⁻* mutations that were apparently created during transfection, the majority of the lines have a very low mutation frequency that has not detectably increased over time. Therefore, the EBV-based vectors may fulfill the theoretical promise of stably replicating plasmid systems for alleviating the problem of the high mutation frequency associated with transfection. In this regard, they appear to be suitable vector systems for the study of rare classes of mutations in mammalian cells.

3. INDUCED MUTATION OF AUTONOMOUSLY REPLICATING PLASMIDS

The finding of efficient replication of simian virus 40 (SV40)-based vectors in 293 cells accompanied by a relatively low spontaneous mutation frequency (Lebkowski *et al.*, 1984) has allowed this shuttle-vector system to be used to study induced mutation. Determination of the mutagenic specificity of one of the most widely studied and most environmentally important mutagens, ultraviolet (UV) light, was the first application of the *lacI* shuttle system (Lebkowski *et al.*, 1985b).

The SV40-*lacI* vector pJYMib (see Fig. 1) was transfected into 293 cells. At 48 hr after transfection, when the vector was established and replicating in the human nuclei, a UV-light dose was given. The vector DNA was allowed to replicate in the human cells for another day to fix mutations. Then vector DNA was purified and returned to *E. coli* for analysis of mutations. A dose–response curve for various UV-light doses was constructed that showed a rising mutation frequency with increasing UV-light dose. A dose of 50 J/m² was used to generate a large collection of *I⁻* UV-light-induced mutations. Most of the UV-light-induced mutations were base substitutions, and the DNA-sequence changes involved were determined for 53 of the mutations using the *lacI* genetic system (Lebkowski *et al.*, 1985b). These data are summarized in Fig. 3. The nature of 32 spontaneous base substitutions that occurred in the human cells is also shown.

The most striking characteristic of the mutagenic specificity of UV light in human cells is its close similarity to the pattern obtained in *E. coli* (Coulondre and Miller, 1977b; J. H. Miller, 1985). This result suggests that human and bacterial cells may react to UV-light damage in similar ways. The majority of the mutations analyzed (81%) were G : C to A : T transitions. In *E. coli*, 62% of all UV-light-induced amber and ochre mutations are due to this change. If mutations were occurring at random in the *lacI* nonsense system, 57% of the G : C to A : T transitions would be at pyrimidine–pyrimidine sequences. Therefore, it is highly significant that in the human system, as in *E. coli*, the great

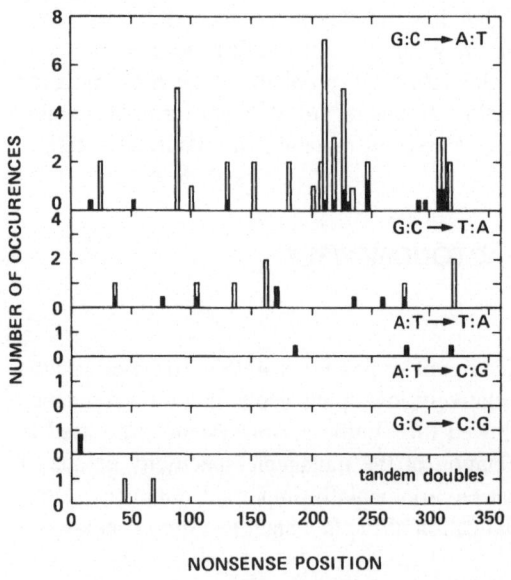

NONSENSE POSITION

FIGURE 3. Mutagenic specificity of UV-light-induced and spontaneous mutations in 293 cells. (□) UV-light-induced mutations; (■) spontaneous mutations. The nature of the mutational change is shown in the upper right corner of each box. The position of the mutation along the 360-amino-acid sequence of the *lac* repressor (coded for by *lacI*) is shown on the abscissa. The height of each bar represents the number of mutations at that site. The heights were normalized to correctly reflect mutation frequency as follows: For UV-light-induced mutations, 1 mutation = 1 unit. For spontaneous mutations, 1 mutation = 0.42 units, to take into account a smaller sample size: (53 UV-induced/32 spontaneous) = 1.7 and a 4-fold lower mutation frequency (0.25). Thus, 1.7 × 0.25 = 0.42. For UV mutations, the number of mutations equals final peak height.

majority (>90%) of the UV-light-induced transition mutations involve pyrimidine–pyrimidine sequences. Since the two photoproducts most closely associated with UV-light mutagenesis, the (6-4) and cyclobutane pyrimidine dimers, occur at pyrimidine–pyrimidine sequences, the localization of the majority of the mutations opposite these sequences argues that UV-light mutagenesis is targeted to premutational lesions in human cells.

The DNA sequence changes induced in human cells by the alkylating agent ethylmethanesulfonate (EMS) were also determined using the *lacI* shuttle (Lebkowski *et al.*, 1986). EMS is representative of a large and important class of compounds that alkylate DNA and cause mutations. The DNA sequence changes for 54 mutations induced by EMS in human cells were determined using the *lacI* genetic system. EMS induced predominantly base substitutions that occurred at a frequency 6 times above background. Of the 54 mutations, 53 (98%) were G : C to A : T transitions. This figure is in accord with the data obtained for *lacI* in *E. coli*, in which 98% of the mutations were also due to this change (Coulondre and Miller, 1977b). The similarity in the nature of the mutations formed by EMS in the human and bacterial systems is consistent with the idea that the mechanism of mutagenesis by EMS is similar in the two systems. Addition of an alkyl group to the 0-6 position of guanine is thought to be the main mutagenic lesion (Drake and Baltz, 1976). If the 0-6 alkyl group is not removed before replication, mispairing of the alkyl-guanine with thymine can result, leading to a full G : C to A : T transition on the next round of replication. As expected, and unlike the result with UV light, for EMS the percentage of mutations that involve pyrimidine dimers is only 53%, close to what would be obtained at random. These results with UV light and EMS indicate that the *lacI* shuttle system is suitable for determining the mutagenic specificity of a range of mutagens and carcinogens in human cells.

As an alternative to performing mutagenesis in a cell line such as 293 that has a naturally lower spontaneous background per base pair for transfected DNA, several groups have taken a different approach. By reducing the size of the target gene, the background of spontaneous mutations per target gene can be reduced artificially. For example, by using a suppressor transfer RNA (tRNA) gene, *supF*, as the target, a spontaneous background of 0.2% was obtained in COS-7 cells by Sarkar *et al.* (1984). The same gene, closely flanked by selected sequences, has been used by Seidman *et al.* (1985). Their vector has a background of 0.04% in CV-1 cells. Of course, reducing the target size does not actually reduce the spontaneous background per base pair. The tRNA coding sequence is only 85 bp long, compared to the more than 1000 bp of *lacI*. Therefore, on a per-base-pair basis, the spontaneous mutation frequencies obtained for the tRNA target in simian cells (approximately 4×10^{-6}/base pair) are still well above the background frequency for the *lacI* gene in human cells (3.5×10^{-7}/base pair).

In the UV and EMS studies described (Lebkowski *et al.*, 1985b, 1986), mutagenesis was performed after the transfected DNA was inside the nuclei of the human cells, leading to typical increases of 4- to 6-fold in mutation frequency over background. Treatment of the vector DNA with high doses of mutagens *in vitro* before transfection can produce larger increases in mutation frequency over background. Use of these same doses *in vivo* would be lethal to the mammalian cells. For example, by treating vector DNA with high doses of UV light (500 J/m^2) *in vitro* before transfection, tRNA mutants were observed at frequencies about 20-fold above the spontaneous background (Seidman *et al.*, 1985). *In vitro* treatment of *lacI* vectors also gives this effect. For example, exposure of the vector DNA to a pH of 5.0 for 1–2 hr before transfection leads to increases of 10- to 20-fold in mutation frequencies. However, a concern in these types of experiments is that treatment of naked DNA *in vitro* with severe mutagenic doses may not adequately represent the behavior of DNA treated with more realistic doses while inside the cell.

It should be clear from the foregoing discussion that the promise of shuttle-vector systems as powerful tools to study mutagenesis in mammalian cells is beginning to be realized. These developments are occurring despite the high spontaneous mutation background associated with transfection that the shuttle systems were instrumental in elucidating.

4. SUMMARY

Studies with autonomously replicating plasmids in mammalian cells have unequivocally shown that there is a high mutation frequency associated with transfection of these vectors into mammalian cells. The mutations encompass point mutations such as base substitutions as well as rearrangements, predominantly deletions. The data indicate that these mutations occur in the mammalian cell and require the nucleus for completion. Overall replication of the vector does not appear to be required to produce the majority of the mutations. This finding implicates intracellular damage as the major cause of the mutations. It also cautions that all transfected DNA is probably subject to an elevated mutation frequency. The deletions and other rearrangements may occur as the result of breakage, degradation, and ligation of the DNA, while base damage may be involved in the base substitutions.

The question of why transfected DNA is subject to attack is an open one. The answer may be that transfected DNA is especially reactive because it enters the cells as naked DNA, uncomplexed into chromatin. It is notable that newly transfected DNA is also subject to very high recombination frequencies, perhaps for the same reasons (see Chapter 13). The phenomenon of elevated mutation frequency on transfection appears to be ubiquitous among mammalian cells and may be present in all eukaryotic cells. Our studies with yeast demonstrate a high mutation frequency in DNA introduced into *Saccharomyces cerevisiae* by the lithium acetate technique (Clancy *et al.*, 1984). The variation among different species in the frequency and nature of mutations induced in transfected DNA presumably reflects differences in the kind of damage and/or repair systems present.

With regard to the use of autonomously replicating plasmids to study mutagenesis in human cells, the high spontaneous mutation frequency encountered has proven to be a stumbling block, but not a barrier. Already, several shuttle systems are beginning to provide novel information about the action of mutagens in human cells. This information will help, for example, to elucidate the early and crucial steps in carcinogenesis. Future developments with stably replicating plasmid systems should carry inquiry far beyond the questions that can now be posed and lead to a much fuller knowledge of the nature of mutations, and of replication itself, in human cells.

ACKNOWLEDGMENTS. I thank the members of my laboratory, especially J. S. Lebkowski, S. Clancy, and R. B. DuBridge, for unpublished data. Work in my laboratory was supported by Grants CA-33056 from the National Cancer Institute and 83-E-101 from the Chicago Community Trust/Searle Scholars Program.

REFERENCES

Ashman, C. R., and Davidson, R. L., 1984, High spontaneous mutation frequency in shuttle vector sequences recovered from mammalian cellular DNA, *Mol. Cell. Biol.* **4**:2226–2272.

Calos, M. P., Lebkowski, J. S., and Botchan, M. R., 1983, High mutation frequency in DNA transfected into mammalian cells, *Proc. Natl. Acad. Sci. U.S.A.* **82**:3015–3019.

Clancy, S., Mann, C., Davis, R. W., and Calos, M. P., 1984, Deletion of plasmid sequences during *Saccharomyces cerevisiae* transformation, *J. Bacteriol.* **159**:1065–1067.

Coulondre, C., and Miller, J. H., 1977a, Genetic studies of the *lac* repressor. III. Additional correlation of mutational sites with specific amino acid residues, *J. Mol. Biol.* **117**:525–567.

Coulondre, C., and Miller, J. H., 1977b, Genetic studies of the *lac* repressor. IV. Mutagenic specificity in the *lacI* gene of *Escherichia coli*, *J. Mol. Biol.* **117**:577–606.

DiMaio, D., Treisman, R., and Maniatis, T., 1982, Bovine papillomavirus vector that propagates as a plasmid in both mouse and bacterial cells, *Proc. Natl. Acad. Sci. U.S.A.* **79**:4030–4034.

Drake, J. W., and Baltz, R. H., 1976, The biochemistry of mutagenesis, *Annu. Rev. Biochem.* **45**:11–37.

DuBridge, R. B., Lusky, M., Botchan, M. R., and Calos, M. P., 1985, Amplification of a bovine papilloma virus–simian virus 40 chimaera, *J. Virol.* **56**:625–627.

Folger, K. R., Wang, E. A., Wahl, G., and Capecchi, M. R., 1982, Patterns of integration of DNA microinjected into cultured mammalian cells: Evidence for homologous recombination between injected plasmid DNA molecules, *Mol. Cell. Biol.* **2**:1372–1387.

Gluzman, Y., 1981, SV40-transformed simian cells support the replication of early SV40 mutants, *Cell* **23**:175–182.

Graham, F. L., Smiley, J., Russell, W. C., and Nairn, R., 1977, Characteristics of a human cell line transformed by DNA from human adenovirus type 5, *J. Gen. Virol.* **36**:59–72.

Hirt, B., 1967, Selective extraction of polyoma DNA from infected mouse cultures, *J. Mol. Biol.* **26**:365–369.

Kelly, T. J., and Nathans, D., 1977, The genome of simian virus 40, *Adv. Virus Res.* **21**:85–173.

Law, M.-F., Lowy, D. R., Dvoretzky, I., and Howley, P. M., 1981, Mouse cells transformed by bovine papillomavirus contain only extrachromosomal viral DNA sequences, *Proc. Natl. Acad. Sci. U.S.A.* **78**:2727–2731.

Lebkowski, J. S., DuBridge, R. B., Antell, E. A., Greisen, K. S., and Calos, M. P., 1984, Transfected DNA is mutated in monkey, mouse and human cells, *Mol. Cell. Biol.* **4**:1951–1960.

Lebkowski, J. S., Clancy, S., and Calos, M. P., 1985a, SV40 replication in adenovirus-transformed human cells antagonizes gene expression, *Nature (London)* **317**:169–171.

Lebkowski, J. S., Clancy, S., Miller, J. H., and Calos, M. P., 1985b, The *lacI* shuttle: Rapid analysis of the mutagenic specificity of ultraviolet light in human cells, *Proc. Natl. Acad. Sci. U.S.A.* **82**:8606–8610.

Lebkowski, J. S., Miller, J. H., and Calos, M. P., 1986, DNA sequence changes induced in human cells by the alkylating agent ethyl methanesulfonate, *Mol. Cell. Biol.* **6**:1838–1842.

Lewin, B., 1980, *Gene Expression: Eukaryotic Chromosomes*, Vol. 2, 2nd ed., Wiley, New York.

Lusky, M., and Botchan, M. R., 1984, Characterization of the bovine papillomavirus plasmid maintenance sequences, *Cell* **36**:391–401.

Lusky, M., and Botchan, M. R., 1985, Genetic analysis of bovine papillomavirus type 1 trans-acting replication factors, *J. Virol.* **53**:955–965.

Mangin, M., Ares, M., and Weiner, A. M., 1985, U1 small nuclear RNA genes are subject to dosage compensation in mouse cells, *Science* **229**:272–275.

Miller, C. K., and Temin, H. M., 1983, High-efficiency ligation and recombination of DNA fragments by vertebrate cells, *Science* **220**:606–609.

Miller, J. H., 1978, The *lacI* gene; Its role in *lac* operon control and its use as a genetic system, in: *The Operon* (J. H. Miller and W. S. Reznikoff, eds.), pp. 31–88, Cold Spring Harbor Press, Cold Spring Harbor, New York.

Miller, J. H., 1985, Mutagenic specificity of ultraviolet light, *J. Mol. Biol.* **182**:45–68.

Miller, J. H., Ganem, D., Lu, P., and Schmitz, A., 1977, Genetic studies of the *lac* repressor. I. Correlation of mutational sites with specific amino acid residues; construction of a colinear gene-protein map, *J. Mol. Biol.* **109**:275–301.

Miller, J. H., Lebkowski, J. S., Greisen, K. S., and Calos, M. P., 1984, Specificity of mutations induced in transfected DNA by mammalian cells, *EMBO J.* **13**:3117–3121.

Mulligan, R. C., and Berg, P., 1980, Expression of a bacterial gene in mammalian cells, *Science* **209**:1422–1427.

Razzaque, A., Mizusawa, H., and Seidman, M. M., 1983, Rearrangement and mutagenesis of a shuttle vector plasmid after passage in mammalian cells, *Proc. Natl. Acad. Sci. U.S.A.* **80**:3010–3014.

Razzaque, A., Chakrabarti, S., Joffee, S., and Seidman, M., 1984, Mutagenesis of a shuttle vector plasmid in mammalian cells, *Mol. Cell. Biol.* **4**:435–441.

Santangelo, G. M., and Cole, C. N., 1983, Preparation of a "functional library" of African green monkey DNA fragments which substitute for the processing/polyadenylation signal in the herpes simplex virus type I thymidine kinase gene, *Mol. Cell. Biol.* **3**:643–653.

Sarkar, S., Dasgupta, U. B., and Summers, W. C., 1984, Error-prone mutagenesis detected in mammalian cells by a shuttle vector containing the *supF* gene of *Escherichia coli*, *Mol. Cell. Biol.* **4**:2227–2230.

Sarver, N., Byrne, J. C., and Howley, P. M., 1982, Transformation and replication in mouse cells of a bovine papillomavirus-pML2 plasmid vector that can be rescued in bacteria, *Proc. Natl. Acad. Sci. U.S.A.* **79**:7147–7151.

Schenborn, E. T., Lund, E., Mitchen, J. L., and Dahlberg, J. E., 1985, Expression of a human U1 RNA gene introduced into mouse cells via bovine papillomavirus DNA vectors, *Mol. Cell. Biol.* **5**:1318–1326.

Seidman, M. M., Dixon, K., Razzaque, A., Zagursky, R. J., and Berman, M. L., 1985, A shuttle vector plasmid for studying carcinogen-induced point mutations in mammalian cells, *Gene* **38:**233–237.

Sugden, B., Marsh, K., and Yates, J., 1985, A vector that replicates as a plasmid and can be efficiently selected in B-lymphoblasts transformed by Epstein–Barr virus, *Mol. Cell. Biol.* **5:**410–413.

Tooze, J., 1981, *DNA Tumor Viruses,* 2nd ed., Cold Spring Harbor Press, Cold Spring Harbor, New York.

Wake, C. T., Gudewicz, T., Porter, T., White, A., and Wilson, J. H., 1984, *Mol. Cell. Biol.* **4:**387–398.

Wigler, M., Pellicer, A., Silverstein, S., and Axel, R., 1978, Biochemical transfer of single-copy eucaryotic genes using total cellular DNA as donor, *Cell* **14:**725–731.

Wigler, M., Sweet, R., Sim, G. K., Wold, B., Pellicer, A., Lacy, E., Maniatis, T., Silverstein, S., and Axel, R., 1979, Transformation of mammalian cells with genes from procaryotes and eucaryotes, *Cell* **16:**777–785.

Wilson, J. H., Berget, P. B., and Pipas, J. M., 1982, Somatic cells efficiently join unrelated DNA segments end-to-end, *Mol. Cell. Biol.* **2:**1258–1269.

Winocour, E., and Keshet, J., 1980, Indiscriminate recombination in simian virus 40-infected monkey cells, *Proc. Natl. Acad. Sci. U.S.A.* **77:**4861–4865.

Yates, J., Warren, N., Reisman, D., and Sugden, B., 1984, A *cis*-acting element from the Epstein–Barr viral genome that permits stable replication of recombinant plasmids in latently infected cells, *Proc. Natl. Acad. Sci. U.S.A.* **81:**3806–3810.

Yates, J. L., Warren, N., and Sugden, B., 1985, Stable replication of plasmids derived from Epstein–Barr virus in various mammalian cells, *Nature (London)* **313:**812–815.

10

GENE PURIFICATION BY TRANSFECTION METHODS

Angel Pellicer

1. INTRODUCTION

The transfer of genetic information is relatively common in prokaryotes, but its incidence in mammalians seems to be negligible. Nevertheless, the transfer of genetic traits from one generation to the next is manipulated by farmers to obtain better plant and animal specimens. This primitive transfer has enabled the identification of specific traits and their assignment to a genetic entity.

The isolation of mutants deficient in a particular function has permitted the design of complementation experiments by adding the missing or altered genes. This use of gene transfer immediately suggests that if the piece of genetic material that is complementing the function is identified and separated from the others, gene isolation can be effected. To succeed in this task, two main challenges have to be met: (1) Gene transfer has to be efficient enough so that, using total DNA from an organism for the detection of transformants, large amounts of starting material and recipient cells are not needed. (2) Techniques have to be devised to accomplish the isolation of different DNA pieces in sufficient quantity to be manipulated.

The first requirement was initially met by using DNA from low-complexity organisms, namely, viruses. One of the first examples of this approach was the identification and isolation of the transforming genes from adenoviruses (Graham *et al.*, 1974). Later, the thymidine kinase gene from herpes simplex was also isolated (Wigler *et al.*, 1977; Bacchetti and Graham, 1977; Maitland and McDougall, 1977). The isolation required fractionation of DNA by digesting it into specific fragments with restriction endonucleases and separating them by agarose gel electrophoresis. After the run, individual fragments or pools of them were extracted from the gel and assayed by gene transfer. The operation was repeated

Angel Pellicer • Department of Pathology and Kaplan Cancer Center, New York University Medical Center, New York, New York 10016.

until a single pure fragment was obtained. It is worth noting that these genes were isolated without any involvement of genetic engineering.

This kind of simple sib-selection process is not feasible in more complex organisms. The solution to that challenge was the recombinant-DNA technology.

As could have been predicted, bacteria and yeast were the first organisms from which genes were isolated using gene transfer and recombinant DNA. With a colony bank of *Escherichia coli* DNA fragments inserted in the plasmid Col El, Clarke and Carbon (1976) were able to identify around 40 genes using complementation of different auxotrophs with the bank. The next step was to cross the species barrier and to complement some *E. coli* mutations with similar banks containing yeast DNA; the same investigators obtained the argininosuccinate lyase (argH) yeast sequences with these methods (Clarke and Carbon, 1978). A similar protocol was used to isolate the yeast galactokinase gene (Schell and Wilson, 1979).

With improvements in gene-transfer techniques, the efficiencies in yeast transformation became high enough to try the complementation experiments in an eukaryotic host. The first gene isolated this way was the yeast arginine permease (Broach *et al.*, 1979); it was done by constructing the bank of yeast DNA fragments in a plasmid that contained pBR322 sequences, with the origin of replication from the yeast 2μ circle and the marker gene *Leu-2*. Several other genes have been subsequently isolated this way (Williamson *et al.*, 1980; Taylor *et al.*, 1982). The temptation to complement mutant yeast with DNA from higher organisms resulted in the isolation of a *Drosophila* gene able to cure an adenine 8 mutation in yeast (Henikoff *et al.*, 1981). The focus of this chapter, however, is the complementation of mammalian cells, and it is necessary to explain the conditions that make possible these experiments.

As mentioned earlier, until 1977, the methods of gene transfer were successful only when DNA of low complexity was used (Graham *et al.*, 1974; Wigler *et al.*, 1977; Bacchetti and Graham, 1977; Maitland and McDougall, 1977). Subsequent improvements made the transfer feasible using total vertebrate and mammalian DNA (Wigler *et al.*, 1978, 1979a). It is relevant here to note that of the several gene-transfer methods (e.g., DEAE–dextran, microinjection, calcium phosphate, liposomes, protoplast fusion, electroporation), so far the only method successfully used for transfer of total cellular DNA has been the calcium phosphate coprecipitation technique, inititally described by Graham and van der Eb (1973) and subsequently modified by Wigler *et al.* (1979a).

To accomplish the first step in all the isolation protocols that will be described later in the chapter, one must be able to detect successful transfer, which is a rare event. There are several approaches to this problem, depending on the phenotype conferred by the gene of interest. Figure 1 shows the three main types of detection–selection methods. (1) Metabolic: The new gene confers on the cell the ability to grow in an environment in which the parental cells die, so that

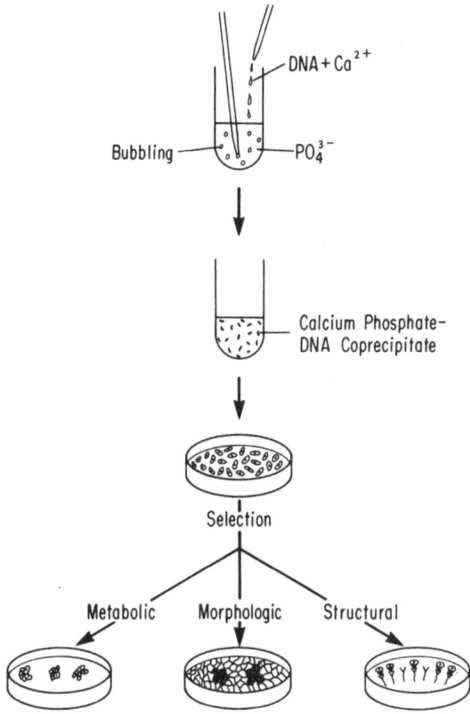

FIGURE 1. Types of selection commonly used after DNA-mediated gene transfer.

primary transformants are identified and isolated as colonies that survive (Wigler *et al.*, 1977; Pellicer *et al.*, 1980a). (2) Morphologic: The cells that have obtained the gene of interest change their phenotype, usually becoming rounder and more refractile. These cells also grow faster than their untransformed neighbors. These changes are sufficient to allow the detection of the altered cells, first under the microscope and later with the naked eye (Shih and Weinberg, 1982; Guerrero *et al.*, 1984). This detection is the one employed for oncogenes. (3) Structural: Here the handle is the expression of a molecule that can be recognized by a specific reagent (usually an antibody), and sometimes the specific recognition is also used to isolate the transformed cells. Examples would be the detection of *T8* (Kavathas and Herzenberg, 1983) and *T9* (Kuhn *et al.*, 1984) by flow cytometer or of *Thy-1* by panning (Berman *et al.*, 1984).

Another important factor for successful gene transfer and isolation is the source of DNA used. Initially, the utilization of DNA from cells or tissues that were expressing the gene of interest was the rule. More recently, a report has

TABLE I. Gene-Isolation Strategies after DNA Transfer

Ref No.[a]	Genes	Source of material		Mode of detection				Methods of isolation			
				Selection (stable)							
		Genomic	cDNA	Metabolic	Morphologic	Structural	No selection (transient)	Experimental linkage	Natural linkage	Subtraction hybridization	Sib selection
1	Chicken *Tk*	+		+			+				
2	Hamster *Aprt*	+		+							
3	Human c-H-*ras*	+			+			+	+		
4	Human c-N-*ras*	+			+			+			
5	Human *Tk*	+		+				+	+		
6	Hamster *Tk*	+		+	+						
7	*Blym-1*	+									
8	Promoter sequences	+		+				+			
9	Human c-K-*ras*	+			+				+		+
10	*Tlym-1*	+			+						
11	*T9*	+			+	+			+		
12	Oncogene in nude mice	+		+	+			+	+		
13	*T8*	+			+	+			+	+	
14	Metastasis gene	+			+			+			
15	GM-CSF-1		+				+				+
16	Enhancer sequences	+		+				+			
17	*IL-2*		+				+				+
18	*T4*	+				+				+	
19	Human glucosidase *AB*	+					+				+

[a] References: (1) Perucho *et al.* (1980a); (2) Pellicer *et al.* (1980b), Lowy *et al.* (1980); (3) Goldfarb *et al.* (1982), Shih and Weinberg (1982), Pulciani *et al.* (1982); (4) Shimizu *et al.* (1983a); (5) Bradshaw (1983), Lin *et al.* (1983), Stuart *et al.* (1985); (6) Lewis *et al.* (1983); (7) Goubin *et al.* (1983); (8) Fried *et al.* (1983); (9) McGrath *et al.* (1983), Shimizu *et al.* (1983b), Nakano *et al.* (1984); (10) Lane *et al.* (1984); (11) Kuhn *et al.* (1984); (12) Fasano *et al.* (1984); (13) Kavathas *et al.* (1984), Littman *et al.* (1985); (14) Bernstein and Weinberg (1985); (15) Lee *et al.* (1985); (16) Ford *et al.* (1985); (17) Yokota *et al.* (1985); (18) Maddon *et al.* (1985); (19) Martiniuk *et al.* (1985).

appeared showing transfer of genes that encode hematopoietic surface markers with DNAs from nonexpressing cells (Hsu *et al.*, 1984). Nevertheless, it is advisable to keep in mind that there is substantial evidence demonstrating that methylation will inhibit a gene in a DNA-mediated-transfer assay (Pellicer *et al.*, 1980b; Wigler *et al.*, 1981), and it is therefore still advisable to use DNA from expressing cells, if they are available.

It is important to mention here a few developments that have contributed to making detection and isolation of transformants easier. One is the discovery that eukaryotic cells tend to incorporate rather indiscriminately the different DNA segments that are added to the tissue-culture media (Wigler *et al.*, 1979b; Pellicer and Esteban, 1982; Boni *et al.*, 1984). This, together with the design of dominant selective schemes (Wigler *et al.*, 1980; Mulligan and Berg, 1981; Colbere-Garapin *et al.*, 1981), has permitted the use of many cells as recipients. Since only approximately 1% of the cells incorporate DNA (Perucho *et al*, 1980b), cotransformation allows an enrichment of two orders of magnitude and increases the chances of finding a transformant with any particular gene to a range between 1/500 and 1/10,000, depending on the system used (Warrick *et al.*, 1980; Kuhn *et al.*, 1984; Martiniuk *et al.*, 1986). The second technical development that fits nicely with that described above is the implementation of methods for replica plating of adherent cultured cells (Stamato and Hohmann, 1975). With replica plating, one can look for a particular feature that can be assayed only by killing the cell. Once the positive transformants are identified, the replica can be used to isolate the clone. Before going into the details of the different gene-isolation protocols, it is worth mentioning that although most reported cases use genomic DNA for gene transfer, there is an increasing trend toward the utilization of complementary DNA (cDNA) that reflects the dramatic improvements during the last few years in obtaining full-length cDNAs (Okayama and Berg, 1982). Another aspect of interest is the mode of expression chosen for the experiments; most genes have been isolated so far through the use of stable transformants, but transient expression has now become a reasonable alternative.

These aspects, together with the methods of isolation, the sources of DNA, and the modes of detection for several representative genes are summarized in Table I.

2. METHODS OF ISOLATION

2.1. Experimental Linkage

As indicated by the title, this method is the linkage of a specific sequence to a restriction-digested DNA prior to gene transfer. This added sequence will eventually allow the retrieval of the gene being sought. There are several ways of achieving this retrieval.

2.1.1. Plasmid Rescue

This method permitted the isolation of the chicken thymidine kinase gene, the first vertebrate gene to be isolated by pure genetic means (Perucho *et al.*, 1980a). A scheme showing some of the details is presented in Fig. 2. It included the screening for several enzymes that do not inactivate the gene in the transfection assay. Since the gene was *Tk,* the selection consisted in the ability to grow in hypoxanthine–aminopterin–thymidine (HAT) conferred by the newly introduced DNA. When an appropriate enzyme that linearized pBR322 and did not cut the gene was identified (*Hind*III), digests of chicken DNA with that enzyme were ligated to *Hind*III-linearized pBR322 in a 1:1 molar ratio. The mixture was used to transform Ltk⁻ cells that were subsequently selected in HAT. The "primary transformants," as these initial transferrants are called, were isolated and grown into mass culture, and DNA was extracted from them to be analyzed in a Southern blot with pBR322 as probe. Due to the phenomenon of cotransformation, the primary transformants usually showed too many bands, and their DNA was used in a second round of transformation. The secondary transformants had few copies of pBR322, and DNA from one of them was digested with a restriction enzyme known not to cut either in pBR or in the *Tk* gene (*Eco*RI and *Bam*HI). The molecules obtained were left to recircularize at low concentration and subsequently used to transform *E. coli* selecting for ampicillin resistence that should be conferred by the β-lactamase gene of pBR322. One of the rescuants obtained was able to transform Ltk⁻ into Tk⁺ cells that grew in HAT. This plasmid contained, linked to the pBR sequences, a fragment of avian DNA that encompassed the chicken thymidine kinase gene. This method can in theory be used to isolate any gene that will give selective advantage to a cultured cell, but it has a few limitations: (1) The gene has to be small, since the plasmid cannot comfortably accommodate large pieces of DNA. (2) It is impossible to be sure that the enzymes utilized do not cut between the plasmid sequence and the gene of interest. Nevertheless, the isolation of the chicken thymidine kinase gene opened the door for the use of gene transfer to isolate genes.

A modification to allow the cloning of longer fragments was reported in 1982 (Lund *et al.*, 1982). It consisted of the use of a cosmid vector that is resistant to tetracycline but sensitive to ampicillin. Following the different steps described, the rescue can be effected through cosmid vectors that will accommodate up to 50 kilobases (kb) of genetic material. A further simplification was designed by Lau and Kan (1984) to isolate the human thymidine kinase gene as a model system. They used a previously prepared cosmid library to tranfect Ltk⁻ cells and to select in HAT for Tk⁺ transformants. Since the cosmids contained the cos sites necessary for packaging, the next step was just to take total DNA from the transformant and package it with lysogenic bacterial extracts as usual.

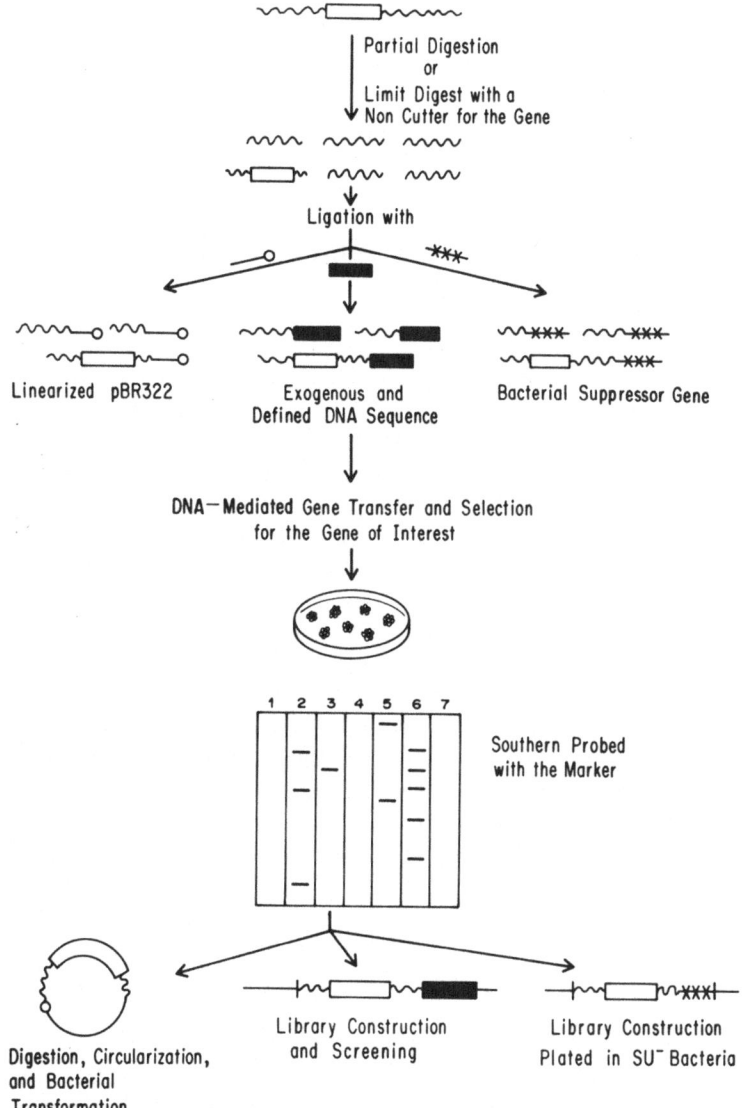

FIGURE 2. Strategies to isolate a gene through linkage to a foreign piece of DNA.

The cosmid colonies obtained were checked by their ability to convert Ltk⁻ cells in Tk⁺ with high efficiency.

2.1.2. Marker Rescue

This method was used as an alternative when plasmid rescue failed, probably due to the absence in the secondary transformant of a complete molecule of pBR322.

The hamster adenine phosphoribosyltransferase (*aprt*) gene was isolated using this approach. It represented the first mammalian gene isolated through DNA-mediated transformation (Pellicer *et al.*, 1980b; Lowy *et al.*, 1980). This gene, like that for Tk, also confers on the recipient aprt⁻ cell the ability to grow in the presence of a selective medium (containing adenine and azaserine), a property lacking in the parental cell.

Here, the initial steps were the same (Fig. 2), i.e., digestion with an enzyme that did not cut the gene (assayed by gene transfer) and its ligation to pBR322 linearized with the same enzyme. Primary and secondary transformants were obtained, and when DNA from the secondaries failed in the plasmid-rescue scheme, a genomic library was constructed from the transformant in λ Charon 4A. This library was screened with P³²-labeled pBR322. The isolated clone was shown to contain a defective pBR molecule and a piece of hamster DNA. When this phage was used to complement the aprt⁻ mutation, its efficiency was 5 orders of magnitude higher than the original hamster DNA, demonstrating the purification of the *aprt* sequences. This method has a more general application, since it allows the gene to be larger and can be improved by performing the initial digestion as a partial with Sau 3A, which gives additional flexibility. Moreover, it does not require the conservation of an intact functional copy of the marker throughout the transformation passages. The retention of a fragment, detectable by hybridization, is sufficient.

2.1.3. tRNA Suppressor Rescue

This method is a convenient modification of the other schemes that permits the use of phages as vector, thereby increasing the size of the insert compared to plasmid rescue, but keeping the advantage of a biological selection to screen the final recombinants.

The genes that were initially cloned using this method are the human activated c-Ha-*ras* (Goldfarb *et al.*, 1982) and N-*ras* (Shimizu *et al.*, 1983b). These are oncogenes for which the method of selection after gene transfer, as described earlier, consists of the ability of the cells that have incorporated and express the oncogene to change their morphology and growth properties in a way that makes it easy to identify them over the background of arrested or moribund cells.

Here, the linkage to the eukaryotic DNA is a 1-kb *Hind*III fragment that contains the transfer (tRNA) suppressor *SupF*, which suppresses amber mutations (Fig. 2).

After secondary transformants were obtained, a genomic library was prepared using the λ phage 1059 with an amber mutation that had been previously introduced by recombination in the right arm. The library was plated in an SU⁻ *E. coli* strain, and therefore only the recombinants that contained the suppressor gene were able to grow. The accomplishment of the oncogene cloning was verified by obtaining very high efficiency of foci formation with the isolated clone. This method has versatility, since it joins the 20-kb capacity with the simplicity of the screening procedure. It has also been used to isolate the hamster *Tk* gene (Lewis *et al.*, 1983).

2.1.4. Function Rescue

The sequences linked to the eukaryotic DNA in the previous examples were used to provide a hybridization handle or to give a biological advantage at the end of the cloning process. Here, the method that will be discussed utilizes the linked sequences for both hybridization and identification of the segment carrying the function that is being investigated.

The best example is the use by Mike Fried's group of polyoma fragments to isolate regulatory sequences from the mammalian genome. In essence, the experiment was to use a *Bam*HI–*Hae*II polyoma fragment that had 89% of the total information but lacked the ability to transform rat cells because part of the promoter region was missing. Mouse DNA was digested with *Hae*II and ligated to the polyoma fragment in a 1 : 4 mass ratio (polyoma/mouse). The mixture was used to transform rat-1 cells. The transformants obtained were analyzed for the presence of polyoma sequences. One cellular clone that contained a single polyoma band was chosen to be used for the isolation procedure. A genomic library was constructed in λgtWESλB and screened with polyoma. The recombinant clone was carefully analyzed to study the role of the mouse sequences linked to the polyoma fragment, and after a Bal-31 deletion scanning, it was concluded that only 58 bases from the mouse DNA were needed to restore the promoter function to the polyoma early region (Fried *et al.*, 1983). The same group has recently communicated the isolation of a sequence with enhancer properties using a very similar strategy, but with F9 DNA as the source (Ford *et al.*, 1985).

In a parallel approach, it has been reported that cotransfection of 3T3 DNA with a cloned 3' long terminal repeat (LTR) from Moloney murine leukemia virus can result in morphological transformation, giving rise to tumorigenic cell lines (Muller and Muller, 1984). The malignant phenotype was found to be associated with specific copies of the LTR through serial passages. In one clone, the LTR was found to be linked to the *c-raf* oncogene, indicating that this might be a feasible approach to clone oncogenes.

2.2. Natural Linkage

This method is based on the fact that mammalian genomes contain sequences that are highly repeated. The most prevalent and best studied is the Alu family in humans. There exist at least 300,000 representatives of this family scattered throughout the genome, so it is likely to be associated with most genes (Jelinek *et al.*, 1980). The members of the family are not identical, but are sufficiently similar to cross-hybridize. Due to its abundance, clones containing the sequence will hybridize to total labeled human DNA, but the most common probe used is a clone, BLUR 8 (Jelinek *et al.*, 1980), that has the sequence of 300 bases cloned into pBR322. The use of this plasmid or its isolated insert increases the sensitivity of detection.

2.2.1. Direct Transformation

This approach was employed for the first time to isolate the H-*ras* activated oncogene by using its focus-forming ability in NIH 3T3 (morphological selection) (Shih and Weinberg, 1982; Pulciani *et al.*, 1982). The scheme is analogous to the one described in Section 2.1, except that here it is not necessary to perform any DNA manipulation prior to gene transfer. The primary transformants, when analyzed by Southern hybridization, present a pattern of intense signal throughout the lane. This is due to the cotransformation phenomenon mentioned previously, by which a cultured cell can incorporate as much as 1% of a genome equivalent when exposed to the DNA–calcium phosphate coprecipitate.

After a second and third passage, the number of bands decreases dramatically, since there is a selection process for the ones that segregate with the transformed phenotype (Fig. 3).

The number of bands associated with a gene will be roughly correlated with the size of the gene, since most of the Alu sequences are in the introns and large genes are usually made at the expense of large and numerous introns. This property supplies useful information for the analyisis of the tertiary transformants: (1) Comparative examination of several transformants (4–6) from the same parental DNA should provide a common pattern of bands that is repeated in all of them, defining the minimal sequences contained in the gene, and some noncommon additional bands different in every clone originated from the integration points or by cotransformation (Pellicer *et al.*, 1978). (2) Addition of the molecular

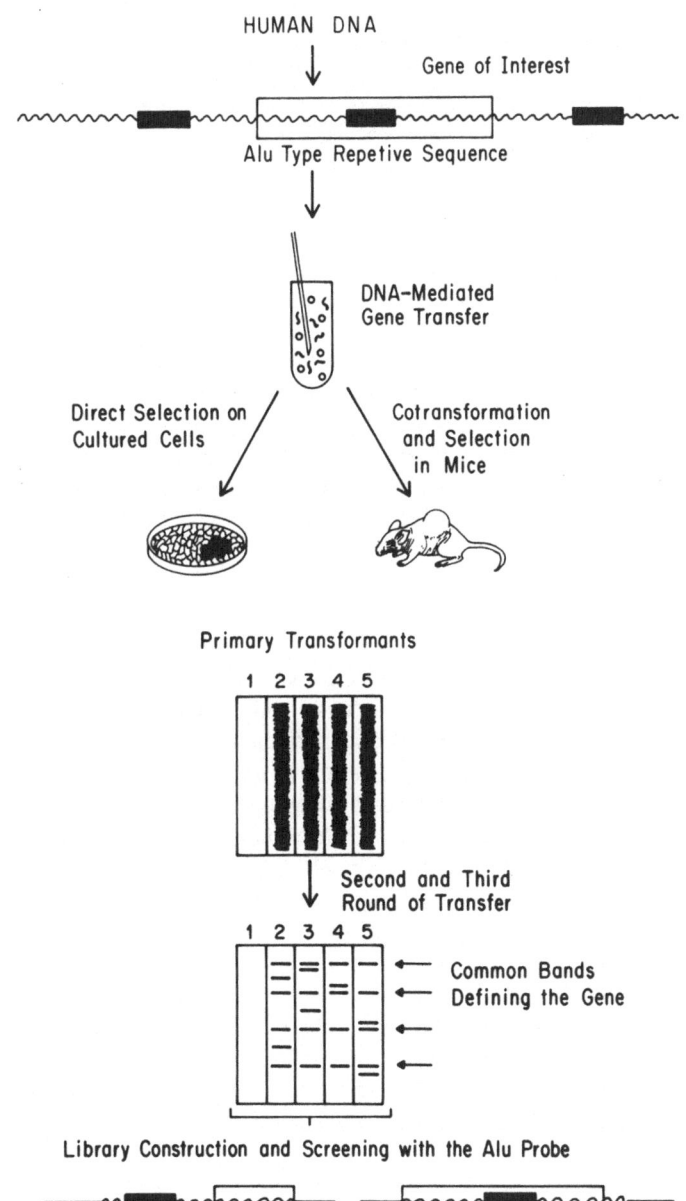

FIGURE 3. Outline of the protocol to isolate genes using highly repetitive sequences.

weights form all common bands will provide an estimate of the minimal size of the gene.

When there are only a few noncommon bands, a recombinant library is made from the transformant and screened with the Alu probe. The isolated clone should contain an Alu sequence inside or adjacent to the gene. The last scenario was observed in the cloning of the activated human H-*ras* gene from the T-24 line (Shih and Weinberg, 1982; Pulciani *et al.*, 1982). Since the gene is small, the repetitive sequences are actually located outside the gene, but close enough to be part of the same recombinant clone. In the case of the human K-*ras* gene, since the number of common bands predicted a large size, the strategy was to clone it in successive walking steps (Shimizu *et al.*, 1983b; McGrath *et al.*, 1983; Nakano *et al.*, 1984). Another interesting gene cloned with this approach is *T9*, the transferrin receptor, in which the gene of interest was not selected but cotransformed with a selectable vector. The surviving colonies were now obtained through structural selection by their surface expression and, after several transfers, were cloned using the human repetitive sequences as probe (Kuhn *et al.*, 1984). Three different teams have used the repetitive sequences to isolate the human *Tk* gene (Bradshaw, 1983; Lin *et al.*, 1983; Stuart *et al.*, 1985).

2.2.2. Cotransformation and Nude Mice

Although based on the same principle of detecting human repetitive sequences on a rodent background, this scheme adds a different approach with the use of cotransformation and the assay of tumorigenesis in nude mice. This method has been devised by Wigler's group (Fasano *et al.*, 1984), modifying an initial version by Blair *et al.* (1982). It takes advantage of the fact that only a subpopulation of cells is usually competent in uptake and expression of foreign DNA. This fraction of cells is specifically selected with the addition of a dominant acting vector together with the DNA to be tested.

In Wigler's experiments, the aim was to screen for and isolate oncogenes that were negative in the 3T3 transformation assay. Using the human tumor cell line mcf-7 as DNA source, its genomic material was cotransfected into NIH3T3 with the *neo^r* gene, and the cells were subsequently selected in the antibiotic G418. After 2 weeks, the surviving colonies (300/plate, 4 plates) were pooled, and 10^7 cells were injected subcutaneously into nude mice. After a latency period of 3–4 weeks, tumors arose that were collected for analysis and subsequent transfers.

Secondary tumors were tested with the Alu probe, and the DNAs could be grouped into three classes by their Southern-blot patterns, which indicated that three independent genes were responsible for nude mice tumorigenesis. One was identified as an amplified normal N-*ras* gene that seems to be sufficient to induce tumors but not to form foci in 3T3. The other two were not recognized by the available probes, so it was concluded that they represented new oncogenes. Recombinant libraries were made from the secondary tumors, and the Alu-containing fragments were isolated using labeled human placental DNA.

This ingenious protocol should facilitate the isolation of new oncogenes with tumorigenic potential.

2.2.3. Metastasis Assay

Using a similar type of approach, Weinberg's group has identified and is isolating a human gene present in the tumor cell line ME-180 from a cervical carcinoma that is capable of inducing the formation of metastases in mice (Bernstein and Weinberg, 1985).

The recipient cells were NIH3T3 transformed with the activated H-*ras* gene and shown not to be metastatic when injected subcutaneously into immunocompetent NSF mice. These cells were cotransfected with DNA from ME-180 plus the *neo*^{r} gene and selected in G418. After 2 weeks, colonies were pooled, and 10^6 cells per animal were injected. One mouse developed metastasis in the lung, and the metastatic phenotype was transferred several times using the protocol described. The molecular analysis of secondary metastases revealed common bands among them, as would be expected if a specific human gene was responsible for the metastatizing behavior. The isolation process is in progress, and this again is a new avenue that might prove to be very useful in the future.

2.3. Sib Selection

This method is based on the ability to individualize the different genes and to assay them for the appropriate phenotype. The first part is feasible with the recombinant-DNA techniques that compartmentalize the genome into many independent clones. The assay is going to be gene transfer.

2.3.1. Stable Transfer of Genomic DNA

The two initial reports were the isolation of chicken *Blym-1* (Goubin et al., 1983) and murine *Tlym-1* (Lane et al., 1984). These are genes that produce morphological transformation in rodent fibroblasts. The assay is simply focus formation as explained above, scored after 2–3 weeks of the transfer and therefore detecting a stable phenotype.

The protocol is easy but laborious, consisting of the isolation of primary transformants in 3T3 from the original tumor DNA. The transformant was used to construct a recombinant library using a λ phage vector. In the case of the

Tlym gene, an initial enrichment of the DNA to be cloned was obtained by sucrose-gradient separation after digesting the DNA with an enzyme known not to inactivate the gene. In both instances, libraries were obtained with 200,000 (*Blym*) and 20,000 (*Tlym*) different clones. The libraries were divided into 10 different pools, which were subsequently amplified to obtain sufficient genetic material for the assay. The foci-formation experiments revealed which one of the pools contained the gene (in some cases, more than one pool is positive and the one with higher efficiency is chosen) (Fig. 4).

The more difficult part is already accomplished at this point, since the critical step is to be able to transform with the library DNA. The difficulties reside in the inhibition that λ DNA produces in DNA-mediated gene transfer and in the risk that the complete gene might not fit in the vector.

The protocol follows with a repetition of the subdivision into smaller pools that are assayed in the same manner. After four or five of these rounds of

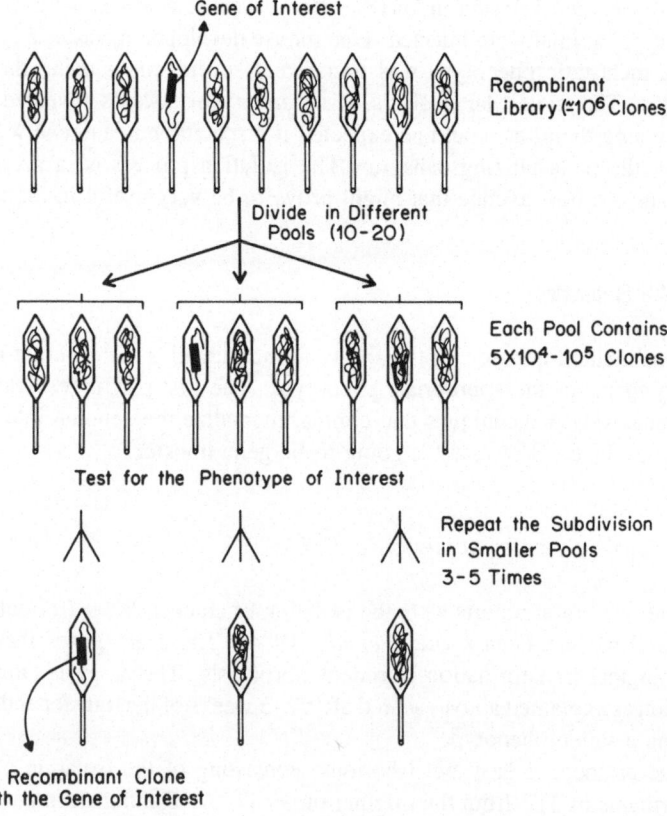

FIGURE 4. Sib-selection method for gene isolation.

fractionation, individual phages can be isolated that can induce the appropriate phenotype. At that point, the efficiency should be five orders of magnitude higher than the original total library. It can be inferred from the foregoing description that this is a reliable way to isolate a gene as long as the assay is sensitive enough to detect activity using the complete library and the gene is small enough to fit in the clones. However, this approach has seldom been used because of the tedious manipulations required.

2.3.2. Transient Expression of cDNAs

This protocol has recently been used to isolate the sequences that code for GM-CSF (Lee *et al.*, 1985; Wong *et al.*, 1985). This molecule stimulates the proliferation of precursor cells into both granulocytic and monocytic lineages. The novelty of this approach consists in the utilization of cDNA cloned in an expressor plasmid as a source of genetic material and the exploitation of transient expression to obtain the product that will be subsequently assayed. In one particular case, cDNAs from a producer line were cloned in the plasmid p91203(b) (Wong *et al.*, 1985). This potent expressor plasmid had previously been shown to produce enough material to allow its detection even when it was diluted 500- to 1000-fold with other plasmids.

A library of 60,000 independent clones was constructed, and 200 pools, each containing 200–500 clones, were prepared for the first screening. The assay of the individual pools was made by transfection of COS-1 cells and collection of the supernatants after transient expression. Using the same sib selection mentioned before, the pools were narrowed down to single clones that were shown to contain the GM-CSF sequences. More recently, cDNA clones for IL-2 have been isolated by the same procedure (Yokota *et al.*, 1985).

The interest of this technique is the use of cDNA in mammalian gene transfer and isolation, since cDNAs have so far been used for isolation by hybridization or bacterial expression in λgtll vectors (Young and Davis, 1983). Additionally, the use of transient expression in an isolation scheme is novel.

We have recently combined the use of a genomic library in concert with transient expression to detect transfer of the human α-glucosidase *AB* (Martiniuk *et al.*, 1985), and experiments are in progress toward the gene isolation.

2.4. Subtraction Hybridization

The principle of this approach is based on the fact that after a recipient cell has been subjected to DNA-mediated gene transfer and screened for whichever phenotype is chosen, the transcripts present in the transformant will be the same as the parental cell except for a few coming from the newly introduced DNA. Since only the cells expressing the gene that is being investigated are chosen, it presents an ideal background to specifically select for the desired message.

This approach has been successfully used for the isolation of the *T8* (Kavathas *et al.*, 1984; Littman *et al.*, 1985) and *T4* (Maddon *et al.*, 1985) genes.

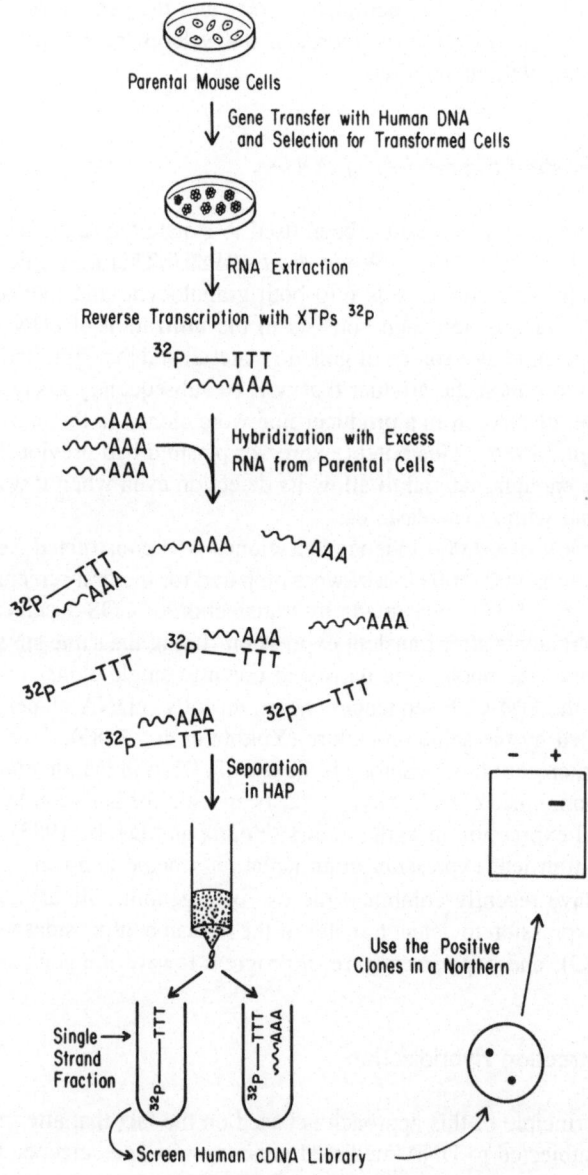

FIGURE 5. Subtraction hybridization as an enriching step in gene purification after gene transfer.

T8 and *T4* are surface markers of the human T-cell lineage that characterize the cytotoxic–suppressor cells (*T8*) or the helper cells (*T4*). To enrich for the subpopulation of cells that takes up DNA, Ltk⁻ cells were cotransfected with human DNA plus the *HSVTK* gene. After colonies appeared, two different approaches were taken. Littman *et al.* (1985) used a rossetting assay employing red cells coupled with a secondary antibody that reacted with the target cells in the presence of the primary antibody. The positive clones were visible with the naked eye, displaying a red color, and they could be isolated and grown in mass culture for analysis and nucleic-acid isolation. On the other hand, Kavathas and Herzenberg (Kavathas *et al.*, 1984) chose to select their positive cells by pooling many Tk⁺ colonies and running the cells through a cell sorter after the cells had been stained for the antigen of interest. In any case, both teams discovered to their dismay that the secondary transformants lacked Alu sequences, and they had to resort to another isolation scheme. Subtraction hybridization had been elegantly used by Mark Davis to isolate the T-cell receptor (Davis *et al.*, 1984) genes and was the avenue chosen. When [p³²]-cDNA made from the transformants was hybridized to a 20-fold excess of messenger RNA (mRNA) from the Ltk⁻ cells, the enrichment obtained was between 10 and 15 times after two rounds of subtraction. This enriched probe was used to screen a human cDNA library, and positive clones were rescreened and confirmed by differential hybridization and by using the isolated clone in Northern analysis comparing RNAs from different cells, some that expressed antigen and others negative for it (Fig. 5).

In the *T8* isolation, the transformants had been selected several times and *T8* expression was amplified, facilitating the enrichment after subtraction. However, in *T4*, no amplification was reported, and despite the low abundance of this mRNA in the transformants, estimated around 5–10 copies per cell, the isolation was accomplished. This indicates that it is a general approach that can be used in a variety of transferred genes. It is important to note the significant improvement represented by using transformants to subtract with parental cells in which most of the RNAs are the same except for the transcripts coming from the transfected genes. Even more important, since heterospecific transfers are commonly used, the RNAs used in the labeled cDNA and subtraction are from one species (mouse), and the gene of interest can be isolated by screening a cDNA library from another (human). This strategy diminishes enormously the background hybridization that occurs by insufficiently subtracted cDNA when it is used to screen a homologous library.

3. DEVELOPING APPROACHES

This section will be dedicated to methods that have been used so far as model systems or that are not yet being generally used. A discussion of related subjects, such as viruses, hybrids, and transgenic mice, will complete the overview.

3.1. Complementation in Bacteria

Although some examples of yeast genes complementing bacterial functions were mentioned at the beginning of the chapter, here the subject is a human gene much more distantly related to the prokaryotic genome. This original approach has been taken to isolate the human purine nucleoside phosphorylase (PNP) gene (Goddard *et al.*, 1983). After an initial enrichment of the mRNA by immunoadsorption of the polysomes with a specific antibody, a cDNA was constructed and cloned into the *Pst*I site of pBR322. This site is close to the β-lactamase promoter, and if the inserted sequences are oriented properly and in frame with the bacterial gene, they can be transcribed and translated in the prokaryotic host and give a functional product. Therefore, the enriched cDNA library obtained was used to transform the PNP-deficient S0312 bacteria. Initially, the transformants were selected for tetracycline resistance and the positive ones replica-plated in medium with deoxyinosine as the sole purine source, requiring the presence of PNP to grow in such a medium. The clones that were positive in this assay were used directly to transform PNP⁻ bacteria on PNP selection medium, and all except one were positive. The bacteria were also shown to produce the human protein by labeling with [^{35}S]-methionine and immunoprecipitation of the extracts with specific antibody. Therefore, a human cDNA with the appropriate bacterial regulatory and signal sequences can complement the prokaryotic mutation. Although this method has not been widely used, it can potentially be applied to isolate mammalian genes, as long as they code for proteins with homologous functions in bacteria.

3.2. Replication Origins

Several observations show that sequences isolated from different species can act as an origin of replication in yeast. DNAs of several species have been used to find autonomously replicating sequences (ARSs) in yeast. ARSs in yeast are assumed to serve as origins of DNA replication, and there is a good and sensitive assay for them based on the increased frequency of transformation efficiency that these sequences confer on yeast plasmids. Using this approach, mouse sequences with these properties have been isolated. Fragments of 4–6 kb resulting from the digestion of mouse DNA with *Bam*HI and *Hind*III were inserted into the yeast plasmid YIp5 (Roth *et al.*, 1983). This plamid contains pBR322 and a yeast gene, *URA3*, that can be used as a selectable marker. Using several of these hybrid plasmids with bacterial, yeast, and mouse DNA, URA3⁻ yeast cells were transformed and selected in media without uracil. The irrelevant sequences did not provide any advantage to the vector, and its efficiency was similar to that of the control, but five of the plasmids assayed increased the efficiency up to three orders of magnitude, indicating that they contained ARS-like sequences. The increased efficiency is explained as a result of the gene-transfer mechanisms: When the foreign DNA is taken up by the yeast, it does not necessarily integrate immediately into the host chromosomes, remaining unintegrated for a variable period of time. If the exogenous DNA is not able to replicate, it is likely to be lost during cell proliferation and division. In contrast, a replication origin will allow the plasmids to muliply, providing to them a high probability of staying in most of the cells that were initially transformed. A similar type of experiment has also been done with *Physarum* DNA with analogous results (Gorman *et al.*, 1981). Going one step further, a Colorado group has recently used a similar approach to isolate what they call "segregators" (Stinchcomb *et al.*, 1985). The plasmids mentioned before with ARS and a selectable marker produce high transformation efficiency, but the plasmid eventually integrates into the chromosome or is lost. An additional sequence, probably of centromeric origin, would stabilize the plasmid as an episome by allowing proper mitotic segregation. Using an *Eco*RI digest of *Caenorhabditis elegans* DNA cloned into a plasmid with pBR322, the *URA3* gene, and ARS1 sequences, it was possible to estimate the presence of around 30 segregators in the *C. elegans* genome, and 7 of them were isolated.

3.3. Retroviruses as a Genetic Bullet

One elegant scheme that has been developed at Columbia University during the last few years is the use of retroviruses as a sort of boomerang that can hit the appropriate gene and also be retrieved (King *et al.*, 1985). It has been used to isolate the *hprt* locus as a model system. The retrovirus employed in the experiments contains the bacterial tRNA suppressor gene *SuIII*. The *hprt* is a convenient locus to choose, since it is haploid in male cells. The infection of the target cells with the retrovirus is followed by the integration of the viral genome in multiple locations. Acting as an insertion mutagen, the integrations will affect the loci where the integrations occur and may also influence the ones nearby. Therefore, the cells are selcted in 8-azaguanine, which will allow the growth of only hprt⁻ variants. It is necessary to calculate the spontaneous mutation frequency of the parental cell line, but in this protocol, the number of mutants obtained with the retroviral treatment was at least one order of magnitude higher than the spontaneous mutations. Once the mutants are obtained, they can be analyzed for the number of copies of the bacterial supressor gene. The line with fewer copies is used to construct a genomic library in a vector with the appropriate mutation, and the library is then plated in SU⁻ bacteria as described earlier. In this way, several true *hprt* clones were isolated and subsequently analyzed.

3.4. Hybrid Enrichment

This is a method that, although it does not use DNA-mediated gene transfer *per se*, is based on the strategies described in previous sections. The principle is to have some selectable function, coded for by the gene chosen for isolation, and to use a cell with that function to make somatic-cell hybrids with another species that is dominant in keeping its chromosomes over the first one (Cirullo *et al.*, 1983). Moreover, the second cells have to lack the selectable function so that only the hybrids that contain the few chromosomes from the first genome will be selected. The innovation is the treatment of these hybrids with high doses of γ-irradiation to extensively fragment their genome and immediately fuse them again with the chromosomally dominant species with subsequent selection for the appropriate function. Due to the extensive fragmentation, the amount of information retained from the first cell type is considerably reduced. Repetition of this cycle two or three times will produce a hybrid that can be analyzed at the molecular level. The technique was used to enrich for the human leucyl and asparaginyl tRNA synthetase in a hamster background. After secondary hybrids were obtained, they were analyzed in a Southern blot with human repetitive sequences. Only a few Alu bands had segregated together with the chosen phenotype; therefore, these hybrid lines can be used to construct a genomic library that on subsequent screening with Alu can identify recombinant clones with the gene of interest.

3.5. Transgenic Mice

Most gene-transfer experiments are carried out in tissue culture with the understanding that it is a model system for live organisms. It is therefore not surprising that eventually some researchers have gone straight for the animal target, transferring genes into fertilized mouse eggs that eventually end up in every cell in around 20% of the born mice (Brinster *et al.*, 1985). This gene-transfer technique can also be used to isolate genes, and it seems to be very promising to study genetic steps involved in mouse development. One of the first such observations was not obtained with pure DNA, but by infecting early embryos with retroviruses. It was observed by Jaenisch's group that some of the mice with newly integrated viruses were producing offspring that died *in utero* at day 13 of development (Schnieke *et al.*, 1983). It was inferred that the viral molecule had integrated at a locus important for embryonic development and that when both alleles were altered, a lethal effect was the consequence.

Since they had the retroviral genome inserted near or at the locus of developmental importance, they could use it as a probe to retrieve that gene by standard recombinant-DNA techniques. In this particular case, the affected gene coded for $\alpha 1$ collagen. Therefore, the insertion of foreign DNA segments in the mouse genome can be used as a mutagen to study complex developmental processes. In fact, Leder's group (Woychik *et al.*, 1985) has recently obtained another developmental mutant that alters limb formation; the interesting thing is the realization that the same mutant phenotype had been previously reported and isolated as a spontaneous event. They are now isolating the involved gene using the foreign DNA as a probe to pinpoint the sequence of interest.

4. CONCLUSIONS

Gene transfer has come of age. During the last ten years, its use has increased enormously, making it today one of the necessary tools for a molecular biologist. The diversification of gene transfer to tackle several genetic problems has had an important impact on gene isolation. The year 1980 was the year in which the first vertebrate genes were isolated by gene-transfer means, and these techniques have developed at a rapid pace, resulting in the isolation of a substantial number of genes.

Today, any gene can be isolated using gene-transfer technology, provided there is a way to identify and isolate the transferrants after transfection:

- The source can be genomic or cDNA.
- Transient and stable expression may be used.
- Different selection schemes can be implemented: metabolic, morphologic, and structural.
- No RNA or protein data are necessary if a selection for gene function is available. Some schemes are based on pure genetic means.
- The method permits isolation of functional sequences that are not genes, e.g., ARS, promoters.

In summary, the combination of gene transfer with recombinant DNA has certainly been synergistic, and it has widened, through the isolation of many interesting genes, our vision of gene expression and cell function.

ACKNOWLEDGMENTS. I thank Joan W. Berman for the critical reading of the manuscript and L. Ramos for its preparation. The experiments described as coming from my laboratory have been supported by grants from the NIH (CA36327 and CA16239) and the March of Dimes (5-380 and ACS 197). Angel Pellicer is a recipient of an Irma Hirschl-Monique Weill/Caulier award.

REFERENCES

Bacchetti, S., and Graham, F. L., 1977, Transfer of the gene for the thymidine kinase to thymidine kinase-deficient human cells by purified herpes simplex viral DNA, *Proc. Natl. Acad. Sci. U.S.A.* **74:**1590–1594.

Berman, J., Basch, R., and Pellicer, A., 1984, Gene transfer in lymphoid cells: Expression of the Thy-1.2 antigen by Thy-1.1 BW5147 lymphoma cell transfected with unfractionated cellular DNA, *Proc. Natl. Acad. Sci. U.S.A.* **81:**7176–7179.

Bernstein, S. C., and Weinberg, R. A., 1985, Expression of the metastatic phenotype in cells transfected with human metastatic tumor DNA, *Proc. Natl. Acad. Sci. U.S.A.* **82:**1726–1730.

Blair, D. G., Cooper, C. S., Oskarson, M. K., Eader, L. A., and Vande Woude, G. F., 1982, New method for detecting cellular transforming genes, *Science* **281:**1122–1125.

Boni, C., Esteban, M., and Pellicer, A., 1984, Expression of cloned vaccinia DNA sequences introduced into animal cells, *J. Gen. Virol.* **65:**1245–1251.

Bradshaw, H. D., 1983, Molecular cloning and cell cycle-specific regulation of a functional human thymidine kinase gene, *Proc. Natl. Acad. Sci. U.S.A.* **80:**5588–5591.

Brinster, R. L., Chen, H. Y., Trumbauer, M. E., Yagle, M. K., and Palmiter, R. D., 1985, Factors affecting the efficiency of introducing foreign DNA into mice by microinjecting eggs, *Proc. Natl. Acad. Sci. U.S.A.* **82:**4438–4442.

Broach, J. R., Strathern, J. N., and Hicks, J. B., 1979, Transformation in yeast: Development of a hybrid cloning vector and isolation of the CAN1 gene, *Gene* **8:**121–133.

Cirullo, R. E., Dana, S., and Wasmuth, J. J., 1983, Efficient procedure for transferring specific human genes into Chinese hamster cell mutants: Interspecific transfer of the human genes encoding leucyl- and asparaginyl-tRNA synthetases, *Mol. Cell. Biol.* **3:**892–902.

Clarke, L., and Carbon, J., 1976, A colony bank containing synthetic Col E1 hybrid plasmids representative of the entire *E. coli* genome, *Cell* **9:**91–99.

Clarke, L., and Carbon, J., 1978, Functional expression of cloned yeast DNA in *Escherichia coli:* Specific complementation of argininosuccinate lyase (arg H) mutations, *J. Mol. Biol.* **120:**517–532.

Colbere-Garapin, F., Horodniceanu, F., Kourilsky, P., and Garapin, A. C., 1981, A new dominant hybrid selective marker for higher eukaryotic cells, *J. Mol. Biol.* **150:**1–14.

Davis, M. M., Cohen, D. I., Nielsen, E. A., Steinmetz, M., Paul, W. E., and Hood, L., 1984, Cell-type-specific cDNA probes and the murine I region: The localization and orientation of Ad alpha, *Proc. Natl. Acad. Sci. U.S.A.* **81:**2194–2198.

Fasano, O., Birnbaum, D., Edlund, L., Fogh, S., and Wigler, M., 1984, New human transforming genes detected by a tumorigenicity assay, *Mol. Cell. Biol.* **4:**1695–1705.

Ford, M., Davies, B., Griffiths, M., Wilson, J., and Fried, M., 1985, Isolation of a gene enhancer within an amplified inverted duplication after "expression selection," *Proc. Natl. Acad. Sci. U.S.A.* **82:**3370–3374.

Fried, M., Griffiths, M., Davies, B., Bjursell, G., LaMantia, G., and Lania, L., 1983, Isolation of cellular DNA sequences that allow expression of adjacent genes, *Proc. Natl. Acad. Sci. U.S.A.* **80:**2117–2121.

Goddard, J. M., Caput, D., Williams, S. R., and Martin, D. W., 1983, Cloning of human purine-nucleoside phosphorylase cDNA sequences by complementation in *Escherichia coli, Proc. Natl. Acad. Sci. U.S.A.* **80:**4281–4285.

Goldfarb, M., Shimizu, K., Perucho, M., and Wigler, M., 1982, Isolation and preliminary characterization of a human transforming gene from T24 bladder carcinoma cells, *Nature (London)* **296:**404–409.

Gorman, J. A., Dove, W. F., and Warren, N., 1981, Isolation of *Physarum* DNA segments that support autonomous replication in yeast, *Mol. Gen. Genet.* **183:**306–313.

Goubin, G., Goldman, D. S., Luce, J., Neiman, P. E., and Cooper, G. M., 1983, Molecular cloning and nucleotide sequence of a transforming gene detected by transfection of chicken B-cell lymphomas DNA, *Nature (London)* **302**:114–119.

Graham, F. L., and van der Eb, A. J., 1973, A new technique for the assay of infectivity of human adenovirus 5 DNA, *Virology* **52**:456–467.

Graham, F. L., Abrahams, P. J., Mulder, C., Heijneker, H. L., Warnaar, S. O., de Vries, F. A. J., Fiers, W., and van der Eb, A. J., 1974, *Cold Spring Harbor Symp. Quant. Biol.* **39**:637–650.

Guerrero, I., Calzada, P., Mayer, A., and Pellicer, A., 1984, A molecular approach to leukemogenesis: Mouse lymphomas contain an activated c-ras oncogene, *Proc. Natl. Acad. Sci. U.S.A.* **81**:202–205.

Henikoff, S., Tatchell, K., Hall, B. D., and Nasmyth, K. A., 1981, Isolation of a gene from *Drosophila* by complementation in yeast, *Nature (London)* **289**:33–37.

Hsu, C., Kavathas, P., and Herzenberg, L. A., 1984, Cell-surface antigens expressed on L-cells transfected with whole DNA from non-expressing and expressing cells, *Nature (London)* **312**:68–69.

Jelinek, W. R., Toomey, T. P., Leinwand, L., Duncan, C. H., Biro, P. A., Choudary, P. V., Weissman, S. M., Rubin, C. M., Houck, C. M., Deininger, P. L., and Schmid, C. W., 1980, Ubiquitous, interspersed repeated sequences in mammalian genomes, *Proc. Natl. Acad. Sci. U.S.A.* **77**:1398–1402.

Kavathas, P., and Herzenberg, L. A., 1983, Stable transformation of mouse L-cells for human membrane T-cell differentiation antigens, HLA and beta 2-microglobulin: Selection by fluorescence-activated cell sorting, *Proc. Natl. Acad. Sci. U.S.A.* **80**:524–528.

Kavathas, P., Sukhatme, V. P., Herzenberg, L. A., and Parnes, J. R., 1984, Isolation of the gene encoding the human T-lymphocyte differentiation antigen Leu-2(T8) by gene transfer and cDNA subtraction, *Proc. Natl. Acad. Sci. U.S.A.* **81**:7688–7692.

King, W., Patel, M. D., Lobel, L. I., Goff, S. P., and Nguyen-Huu, M. C., 1985, Insertion mutagenesis of embryonal carcinoma cells by retroviruses, *Science* **228**:554–558.

Kuhn, L. C., McClelland, A., and Ruddle, F. H., 1984, Gene transfer, expression, and molecular cloning of the human transferrin receptor gene, *Cell* **37**:95–103.

Lane, M. A., Sainten, A., Doherty, K. M., and Cooper, G. M., 1984, Isolation and characterizaton of a stage-specific transforming gene, Tlym-1, from T-cell lymphomas, *Proc. Natl. Acad. Sci. U.S.A.* **81**:2227–2231.

Lau, Y.-F., and Kan, Y. W., 1984, Direct isolation of the functional human thymidine kinase gene with a cosmid shuttle vector, *Proc. Natl. Acad. Sci. U.S.A.* **81**:414–418.

Lee, F., Yokota, T., Otsuka, T., Gemmell, L., Larson, N., Luh, J., Arai, K.-I., and Rennick, D., 1985, Isolation of cDNA for a human granulocyte-marcrophage colony-stimulating factor by functional expression in mammalian cells, *Proc. Natl. Acad. Sci. U.S.A.* **82**:4360–4364.

Lewis, S. A., Shimizu, K., and Zipser, D., 1983, Isolation and preliminary characterization of the Chinese hamster thymidine kinase gene, *Mol. Cell. Biol.* **3**:1815–1823.

Lin, P. F., Zhao, S. Y., and Ruddle, F. H., 1983, Genomic cloning and preliminary characterization of the human thymidine kinase gene, *Proc. Natl. Acad. Sci. U.S.A.* **80**:6528–6532.

Littman, D. R., Thomas, Y., Maddon, P. J., Chess, L., and Axel, R., 1985, The isolation and sequence of the gene encoding T8: A molecule defining functional classes of T lymphocytes, *Cell* **40**:237–246.

Lowy, I., Pellicer, A., Jackson, J. F., Sim, G. K., Silverstein, S., and Axel, R., 1980, Isolation of transforming DNA: Cloning the hamster aprt gene, *Cell* **22**:817–823.

Lund, T., Grosveld, F. G., and Flavell, R. A., 1982, Isolation of transforming DNA by plasmid rescue, *Proc. Natl. Acad. Sci. U.S.A.* **79**:520–524.

Maddon, P. J., Littman, D. R., Godfrey, M., Maddon, D. E., Chess, L., and Axel, R., 1985, The isolation and nucleotide sequence of a cDNA encoding the T cell surface protein T4: A new member of the immunoglobulin gene family, *Cell* **42**:93–104.

Maitland, N. J., and McDougall, J. K., 1977, Biochemical transformation of mouse cells by fragments of herpes simplex virus DNA, *Cell* **11**:233–241.

Martiniuk, F., Pellicer, A., and Hirschhorn, R., 1985, Transient expression of human neutral alpha glucosidase AB (glucosidase II) in enzyme deficient mouse lymphoma cells, *J. Biol. Chem.* **260**:14351–14354.

Martiniuk, F., Pellicer, A., Mehler, M., and Hirschhorn, R., 1986, Detection, frequency and stability of cotransformants expressing non-selectable human enzymes, *Somat. Cell Mol. Genet.* **12**:1–12.

McGrath, J. P., Capon, D. J., Smith, D. H., Chen, E. Y., Seeburg, P. H., Goeddel, D. V., and Levinson, A. D., 1983, Structure and organization of the human Ki-ras proto-oncogene and a related processed pseudogene, *Nature (London)* **304**:501–506.

Muller, R., and Muller, D., 1984, Co-transfection of normal NIH/3T3 DNA and retroviral LTR sequences: A novel strategy for the detection of potential c-onc genes, *EMBO J.* **3**:1421–1427.

Mulligan, R. C., and Berg, P., 1981, Selection for animal cells that express the *Escherichia coli* gene coding for xanthine-guanine phosphoribosyl-transferase, *Proc. Natl. Acad. Sci. U.S.A.* **78**:2072–2076.

Nakano, H., Yamamoto, F., Neville, C., Evans, D., Mizuno, T., and Perucho, M., 1984, Isolation of transforming sequences of two lung carcinomas: Structural and functional analysis of the activated c-K-ras oncogenes, *Proc. Natl. Acad. Sci. U.S.A.* **81**:71–75.

Okayama, H., and Berg, P., 1982, High efficiency cloning of full length cDNA, *Mol. Cell, Biol.* **2**:161–169.

Pellicer, A., and Esteban, M., 1982, Gene transfer stability and biochemical properties of animal cells transformed with vaccinia DNA, *Virology* **122**:363–380.

Pellicer, A., Wigler, M., Axel, R., and Silverstein, S., 1978, The transfer and stable integration of the HSV thymidine kinase gene into mouse cells, *Cell* **14**:133–141.

Pellicer, A., Wagner, E. F., El Kareh, A., Dewey, M. J., Reuser, A. J., Silverstein, S., Axel, R., and Mintz, B., 1980a, Introduction of a viral thymidine kinase gene and the human β-globin gene into developmentally multipotential mouse teratocarcinoma cells, *Proc. Natl. Acad. Sci. U.S.A.* **77**:2098–2102.

Pellicer, A., Robins, D., Wold, B., Sweet, R., Jackson, J., Lowy, I., Roberts, J. M., Sim, G.-K., Silverstein, S., and Axel, R., 1980b, Altering genotype and phenotype by DNA mediated gene transfer, *Science* **209**:1414–1422.

Perucho, M., Hanahan, D., Lipsich, L., and Wigler, M., 1980a, Isolation of the chicken thymidine kinase gene by plasmid rescue, *Nature (London)* **285**:207–210.

Perucho, M., Hanahan, D., and Wigler, M., 1980b, Genetic and physical linkage of exogenous sequences in transformed cells, *Cell* **22**:305–317.

Pulciani, S., Santos, E., Lauver, A. V., Long, L. K., Robbins, K. C., and Barbacid, M., 1982, Oncogenes in human tumor cell lines: Molecular cloning of a transforming gene from human bladder carcinoma cells, *Proc. Natl. Acad. Sci. U.S.A.* **79**:2845–2849.

Roth, G. E., Blanton, H. M., Hager, L. J., and Zakian, V. A., 1983, Isolation and characterization of sequences form mouse chromosomal DNA with ARS function in yeast, *Mol. Cell Biol.* **3**:1898–1909.

Schell, M. A., and Wilson, D. B., 1979, Cloning and expression of the yeast galactokinase gene in an *Escherichia coli* plasmid, *Gene* **5**:291–303.

Schnieke, A., Harbers, K., and Jaenisch, R., 1983, Embryonic lethal mutation in mice induced by retrovirus insertion into the alpha 1(I) collagen gene, *Nature (London)* **304**:315–320.

Shih, C., and Weinberg, R. A., 1982, Isolation of a transforming sequence from a human bladder carcinoma cell line, *Cell* **29**:161–169.

Shimizu, K., Goldfarb, M., Perucho, M., and Wigler, M., 1983a, Isolation and preliminary characterization of the transforming gene of a human neuroblastoma cell line, *Proc. Natl. Acad. Sci. U.S.A.* **80**:383–387.

Shimizu, K., Birnbaum, D., Ruley, M. A., Fasano, O., Suard, Y., Edlund, L., Taparowski, E., Goldfarb, M., and Wigler, M., 1983b, Structure of the Ki-ras gene of the human lung carcinoma cell line Calu-1, *Nature (London)* **304**:497–500.

Stamato, T. D., and Hohmann, C. K., 1975, A replica plating method for CHO Cells using nylon cloth, *Cytogent. Cell Genet.* **15**:372–379.

Stinchcomb, D. T., Mello, C., and Hirsch, D., 1985, *Caenorhabditis elegans* DNA that directs segregation in yeast cells, *Proc. Natl. Acad. Sci. U.S.A.* **82**:4167–4171.

Stuart, P., Ito, M., Stewart, C., and Conrad, S. E., 1985, Induction of cellular thymidine kinase occurs at the mRNA level, *Mol. Cell. Biol.* **5**:1490–1497.

Taylor, G. R., Barclay, B. J., Storms, R. K., Friesen, J. D., and Haynes, R. H., 1982, Isolation of the thymidilate synthetase gene (TMP1) by complementation in *Saccharomyces cerevisiae, Mol. Cell. Biol.* **2**:437–442.

Warrick, H., Hsiung, N., Shows, T., and Kucherlapati, R., 1980, DNA mediated cotransfer of unlinked mammalian cell markers into mouse L cells, *J. Mol. Biol.* **86**:341–346.

Wigler, M., Silverstein, S., Lee, L.-S., Pellicer, A., Cheng, Y.-C., and Axel, R., 1977, Transfer of purified herpes virus thymidine kinase gene to cultured mouse cells, *Cell* **11**:223–232.

Wigler, M., Pellicer, A., Silverstein, S., and Axel, R., 1978, Transfer of single copy genes using total cellular DNA as donor, *Cell* **14**:725–731.

Wigler, M., Pellicer, A., Silverstein, S., Axel, R., Urlaub, G., and Chasin, L., 1979a, Transformation of the aprt locus in mammalian cells, *Proc. Natl. Acad. Sci. U.S.A.* **76**:1373–1376.

Wigler, M., Sweet, R., Sim, G. K., Wold, B., Pellicer, A., Lacy, E., Maniatis, T., Silverstein, S., and Axel, R., 1979b, Transformation of mammalian cells with genes from procaryotes and eucaryotes, *Cell* **16**:777–785.

Wigler, M., Perucho, M., Kurtz, D., Dana, S., Pellicer, A., Axel, R., and Silverstein, S., 1980, Transformation of mammalian cells with an amplifiable dominant-acting gene, *Proc. Natl. Acad. Sci. U.S.A.* **77**:3567–3570.

Wigler, M., Levy, D., and Perucho, M., 1981, The somatic replication of DNA methylation, *Cell* **24**:33–40.

Williamson, V. M., Bennetzen, J., Young, E. T., Nasmyth, K., and Hall, B. D., 1980, Isolation of the structural gene for alcohol dehydrogenase by genetic complementation in yeast. *Nature (London)* **283**:214–216.

Wong, G. G., Witek, J. S., Temple, P. A. Wilkens, K. M., Levy, A. C., Luxenberg, D. P., Jones, S. S., Brown, E. L., Kay, R. M., Orr, E. C., Shoemaker, C., Golde, D. W., Kaufman, R. J., Hewick, R. M., Wang, E. A., and Clark, S. C., 1985, Human GM-CSF: Molecular cloning of the complementary DNA and purification of the natural recombinant proteins, *Science* **228**:210–214.

Woychik R. P., Stewart, T. A., Davis, L. G., D'Eustachio, P., and Leder, P., 1985, An inherited limb deformity created by insertional mutagenesis in a transgenic mouse, *Nature* **318**:36–40.

Yokota, T., Arai, N., Lee, F., Rennick, D., Mosmann, T., and Arai, K., 1985, Use of a cDNA expression vector for isolation of mouse interleukin 2 cDNA clones: Expression of T-cell growth-factor activity after transfection of monkey cells, *Proc. Natl. Acad. Sci. U.S.A.* **82**:68–72.

Young, R. A., and Davis, R. W., 1983, Efficient isolation of genes by using antibody probes, *Proc. Natl. Acad. Sci. U.S.A.* **80**:1194–1198.

11

APPLICATIONS OF GENE TRANSFER IN THE ANALYSIS OF GENE AMPLIFICATION

G. M. Wahl, S. Carroll, P. Gaudray, J. Meinkoth, and J. Ruiz

1. INTRODUCTION

Gene amplification is a ubiquitous process that occurs under conditions in which elevated levels of specific gene products provide a cell with a growth advantage or are required for cellular differentiation to proceed. It has been shown to mediate the resistance of prokaryotes and eukaryotes to antiproliferative agents and adverse environmental conditions (for reviews, see Anderson and Roth, 1977; Biedler *et al.*, 1983; Cowell, 1982; Hamlin *et al.*, 1984; Schimke, 1984; Stark and Wahl, 1984) and to allow the accumulation of structural RNAs and specific proteins during development (e.g., see Hamlin *et al.*, 1984; Stark and Wahl, 1984). A role for gene amplification in the development or progression of malignancy has also been proposed to account for the frequent amplification and overexpression of protooncogenes in a variety of tumor cells grown *in vitro* or isolated from patients *in vivo* (for reviews, see Bishop, 1983; Schimke, 1984).

Molecular and cytogenetic analyses have revealed that the amplification process generally results in the coamplification of 150–3000 kilobases (kb) of DNA with the target gene (Cowell, 1982; Biedler *et al.*, 1983; Stark and Wahl, 1984). The coamplification of such large regions of DNA within a chromosome results in the appearance of expanded chromosomal regions (ECRs), which are often referred to as homogeneously staining regions (HSRs), since they often do not exhibit the typical pattern observed with the trypsin–Giemsa banding method. Alternatively, autonomously replicating acentric chromosomes called single or double minute chromosomes (SMs or DMs) are also frequently observed in some types of cells (for reviews, see Cowell, 1982; Biedler *et al.*, 1983; see

G. M. Wahl, S. Carroll, P. Gaudray, J. Meinkoth, and J. Ruiz • Gene Expression Laboratory, The Salk Institute, La Jolla, California 92037.

also Kaufman et al., 1979; Hamkalo et al., 1985). Several independent studies using different methods have shown that the DNA that flanks the target gene is often rearranged after multiple rounds of drug selection (e.g., see Ardeshir et al., 1983), although one report failed to find evidence of significant rearrangements over a 150-kb region in a Chinese hamster ovary cell line containing more than 1000 copies of the dihydrofolate reductase gene (Montoya-Zavala and Hamlin, 1985). It has also been observed that amplification can proceed from the site of the resident gene in some cases (Wahl et al., 1982, 1983), while in others, the selected gene is amplified at or near the site of a translocation (e.g., see Flintoff et al., 1984; Stallings et al., 1984).

Two of the most commonly accepted models hypothesize that at least two steps occur to generate amplified sequences. First, multiple copies of a DNA segment including the selected gene must be synthesized, possibly by multiple unscheduled rounds of DNA replication within one S phase (for reviews, see Hamlin et al., 1984; Schimke, 1984; Stark and Wahl, 1984). Second, these overreplicated sequences must be recombined to form either the linear array found in ECRs (HSRs) or the presumptive large circular molecules that comprise SMs and DMs (Hamkalo et al., 1985). The combined frequencies of these events in drug-selected cells (around 10^{-5}–10^{-6}) are generally far higher than those observed for spontaneous point mutations (typically 10^{-6}–10^{-8}). An analysis capable of measuring the frequency of the first duplication event suggests that the duplication frequency is substantially higher than 10^{-5} [around 10^{-3} in one example (Johnston et al., 1983)].

While the frequency of gene amplification is high in genetic terms, it is still too low to enable molecular studies to investigate the nature of the intermediates involved in the first step of the amplification process. Due to the large sizes and complex structures of amplified "units," it has been difficult to answer several important questions about the amplification process. For example, it has not been possible to assess whether there are discrete sequences within the unit that serve as the initiation and termination points, whether units are linked in precise tandem arrays, or which sequences participate in the recombination steps that resolve the putative overreplicated structure. It is also unclear whether a gradient of amplified sequences occurs in mammalian cells as it does in Drosophila, in which overreplication is known to mediate amplification (Spradling, 1981; Spradling and Mahowald, 1981; Osheim and Miller, 1983; Kalfayan et al., 1985). The structures of amplified regions are also generally deduced in cell lines with numerous copies of amplified DNA to facilitate the analysis. However, the generation of such lines requires multiple rounds of selection, and rearrangements that may be unrelated to the initiation or resolution events are likely to occur and further complicate the analysis. Sections 4 and 5 show how gene-transfer technology can be used to gain insight into some of these crucial issues.

Although the large size of amplified structures presents an obstacle to detailed molecular analysis, it has enabled coamplification of genetically engineered or cotransferred sequences to provide a method for the overproduction of products from genes that encode biologically important proteins. Progress in this research and the insights it has given into some of the factors that influence the fidelity of the amplification process are summarized in Sections 2 and 3.

2. SELECTION OF TRANSFORMANTS WITH AMPLIFIED SEQUENCES

2.1. Selection Strategies

The isolation of cell lines with highly amplified sequences generally requires multiple rounds of selection with increased stringency of selection at each step. One highly successful strategy for obtaining amplified lines has employed sequential selection with increasing concentrations of tight-binding and specific enzyme inhibitors. Two such inhibitors are methotrexate (MTX), an inhibitor of dihydrofolate reductase (DHFR) (Alt *et al.*, 1978), and N-(phosphonacetyl)-L-aspartate (PALA), an inhibitor of aspartate transcarbamylase (part of the trifunctional CAD protein that catalyzes the first three steps of *de novo* uridine biosynthesis (Kempe *et al.*, 1976; Wahl *et al.*, 1979). Resistance to a broad range of agents has now been shown to be mediated by gene amplification (e.g., see Tables 1 and 2 of Stark and Wahl, 1984). However, for some agents, the proportion of clones that resist the drug because of target-gene amplification is highly dependent on the protocol used for selection. For example, resistance to MTX can be achieved by at least two mechanisms other than amplification: (1) mutations that alter the binding of MTX to DHFR (Flintoff *et al.*, 1976; Haber *et al.*, 1981) and (2) decreased transport of MTX into the cell (Sirotnak *et al.*, 1981). Selection with high levels of MTX in one step generally isolates cells by either of these two mechanisms, presumably because a gene increase of sufficient magnitude to engender resistance to high drug concentrations in one step is highly unlikely. CAD gene amplification, however, is the only mechanism of resistance to PALA in mouse, Syrian hamster, rat, and two lines of Chinese hamster cells (Kempe *et al.*, 1976; Wahl *et al.*, 1979; J. Meinkoth and G. M. Wahl, in prep.). Thus, selection for resistance to high levels of PALA in one step sharply decreases the frequency of clones obtained, but the drug is resisted only by amplification of the CAD gene.

In a second selective strategy, target enzymes that are not essential for cell survival under normal conditions become essential in special media. For example, Kellems and collaborators used a medium that contained toxic concentrations of adenosine, alanosine (an inhibitor of *de novo* AMP biosynthesis, included to prevent selection of adenosine kinase mutants), deoxycoformycin (to inhibit adenosine deaminase), and uridine (to bypass the adenosine block of UMP biosynthesis) (Yeung *et al.*, 1983). In this medium, the cell must produce sufficient ADA to detoxify the adenosine (converting it to inosine) and to overcome inhibition by deoxycoformycin. Sequential selection of cells with increasing concentrations of deoxycoformycin (in the presence of adenosine, alanosine, and uridine) then yields mutant cells with up to a 10,000-fold overproduction of ADA (Yeung *et al.*, 1983; R. Kellems, personal communication). The availability of these lines has greatly facilitated the isolation of full-length ADA complementary DNA (cDNA) clones, and complementation of ADA-deficient bacteria by these clones has enabled the isolation of functional ADA cDNA clones (Yeung *et al.*, 1985).

A third strategy involves the selection of phenotypically normal revertants from cell lines that contain a defective endogenous or transfected gene. The defective genes that have been used encode either kinetically altered or thermosensitive enzymes, or they are expressed at a low level because of weak, down-regulated, or mutationally altered promoters. While reversion of such genes could be achieved by a second site mutation within the gene or within the transcriptional regulatory sequences, amplification of the defective gene has been shown to be the predominant mechanism in all cases analyzed thus far. Examples include the amplification of hypoxanthine-guanine phosphoribosyltransferase (HGPRT) genes that encode thermosensitive (Fuscoe *et al.*, 1983) or kinetically altered enzymes (Brennand *et al.*, 1982), of integrated simian virus 40 (SV40) genomes that encode a temperature-sensitive large tumor (large-T) antigen (Hiscott *et al.*, 1980), of transfected thymidine kinase (TK) (Roberts and Axel, 1982) or *v-scrc* (Glanville, 1985) genes that contain weak promoters, and of an altered endogenous DHFR gene in mouse cells (Haber *et al.*, 1981). A 10-fold (Fuscoe *et al.*, 1983) to 50-fold (Brennand *et al.*, 1982; Roberts and Axel, 1982) increase in gene number has been observed using the "phenotypic reversion" approach, but higher levels of amplification may be difficult to achieve, since it is difficult to regulate the stringency of this type of selection.

In a variation of the phenotypic reversion approach, Ringold and his colleagues (Chapman *et al.*, 1983) have constructed a recombinant with an *Escherichia coli* xanthine-guanine phosphoribosyltransferase gene (*xgprt* or *gpt*) under the control of the murine mammary tumor virus long terminal repeat (MMTV-LTR). In the presence of subinducing steroid concentrations, this gene is expressed at a very low level, and the cells can be made steroid-dependent for growth in an appropriate medium. Selection of cells to grow in the presence of mycophenolic acid (which inhibits the endogenous conversion of IMP to XMP) and limiting quantities of xanthine and steroids could therefore yield two types of mutants that could grow under these conditions: (1) those capable of inducing the transfected gene at low steroid concentrations due either to a promoter lesion in the MMTV-LTR or to an elevated level of expression of the unlinked endogenous glucocorticoid receptor or (2) amplification of the transfected genes resulting in sufficient *gpt* expression to allow for cell survival. Under these selective conditions, most of the cells that were isolated showed an increased copy number of the transfected *gpt* gene. However, two clones displayed a reproducible 2- to 3-fold increase in the specific activity of the glucocorticoid receptor. While neither the mechanism nor the physiological consequences of this small increase in the receptor were ascertained, the data suggest that it may be possible to tailor such an approach to achieve overproduction of unlinked genes for which direct selections cannot be easily designed.

2.2. Cloned Amplifiable Genes

Functional cloned genes for which convenient amplification selections are available are listed in Table I. Genomic clones of the DHFR and CAD genes have been isolated in cosmids, since they are more than 30 kb in length (Robert de Saint Vincent *et al.*, 1981; Milbrandt *et al.*, 1983a,b). Introduction of molecules this large can be achieved with high efficiencies by fusing bacterial protoplasts harboring the cosmids directly to recipient mammalian cells (Robert de Saint Vincent *et al.*, 1981; Milbrandt *et al.*, 1983a,b) or by the calcium phosphate coprecipitation method (e.g., see Wahl *et al.*, 1984a,b). While it is difficult to link other sequences to these genes using *in vitro* molecular cloning methods due to their large size and consequent lack of unique restriction sites, it is possible to link sequences together *in vivo* because of the highly efficient mammalian-cell recombination machinery (e.g., see Folger *et al.*, 1982; Robert de Saint Vincent and Wahl, 1983; Kucherlapati *et al.*, 1984; Rubnitz and Subramani, 1984). These genes are generally introduced into mutant cells that are DHFR-deficient [DXB11 (Urlaub and Chasin, 1980) or DG44, a CHO-Pro3 cell with a double deletion of the DHFR locus that is similar to other DG cells described by Urlaub *et al.*, (1983)] or CAD-deficient [Urd-A (Patterson and Carnright, 1977)]. Transformants that express these genes are readily identified by their ability to grow in a medium that lacks thymidine, glycine, and hypoxanthine (for DHFR) or uridine (for CAD). However, one can also introduce these genes into wild-type cells and select for transformants that resist higher levels of MTX or PALA than can be achieved by amplification of the endogenous gene in one step. This results in the isolation of transformants with many copies of the donated sequences, generally arrayed as a head-to-tail tandem repeat. Simultaneous introduction of two different plasmids (or cosmids) can be achieved either by mixing bacterial protoplasts containing the desired molecules prior to fusion with the recipient animal cell or by fusing protoplasts that harbor both molecules (Wahl *et al.*, 1983). Alternatively, cointroduction can be achieved by the calcium phosphate transfection procedure by mixing a high ratio of the unselected gene with the selected amplifiable gene prior to precipitation. These procedures generally result in the introduction of multiple copies of both sequences, usually at one chromosomal site, although multiple integrations are seen occasionally. It should be possible to introduce many copies of both the unselected and selected

TABLE I. Amplifiable Markers

Selectable marker	Size of selectable gene (kb)	Selection	Amplification level[a]	Reference
I. DHFR				
A. Genomic clones				
1. Chinese hamster	24.5	MTX[b]	500	Milbrandt et al. (1983a,b)
2. Syrian hamster	≥10.5[c]	MTX	?	Schoner and Littlefield (1981)
3. Mouse	31.5	MTX	?	Crouse et al. (1982)
B. Modular cDNA clones				
1. MTX[r]	≈1.2	High MTX concentration[b]	50–100	Simonsen and Levenson (1983)
2. Wild-type	21.2–3.5	MTX[b]	1000–2000	Subramani et al. (1981) Crouse et al. (1983) Murray et al. (1983)
II. CAD				
A. Genomic clone	≈35	PALA[b]	250	Robert de Saint Vincent et al. (1981) Wahl et al. (1984a,b)
B. Modular gene (E. coli pyr B)	1.3	PALA[b?]	?	J. Ruiz and G. M. Wahl (1986)
III. ADA (Modular gene)	1.1 (human) 1.1 (mouse)	11AAU + dCF[b,d]	10,000 (for endogenous gene)	Orkin et al. (1985)
IV. TK (viral-HSV)	≈3.8	AAT[e]	50	Yeung et al. (1983, 1985)
V. GPT (E. coli)	≈1.2	Limiting xanthine + mycophenolic acid	10	Roberts and Axel (1982) Chapman et al. (1983)

[a] Maximum reported to date.
[b] Selection can be made dominant.
[c] Not tested for function; the complete gene may be larger.
[d] 11AAU = 1.1 mM adenosine, 50 μM alanosine, 1 mM uridine, 2′ deoxycoformycin (variable concentration).
[e] AAT = 15 μg/ml adenine, 1μg/ml amethopterin, 5 μg/ml thymidine

genes by applying a stringent selection at the time of gene transfer. In this case, a high copy number of the unselected gene should be attainable without employing multiple steps of selection. As described below, such selections often result in the rearrangement or loss, or both, of the unselected sequence.

An alternative to using large genomic clones is to employ full-length cDNA clones and attach to them the required 5' transcriptional start and 3' polyadenylation signals. Several strategies have been used to prepare functional DHFR cDNA clones. For example, recombination in bacteria between two λ phage clones containing overlapping genomic and cDNA sequences was used to produce hybrid cDNA–genomic clones with a natural DHFR promoter and 1.5 kb of 5' flanking DNA with or without the first exon (Crouse et al., 1983). While the exon-containing and exon-deficient genes efficiently transform DHFR-Chinese hamster ovary (CHO) cells to the DHFR$^+$ phenotype, all transformants had lower than wild-type DHFR specific activities even though they contained multiple copies of the donated recombinant genes. The lack of essential 3' sequences required for efficient processing and polyadenylation in these genes may account in part for their poor expression (Crouse et al., 1983). On the other hand, it is likely that sequences flanking the donated minigenes can also have profound effects on their expression and cause the same construct to be expressed at different levels in independent transformants. For example, transformants with 5 copies of one integrated minigene have the same level of expression as other transformants with 300 copies of the same minigene (Crouse et al., 1983). Other functional modular DHFR (Gasser et al., 1982; Subramani et al., 1981; Murray et al., 1983; Kaufman and Sharp, 1982; Kaufman et al., 1985) and ADA (Orkin et al., 1985; Yeung et al., 1985) cDNA clones have been placed under the control of a variety of higher-efficiency viral promoters and 3'-polyadenylation signals to achieve higher levels of expression. While detailed data are not available for comparisons of the different constructs, it is generally agreed that the modular DHFR genes are expressed less efficiently on a per-gene basis than are the endogenous CHO or mouse DHFR genes. As described in Section 3, inefficient expression can also result from factors other than those related to transcription and correct processing. To obtain the level of gene expression required for cell growth under selective conditions, multiple tandem copies are often integrated at a single chromosomal site (Crouse et al., 1983; Kaufman et al., 1983).

It is interesting that the number of integrated modular DHFR genes is low when carrier DNA is not used in the transformation (8 of 8 in one study had fewer than 5 copies), but can range between fewer than 5 and more than 400 copies of the same modular genes (in the same study) when carrier DNA is used (Crouse et al., 1983). For unclear reasons, genes within tandem arrays are often expressed inefficiently, and those most poorly expressed may be selected against and lost after gene amplification (Crouse et al., 1983; Southern and Berg, 1982).

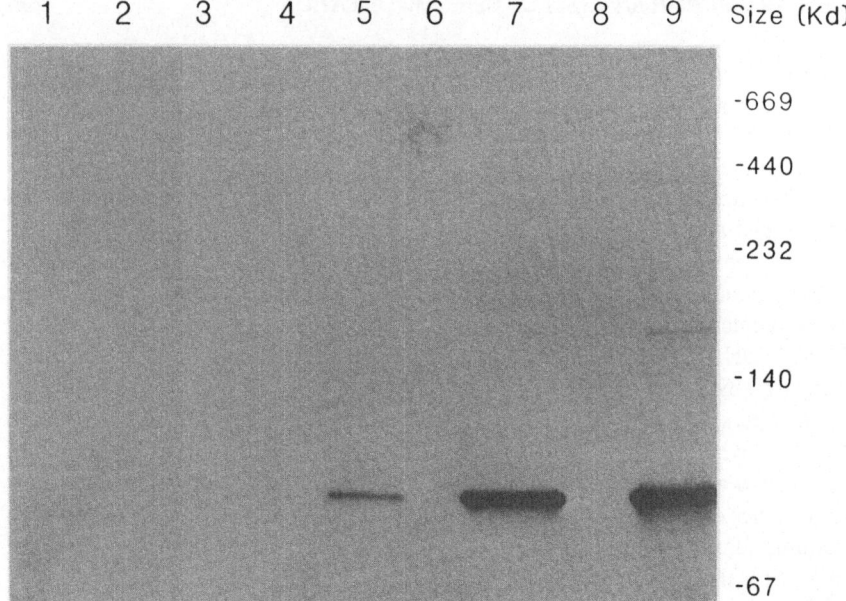

| 1 | 2 | 3 | 4 | 5 | 6 | 7 | 8 | 9 | Size (Kd) |

-669

-440

-232

-140

-67

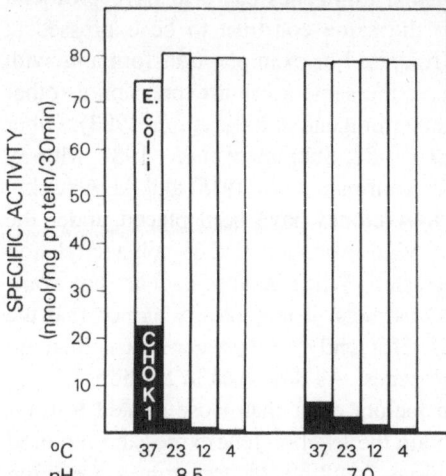

FIGURE 1. Determination of *E. coli* ATCase activity in mammalian cells. (Above) *In situ* ATCase gel assay. Cell-free extract (20 μg protein) was fractionated in a 7.5% acrylamide gel under nondenaturing conditions. The gel was then incubated in a solution containing the ATCase substrates carbamyl phosphate and aspartate. The location of the active ATCase enzyme was visualized by successive immersion of the gel in $Pb(NO_3)_2$ and $(NH_4)_2S$, which precipitate *in situ* the phosphate liberated by the ATCase reaction. Lanes 1–3 and 5 are G418-resistant D20 transformants containing the following recombinant plasmids: (1) pSV2* neo-1 (*neo* transcriptional unit from pSV2neo inserted into pBR-based vector pML); (2) LTRpyrBneo-4 (LTRpyrBneo is *pyrB* placed under transcriptional control of Mo-MSV promoter but without a eukaryotic polyadenylation signal); (3) pSV2pyrBneo-2 (pSV2pyrBneo is *pyrB* inserted into standard pSV2 expression vector); (4) D20 parent line; (5) LTRpyrBneo′-2 (LTRpyrBneo′ is *pyrB* under Mo-MSV transcriptional control and including the SV40 late polyadenylation signal); (6) B5-4 (a Syrian hamster line containing a 100-fold amplification of the CAD gene); (7) *E. coli* strain HS1054 (0.5 μg cell-free extract of this *pyrB*-containing strain); (8) CHO-K1 (the wild-type CHO cell line that was the grandparent of D20); (9) *E. coli* strain HS1054 (2.5 μg cell-free extract). In lane 3 (pSV2pyrBneo-2), the intensity of the ATCase activity band is approximately 10-fold less than that of LTRpyrBneo-2 (lane 5). (Above left) Quantitative ATCase assay: Temperature and pH profiles of *E. coli pyrB* and CHO-K1 ATCase enzymes. ATCase activity was measured in crude extracts as described by Patterson and Carnright (1977). Specific activity values were calculated by quantitating the conversion of [^{14}C]aspartate to carbamyl-[^{14}C]aspartate.

A minigene that engenders PALA resistance has also been constructed (Ruiz and Wahl, 1986), but this required a far different strategy than employed above since it has not been possible to prepare full-length cDNA clones to the 7.9-kb CAD messenger RNA (mRNA) (Padgett *et al.*, 1982; Shigesada *et al.*, 1985), and the numerous small exons and introns in the CAD gene preclude the use of the *in vivo* recombination method used to prepare DHFR minigenes (Crouse *et al.*, 1983). The target of PALA is the aspartate transcarbamylase (ATCase) activity of the trifunctional CAD protein. Since bacterial ATCase also binds PALA (Collins and Stark, 1971), we reasoned that expression of bacterial ATCase in mammalian cells might lead to PALA resistance in a fashion similar to the overproduction of the endogenous CAD protein. To test this idea, the *E. coli pyrB* gene (which encodes the ATCase catalytic subunit) (Pauza *et al.*, 1982) was placed under the control of either an SV40 or Moloney murine sarcoma virus (Mo-MSV) LTR and an SV40 polyadenylation signal. Since it was unclear whether the bacterial gene would function, the dominant selectable neomycin phosphoribosyltransferase gene (*neo*) was inserted into this vector so that transformants containing a single copy of the recombinant molecule could be isolated on the basis of their resistance to the drug G418 (Southern and Berg, 1982). CHO cells, that are highly sensitive to PALA because of a defect in the ATCase function of the CAD protein [D20 (Patterson and Carnright, 1977)] were transformed with the *pyrB-neo* recombinant, selected to resist G418, and then analyzed for PALA resistance. In contrast to the parental D20 cells, which are sensitive to 0.1–0.2 μm PALA [the PALA dose that killed 50% of the cells (LD_{50})], six independently isolated transformants containing the *pyrB* gene displayed a range of LD_{50}'s from 0.5 to 10.0 μm PALA. Figure 1A shows an assay in which cell extracts prepared from transformants containing the *pyrB* gene or wild-type D20 cells were electrophoresed through a nondenaturing gel and the ATCase activity was detected *in situ* in the gel. It is clear from this figure that the protein produced in the *pyrB* gene transformants migrates at the same position as the *E. coli* ATCase. Interestingly, this analysis indicated that the active form of the ATCase catalytic subunit is a trimer in mammalian cells as it is in bacteria. Figure 1B shows that this enzyme can be analyzed in the presence of a functional mammalian CAD protein, since the pH and temperature optima for the bacterial ATCase are very different from those of the mammalian CAD protein. The latter assay provides a rapid and convenient method with approximately the same sensitivity as the widely used chloramphenicol transacetylase assay (Gorman *et al.*, 1982). Figure 2 shows that a typical transformant containing 2–4 copies of the *pyrB* gene resists PALA through the amplification of the donated genes. The novel fragments seen in the Southern blot in Fig. 2 indicate that some rearrangements occurred during gene transfer and integration, but that these rearrangements were included in the amplified unit, since their abundance increased in parallel with that of the selected sequences. The maximal level of amplification that can be attained with the *pyrB* gene has yet to be determined.

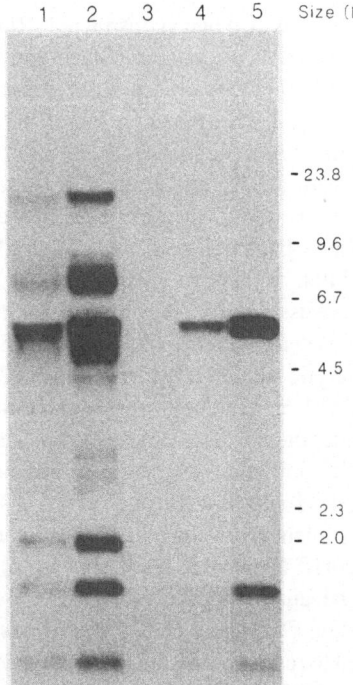

FIGURE 2. Analysis of restriction patterns in a *pyrB* transformant of D20 cells and a PALA-resistant derivative. DNA from the indicated cell lines was digested with endonuclease PstI, fractionated in a 0.7% agarose gel, blotted to nitrocellulose, and hybridized with ^{32}P-labeled pSV2pyrB-neo2. (1) G418-resistant transformant pSV2pyrBneo-2; (2) PALA-resistant derivative pSV2pyrBneo-2 S1-3 (resists 60 μM PALA); (3) D20 parent, (4, 5) a reconstruction experiment containing approximately 1 copy and 10 copies of PstI-restricted pSV2pyrBneo plasmid, respectively.

3. COAMPLIFICATION OF LINKED GENES

The production of biological products from mammalian cells that require specific *in vivo* modifications for biological activity is clearly done best in mammalian cells, since qualitatively and quantitatively equivalent processing cannot generally be achieved in prokaryotes or yeast. Amplification of the genes that encode such proteins in concert with a selectable marker is one means by which large amounts of these products can be produced economically in mammalian cells. In addition to providing a useful tool for production, experiments that have utilized gene coamplification have also provided insight into the molecular biology of the amplification process.

As described in Section 1, amplified units are generally 150–3000 kb long, and functional genes situated within the amplification boundaries are coamplified and generally overexpressed in parallel with the selected gene (Padgett *et al.*, 1982; Debatisse *et al.*, 1984; Wahl *et al.*, 1983). It has been possible to introduce random chromosomal fragments in the presence of pure DNA containing a marker gene of interest and obtain large (up to 2000 kb) DNA insertions consisting of a patchwork arrangement of the introduced chromosomal sequences ligated to the marker gene (Perucho *et al.*, 1980; Roberts and Axel, 1982). When the chromosomal DNA contains a dihydrofolate reductase (DHFR) gene, one can obtain transformants in which the marker gene is ligated sufficiently close to the DHFR gene that both can be coamplified after selection with methotrexate (MTX) (Perucho *et al.*, 1980; Christman *et al.*, 1982). This approach first employed chromosomal DNA obtained from a cell line with a mutant DHFR gene (cell line A29) that encodes a protein that binds MTX inefficiently and thus provides a dominant selectable marker enabling introduction of this gene into cells containing a functional DHFR gene. While the overproduction of hepatitis B surface antigen has been achieved by this approach (Christman *et al.*, 1982), it is no longer used frequently, since the inserted array of molecules can be unstable or prone to frequent rearrangement, and in only a proportion of the transformants obtained are the marker and selectable genes sufficiently close to allow for coamplification (Perucho *et al.*, 1980). However, in addition to demonstrating the feasibility of coamplification for overproducing proteins, this approach also substantiated the claims made from cytogenetic observations that amplified regions could span 1000 kb or more of DNA (e.g., see Cowell, 1982; Biedler *et al.*, 1983; Hamlin *et al.*, 1984; Stark and Wahl, 1984).

The coamplification of unlinked genes following transfection of random chromosomal fragments has now given way to the use of plasmids in which the selected and marker genes are linked *in vitro* by molecular cloning techniques or *in vivo* through the host-cell recombination machinery (Scahill *et al.*, 1983; Kaufman *et al.*, 1985). If the *in vivo* recombination approach is used, the molar ratio of selected to unselected sequences will determine the probability of obtaining transformants in which both sequences are linked (Scahill *et al.*, 1983; Kaufman *et al.*, 1985). While this ratio must be determined empirically in each case, there is a strategy that improves the likelihood of isolating transformants that express both sequences. In this approach, an enhancer is placed on the unselected sequence, and the selected gene is put under the control of an enhancer-dependent promoter (Kaufman *et al.*, 1985). Since the enhancer should function independently of orientation and should be relatively insensitive to the distance from the promoter, *in vivo* recombination events should frequently result in the construction of molecular aggregates in which both genes will be functional. However, such aggregates generally have rearranged sequences, and the distances separating the selected gene and the marker gene cannot be controlled (Kaufman *et al.*, 1985).

The selectable gene employed most frequently has been the DHFR gene, since several functional minigenes have been constructed and mutant Chinese hamster ovary (CHO) cells that are DHFR-deficient have been isolated (Urlaub and Chasin, 1980; Urlaub *et al.*, 1983). The availability of such cells enables a straightforward selection of transformants expressing the donated sequences by isolating clones that survive in the absence of hypoxanthine, thymidine, and glycine. Sequential selection of transformants containing the donated DHFR genes with increasing concentrations of MTX can yield clones with up to a 2000-fold amplification of the introduced DHFR genes and linked marker genes. In some cases, there is an equivalent overproduction of the product encoded by the cotransforming and DHFR genes (see Table II). For example, coamplified human α- and γ-interferons (Haynes and Weissman, 1983; Scahill *et al.*, 1983) have constitutive levels of expression in CHO cells that equal or exceed those attained in the most proficient human cells. In a CHO cell line with coamplified genomic β-interferon genes that can be induced either by polyriboinosinic acid–polyribocytidilic acid or by high cell density in CHO cells, the level of production from the induced amplified genes can be up to 300 times that seen in human fibroblasts (McCormick *et al.*, 1984). Similarly, human tissue-type plasminogen activator represents 5% of the total pulse-labeled protein and 50% of the secreted protein in some CHO cell clones in which it is coamplified over 100-fold with the DHFR gene (Kaufman *et al.*, 1985). In each of these cases, the human proteins have been shown to have biological activities roughly equivalent to those derived from the native sources. However, it is likely that the modifications made in the CHO cells are not identical to those made in the human cells (e.g., see Kaufman *et al.*, 1985). The current data indicate that in this diverse array of proteins, the alternative modifications do not significantly affect their biological activities. Furthermore, it is apparent that the CHO-cell processing machinery is not limiting even when foreign proteins reproduced at such high levels. In addition, while there have been reports that an amplified metallothionein 1 ($Mt - 1$) gene can no longer be regulated by glucocorticoids (Mayo and Palmiter, 1982), the studies described above indicate that loss of gene regulation is not a direct consequence of gene amplification.

In contrast to the preceding examples are cases in which a functional gene is amplified and expressed but no protein is made or in which gene rearrangements occur on amplification. An example of high-level transcription with low-level protein production was observed in the same DHFR⁻ CHO cells as used in the successful coamplification experiments described above. Clones that were selected to contain 500–1000 copies of the DHFR gene and a coamplified human α-globin gene (Lau *et al.*, 1984) produced copious amounts of α-globin mRNA that could be translated into α-globin protein *in vitro*, but no α-globin was synthesized *in vivo*. Whether this reflects an instability of the α-globin proteins in CHO cells, unusual codon usage in the α-globin mRNA and a lack of the corresponding transfer RNA in CHO cells, or other factors has not been determined. However, this example emphasizes that the capacity for a protein to be produced in a heterologous system must be tested empirically. It has also been found that in two clones transfected with the same construct containing a DHFR gene and a functional simian virus 40 (SV40) early region, some amplified lines express large tumor (large-T) antigen but little or no small tumor (small-t) antigen, while others do the opposite, even though the genes that encode these two proteins overlap (Kaufman and Sharp, 1982). Furthermore, large-T antigen production

TABLE II. Gene Product Overexpression and Gene Rearrangements in Coamplified Sequences

Selected gene[a]	Coamplified gene[b]	Fold overproduction			Stability[c]	Rearrangements on insertion/ amplification[c]	Clonal variability of amplification/ gene product[c]	Reference
		Selected gene DNA/RNA[c]	Coamplified DNA/RNA[c]	Coamplified protein[d]				
DHFR	SV40 T, t	100–500/ N.D.	100–50/ N.D.	t = 1% of TSP without alteration; t = 10% of TSP with rearrangement[e]	N.D.	+/+	+/+	Kaufman and Sharp (1982)
DHFR (pAdD26)	Human α-, γ-IFN[f]	5–10/20–80	20–50/12–80	$0.2–1 \times 10^5$ U/ml per day α- or γ-IFN	V	−/−	N.D./N.D.	Haynes and Weissmann (1983)
DHFR [padD26 SV(A)-3]	Human γ-IFN[g]	N.D./N.D.	N.D./N.D.	0.5×10^5 U/ml per day 6× overproduction	V	N.D./N.D	N.D./+	Scahill et al. (1983)
DHFR (pSVM)	Human IFN (genomic clone)	25/N.D.	25/25	10^7 U/ml per day (induced) 5 × 10 U/ml per day (uninduced)	N.D.	+/N.D.	N.D./N.D.	McCormick et al. (1984)
DHFR (pSV2)	α-Globin; α-Globin + Mt-1 promoter	500–1000/ N.D.	500–1000/500–1000	0 (mRNA transcribed but no α-globin detected)	S	−/+	N.D./N.D.	Lau et al. (1984)
DHFR [padD2 SV(A)3]	Human t-PA	100/100	100/100	5% TSP (pulse label); 50% secreted protein	V	+/+	+/+	Kaufman et al. (1985)

[a] The designations in parentheses indicate the modular DHFR gene used.
[b] (IFN) Interferon; (t-PA) tissue-type plasminogen activator.
[c] (N.D.) No data; (V) variable; (S) stable.
[d] (TSP) Total soluble protein.
[e] t Antigen produced was truncated.
[f] Genes introduced on one plasmid.
[g] Genes introduced by cotransformation.

was maximal in the one cell line analyzed, which contained about 100 copies of the transfected DNA. Further selection of these cells to resist higher concentrations of methotrexate yielded cells with 500 copies of the DHFR gene, but no increased synthesis of large-T protein. This was apparently due to a specific loss of SV40 DNA sequences during the amplification process. In a similar fashion, cells selected to contain approximately 500 DHFR genes suffered a rearrangement in the small-t antigen coding sequences, resulting in the production of 10% of the soluble protein as a truncated version of the small-t antigen (Kaufman and Sharp, 1982). These data suggest that these proteins are likely to be cytotoxic when overproduced at such high levels and that rearrangements that either delete some of the sequences or produce inactive gene products are selected for during the amplification process. The overproduction of other proteins such as that encoded by the *E. coli gpt* gene has also been reported to be cytotoxic when produced at high levels (Ringold *et al.*, 1981). Placement of such genes under the control of inducible promoters may enable the isolation of cell lines with higher levels of amplification of unrearranged genes, since the synthesis of the cytotoxic proteins could be regulated more precisely.

The ability to obtain amplification of an unselected gene varies among subclones derived from the same parent and among different parental clones (e.g., see Kaufman *et al.*, 1985). The latter observation probably reflects differences in the arrangement of the inserted sequences, the proximity of the marker to the selected gene, and influences of the different chromosomal environments into which the genes have inserted in different transformants. Differences in coamplification in subclones derived from a single parent are somewhat surprising, since *in situ* hybridization has shown that in amplified CHO-cell transformants, the selected and unselected genes generally reside together in large expanded chromosomal regions (Lau *et al.*, 1984; Kaufman *et al.*, 1983). The large sizes of the amplified regions indicate that the chromosomal sequences flanking the inserted genes must have been involved in the amplification process in these lines. However, the complex DNA blotting patterns of many of the parental transformants indicate that they contain multiple copies of the donated sequences and that some of these copies were rearranged upon integration. Upon selection for amplification, clones are often obtained in which a subset of these bands are increased in abundance and only a fraction of the clones overexpress the unselected marker (Kaufman *et al.*, 1985). It is possible that certain sequence arrangements within the donated array provide favored sites for rearrangements during amplification, or they may participate directly in the amplification process along with flanking chromosomal sequences. Recombination at such sites may disassociate the selected and unselected genes from one another in some lines, while the linkage may be preserved in others. Amplification mediated in part by recombination between plasmid sequences repeated in an array of human clones inserted into a mouse chromosome provides one precedent for the direct participation of the donated DNA in the amplification processes (Roberts and Axel, 1982; Roberts *et al.*, 1983).

Independent clones can also differ over several orders of magnitude in their expression of the unselected gene. Thus, it is essential to screen for clones that not only coamplify this gene, but also overexpress it to an acceptable level. Since in some cases the number of clones that coamplify the unselected gene can be low, it has been necessary to derive an alternative to the arduous procedure of screening individual clones for their levels of expression. One successful strategy has been to identify pools of clones that amplify the selected gene and then screen the different pools of amplified lines for those pools in which overproduction of the unselected product occurs at the highest level (Kaufman et al., 1985). This method screens for those clones that contain either the most favorable sequence arrangements for amplification or an integration site that favors amplification at a high frequency (see Section 4). Since pools with the highest activity for the unselected gene are likely to have the highest proportion of cells in which the unselected gene is overexpressed at the highest level, the task of identifying and purifying subclones is greatly facilitated. Subcloning is essential in such cases, since independent clones picked from such pools have large differences in both the degree of overexpression and the stability of the amplified sequences (Kaufman et al., 1985).

4. POSITION EFFECTS IN GENE AMPLIFICATION

The production of multiple chromosomal or extrachromosomal copies of a DNA sequence could occur by a variety of mechanisms such as: (1) unequal sister chromatid exchange, (2) unscheduled rounds of DNA synthesis within a single S phase followed by recombination within the overreplicated region, and (3) replicative transposition (for reviews of the details of the models, see, for example, Hamlin et al., 1984; Stark and Wahl, 1984; Schimke, 1984). It is reasonable to expect, therefore, that the position of a gene relative to preferred sites for recombination, to replication origins, or to transposable elements could have profound effects on the frequency and/or the cytogenetic consequences of its amplification. The importance of gene position in gene amplification has been directly demonstrated in bacteria, in which the frequency of spontaneous duplication of different regions of the *Salmonella* chromosome was observed to vary over more than three orders of magnitude (Anderson and Roth, 1981). The frequency differences were closely correlated with the position of the selected gene relative to flanking sequences that participate in homologous unequal sister chromatid exchanges.

To investigate the role of gene position in gene amplification in mammalian cells, it is necessary to analyze the amplification of the same gene in different chromosomal environments in a single cell line. In one study, a collection of random but molecularly characterized human clones were cotransfected with an adenine phosphoribosyltransferase (APRT) gene and a promoter-defective thymidine kinase (TK) gene into APRT⁻ TK⁻ mouse L cells. APRT⁺ transformants were then selected for amplification by isolating cells with elevated TK expression. In this case, the amplifiable marker (TK) was embedded in defined carrier DNA so the consequences of amplification could be defined precisely at the molecular level. Each transformant has a unique organization of human DNA flanking the APRT and TK genes. Since independently isolated clones differ in the organization and content of this array as well as in the chromosomal location into which this large molecular patchwork is inserted, it is not surprising that different clones amplify the inserted sequences at rates that differ by more than 100-fold. Interestingly, after amplification, each of eight subclones derived from a single parental clone elicits a unique molecular fingerprint consisting of a common core sequence amplified along with different lengths (40–200 kb) of flanking human DNA. One inference from this result is that the core region contains sequences capable of initiating the amplification process. This result is strikingly similar to the observation in prokaryotes that different amounts of flanking DNA can be coamplified with the same marker gene in independently isolated clones (e.g., see Tlsty *et al.*, 1984). Similar observations have been made for the amplification of endogenous eukaryotic genes (G. Buttin, personal communication; Ardeshir *et al.*, 1983). These results suggest that while there may be alternative sequences that can be used for the termini of amplification units, once the choice is made in the first step of amplification, the same sequences are generally used in succeeding steps. The mechanisms that govern the preferential use of some sites to the exclusion of alternative sites after the first amplification step have not been elucidated.

An analysis of the molar ratios of the coamplified DNA within the amplified domain of one clone in the APRT–TK system was consistent with the existence of an amplification gradient generated by multiple rounds of unscheduled replication, followed by recombination between the plasmid sequences that are the most common repeated element in the array (Roberts *et al.*, 1983). Unscheduled DNA replication has been shown to be the mechanism by which chorion genes are amplified during cogenesis in *Drosophila* (see also deCicco and Spradling, 1984) and has been proposed as a mechanism for the amplification of endogenous genes in mammalian cells (for reviews, see Hamlin *et al.*, 1984; Schimke, 1984; Stark and Wahl, 1984). The DNA within the amplified APRT–TK array is apparently subject to a gene-correction mechanism, since lines with amplified defective APRT genes can be isolated in a single step from cells with amplified wild-type APRT genes at the same frequency at which lines with a single APRT⁺ TK⁺ gene yield APRT⁻ TK⁺ colonies. The frequency change from an amplified APRT⁺ TK⁺ to an APRT⁻ TK⁺ phenotype ranged from 3×10^{-4} to 4×10^{-7} in 12 independent clones that were analyzed. These frequencies are equal to or higher than those observed for gene conversion in other mammalian systems (Liskay and Stachelik, 1983), even though in the APRT–TK system the APRT⁺ gene must first be mutated to APRT⁻ and this change must then be transmitted through the other amplified APRT⁺ genes. It is unclear whether it is the random array of chromosomal sequences in these cell lines that causes this unexpected behavior, and it remains to be established whether the amplification mechanisms deduced from this system will prove to be generally applicable.

The introduction of cloned amplifiable marker genes into apparently random chromosomal locations by DNA transfection in the *absence* of carrier DNA provides another method by which to investigate the influence of gene position on gene amplification (Robert de Saint Vincent *et al.*, 1981; Kaufman and Sharp, 1982; Wahl *et al.*, 1984b). Data from several studies using this approach are consistent with the interpretation that gene position can profoundly influence amplification frequency. For example, when a functional genomic clone of the CAD gene was inserted into the chromosomes of CAD-deficient [Urd⁻A (Patterson and Carnright, 1977)] mutants of Chinese hamster ovary (CHO) cells (Robert de Saint Vincent *et al.*, 1981; Wahl *et al.*, 1984b), one of four transformants was observed to amplify the inserted sequences at a frequency (Wahl *et al.*, 1984b) and rate (G. N. Proctor and G. M. Wahl, unpublished observations) that are at least 1000 times those of the other transformants or of the endogenous gene in wild-type cells. Since the dihydrofolate reductase (DHFR) gene in this transformant (called T5) is amplified at the same frequency as it is in wild-type cells, the high-frequency-amplification phenotype of T5 is not due to a genetic alteration that enables other genes to be amplified at a high frequency. A more likely explanation is that high-frequency amplification in T5 is due either to the organization of the inserted sequences or to an influence of their local genomic environment.

Two observations lead us to the interpretation that high-frequency amplification in T5 is related to the position of integration of the donated sequences. First, the inserted sequences in all the transformants we have analyzed share a common molecular organization consisting of a core of tandemly repeated CAD genes flanked by variable amounts of rearranged CAD sequences. Thus, there are no obvious characteristics of the molecular structure of T5 that could account for the 1000-fold difference in amplification frequency. Second, the gel electrophoretic method described in Fig. 3 has enabled us to determine that the capacity for high-frequency amplification in T5 is related to the production of an extrachromosomal molecule containing the donated CAD sequences (S. Carroll, P. Gaudray, M. L. DeRose, E. Nakkim, and G. M. Wahl, in preparation) (see Fig. 4). Importantly, this molecule could not be detected by the Hirt lysate procedure (Hirt, 1967), perhaps because of its large size or due to a particular characteristic of its molecular structure. We infer that the production of this molecule is related to the high-frequency-amplification phenotype from three observations: (1) Other transformants in which the CAD sequences are amplified at a low frequency do not produce such a molecule; (2) subclones of T5 amplify at different frequencies, and the subclone that has the highest amplification frequency produces the highest quantity of the extrachromosomal molecule; and (3) one *N*-(phosphonoacetyl)-L-aspartate (PALA)-resistant derivative of T5 that has amplified the donated sequences does not amplify these sequences further

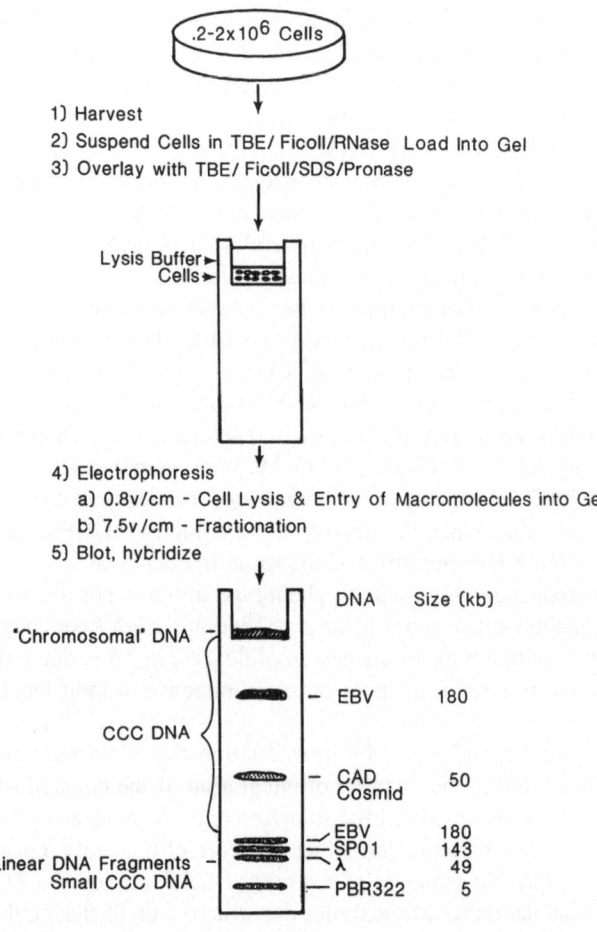

FIGURE 3. A simple gel electrophoresis method to detect high-molecular-weight DNA. Gels were run essentially as described (Eckhardt, 1978; Gardella *et al.*, 1984). Modifications of this technique allow the use of as few as 5×10^4 cells to detect extrachromosomal DNA in T5. Additional information regarding these modifications and the migration of various size markers will be published in S. Carroll, P. Gaudray, M. L. DeRose, E. Nakkim, and G. M. Wahl (in preparation).

at a high frequency, and it also fails to synthesize any of the extrachromosomal molecule. Since *in situ* hybridization has consistently shown integration of CAD cosmids following DNA transfection (Robert de St. Vincent *et al.*, 1981; Wahl *et al.*, 1984a,b) and an analysis of an additional 17 CAD cosmid transformants has not shown any containing extrachromosomal CAD carrying molecules, we propose that the generation of extrachromosomal CAD sequences in T5 derives either from the insertion of the CAD cosmid near a replication origin or from the destabilization of a replication domain as a consequence of insertion of the

200-kb tandem array of CAD sequences. Rereplication in this region could generate extrachromosomal molecules by recombination within the overreplicated region at homologous sequences in the tandem array or by other recombination processes (e.g., see Schimke, 1984). As discussed later in this section, such rereplication could be triggered by DNA damage introduced by the selective agent itself (e.g., see Varshavsky, 1981a,b). Since the donated sequences in T5 appear to be localized to a single chromosome (Wahl *et al.*, 1984b) and the amplified gene copies in two PALA-resistant derivatives of T5 are also chromosomal, we infer either that the extrachromosomal molecules can be reintegrated or that the two amplified lines we have studied were generated by an intrachromosomal expansion of the donated sequences. Analysis of the fate of the extrachromosomal molecules during PALA selection will help to resolve this question. Transient production of extrachromosomal molecules followed by rapid reintegration has previously been postulated as a mechanism of gene amplification in CHO cells (Kaufman and Schimke, 1981).

The migration of the extrachromosomal molecule in the gel system described in Fig. 3 is very similar to that of the 208-kb covalently closed circular plasmid of *Agrobacterium tumefaciens* (see Figs. 3 and 4) and of the covalently closed circular molecule found in some Epstein–Barr-virus-infected cells (Gardella *et*

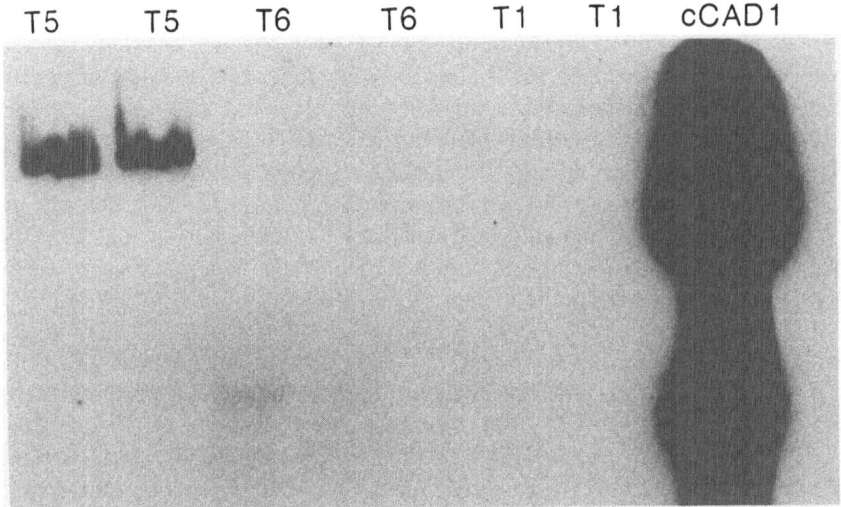

FIGURE 4. Demonstration of the presence of an extrachromsomal molecule in T5. Duplicate samples of 10^6 cells from T5 and from two transformants, T1 and T6, which do not exhibit the hyperamplification phenotype, were run on a vertical Eckhardt gel as described in Fig. 3. To transfer high-molecular-weight DNA efficiently to nitrocellulose, the gel was first treated for 10 min in 0.25 N HCl. A Southern blot was hybridized in the presence of dextran sulfate for 16 hr with 2×10^6 cpm/ml of a nick-translated fragment specific for the 3′ end of the CAD gene. Similar results are also obtained using a probe specific for the 5′ end of the gene. A 52-kb CAD cosmid was run as a size marker. The blot was exposed to Kodak XAR-5 film for 48 hr at $-70°C$.

al., 1984). However, the extrachromosomal molecule in T5 either is not likely
to be covalently closed or is not composed entirely of DNA, since it is quantitatively converted to a lower-molecular-weight species with an approximate
size of 40–80 kb upon alkaline lysis. A reanalysis of this experiment with
mitochondrial DNA probes and visualization of the ethidium-bromide-stained
gel showed that the alkaline lysis procedure did not disrupt the *Agrobacterium*
plasmids or large circular mitochondrial molecules (P. Gaudray, unpublished
data). We infer that the smaller molecules are at least partially double-stranded
after neutralization of the alkaline lysate, since they can be cleaved with restriction enzymes. A model that is consistent with these observations is that the
apparently large extrachromosomal molecule is a replication intermediate consisting of intertwined circles with nicked daughter strands. Alkaline treatment
would denature the nicked strands, leaving two complementary intertwined single-stranded circles that could then partially renature on neutralization of the
alkaline solution. Further characterization of this molecule (which is currently
in progress) is necessary to determine whether this model is correct. Importantly,
if the model is correct, it suggests that the extrachromosomal molecules should
contain a functional replication origin. We are now determining whether the
molecules isolated after alkaline extraction can replicate autonomously after
transfection into Urd⁻ CHO cells and whether they engender a high-frequency-
amplification phenotype if they are integrated.

If the proximity of a gene to origins of DNA replication or highly recombinogenic sites determines the frequency of amplification, then the high-frequency-amplification phenotype should be relatively common. This expectation
is based on the following considerations: There are approximately 50,000 DNA
replication origins in an average-size mammalian genome (Huberman and Riggs,
1968). If such sequences are distributed randomly throughout the genome, if
gene insertion occurs randomly, and if insertion within 1–10 kbp of an origin
could confer a high-frequency-amplification phenotype, then roughly 1 of 10 to
1 of 100 transformants should display this phenotype.

The results of one recent report showed that a site that engenders high-
frequency amplification was present in 1 of 40 transformants of NIH 3T3 cells
transfected with a *v-src* gene placed under the control of a mouse metallothionein
(*Mt − 1*) promoter (Glanville, 1985). In these experiments, the *Mt-1 v-src* plasmid was introduced into NIH 3T3 cells, and clones displaying a transformed
phenotype were selected by growth in soft agar. Of 40 apparently transformed
clones isolated by this procedure, 2 lost the transformed phenotype on passage
at low cell density. Of these 2, 1 regained the transformed phenotype when
allowed to grow to confluence. The reversible gain and loss of the transformed
phenotype were shown to be due to a corresponding amplification and high-level
expression of the transfecting *v-src* sequences at high cell density and a loss of
these sequences at low cell density. Single and double minute chromosomes

(SMs and DMs), hallmarks of gene amplification in mouse cells, were present in this cell line when it was grown at confluence, but were not present when the line was passaged at low density. In addition, SMs and DMs were not present in wild-type NIH 3T3 cells or in NIH 3T3 cells expressing high levels of *v-src* as a consequence of viral infection. Since the *v-src* gene is not normally regulated by confluence, nor is its expression elevated in the other 38 transformants analyzed, the data are consistent with the interpretation that gene integration in this transformant occurred in a position at which overreplication is triggered or "induced" in response to undefined signals elicited at confluence. Since overexpression of the *v-src* gene in a cell line at confluence enables such a cell to continue to grow to form foci, integration of the *v-src* gene into such a position would be a highly selected event. The reversible nature of the amplification in this cell line, and the presence of SMs and DMs, which are thought to be large closed circular molecules (Hamkalo *et al.*, 1985), are consistent with the interpretation presented above that the high-frequency-amplification phenotype can be associated with the generation of extrachromosomal circular molecules. One caution in interpreting this study is that the DNA transfection was done in the presence of salmon sperm carrier DNA. It is unclear, therefore, whether the interesting properties observed are related to sequences in the carrier DNA, to the site of integration in the mouse chromosome, or to chromosome disruption caused by insertion of a large molecular patchwork.

Implicit in the notion that high-frequency gene amplification is associated with the proximity of a gene to a sequence such as a replication origin is the assumption that such a sequence can act in *cis* to affect gene amplification. Results from several experimental systems corroborate this assumption. For example, the chorion genes in *Drosophila* ovarian follicle cells are amplified at a specific time during oogenesis (Spradling, 1981a,b). Previous experiments indicated that two origins of DNA replication, one on chromosome 3 and one on the X chromosome, are stimulated to rereplicate in response to a specific developmental signal leading to the generation of gradients of amplified sequences containing the two chorion genes in the most highly amplified regions (Spradling, 1981; Osheim and Miller, 1983). Regions containing the putative origin sequences were isolated (deCicco and Spradling, 1984) and reintroduced into other chromosomal positions by P-element-mediated gene transfer (deCicco and Spradling, 1984; Levine and Spradling, 1986; Kalfayan *et al.*, 1985). The gene-transfer experiments demonstrated that a 3.8-kb genomic segment could initiate amplification of adjacent sequences at the correct time during development in a variety of genomic positions. However, the level of amplification in each of these positions was variable and significantly lower than that observed in the native position. This suggests either that adjacent sequences could exhibit a suppressing effect or that sequences in addition to the 3.8-kb region are required to regulate the amplitude of the amplification process. Other experiments have provided evidence that *trans*-acting factors are also necessary to control the number of times that these origins stimulate rereplication (Orr *et al.*, 1984; Kalfayan *et al.*, 1985).

A second system that indicates a role for *cis*-acting sequences in gene amplification concerns the bromodeoxyuridine (BrdU)-dependent amplification of prolactin sequences in rat GH pituitary cells (Biswas and Hanes, 1982). Synthesis of prolactin in a mutant of GH cells (GH$_1$ 2C$_1$) that normally does not synthesize prolactin can be achieved when they or a BrdU-resistant derivative are grown in the presence of high concentrations of BrdU (Biswas *et al.*, 1977, 1979; Biswas and Hanes, 1982). The elevated prolactin synthesis is dependent on the incorporation of BrdU into genomic DNA (Biswas *et al.*, 1979) and is due to an approximately 40-fold increase in the number of prolactin-gene sequences (Biswas and Hanes, 1982). The amplification is reversible and is mediated by the production of an approximately 20-kb extrachromosomal molecule (Wilson *et al.*, 1983a,b). The frequency of cells in the population that undergo this BrdU-induced amplification has not been reported. A 10.3-kb segment containing DNA that flanks the 5' and 3' ends of the rat prolactin gene in BrdU-responsive GH cells can apparently mediate the amplification of linked sequences following transfection and integration into the chromosomes of recipient mouse cells (Biswas *et al.*, 1984). Similar to the results described above for *Drosophila*, the level of amplification is less than that seen in the parental cells. It is not yet clear whether BrdU-induced gene amplification in the transfectant is also mediated by production of an extrachromosomal molecule. An equivalent 10.3-kb restriction fragment derived from wild-type (BrdU-nonresponsive) GH cells fails to mediate amplification of linked sequences in the mouse cell transformants. Since a restriction fragment of approximately the same size is present in both the BrdU-responsive and -nonresponsive cells, the data suggest that a rather small alteration occurring within this region is capable of mediating BrdU-dependent amplification of adjacent sequences.

The data indicate that treatment of either rat or mouse cells with BrdU leads to the generation of a similar, reversible signal that results in the overreplication of the prolactin gene and 10–15 kb of flanking DNA. This signal could result from changes caused by BrdU incorporation into the BrdU-responsive sequences that flank the prolactin gene, leading to the generation of a damaged region that is frequently overreplicated when repaired. On the other hand, BrdU incorporation into the total genomic DNA of the cell can also lead to DNA damage. It is conceivable that such generalized DNA damage could evoke a damage response analogous to that elicited in *E. coli* (for a review, see Walker, 1984) or yeast (McClanahan and McEntee, 1984; Ruby *et al.*, 1983) in response to a spectrum of DNA-damaging agents or environmental conditions. It is possible that the genetic alteration or alterations presumed to be present in the region flanking the BrdU-responsive prolactin gene create an origin of replication that can be fired in response to signals evoked by generalized DNA damage caused by BrdU incorporation.

The amplification of some surface-antigen genes after transfer of total genomic DNA from human cells into mouse cells provides another example of high-frequency gene amplification. Cotransfection of mouse LMTK⁻ cells with a herpes simplex virus TK gene and total genomic DNA enables one to isolate Tk⁺ transformants containing $1–10 \times 10^3$ kb of the cotransforming sequences (Perucho et al., 1980; Robins et al., 1980). When human DNA is used as the donor, approximately 1 of 1000 transformants expresses one of a variety of human T-cell differentiation markers (Kavathas and Herzenberg, 1983a; Hsu et al., 1984). These cells could be purified from nonexpressing cells after binding fluorescent monoclonal antibodies and aseptic sorting in the fluorescence-activated cell sorter (FACS) (Kavathas and Herzenberg, 1983a,b; Hsu et al., 1984). Of 20 independent clones that express the human Leu-2 antigen, 10 displayed striking variability in the amount of antigen expressed per cell, and in one of these the mean quantity of Leu-2 antigen per cell was at least 7 times greater than that seen in the others. Repeated rounds of sorting in the FACS yielded cells with a 50-fold increase in Leu-2 antigen per cell, which was shown by DNA and RNA blotting to result from elevated copies of the human Leu-2 gene (Kavathas and Herzenberg, 1983a; Kavathas et al., 1984). The size of the coamplified region was estimated to be approximately 10,000 kb on the basis of the total sizes of amplified human specific DNA detected in DNA blots after hybridization with human repetitive DNA. The high cell-to-cell variability of Leu-2 expression in a clone is probably related to the existence of the Leu-2 gene on DMs in these cells (Kavathas and Herzenberg, 1983a). It is unclear whether the Leu-2 genes in the transformants were originally integrated into chromosomes or whether the transfecting sequences exist as autonomously replicating extrachromosomal elements directly after entry into the mouse cell nucleus. If the latter interpretation is correct, then "high-frequency amplification" in this case would probably mean that the FACS is detecting those cells that have more DMs due to the random partitioning of these structures at mitosis, rather than the de novo amplification, as in the BrdU-induced amplification of prolactin genes, the confluency-dependent amplification of transfected v-src genes, or the amplification of the CAD gene described above.

Since other human surface-antigen sequences introduced into mouse cells do not exhibit variable expression due to amplification, and 10 of 20 Leu-2 transformants exhibit similar properties, it is reasonable to infer that the capacity for "high-frequency amplification" exhibited by the Leu-2 gene derives from sequences linked to it. However, this behavior is not elicited by the Leu-2 gene in its normal chromosomal position. It may thus be that this phenotype is manifested only when the sequences are removed from the content of their normal flanking sequences, when they are placed adjacent to a subset of new sequences, or when they are placed in a new nuclear environment (see Section 6). It is interesting that the independently isolated transformants with *Leu-2* sequences vary in their ability to undergo *Leu-2* "amplification" (replication?). This may be due to the effects of different flanking sequences in each transformant as described above for the amplification of *Drosophila* chorion sequences inserted into different genomic positions.

The preceding discussion has focused on genomic sites associated with high-frequency or programmed gene amplification. On the other hand, some chromosomal positions should not be conducive to gene amplification, either because their reiteration would lead to a structure incompatible with chromosome function or because coamplification of adjacent sequences could lead to over-production of a toxic product (see Section 3). As an example of a potentially lethal site for duplication, the region containing the termination of chromosome synthesis in *Salmonella* apparently cannot be duplicated (Anderson and Roth, 1981). Gene-transfer experiments also reveal such sites in mammalian chromosomes. For example, Gasser and Schimke (1986) have observed that no amplified lines could be obtained from 3 of 12 CHO cell transformants containing donated DHFR genes. It is possible that amplification of the donated genes does occur in these lines, but at frequencies below the level of sensitivity employed (the lower limits of amplification frequency were not given). An interesting outcome of these experiments was the observation that the site of insertion can apparently affect the cell-cycle-specific regulation of the DHFR gene. However, amplification of both unregulated and regulated DHFR genes was observed, suggesting that correct regulation is not a prerequisite for amplification to occur. From the available data, it is not possible to conclude whether there are quantitative differences in the amplification frequencies of regulated and unregulated DHFR genes.

In addition to the effect of gene position on gene-amplification frequency, results from gene-transfer studies indicate that gene position can also influence the cytogenetic consequences of amplification. For example, in several transformants, amplification was not associated with obvious chromosome rearrangements at the position of insertion of donated CAD genes (Wahl *et al.*, 1984b; G. M. Wahl and M. L. DeRose, unpublished observations). However, in one transformant, the CAD genes were inserted very close to a centromere, and gene movement was found in two independently derived PALA-resistant derivatives of this transformant. Interestingly, while the original transformant contains only a single insertion site, the two amplified lines investigated contain two chromosomes with the transferred genes, one of which is the original chromosome with no apparent increase in gene copy number. The amplified sequences are localized to a second chromosomal region, either at the telomere in one line or in a chromosomal fragment or extrachromosomal element in the second line. In the absence of selection, amplification in the former was stable, while it was unstable in the latter. It is important to emphasize that the PALA-resistant lines were studied after the first step of selection, demonstrating that gene movement was an early event in the amplification process. Maintenance of the original chromosome and appearance of amplified copies in the second site is consistent with either of two models. First, if the amplified unit included the centromere, then amplification could create a di- or polycentric chromosome that should be very unstable due to breakage–fusion–bridge cycles (e.g., for a review, see McClintock, 1984). Part of the chromosome could be lost by such a cycle, and this event would probably be lethal. However, nondisjunction could occasionally generate a cell with two copies of the original chromosome, only one of which undergoes amplification and a breakage–fusion–bridge cycle as described above. This would generate chromosome fragments that could either attach to chromosome telomeres, exist as free unstable elements, or integrate into other chromosomal locations. Kaufmann *et al.* (1983) have proposed that the integration of transfected sequences into telomeric regions may disrupt telomere function. They suggest that such disruption may also result in breakage–fusion–bridge cycles, leading to the formation of dicentric chromosomes followed by further rearrangements and the generation of multiple chromosomes with amplified sequences. A second model to explain maintenance of donated sequences at the original site and appearance of amplified sequences at a second site is that the CAD gene inserted near an element capable of transposing large regions of DNA. Sequences capable of transposing as much as 300 kb of DNA have been observed in *Drosophila* (Paro *et al.,* 1983). Replicative transposition (for a review, see Stark and Wahl, 1984) could then generate the observed amplification pattern.

5. THE EFFECT OF "NUCLEAR ENVIRONMENT" ON GENE AMPLIFICATION

Early work on gene amplification indicated that some cell lines (e.g., mouse or human) could amplify a specific gene in either expanded chromosomal regions or extrachromosomal elements (for reviews, see Cowell, 1982; Biedler *et al.,* 1983; Stark and Wahl, 1984), while in other lines (e.g., hamster), the same gene would be amplified only within a chromosome. The data presented above suggest that part of the reason for such differences could relate to the local genomic environment of the selected gene in different cell lines and the extent of flanking DNA coamplified in different mutants. Another possibility is that there is a different capacity for cell lines to produce extrachromosomal elements vs. expanded chromosomal regions or that some cell lines can maintain extrachromosomal elements better than others (Kaufman and Sharp, 1982). These alternatives may be broadly classified as influences of local genomic environment vs. influences of nuclear environment, where the latter could reflect as yet unidentified cell-line differences in recombination, response to DNA damage, ability to rereplicate sequences multiple times within one S phase, or other influences.

One way of distinguishing between local sequence effects and nuclear environmental effects is to transfer whole chromosomes by cell fusion or chromosome-mediated gene transfer from one cell line to another to investigate how the same selected marker in the same local genomic environment behaves in a new nuclear environment. In one application of this strategy, a human neuroblastoma cell line harboring an expanded chromosomal region was fused with mouse fibroblasts. All the resulting clones contained extrachromosomal elements (Jakobsson *et al.*, 1983). A second study employed methotrexate (MTX)-resistant murine SEWA1R TC13 cells, which, when cultured *in vitro*, contain amplified dihydrofolate reductase sequences either in expanded chromosomal regions (40% of the cells) or in extrachromosomal elements (60% of the cells). Following fusion of a population of such cells with Chinese hamster V79 cells, all MTX-resistant hybrids were observed to contain expanded chromosomal regions (Jakobsson *et al.*, 1983). In each case, the difference in genetic background was inferred to influence the presence of extrachromosomal elements or expanded chromosomal regions. These results have been interpreted to suggest that the nuclear environment of the hamster cell does not efficiently support the replication of extrachromosomal elements or that the murine nuclear environment inefficiently aggregates or integrates extrachromosomal elements into a chromosome. However, it is equally plausible that extrachromosomal elements are not transferred in sufficient quantities or maintained during the process of cell fusion or that an expanded extrachromosomal region is fragile and fragmented during the process. In addition, other results are in contrast to these findings. For example, the fusion of mouse neuroblastoma cells containing only extrachromosomal elements with Chinese hamster brain cells produced hybrids that consistently contained extrachromosomal elements (Kanotanaka *et al.*, 1982).

In an attempt to assess whether the nuclear environment of a hamster cell influences the stability of an amplified unit, the CAD gene containing chromosome of wild-type mouse S180 was transferred into CAD-deficient (Urd-A) Chinese hamster ovary (CHO) cells. In mouse S180 cells, the CAD gene is most frequently amplified in an unstable fashion [6 of 8 clones analyzed (J. Meinkoth, Ph.D. thesis, 1985, University of California, San Diego)]. In two of three unstably amplified S180 cell lines, extrachromosomal elements were present, while in one cell line, the unstable CAD genes were chromosomally localized. If the nuclear milieu of the hamster cell influences the stability of amplified sequences, then amplification of the murine CAD gene in the hamster cell nucleus should result in a stable phenotype, and the amplified sequences should be associated with the chromosome. If, on the other hand, the stability of the amplified unit is determined by sequences contained within the amplified unit or adjacent to the CAD gene, then amplification of the mouse CAD gene in the CHO cell should frequently be unstable and generally extrachromosomal. Since amplification of an "endogenous" gene is selected for, this analysis avoids the potential complications associated with the construction of somatic-cell hybrids or the chromosome-mediated transfer experiments described above. In addition, because the entire murine chromosome is transferred to the recipient cell, the sequences flanking the gene under selection are presumably the same as in the parental S180 mouse cell line. Seven independent amplified cell lines derived from the same parental single cell clone were examined. All seven retain the amplified CAD gene copies following 60 days of propagation under nonselective conditions and are still present at the same copy number in three cell lines that were analyzed after 140 days of nonselective growth. Metaphase chromosome spreads prepared from the latter three cell lines contain a single unpaired mouse chromosome, the same chromosome number as the parental cell clone, and no extrachromosomal elements. The results strongly indicate that the nuclear environment of the hamster cell is responsible for the generation of stable units of CAD gene amplification. Moreover, the influence of the hamster-cell nucleus can affect in some as yet unspecified way a murine chromosome previously shown to generate unstably amplified sequences.

6. CONCLUSIONS AND PERSPECTIVES

The number of reports describing gene amplification in bacteria, fungi, protozoa, plants, dipterans, and mammals establish this process as one of fundamental biological importance. Comparisons of the details of amplification in different species reveal a number of common features that suggest that it may be possible to apply to higher eukaryotes some of the mechanistic insights gleaned from simpler organisms. For example, the frequency of amplification in bacteria is highly dependent on the position of a sequence in the genome, which is probably due to the location of the selected gene relative to favorable sites for recombination (Anderson and Roth, 1981). In mammalian cells, the position of a gene relative to recombinational hot spots, replication origins, transposable elements, or other sequences that could initiate or resolve the process would also be expected to influence the frequency or cytogenetic characteristics of the amplification process. However, a comparison of the amplification of different endogenous genes located in different genomic positions cannot be used to analyze the effects of gene position, since such an approach would include several uncontrolled variables. One variable is the selective agent, since there is growing evidence that the selective agents themselves may influence the amplification process (Schimke, 1984). The dissimilar metabolic consequences of different selective agents or protocols could therefore impinge on the amplification process or bias the mechanism in unexpected ways. In addition, it is clear that as yet unidentified characteristics of the "nuclear environment" of different cells can determine whether a sequence is amplified as part of an expanded chromosomal region or as autonomously replicating extrachromosomal elements (see Section 4). Thus, to gain insight into position effects in gene amplification while keeping constant the potential influence of the selective agent and cell line, it is necessary to use a single selective protocol in one cell line to analyze the amplification of the same gene placed in different chromosomal locations.

The introduction of amplifiable genomic clones and minigenes into different chromosomal locations by gene transfer provides a powerful method for determining the effects of gene position on gene amplification. These studies have demonstrated that certain genomic regions enable amplification to occur at high frequency, while amplification in others is either not possible or occurs at frequencies below the limits of detection (see Section 4). These studies, in effect, identify genomic regions that somehow control the amplification of the surrounding sequences. However, one may criticize the gene-transfer approach in that gene insertion not only places a gene in a new genomic context, but also *disrupts* a previously existing chromosomal region. Consequently, position effects may reflect the effects of disruption, adjacent sequences, or both. In this light, it is important to emphasize that an alteration adjacent to the prolactin gene enables it to be amplified at an apparently high frequency after bromodeoxyuridine (BrdU) administration and that the BrdU-inducible phenotype is maintained after transfer into a new genomic position (see Section 4). Gene transfer has also been useful for identifying at least part of the region responsible for the developmentally regulated amplification of *Drosophila* chorion sequences (see Section 4). Furthermore, the ocelliless mutation, an inversion that places new sequences adjacent to the chorion gene-amplification control element, enables the new sequences to be amplified instead of the chorion genes (Spradling, 1981b). These examples provide precedents for the existence of chromosomal sequences that can control or initiate the amplification of adjacent regions. It is reasonable to infer, therefore, that gene transfer should place amplifiable target genes adjacent to such sites and that high-frequency amplification in at least some cases should reflect the favorable positioning of the transferred gene in these regions.

Gene insertion also introduces molecular tags that should enable the isolation of hyperamplifiable regions by molecular cloning techniques. Sufficient quantities of putative "amplification stimulating sequences" will then be available in pure form to allow detailed molecular characterizations and to enable subsequent gene-transfer studies to show that such sequences can act in *cis* when placed in new chromosomal environments. If the hypothesis is correct that hyperamplifiable sites reflect the presence of nearby origins of DNA replication (see Section 4), then hyperamplifiable cell lines will provide a valuable tool for the isolation and comparison of these heretofore elusive genomic structures.

The observation that gene introduction reveals sites that engender high-frequency amplification may provide a clue to the frequent association of translocation with amplification. Many examples have now been reported in which a gene is not amplified from its normal genomic position (e.g., see Cowell, 1982; Biedler *et al.*, 1983; Stark and Wahl, 1984). Two explanations of this observation have been presented: (1) Amplification intermediates are likely to be unstable and prone to breakage, resulting in frequent translocation (Stark and Wahl, 1984); (2) translocation places a gene that is unlikely to be amplified in its native site into a new genomic location in which amplification either occurs more frequently or is inducible (see Section 4) (see also Stark and Wahl, 1984). With regard to the latter model, gene-transfer experiments indicate that both high- and low-frequency sites exist in the genome. The ocelliless inversion provides precedent for gene rearrangements moving previously unamplified sequences into a region where they can be amplified. It is tempting to speculate that the translocations that are associated with amplification reflect the repositioning of genes into "preferred" locales for amplification.

Similar to the amplification of endogenous genes, transferred genes generally comprise only a fraction of the genomic sequences in the "amplification unit." The large size of these units has provided a means for amplifying unselected genes along with the amplifiable marker. This technology has provided an economically feasible means of producing large amounts of correctly processed, biologically active proteins that are otherwise present in only trace quantities. A consideration of the factors that determine the maximum attainable level of amplification of a given gene is of importance for practical reasons to enable the correct choice of amplification systems for a given application and because it gives insight into the dynamics of the amplification process. One factor that has emerged from the coamplification experiments is that high-level expression of genes that encode products that are deleterious to the cell often results in the selection of cells containing rearrangements that prevent expression of the gene or inactivate its product (see Section 3). Thus, the maximum attainable level of amplification for both endogenous and introduced genes is in part related to the maximum tolerable level of gene products encoded by adjacent genes. However, this factor is probably not sufficient to explain why the maximum level of CAD gene amplification is only about 300-fold compared to 2000-fold for the dihydrofolate reductase gene and 10,000-fold for the adenosine deaminase (ADA) gene. Other important factors are the concentration ranges over which the selective agents can be used and the relationship between the level of overproduction and the corresponding increase in drug resistance. For example, while *N*-(phosphonoacetyl)-L-aspartate (PALA) can be used over about a 250-fold concentration range (100 μM–25 mM; higher concentrations are toxic because PALA is a tetravalent anion), methotrexate can be used over more than an 8000-fold range (100 nM–800 μM). In addition, while a 5- to 10-fold increase in CAD activity gives a 100- to 1000-fold increase in resistance to PALA, there is an approximately 1:1 relationship between the level of ADA overproduction and the degree of resistance to deoxycoformycin.

The application of gene transfer to gene amplification is thus divided between two endeavors. On the one hand, coamplification makes feasible the use of mammalian cells as factories for the production of biological products that require specific *in vivo* modifications. On the other hand, it provides a novel means of probing the genome for a diversity of sites that are likely to be involved in the events that initiate and/or resolve the amplification process. While it has not been possible to investigate the molecular changes that occur in the earliest steps involved in gene amplification due to the relatively low frequency of this event, the discovery of hyperamplifiable cell lines now gives a resource with which to analyze these first crucial events. Studies in the immediate future will certainly center around the isolation and functional characterization of such sites. It is also important to stress that a number of studies now indicate that drugs that interfere with DNA synthesis or alter DNA structure may actually induce the amplification process (e.g., see Varshavsky, 1981a,b; Schimke, 1984). It is conceivable, therefore, that gene amplification is one facet of the cellular response to environmental shocks. It will be interesting in this regard to determine whether the hyperamplifiable sites identified by gene transfer prove to be activated in normal and tumor cells in response to environmental stress. If this proves to be true, then amplification would serve as a genomic footprint for the stress response and, in conjunction with transfer of the appropriate vectors, would provide an experimentally manipulable system with which to elucidate its biology.

ACKNOWLEDGMENTS.   We thank Peggy DeRose, Eric Nakkim, and John Emery for their excellent and devoted technical assistance and our colleagues who kindly provided us with manuscripts prior to their publication to enable us to make this chapter as current as possible. This work was supported in part by a grant from the G. Harold and Leila Y. Mathers charitable foundation and by NIH Grant R01 6M 27754. P.G. was supported by a grant from E.M.B.O., and J.R. and J.M. are predoctoral fellows of the Biology Department, University of California, San Diego.

REFERENCES

Alt, F., Kellems, R., Bertino, J. R., and Schimke, R. T., 1978, Selective multiplication of dihydrofolate reductase genes in methotrexate-resistant variants of cultured murine cells, *J. Biol. Chem.* **253:**1357–1370.

Anderson, R. P., and Roth, J. R., 1977, Tandem genetic duplications in phage and bacteria, *Annu. Rev. Microbiol.* **31:**473–504.

Anderson, P., and Roth, J., 1981, Spontaneous tandem genetic duplications in *Salmonella typhimurium* arise by unequal recombination between rDNA (*rrn*) cistrons, *Proc. Natl. Acad. Sci. U.S.A.* **78:**3113–3117.

Ardeshir, F., Giulotto, E., Zeig, J., Brison, O., Liao, W., and Stark, G., 1983, Structure of amplified DNA in different Syrian hamster cell lines resistant to *N*-(phosphonacetyl)-L-aspartate, *Mol. Cell. Biol.* **3:**2076–2088.

Barsoum, J., and Varshavsky, A., 1983, Mitogenic hormones and tumour protomers greatly increase the incidence of colony-forming cells bearing amplified dihydrofoltate reductase genes, *Proc. Natl. Acad. Sci. U.S.A.* **80:**5330–5334.

Biedler, J. L., Meyers, M. B., and Spengler, B. A., 1983, Homogeneously staining regions and double minute chromosomes, prevalent cytogenetic abnormalities of human neuroblastoma cells, *Adv. Cell. Neurobiol.* **4:**267–307.

Bishop, J. M., 1983, Cellular oncogenes and retroviruses, *Annu. Rev. Biochem.* **52:**301–354.

Biswas, D. K., and Hanes, S. D., 1982, Increased level of prolactin gene sequences in bromodeoxyuridine treated GH cells, *Nucleic Acids Res.* **10:**3995–4008.

Biswas, D. K., Lyons, J., and Tashjian, A. H., Jr., 1977, Induction of prolactin synthesis in rat pituitary tumor cells by 5-bromodeoxyuridine, *Cell* **11:**431–439.

Biswas, D. K., Abdulah, K. T., and Brennessel, B. A., 1979, On the mechanism of 5-bromodeoxyuridine induction of prolactin synthesis in rat pituitary tumor cells, *J. Cell. Biol.* **81:**1–6.

Biswas, D. K., Hartigan, J. A., and Pichler, M. H., 1984, Identification of DNA sequence responsible for 5-bromodeoxyuridine-induced gene amplification, *Science* **225:**941–943.

Brennand, J., Chinault, A. C., Konecki, D. S., Melton, D. W., and Caskey, C. T., 1982, Cloned cDNA sequences of the hypoxanthine/guanine phosphoribosyl transferase gene from a mouse neuroblastoma cell line found to have amplified genomic sequences, *Proc. Natl. Acad. Sci. U.S.A.* **79:**1950–1954.

Chapman, A. B., Costello, M. A., Lee, F., and Ringold, G. M., 1983, Amplification and hormone-regulated expression of a mouse mammary tumor virus–*Eco gpt* fusion plasmid in mouse 3T6 cells, *Mol. Cell. Biol.* **3:**1421–1429.

Christman, J. K., Gerber, M., Price, P. M., Flordellis, C., Edelman, J., and Acs, G., 1982, Amplification of expression of hepatitis B surface antigen in 3T3 cells cotransfected with a dominant-acting gene and cloned viral DNA, *Proc. Natl. Sci. U.S.A.* **79:**1815–1819.

Collins, K., and Stark, G. R., 1971, Aspartate Transcarbamylase *J. Biol. Chem.* **246:**6599–6605.

Cowell, J. K., 1982, Double minutes and homogeneously staining regions: Gene amplification in mammalian cells, *Annu. Rev. Genet.* **16:**21–59.

Crouse, G. F., Simonsen, C. C., McEwan, R. N., and Schimke, R. T., 1982, Structure of amplified normal and variant dihydrofolate reductase genes in mouse sarcoma S180 cells, *J. Biol. Chem.* **257:**7887–7897.

Crouse, G. F., McEwan, R. N., and Pearson, M. L., 1983, Expression and amplification of engineered mouse dihydrofolate reductase minigenes, *Mol. Cell. Biol.* **3:**257–266.

Debatisse, M., Robert de Saint Vincent, B., and Buttin, G., 1984, Expression of several amplified genes in an adenylate-deaminase overproducing variant of Chinese hamster fibroblasts, *EMBO J.* **3:**3123–3127.

DeCicco, D. V., and Spradling, A. C., 1984, Localization of a *cis*-acting element responsible for the developmentally regulated amplification of *Drosophila* chorion genes, *Cell* **38:**45–54.

Eckhardt, T., 1978, A rapid method for the identification of plasmic desoxyribonucleic acid in bacteria, *Plasmid* **1:**584–588.

Flintoff, W. E., Davidson, S. V., and Siminovitch, L., 1976, Isolation and partial characterization of three methotrexate-resistant phenotypes from Chinese hamster ovary cells, *Somat. Cell Genet.* **2:**245–261.

Flintoff, W. F., Livingston, E., Duff, C., and Worton, R. G., 1984, Moderate-level gene amplification in methotrexate-resistant Chinese hamster ovary cells is accompanied by chromosomal translocations at or near the site of the amplified DHFR gene, *Mol. Cell. Biol.* **4:**69–76.

Folger, K. R., Wong, E. A., Wahl, G., and Capecchi, M. R., 1982, Patterns of integration of DNA microinjected into cultured mammalian cells: Evidence for homologous recombination between injected plasmid DNA molecules, *Mol. Cell. Biol.* **2:**1372–1387.

Fuscoe, J. C., Fenwick, R. G., Jr., Ledbetter, D. H., and Caskey, C. T., 1983, Detection and amplification of the HGPRT locus in Chinese hamster cell, *Mol. Cell. Biol.* **3:**1086–1096.

Gardella, T., Medveczky, P., Sairenji, T., and Mulder, C., 1984, Detection of circular and linear herpesvirus DNA molecules in mammalian cells by gel electrophoresis, *J. Virol.* **50**:248–254.

Gasser, C. S., and Schimke, R. T., 1986, Regulation of transfected murine dihydrofolate reductase genes (submitted).

Gasser, C., Simonsen, C. C., Schilling, J. W., and Schimke, R. T., 1982, Expression of abbreviated mouse dihydrofolate reductase genes in cultured hamster cells, *Proc. Natl. Acad. Sci. U.S.A.* **79**:6522–6526.

Glanville, N., 1985, Unstable expression and amplification of a transfected oncogene in confluent and subconfluent cells, *Mol. Cell. Biol.* **5**:1456–1464.

Gorman, C. M., Moffat, L. F., and Howard, B. H., 1982, Recombinant genomes which express chloramphenicol acetyltransferase in mammalian cells, *Mol. Cell. Biol.* **2**:1044–1051.

Haber, D. A., Beverley, S. M., Kiely, M. L., and Schimke, R. T., 1981, Properties of an altered dihydrofolate reductase encoded by amplified genes in cultured mouse fibroblasts, *J. Biol. Chem.* **256**:9501–9510.

Hamkalo, B. A., Farnham, P. J., Johnston, R., and Schimke, R. T., 1985, Ultrastructural features of minute chromosomes in a methotrexate-resistant mouse 3T3 cell line, *Proc. Natl. Acad. Sci. U.S.A.* **82**:1126–1130.

Hamlin, J. L., Milbrandt, J. D., Heintz, N. H., and Azizkhan, J. C., 1984, DNA sequence amplification in mammalian cells, *Int. Rev. Cytol.* **90**:31–82.

Haynes, J., and Weissman, C., 1983, Constitutive, long-term production of human interferons by hamster cells containing multiple copies of a cloned interferon gene, *Nucleic Acids Res.* **11**:687–706.

Hirt, B., 1967, Selective extraction of polyoma DNA from infected mouse cell cultures, *J. Mol. Biol.* **26**:365–369.

Hiscott, J., Murphy, D., and Defendi, V., 1980, Amplification and rearrangement of integrated SV40 DNA sequences accompany the selection of anchorage-independent transformed mouse cells, *Cell* **22**:535–543.

Hsu, C., Kavathas, P., and Herzenberg, L. A., 1984, Cell-surface antigens expressed on L-cells transfected with whole DNA from non-expressing and expressing cells, *Nature (London)* **312**:68–69.

Huberman, J. A., and Riggs, A. D., 1968, On the mechanism of DNA replication in mammalian chromosomes, *J. Mol. Biol.* **32**:327–341.

Jakobsson, A., Dahllof, B., Martinsson, T., and Levan, G., 1983, Transfer of methotrexate resistance with somatic cell hybridization, *Hereditas* **99**:293–302.

Johnston, R. N., Beverley, S. M., and Schimke, R. T., 1983, Rapid spontaneous dihydrofolate reductase gene amplification shown by fluorescence-activated cell sorting, *Proc. Natl. Acad. Sci. U.S.A.* **80**:3711–3715.

Kalfayan, L., Levine, J., Orr-Weaver, T., Parks, S., Wakimoto, B., de Cicco, D., and Spradling, A., 1985, Localization of sequences regulating *Drosophila* chorion gene amplification and expression, *Cold Spring Harbor Symp. Quant. Biol.* **50**:527–535.

Kanotanaka, K., Higashida, H., Fukami, H., and Tanaka, T., 1982, Double minutes in mouse neuroblastoma cells and their hybrids, *Cancer Genet. Cytogenet.* **5**:51–62.

Kaufman, R. J., and Schimke, R. T., 1981, Amplification and loss of dihydrofolate reductase genes in a Chinese hamster ovary cell line, *Mol. Cell. Biol.* **1**:1069–1076.

Kaufman, R. J., and Sharp, P. A., 1982, Amplification and expression of sequences cotransfected with a modular dihydrofolate reductase complementary DNA gene, *J. Mol. Biol.* **159**:601–621.

Kaufman, R. J., Brown, P. C., and Schimke, R. T., 1979, Amplified dihydrofolate reductase genes in unstably methotrexate-resistant cells are associated with double minute chromosomes, *Proc. Natl. Acad. Sci. U.S.A.* **76**:5669–5673.

Kaufman, R. J., Sharp, P. A., and Latt, S. A., 1983, Evolution of chromosomal regions containing transfected and amplified dihydrofolate reductase sequences, *Mol. Cell. Biol.* **3**:699–711.

Kaufman, R. J., Wasley, L. C., Spiliotes, A. J., Gossels, S. D., Latt, S. A., Larsen, G. R., and Kay, R. M., 1985, Coamplification and coexpression of human tissue-type plasminogen activator and murine dihydrofolate reductase sequences in Chinese hamster ovary cells, *Mol. Cell. Biol.* **5**:1750–1759.

Kavathas, P., and Herzenberg, L. A., 1983a, Amplification of a gene coding for human T-cell differentiation antigen, *Nature (London)* **306**:385–387.

Kavathas, P., and Herzenberg, L. A., 1983b, Stable transformation of mouse L cells for human membrane T-cell differentiation antigens, HLA and β2-microglobulin: Selection by fluorescence-activated cell sorting, *Proc. Natl. Acad. Sci. U.S.A.* **80**:524–528.

Kavathas, P., Sukhatme, V. P., Herzenberg, L. A., and Parnes, J. R., 1984, Isolation of the gene encoding the human T-lymphocyte differentiation antigen Leu-2 (T8) by gene transfer and cDNA subtraction, *Proc. Natl. Acad. Sci. U.S.A.* **81**:7688–7692.

Kempe, T. D., Swyryd, E. A., Bruist, M., and Stark, G. R., 1976, Stable mutants of mammalian cells that overproduce the first three enzymes of pyrimidine nucleotide biosynthesis, *Cell* **9**:541–550.

Kohl, N. E., Kanda, N., Schreck, R. R., Bruns, G., Latt, S. A., Gilbert, F., and Alt, F. W., 1983, Transposition and amplification of oncogene-related sequences in human neuroblastomas, *Cell* **35**:359–367.

Kucherlapati, R. S., Eves, E. M., Song, K. Y., Morse, B. S., and Smithies, O., 1984, Homologous recombination between plasmids in mammalian cells can be enhanced by treatment of input DNA, *Proc. Natl. Acad. Sci. U.S.A.* **81**:3153–3157.

Lau, Y., Lin, C. C., and Kan, Y. W., 1984, Amplification and expression of human α-globin genes in Chinese hamster ovary cells, *Mol. Cell. Biol.* **4**:1469–1475.

Levine, J., and Spradling, A., 1986, DNA sequence of a 3.8 kilobase pair region controlling *Drosophila* chorion gene amplification, *Chromosoma* **92**:130.

Liskay, R. M., and Stachelek, J. L., 1983, Evidence for intrachromosomal gene conversion in cultured mouse cells, *Cell* **35**:157–165.

Mayo, K. E., and Palmiter, R. D., 1982, Altered regulation of the mouse metallothionein-1 gene following gene amplification or transfection, in: *Gene Amplification* (R. T. Schimke, ed.), pp. 67–73, Cold Spring Harbor Press, Cold Spring Harbor, New York.

McClanahan, T., and McEntee, K., 1984, Specific transcripts are elevated in *Saccharomyces cerevisiae* in response to DNA damage, *Mol. Cell. Biol.* **4**:2356–2363.

McClintock, B., 1984, The significance of responses of the genome to challenge, *Science* **226**:792–801.

McCormick, F., Trahey, M., Innis, M., Dieckmann, B., and Ringold, G., 1984, Inducible expression of amplified human beta interferon genes in CHO cells, *Mol. Cell. Biol.* **4**:166–172.

Milbrandt, J. D., Azizkhan, J. C., Greisen, K. S., and Hamlin, J. L., 1983a, Organization of a Chinese hamster ovary dihydrofolate reductase gene identified by phenotypic rescue, *Mol. Cell. Biol.* **3**:1266–1273.

Milbrandt, J. D., Azizkhan, J. C., and Hamlin, J. L., 1983b, Amplification of a cloned Chinese hamster dihydrofolate reductase gene after transfer into a dihydrofolate reductase-deficient cell line, *Mol. Cell. Biol.* **3**:1274–1282.

Montoya-Zavala, M., and Hamlin, J. L., 1985, Similar 150-kilobase DNA sequences are amplified in independently derived methotrexate-resistant Chinese hamster cells, *Mol. Cell. Biol.* **5**:619–627.

Mulligan, R., and Berg, P., 1981, Selection for animal cells that express the *Escherichia coli* gene coding for xanthine-guanine phosphoribosyl transferase, *Proc. Natl. Acad. Sci. U.S.A.* **78**:2072–2076.

Murray, M. J., Kaufman, R. J., Latt, S. A., and Weinberg, R. A., 1983, Construction and use of a dominant, selectable marker: A Harvey sarcoma virus–dihydrofolate reductase chimera, *Mol. Cell. Biol.* **3**:32–43.

Orkin, S. H., Goff, S. C., Kelley, W. N., and Daddona, P. E., 1985, Transient expression of human adenosine deaminase cDNAs: Identification of a nonfunctional clone resulting from a single amino acid substitution, *Mol. Cell. Biol.* **5**:762–767.

Orr, W. K., Komitopoulou, K., and Kafatos, F. C., 1984, Mutants suppressing in *trans* chorion gene amplification in *Drosophila*, *Proc. Natl. Acad. Sci. U.S.A.* **81**:3773–3778.

Osheim, Y. N., and Miller, O. L., Jr., 1983, Novel amplification and transcriptional activity of chorion genes in *Drosophila melanogaster* follicle cells, *Cell* **33**:543–553.

Padgett, R. A., Wahl, G. M., and Stark, G. R., 1982, Structure of the gene for CAD, the multifunctional protein that initiates UMP synthesis in Syrian hamster cells, *Mol. Cell. Biol.* **2**:293–301.

Paro, R., Goldberg, M. L., and Gehring, W. J., 1983, Molecular analysis of large transposable elements carrying the white locus of *Drosophila*-melanogaster *EMBO J.* **2**:835–860.

Patterson, D., and Carnright, D. V., 1977, Biochemical genetic analysis of pyrimidine biosynthesis in mammalian cells. I. Isolation of a mutant defective in the early steps of *de novo* pyrimidine synthesis, *Somat. Cell Genet.* **3**:483–495.

Pauza, C., Karels, M. J., Navre, M., and Schachman, H. K., 1982, Gene encoding *Escherichia coli* aspartate transcarbamylase: The *pyrB-pyrI* operon, *Proc. Natl. Acad. Sci. U.S.A.* **79**:4020–4024.

Perucho, M., Hanahan, D., and Wigler, M., 1980, Genetic and physical linkage of exogenous sequences in transformed cells, *Cell* **22**:309–317.

Ringold, G., Dieckmann, B., and Lee, F., 1981, Co-expression and amplification of dihydrofolate reductase cDNA and the *Escherichia coli* XGPRT gene in Chinese hamster ovary cells, *J. Mol. Appl. Genet.* **1**:165–175.

Robert de Saint Vincent, B., and Wahl, G. M., 1983, Homologous recombination in mammalian cells mediates formation of a functional gene from two overlapping gene fragments, *Proc. Natl. Acad. Sci. U.S.A.* **180**:2002–2006.

Robert de Saint Vincent, B., Delbruck, S., Eckhart, W., Meinkoth, J., Vitto, L., and Wahl, G., 1981, The cloning and reintroduction into animal cells of a functional CAD gene, a dominant amplifiable genetic marker, *Cell* **27**:267–277.

Roberts, J. M., and Axel, R., 1982, Gene amplification and gene correction in somatic cells, *Cell* **29**:109–119.

Roberts, J. M., Buck, L. B., and Axel, R., 1983, A structure for amplified DNA, *Cell* **33**:53–63.

Robins, D. M., Ripley, S., Henderson, A., and Axel, R., 1980, Transforming DNA integrates into the host chromosome, *Cell* **23**:29–39.

Rubnitz, J., and Subramani, S., 1984, The minimum amount of homology required for homologous recombination in mammalian cells, *Mol. Cell. Biol.* **4**:2253–2258.

Ruby, S. W., Szostak, J. W., and Murray, A. W., 1983, Cloning regulated yeast genes from a pool of *lacZ* fusions, *Methods Enzymol.* **101**:253–269.

Ruiz, J. C., and Wahl, G. M., 1986, *Escherichia coli* aspartate transcarbamylase: A novel marker for gene amplification and expression studies in mammalian cells, *Mol. Cell. Biol.* **6** (in press).

Scahill, S. J., Devos, R., Van der Heyden, J., and Fiers, W., 1983, Expression and characterization of the product of a human immune interferon cDNA gene in Chinese hamster ovary cells, *Proc. Natl. Acad. Sci. U.S.A.* **80**:4654–4658.

Schimke, R. T., 1984, Gene amplification, drug resistance, and cancer, *Cancer Res.* **44**:1735–1742.

Schoner, R. G., and Littlefield, J. W., 1981, The organization of the dihydrofolate reductase gene of baby hamster kidney fibroblasts, *Nucleic Acids Res.* **9**:6001–6014.

Shigesana, K., Stark, G. R., Maley, J. A., Niswander, L. A., and Davidson, J. N., 1985, Construction of a cDNA to the hamster CAD Gene and its application toward defining the domain for aspartate transcarbamylase, *Mol. Cell. Biol.* **5**:1735–1742.

Simonsen, C. C., and Levinson, A. D., 1983, Isolation and expression of an altered mouse dihydrofolate reductase cDNA, *Proc. Natl. Acad. Sci. U.S.A.* **80**:2495–2499.

Sirotnak, F. M., Moccio, D. M., Kelleher, L. E., and Goutas, L. J., 1981, Relative frequency and kinetic properties of transport-defective phenotypes among methotrexate-resistant L1210 clonal cell lines derived *in vivo*, *Cancer Res.* **41**:4447–4452.

Southern, P. J., and Berg, P., 1982, Transformation of mammalian cells to antibiotic resistance with a bacterial gene under control of the SV40 early region promoter, *J. Mol. Appl. Genet.* **1**:327–341.

Spradling, A. C., 1981, The organization and amplification of two chromosommal domains containing *Drosophila* chorion genes, *Cell* 27:193–201.

Spradling, A. C. and Mahowald, A. P., 1981, A chromosome inversion alters the pattern of specific DNA replication in *Drosophila* follicle cells, *Cell* 27:203–209.

Stallings, R. L., Munk, A. C., Longmire, J. L., Hildebrand, C. E., and Crawford, B. D., 1984, Assignment of genes encoding metallothioneins I and II to Chinese hamster chromosome 3: Evidence for the role of chromosome rearrangement in gene amplification, *Mol. Cell. Biol.* 4:2932–2936.

Stark, G. R., and Wahl, G. M., 1984, Gene amplification, *Annu. Rev. Biochem.* 53:447–491.

Subramani, S., Mulligan, R., and Berg, P., 1981, Expression of the mouse dihydrofolate reductase complementary deoxyribonucleic acid in simian virus 40 vectors, *Mol. Cell. Biol.* 1:854–864.

Tlsty, T. D., Albertini, A. M., and Miller, J. H., 1984, Gene amplification in the *lac* region of *E. coli*, *Cell* 37:217–224.

Urlaub, G., and Chasin, L. A., 1980, Isolation of Chinese hamster cell mutants deficient in dihydrofolate reductase activity, *Proc. Natl. Acad. Sci. U.S.A.* 77:4216–4220.

Urlaub, G., Kas, E., Carothers, A. M., and Chasin, L. A., 1983, Deletion of the diploid dihydrofolate reductase locus from cultured mammalian cells, *Cell* 33:405–412.

Varshavsky, A., 1981a, On the possibility of metabolic control of replicon "misfiring": Relationship to emergence of malignant phenotypes in mammalian cell lineages, *Proc. Natl. Acad. Sci. U.S.A.* 78:3673–3677.

Varshavsky, A., 1981b, Phorbol ester dramatically increases incidence of methotrexate-resistant mouse cells: Possible mechanisms and relevance to tumor promotion, *Cell* 25:561–572.

Wahl, G. M., Padgett, R. A., and Stark, G. R., 1979, Gene amplification causes overproduction of the first three enzymes of UMP in *N*-(phosphonacetyl)-L-aspartate resistant hamster cells, *J. Biol. Chem.* 254:8679–8689.

Wahl, G. M., Vitto, L., Padgett, R. A., and Stark, G. R., 1982, Single-copy and amplified CAD genes in Syrian hamster chromosomes localized by a highly sensitive method for *in situ* hybridization, *Mol. Cell. Biol.* 2:308–319.

Wahl, G. M., Vitto, L., and Rubnitz, J., 1983, Co-amplification of rRNA genes with CAD genes in *N*-(phosphonacetyl)-L-aspartate-resistant Syrian hamster cells, *Mol. Cell. Biol.* 3:2066–2075.

Wahl, G. M., Allen, V., Delbruck, S., Eckhart, W., Meinkoth, J., Robert de Saint Vincent, B., and Vitto, L., 1984a, Analysis of CAD gene amplification using a combined approach of molecular genetics and cytogenetics, in: *Advances in Experimental Medicine and Biology,* Vol. 172, *Eukaryotic Cell Cultures: Basics and Applications* (R. T. Acton and J. D. Lynn, eds.), pp. 319–345, Plenum Press, New York.

Wahl, G. M., Robert de Saint Vincent, B., and DeRose, M. L., 1984b, Effect of chromosomal position on amplification of transfected genes in animal cells, *Nature (London)* 307:516–520.

Walker, G., 1984, Mutagenesis and inducible responses to deoxyribonucleic acid damage in *Escherichia coli*, *Microbiol. Rev.* 48:60–93.

Wilson, D. J., Hanes, S. D., Pichler, M. H., and Biswas, D. K., 1983a, 5-Bromodeoxyuridine-induced amplification of prolactin gene in GH cells is an extrachromosomal event, *Biochemistry* 22:6077.

Wilson, D. J., Pichler, M. H., and Biswas, D. K., 1983b, Extent of BrdUrd-induced prolactin gene amplification in GH cells, *DNA* 2:237–242.

Yeung, C., Ingolia, D. E., Bobonis, C., Dunbar, B. S., Riser, M. E., Siciliano, M. J., and Kellems, R. E., 1983, Selective overproduction of adenosine deaminase in cultured mouse cells, *J. Biol. Chem.* 258:8338–8345.

Yeung, C., Ingolia, D. E., Roth, D. B., Shoemaker, C., Al-Ubaidi, M. R., Yen, J., Ching, C., Bobonis, C., Kaufman, R. J., and Kellems, R. E., 1985, Identification of functional murine adenosine deaminase cDNA clones by complementation in *Escherichia coli*, *J. Biol. Chem.* 260:10,299–10,307.

RETROVIRAL INTEGRATION AND INSERTIONAL MUTAGENESIS

Stephen P. Goff

1. INTRODUCTION

In the last decade, technology for the detection and isolation of specific genes has improved so dramatically that the DNA that encodes virtually any known protein can be obtained with modest effort and expense. Furthermore, we can manipulate cloned genes with precision, making specific alterations that might affect either coding or regulatory sequences of these genes, and we can reinstate the DNAs in cells in a functional form to assess the consequences of these alterations. Because of these technical advances, our understanding of the structural determinants of gene function has grown enormously.

The key step in the isolation of a particular gene is usually devising a scheme for the detection of that gene. The majority of the genes from higher eukaryotes cloned to date were isolated by virtue of a biochemical handle on the gene products. If a messenger RNA (mRNA) transcribed from a gene can be isolated in moderately pure form, then nucleic acid probes can be prepared from that mRNA (as was used for the isolation of the globin genes); if antibodies specific for the protein product of the gene are available, they can be used to screen cultures expressing proteins from the cloned DNAs (as in the cloning of the gene for terminal deoxynucleotidyl transferase); if amino acid sequences of the protein are known, the information can be used to synthesize oligonucleotide probes homologous to the gene (as in the cloning of the human insulin receptor). More challenging problems are presented by the isolation of genes for which there is no biochemical handle and for which no protein product is known. If a gene is known only by the phenotype of a particular allele, functional assays must be used for the detection of the gene after transfer back into a eukaryotic host. The isolation of several genes responsible for eliciting the morphological

Stephen P. Goff ● Department of Biochemistry and Molecular Biophysics, College of Physicians and Surgeons, Columbia University, New York, New York 10032.

transformation of mammalian cells has required these more complex assays, and Chapters 1 and 10 describe such procedures. The genes are detected by virtue of their ability, as dominant genetic markers, to induce growth of contact-inhibited cells after the transfer of total genomic DNA into recipient cells. Such procedures depend on the dominant action of the transferred gene and on the existence of a readily observable phenotype in the recipient cells. Once transferred to new cells, the genes have been effectively separated from all other DNA of the cell of origin; they have been added to the DNA in the recipient cell. The final isolation step usually depends on establishing linkage between the transferred gene and a known, additional sequence or on the preexisting linkage of the gene with repetitive sequences specifically present in the genome of the donor cell and absent in the recipient.

An alternative approach to the detection and isolation of genes is the generation of new alleles with a desired phenotype, and such that the mutant gene is marked with a recognizable sequence. This is achieved by insertional mutagenesis: the insertion of a known genetic element into an unknown gene, disrupting or altering the expression of that gene and inducing a desired phenotype. Thereafter, the gene is marked by its new linkage to a defined segment of DNA and, if the mutagenic element carries its own genes, by its linkage to the markers on the element. The isolation of the gene is then straightforward. The mutagenic element is isolated along with portions of the flanking DNA, and the affected gene is sought within these adjacent regions. The wild-type allele can be recovered by using the mutant allele as a nucleic-acid probe.

This chapter will discuss progress toward the use of insertional mutagenesis to alter, detect, and clone genes from mammalian cells. The genetic elements that we will describe to be used as mutagens are the retroviruses: RNA viruses that induce the formation of an integrated proviral DNA as a normal part of their life cycle. We will describe what is known about the integration reaction, and we will summarize the studies of the consequences of retroviral insertions into host genes.

1.1. Insertional Mutagenesis

Insertional mutagenesis has long been recognized as a powerful method for the inactivation and identification of genes (Bukhari et al., 1977; Kleckner, 1977). Transposable elements of prokaryotes are widely used to create mutations; those elements that carry genes that confer antibiotic resistances are particularly useful in tagging a gene and linking a mutant allele to a selectable marker. The subsequent movement of the mutant allele can be followed by tests for the selectable drug resistance. The mutator phage μ is also of significant utility in generating insertions and is even more closely analogous to the retroviruses of eukaryotic cells. This DNA virus, like phage λ, forms an integrated prophage when converting a bacterial host to a lysogen; unlike λ, however, phage μ inserts its genome at nearly random sites on the host chromosome and can therefore induce mutations in virtually any gene.

Related elements have now been found and exploited in a number of eukaryotic hosts; the similarity of these elements to proviral DNAs is striking and has been noted by many investigators (Shimotohno *et al.*, 1980; Varmus, 1983). Transposable elements termed Ty elements have been found in a broad range of yeast strains. These elements, present in high copy number in most laboratory strains, ordinarily move at such low frequency to new sites that mutations are rarely induced. Only a few insertions have been studied (Roeder and Fink, 1980; Silverman and Fink, 1984). The size and structure of these elements are strikingly similar to those of the retroviruses in higher cells, and very recently, the relationship between these elements and the retroviruses has been extended (Boeke *et al.*, 1985). The transposition of the Ty elements occurs by the formation of an RNA intermediate; this RNA is copied into double-stranded DNA (dsDNA) forms by a reverse transcriptase encoded by the element and the DNA is inserted into the host by an integrase function, similarly encoded in the element.

Similar mobile elements have been found and exploited in *Drosophila melanogaster*. The *white* locus of *Drosophila* was first cloned in the form of sequences adjacent to a *copia* element that was responsible for the insertional inactivation of the gene and had resulted in the formation of the *white*[a] allele (Bingham *et al.*, 1981). Further analysis of the locus showed that an astonishing number of the known alleles were formed by the insertion of various transposable elements, and similar results have been found for mutations in the *bithorax* and *notch* loci. The most sophisticated use of transposable elements in this organism has been the use of the P elements to generate mutations (Searles *et al.*, 1982; O'Hare and Rubin, 1983). These elements are repressed in the cytotype of their normal host, but are activated to transpose soon after zygote formation between a male gamete carrying the element and an egg lacking them. The resulting progeny suffer massive insertional mutagenesis, the consequences of which are readily detected in the next generation. An insertion that produces an interesting phenotype facilitates the cloning of the responsible gene: After suitable backcrosses to remove irrelevant elements, the gene is cloned by simply cloning the element. Germ-line insertions of cloned copies of these elements can also be engineered (Spradling and Rubin 1982).

1.2. Retroviral Life Cycle

The retroviruses, originally termed RNA tumor viruses, are RNA viruses with an unusual life cycle (Varmus, 1983). The virion particles contain two copies of a genomic RNA, annealed in a dimer by noncovalent interactions, wrapped in a protein core, and surrounded by a bilayer envelope. On infection, the RNA serves as template for the formation of a linear dsDNA molecule by the enzyme reverse transcriptase; this enzyme is contained inside the virion and is brought into the cell by the infecting particle. The reaction is complex (Gilboa *et al.*, 1979). The product DNA is collinear with the genomic RNA, but contains added duplicated sequences termed long terminal repeats (LTRs) at the ends. This DNA is next transported by unknown mechanisms to the nucleus, where two double-stranded circular species appear. One circle contains only one copy of the LTR sequences and has a structure consistent with its formation by homologous recombination between the two LTRs of the linear DNA; the other circle contains two tandem LTRs and probably arises by the blunt-end ligation of the termini of the linear molecule.

In the important step for our consideration, the viral DNA is integrated into the host genome by an efficient and specific reaction. The recombination site on the viral DNA is exceedingly specific: It is at the edges of the LTR sequences, opening up the circular DNA such that the integrated proviral DNA is similar in organization to the initial unintegrated linear structure. Thus, the integrated sequences are flanked by LTRs and the rest of the viral DNA lies between them. The sites in the host genome are, to a first approximation, arbitrarily selected. The integrated DNA is quite stable, as we will see from the work discussed below. In the case of the replication-competent viruses, the integrated DNA serves as the template for transcription by the RNA polymerase II system of the cell to form viral mRNAs; transcription of these messengers begins in the 5' LTR and terminates in the 3' LTR. These RNAs are translated to form the viral proteins, and these proteins assemble with virion RNA to form the progeny virions. Virions are shed from the cell surface through the plasma membrane, and the mature virions are surrounded by a lipid bilayer from the host membrane. The production can continue unabated indefinitely because the release of virions does not permanently damage the cell membrane, nor does the production represent a serious drain on the resources of the cell. The released progeny can spread to new target cells.

The life cycle immediately suggests that these agents are ideally suited for use as insertional mutagens of vertebrate cells. The insertion events can be induced at will simply by applying virions to cells, the number of insertion events can be controlled by choosing the appropriate titer of virus, and the cells are not killed by the insertion of the viral DNA, even though viral gene products and progeny virus are produced.

2. INTEGRATION REACTION: SEQUENCES AND STRUCTURES

Most of our information about the integration reaction by which retroviral DNAs are inserted into the host genome is highly inferential, derived from analyses of DNA sequences before and after the event. This paucity of information is due to the simple fact that no system other than an actual viral infection, either an *in vitro* reaction or an *in vivo* circumstance, has been found that can reproduce the recombination reaction. Transfection and microinjection of naked DNA, for example, do not result in recombination at the appropriate site (Folger *et al.*, 1982; Kriegler and Botchan, 1983; Luciw *et al.*, 1984). Fortunately, there are now considerable sequence data on the reactants and products for a number of viral systems, and the rules of the reaction are by now quite well defined. Recently, the powerful techniques of *in vitro* mutagenesis have been applied to the viral DNA, and the ability of the viral genomes recovered from these manipulations to integrate has been determined. These results have helped to define the nature of the immediate precursor of the integrated provirus, the sequences on the virus that are required for the event, and the determinants of the precise sites of recombination to the host DNA. These points have been summarized recently (Panganiban, 1985).

2.1. Structure of the Viral DNA before and after Integration

As noted in Section 1.2, the initial product of reverse transcription is a full-length linear dsDNA flanked by long terminal repeat (LTR) sequences (Hughes *et al.*, 1978). The LTRs, ranging from 300 to 1300 base pairs (bp) in size in different viruses, are in the same, or direct, orientation. We will refer to the LTRs by the usual convention: The left or 5' LTR is the one that forms the 5' end of the viral mRNA during viral gene expression, and the right or 3' LTR forms the 3' end of the transcripts. Each LTR sequence contains short inverted repeats at its edges, ranging from as few as 2 bp to as many as 13 bp; in some cases, the length of the inverted repeats can be extended by allowing a few mismatches. Since each LTR contains two copies of these inverted repeat sequences, and since there are two LTRs in the full-length linear DNA, there are four copies of the inverted repeats in the molecule. The topology also dictates that the full-length molecule has these inverted repeats at its termini: The left terminus of the molecule is contributed by the left edge of the 5' LTR and the right terminus of the molecule by the right edge of the 3' LTR (Swanstrom *et al.*, 1981; Temin, 1981).

The initial product of reverse transcription of the retroviral RNA is the linear DNA, and at later times there appear two distinct circular species, bearing either one copy, or two tandem copies, of the LTR sequences. All three of these forms coexist in the infected cell for an extended period of time, ranging from one to a few days after infection. The precursor–product relationships among these species are not completely certain. The linear form is definitely the first discrete species to appear after infection and is presumed to be the precursor for the circular forms on this basis. The circular forms appear at about the same time, and there is no dramatic change in the mass ratio between them during the course of the infection; there is no evidence that one serves as precursor to the other.

There is some variation from virus to virus, and from cell line to cell line, in the characteristic ratio between the circles, such that under some conditions the smaller circle may be more abundant, while under other circumstances the larger may be more abundant. Rarely do the concentrations of the two circles differ by more than a factor of 2. The maximum concentration of the circles never approaches that of the linear DNA, and typically the circles are only one tenth as abundant as the linears. The simplest assumption from these observations is that the linear form is made first and that it serves simultaneously as the precursor for the formation of both circles, yielding the smaller circle by homologous recombination between the direct repeats of the LTRs and the larger circle by blunt-end joining of the termini. This assumption is far from proven.

The integrated proviral DNA appears soon after the circular molecules, and because of their nuclear localization, it is reasonable to assume that one or both of these circular species serve as the precursor. Because all three DNA forms coexist in the infected cell, however, it has been uncertain for many years which form(s) can serve in this critical role. If the linear form were the precursor, the termini would presumably be recognized by the integration enzymes. If the single-LTR circle were the precursor, the edges of the LTR would be recognized; there would have to be a duplication of the LTR during the integration reaction to account for the ultimate presence of the two LTRs flanking the integrated provirus. There is in fact a direct precedent for this mechanism in the formation of cointegrants of two prokaryotic plasmids mediated by insertion elements (Shapiro, 1979). If the double-LTR circle were the precursor, the junction between the LTRs would serve as the recognized sequence. There is now compelling evidence that this last possibility is the correct one and that the novel joint formed by the joining of the LTRs is at least the major integration site for the retroviruses. Panganiban and Temin (1984a) isolated a small DNA restriction fragment containing this sequence from a cloned double-LTR circle of spleen necrosis virus (SNV) and inserted this fragment into the central portion of a defective SNV genome. This DNA was applied to permissive cells along with helper virus genomes, the mixed virus population was allowed to spread through the culture, and the structure of the integrated DNAs was determined by Southern blots. The results showed that either of two sequences in the defective genome was utilized for integration reactions: either the normal sequences of the LTRs or the novel insert containing the LTR–LTR junction. The two sequences were used at equal frequency, suggesting that the LTR–LTR junction might be the only sequence recognized out of the possible structures formed by the terminal LTRs.

There is another important reaction product besides the integrated provirus, first uncovered in the analysis of cloned copies of circular retroviral DNAs (Shoemaker *et al.*, 1980, 1981a,b). It was found that a surprisingly high proportion of the circular molecules had suffered an intramolecular recombination reaction, resulting in either a deletion or an inversion of portions of the genome. In most cases, an end point of these rearrangements occurred at an LTR edge, suggesting that the events were mediated by the integration machinery intended to insert the virus into the host genome. Sequence analysis supported this idea, since all the hallmarks of an integration event were reproduced in these intramolecular events.

2.2. Sequence Features of the Reaction

What are the nucleotide sequences at the LTR–LTR junction? A summary of known sequences from a variety of retroviruses is given in Table I. In most cases, the actual junction has not been sequenced and the results shown are as predicted from the sequence of linear or integrated genomes, but our knowledge of the reverse transcription process allows such predictions to be made with some confidence. The sequences share several features. First, they contain inverted repeats of varying lengths; some are long perfect inverted repeats, while others show only imperfect repeats. Second, they contain a pair of conserved dinucleotide sequences, . . .CA. . . .TG. . ., flanking the center of symmetry of the inverted repeat, with only a few exceptions. Usually, there are 4 bp between these bases, with the possible exception of the human viruses, predicted to contain only 3 bp between the conserved bases. Included in Table I are the sequences predicted for the joining of the termini of a few transposable elements that move by reverse transcription of RNA; these sequences share many of the features of the retroviral LTR–LTR junctions, suggesting that the integration of these elements occurs via related mechanisms (Farabaugh and Fink, 1980; Dunsmuir et al., 1980).

Comparisons of the sequences at the termini of integrated proviruses with those of the LTR–LTR junction have shown that retroviral integration is not like that of phage λ (Mizuuchi et al., 1981). The reaction is not a simple recombination in which the viral DNA is cut, the host DNA is cut, and the ends are exchanged (Dhar et al., 1980; Shoemaker et al., 1980, 1981a,b; Van Beveren et al., 1981). The break points in the viral DNA are not precisely at the center of symmetry of the LTR–LTR junction, but are after the conserved CA and TG dinucleotides; the central base pairs between the dinucleotides are lost during the reaction and do not appear in the integrated provirus. Furthermore, a small number of bases of the host DNA are duplicated during the recombination (Hughes et al., 1981a; Majors and Varmus, 1981); the number ranges from 4 to 6 bp among different retroviruses. These bases are found in only single copy at the target site before infection, but are found twice, flanking the integrated provirus, after the insertion event. A plausible mechanism to account for this observation has been proposed (Shoemaker et al., 1980, 1981a,b). Stripped to its minimal content, the idea is that cleavage enzymes make staggered cuts in both the viral and host DNAs; the protruding ends of the viral termini are removed, while the protruding ends of the host termini are "filled in" by extension

TABLE I. Sequences of LTR–LTR Junctions

Virus	Predicted LTR–LTR junction[a]	Reference
M-MuLV	*********** **** *********** CGGGGGTCTTTCA'TTAA'TGAAAGACCCCAC	Shinnick et al. (1981)
FeLV	*********** **** *********** CGGGGGTCTTTCA'TTAA'TGAAAGACCCCCT	Hampe et al. (1983)
IAP	* *** *** **** *** * ** * CTAGCGCGGGACA'TCGA'TGTGAGGAGCCGC ** * * **** **** ** * **	Ono et al. (1985)
SNV	CTTCGGTACAAACA'TTAA'TGTGGGAGGGAGC *********** **** ***********	Shimotohono et al. (1980)
MMTV	GCCGACTGCGGCA'GCAA'TGCCGCGCCTGCA *** *** ** ** **** *** *** ***	Majors and Varmus (1983)
ASV	GCAGAAGGCTTCA'TTAA'TGTAGTCTTATGC * * ** * ** * * *	Swanstrom et al. (1981)
HTLV I	AATTTAGTACACA'GTGA'TGACAATGACCAT * * ** ** ** ** ** * *	Seiki et al. (1983)
HTLV III	AAAATCTCTAGCA'GAC'TGGAAGGGCTAAT ** *** ****** ***** *** **	Wain-Hobson et al. (1985)
Copia	TAAATTACAAACA'?'TGTTGGAATATA * * * * * *	Mount and Rubin (1985)
Tyl	TTCACCCATTTCTCA'?'TGTTGGAATAGAAAT	Farabaugh and Fink (1980)

[a] Bases that contribute to the inverted repeats are marked by asterisks. The sites of cleavage and joining to host DNA are marked by apostrophes. Most sequences are predicted only from the sequences of fragments of the viral genomes and the positions of probable sites for initiation of DNA synthesis of each DNA strand. The sites for initiation of plus strands are very speculative in most cases.

of the recessed ends. This idea would account for the loss of the central 4 bases of the viral DNA and for the duplication of 4–6 bp of host DNA. The order of the events is unclear, and models can be built in which the removal and filling in of the protruding ends follow the actual joining of one of the viral and host strands. It is also uncertain whether the termini would be 3' or 5' protruding ends. In the simplest versions, the viral and host termini are of the same type: Either both have 3' protruding ends or both have 5' protruding ends.

2.3. Essential Sequences at the Viral Integration Site

Several mutational studies have been carried out to determine the portion of the junction sequences that are required for the integration reaction. Early studies (Panganiban and Temin, 1983) showed that deletions of SNV impinging on the junction sequences from within the LTR could approach to within 8–12 bp of the center of symmetry without abolishing the ability of the virus to replicate and integrate. Deletions that extended closer to the inverted repeat caused severe decreases in the ability of the virus to replicate: Titers dropped 10^3-fold below normal, and integration, as judged by Southern blots, dropped 10^5-fold relative to the wild-type. These results suggest that the inverted repeats are essential for the reaction and that roughly 10 bp from the center of symmetry may be required.

Our laboratory has carried out related studies (Colicelli and Goff, 1985), using the Moloney murine leukemia virus (M-MuLV). A single base substitution was generated at the right edge of the LTRs, changing 1 of the 4 bp at the center of symmetry of the LTR–LTR junction. This virus was fully replication-competent, demonstrating that perfect symmetry is not essential for recognition of the site; the enzymes responsible for cleavage must resemble some restriction enzymes (e.g., *BglI*, *XmnI*) in utilizing a fragmented recognition site, with unimportant bases in the center. Deletions were also generated, extending from the edge of the LTR varying distances into the LTR sequences; the resulting deletion mutants could still carry out the formation of all three of the unintegrated DNA forms. The LTR–LTR junction was still formed, but the deletions abolished the perfect symmetry, removing bases asymmetrically from one side of the normal junction. As bases were removed from the center, the outer portions of the inverted repeat sequences were brought closer to the center. The most interesting mutation was the deletion of a single base pair at the exact edge of the LTR. The removal of this one base pair had no effect on the efficiency of replication or integration. This virus integrated, however, with a difference: 3 bp rather than the usual 4 bp were lost during the reaction. The sites of joining to the host DNA, in other words, had moved inward one base; the cleavages were still made after the conserved CA and TG dinucleotides. The result suggests that the point of joining is not determined by the distance from the center of symmetry of the LTR–LTR junction or from the edges of the LTR; rather, the cleavages on the two sides are determined by an "outside-in" measurement and are made relative to the outer portions of the inverted repeat. The protein subunits that must be responsible for the two cleavages are at least partially independent; there is no precise interaction between the enzymes on the two sides, since the repeats can be drawn together by 1 bp (3.4 Å) without deleterious effect. The simplest model for the recognition and cleavage steps is that protein factors bind to the outer portions of the inverted repeat sequence; these factors reach inward, toward the center of the inverted repeat, and cleave after the CA and TG dinucleotides. The central bases, whether 3 or 4, are removed and lost.

These studies also showed that larger deletions, extending from the center of symmetry outward into the rest of the inverted repeat sequence, do abolish integration. Deletion of even 2 bp blocks the reaction; only revertant viruses, arising by the reinsertion of one of the two missing bases, are recovered from such a mutant. A larger deletion of 8 bp could not integrate at all, and no replicating virus could be recovered in an appropriate host (Colicelli and Goff, 1986). These results suggest that the recognition region extends close to the LTR edge; alternatively, bringing the two recognition regions too close together may block the reaction.

2.4. Sequences at the Target Site in the Host Genome

Each virus induces the duplication of a characteristic number of base pairs of host DNA on integration (Dhar *et al.*, 1980; Shoemaker *et al.*, 1980, 1981a,b; Fitts and Temin, 1983; Majors and Varmus, 1981; Hughes *et al.*, 1981a,b). The murine leukemia viruses duplicate 4 bp; spleen necrosis virus duplicates 5 bp; the mouse mammary tumor virus duplicates 6 bp. The fact that the number duplicated is characteristic of the virus and not of the host suggests that the virus encodes a function involved in the duplication, but it is not known whether the function is a viral gene product or simply a feature of the sequences at the viral integration site. Recently, there have been constructed in M-MuLV mutations of both types that have altered the number of bases of host DNA duplicated, albeit only occasionally. A deletion of 1 bp at the LTR edge [mentioned in Section 2.3 (Colicelli and Goff, 1985)] induced the duplication of 5 bp rather than the usual 4 bp in one particular integration event; a very different mutation in the *pol* gene [to be described in Section 3.2 (Donehower and Varmus, 1984)] did the same. It is not clear whether these mutations specifically changed the number of bases duplicated to 5 bp or simply made the reaction more error-prone. In fact, they may not even have increased the frequency of such events; one wild-type provirus has been reported to have duplicated 5 bp (Van Beveran *et al.*, 1981), and the number of mutant proviruses examined is too small to obtain meaningful estimates of frequency.

A large number of wild-type integrated proviruses have been cloned, and the junction fragments of many have been sequenced. Only rarely have the preintegration sites been cloned from uninfected cells, using as probes the cellular DNA adjacent to a cloned provirus; these were the studies that demonstrated conclusively that bases are duplicated during the reaction (Hughes *et al.*, 1981a). Although few preintegration sites have been cloned, their sequence can be reconstructed from the sequence of the proviral junctions by simply assuming that the usual number of bases were duplicated. Examination of large numbers of such sequences for conserved features has revealed no consensus at all; the weak consensus proposed from limited data (Shoemaker *et al.*, 1981b) was not supported by subsequent results. Thus, the local environment is not important for the reaction, and virtually any sequence can be a target. There is no tendency for integration to occur into AT-rich or AT-poor sequences. This fact is important in the use of the viruses for mutagenesis because it bears on the frequency of recovery of mutations in a particular target gene. As we shall see below, there may well be global determinants of target preference, but there seems to be no direct preference for any DNA sequence. In other systems, there may be specificity. For example, it has been suggested that *Drosophila* elements may show preferred targets (Inouye *et al.*, 1984); certainly many insertions have been found in promoter regions of genes (Ikenaga and Saigo, 1982). We should anticipate the possibility that the retroviruses may show a similar preference for such sequences, perhaps because of "open" chromatin configurations in the vicinity of these sequences.

3. INTEGRATION ENZYMES AND FACTORS

Because there is no *in vitro* system that performs the retroviral integration reaction, there is little biochemical information on the enzymes that catalyze it. The bulk of the data are genetic, suggesting that both host and viral functions are involved. In the murine system, there exists a single locus, termed *Fv-1*, that affects retroviral infection intracellularly; this gene can block the formation of the retroviral DNA soon after infection or its integration into the genome. We will briefly review its properties below. In several systems, there has been identified a viral gene essential for integration: The 3′ portion of the *pol* gene clearly encodes a *trans*-acting function, distinct from reverse transcriptase, that is needed for establishment of the provirus. This gene is highly conserved among the retroviruses (Chiu *et al.*, 1984; Toh *et al.*, 1983; Patarca and Haseltine, 1984) and is even homologous to genes of transposable elements (Mount and Rubin, 1985; Boeke *et al.*, 1985). But only in the avian system has elucidation of the biochemistry of the protein products of this portion of the *pol* gene begun.

Here, proteins have been shown to specifically bind and endonucleolytically nick the viral sequences of the long terminal repeat (LTR)–LTR junction. These important experiments are summarized below.

3.1. Host Factors: *Fv-1* Gene

The *Fv-1* gene was originally defined by a difference between strains of mice in their sensitivity to erythroleukemia induced by the Friend murine leukemia virus (Lilly and Pincus, 1973; Jolicoeur, 1979). Strains bearing a particular allele, now termed $Fv-1^n$, were sensitive to the virus, and strains bearing a different allele, now termed $Fv-1^b$, were resistant. The resistance was dominant over sensitivity in the heterozygous animals, precluding the possibility that the resistance was due to the lack of a viral receptor. Later analysis revealed that the resistance could be reproduced in tissue culture; lymphoid or even fibroblastic cells from the resistant mice were also resistant to infection by the virus (Decleve *et al.*, 1975).

Further analysis of other strains of murine leukemia viruses revealed that three types could be distinguished on the basis of their ability to replicate on cells derived from mice of different *Fv-1* genotype. One type of virus, including the original Friend virus isolate, was termed N-tropic and replicated well on $Fv-1^n$ cells, but not on $Fv-1^b$ cells; a second type termed B-tropic replicated on $Fv-1^b$ but not $Fv-1^n$ cells; a third type termed NB-tropic could replicate well on either type of cell. Mixed virus particles containing proteins of both N- and B-types were found to be restricted by both $Fv-1^n$ and $Fv-1^b$ alleles (Rein *et al.*, 1976). The *Fv-1* gene itself was mapped by classic genetic means to mouse chromosome 4.

The block to infection in nonpermissive pairings of virus and host was shown to occur at an interesting stage in the life cycle by direct biochemical analyses (Jolicoeur and Baltimore, 1976; Jolicoeur and Rassart, 1980; Yang *et al.*, 1980). The block occurred early in the infection and prevented the successful formation of the integrated provirus. In some experimental conditions, there was little or no effect on the formation of any of the three unintegrated DNAs; in other conditions, the formation of the linear DNA was normal, but the level of the circular forms was depressed; in still other conditions, even the level of the linear DNA was lowered. These analyses suggested that the resistant allele encoded a function that blocked some step in the processing of the viral DNA; perhaps there was some effect on the late steps of reverse transcription, or on the transport of the DNA to the nucleus, or on the actual integration step.

The magnitude of the effect—i.e., the ratio of the apparent titer of virus on sensitive cells to that on resistant cells—ranges from as little as 10 to a maximum of 1000, depending on the experimental conditions and the choice of virus and cell (Chinsky and Soeiro, 1981). Thus, the resistance is never an absolute block. There is a curious dependence on multiplicity of infection: At high multiplicity, the resistance is overcome, and virus successfully infects even nonpermissive cells (Pincus *et al.*, 1975; Duran-Troise *et al.*, 1977). Under these conditions of high multiplicity of infection, there are extremely high levels of virion particles impinging on the cell; the particle/plaque-forming-unit (PFU) ratio for the retroviruses is high, and thus thousands of physical particles may attach to each cell when the calculated applied ratio is only 10 PFU/cell. It seems that under these conditions, the resistance is ineffective, and it may be that the incoming virions have swamped out some blocking enzymes, present in only finite numbers.

Is the block directly at the stage of integration? There is little evidence in support of this possibility. The most severely affected infections tend to be the ones that show a reduction in all the DNA forms including the linear DNA. This would suggest that the primary block in all the resistant cells may be at this early step; the reduction in all the subsequent DNA forms could be a cascade effect of the reduced amount of linear DNA. Those weaker blocks, in which the amount of linear DNA is hardly affected at all, might show reduced levels of circles simply because the cascade amplifies the minimal effect on the linears. Nevertheless, we should anticipate that there may be many cellular gene products involved in multiple steps of the processing of the viral DNA (Chinsky and Soiero, 1982). Cellular enzymes may be required for the completion of synthesis of the linear DNA (Jha *et al.*, 1980; Richter *et al.*, 1984), for the nuclear localization of the DNA, for the ligation of the termini to form the double-LTR circle, and for several steps in the integration process. There is evidence that these host proteins may be expressed in a cell-cycle-dependent way. Infection can proceed well only when the cells proceed through S phase; viral DNA apparently integrates poorly when the cells are arrested by contact inhibition at the time of infection. It will be important to determine what these host factors are and how they interact with the viral DNA. The *Fv-1* gene is the first one to be identified even indirectly to affect these events.

The viral gene responsible for the tropism of a given strain of murine leukemia virus has been unequivocally identified. Early work showed that viruses of different tropism showed variations in two-dimensional fingerprints of the genomic RNA, with the variant oligonucleotide mapping in the *gag* gene; analysis of the *gag* proteins showed that the P30 cleavage product had an electrophoretic mobility that was characteristic of the tropism (Hopkins *et al.*, 1977; Gautsch *et al.*, 1978, 1980). Furthermore, mutant viruses could be isolated by selection for growth on nonpermissive hosts; the changes in tropism were associated with changes in both the oligonucleotide and the mobility of the P30 protein. These data argued that the significant changes were closely linked to the P30 *gag* domain. Recently, DNA clones of the genomes of N- and B-tropic viruses were obtained, and DNA-fragment exchanges were performed to generate chimeric viral genomes (DesGroseillers and Jolicoeur, 1983; Ou *et al.*, 1983; Boone *et al.*, 1983). Virus was recovered from each chimera and typed for tropism. The results showed that the tropism mapped to the P30 region and that in fact changes in a short stretch of amino acids were sufficient to change the tropism of the entire virus. This implies that P30 can interact in some way with the *Fv-1* gene product.

The fact that P30 determines the *Fv-1* tropism of the murine leukemia viruses suggests that P30 is involved in steps of reverse transcription or integration, but it is not clear how it is involved. The P30 protein is the major capsid protein of the virion core and is essential for the assembly of virions. Perhaps its job in early steps is simply to protect the incoming RNA and DNA; perhaps the *Fv-1* gene product of a resistant cell can attack the virion and can block complete reverse transcription. Alternatively, P30 may have a more active role in early events. The protein does bind to reverse transcriptase, and it may enhance its activity; it could be that the *Fv-1* protein blocks this enhancement.

3.2. Genetics of Viral Integration Enzymes

Many prokaryotic transposable elements have been shown to encode functions necessary for their own transposition (Gill *et al.*, 1979), and the retroviruses also encode such functions. The *in vitro* mutagenesis of retroviral genomes has been used to define the functions of several viral proteins. The best evidence that these viruses encode a function directly necessary for the establishment of the integrated provirus has come from analysis of mutants with defined lesions at various positions in the genome.

In the first such study, small deletion mutations were generated in a cloned DNA copy of the genome of Moloney murine leukemia virus (M-MuLV) (Schwartzberg *et al.*, 1984). Because no system other than a normal infection results in the correct integration of viral DNAs, an indirect approach to analyzing such mutants had to be employed. The altered DNAs were first introduced stably into recipient cells by calcium-phosphate-mediated cotransformation with the selectable pSV2neo DNA, and transformants were selected by exposure of the cells to the antibiotic G418. This transformation event results in the formation of integrated viral DNA, but the process does not occur by the efficient viral machinery: No viral proteins are required, and the viral DNA is integrated aberrantly, with no recombinations occurring at the edges of the LTR. Nevertheless, a complete provirus can be introduced simply by using a full-length clone.

Mutant genomes bearing deletions at a variety of positions in the viral *pol* gene were all able to induce the assembly and release of virion particles by the transformants. Mutations in the major portion of the *pol* gene abolished reverse-transcriptase activity as measured on synthetic templates (Goff *et al.*, 1981), and the virions were inert and uninfectious; no viral DNA could be produced after infection of fresh recipient cells by these viruses. Mutations in the 3' portion of the genome, however, induced a very different phenotype. These mutations had no effect on reverse transcriptase, and the virions could direct the formation of all three DNA forms after infection of fresh cells. The viral DNA did not integrate in these cells, however, and the infection aborted at this stage. No integrated proviruses could be detected in the cells. These results suggested that the 3' portion of the *pol* gene was required for integration and that integration was essential for productive infection; the unintegrated DNAs apparently could not serve as templates sufficiently well to produce progeny virus. There were other quantitative effects on the rate of replication that were caused by these mutations. There was a decrease in the level of production of virus particles, suggesting that the aberrant *pol* product had some role in virion assembly; there was a decrease in the rate and extent of cleavage of the *gag* precursor protein in the virions, suggesting that the aberrant *pol* protein might be involved in maturation of the virus. We now know that a third domain of the *pol* gene is in fact necessary for the processing of the *gag* protein, and so it is reasonable that the 3' deletions have some minor effect on the other function of the *pol* precursor.

A similar and more detailed study has been performed with a single amino acid substitution in the same domain (Donehower and Varmus, 1984). A point mutation in the 3' portion of the *pol* gene of M-MuLV was generated and was found to produce a similar phenotype to the deletions: Virions were produced by transformants, and these virions were unable to establish integrated proviruses in infected cells. The ability of this mutant to serve as helper virus for the transmission of replication-defective, genetically marked, viral constructs was also tested. Cells were formed expressing both the mutant genome and a viral construct carrying the herpes virus gene that encodes thymidine kinase (TK). The mutant could encapsidate and transmit the TK construct to new cells, but there was a major defect in the efficiency with which the construct was integrated into the recipient cells. Rare cells that received the gene could be isolated, and many were integrated aberrantly, at sites other than the usual one formed by the joining of the LTRs. These results confirmed that the 3' portion of the *pol* gene was indeed needed in *trans* for the establishment of the integrated DNA and showed that DNAs introduced into cells by infection could integrate aberrantly, at very low efficiency (perhaps 100-fold less efficiently than normal), when the viral function was disrupted.

Recent work has shown that occasionally the virus bearing the point mutation could transfer viral genomes such that integration did occur at the correct site. Analysis of these integrants has shown that the normal sites on the viral genome were utilized. The point mutant is apparently slightly leaky, and the defective protein may rarely be able to catalyze the recombination; alternatively, there may be functions in the cell that can replace the viral protein. Of two provirus clones sequenced, one insertion was associated with the duplication of the usual 4 bp of host DNA, and one was associated with duplication of 5 bp; it is not clear whether the mutant protein is specifically responsible for a duplication of 5 bp part of the time or whether inefficiency of the reaction simply causes other proteins to occasionally make mistakes (H. E. Varmus, personal communication). A mutation in the integration site was also found to result in the occasional duplication of 5 bp. As noted above, it is not even clear whether the frequency of such "errors" is high in the case of the mutants; one wild-type integrant has been reported to have duplicated 5 bp (Van Beveran *et al.*, 1981).

Another study of a naturally occurring mutant of M-MuLV (Gerwin *et al.*, 1979) offered support for the mapping of the functions of the *pol* gene; the reverse transcriptase function was shown to be encoded in the center of the *pol* region, leaving a large portion of the gene at the 3' end unaccounted for (Levin *et al.*, 1984). Other mutants that potentially affect this function have not been fully characterized (Rein *et al.*, 1978).

Analogous mutational experiments have also been performed in the spleen necrosis virus (SNV) system (Panganiban and Temin, 1984b). A total of five mutations in the 3' portion of the *pol* gene of SNV were generated, and the altered DNAs were introduced into chicken cells. A single one of the altered DNAs was able to replicate and spread throughout the cultures; no virus could be recovered from the other mutant DNAs. Some of the inactive clones contained mutations at the same position, and of similar size, as the active clone, and it is unclear at this point why the biological properties of the clones are so different; presumably, the amino acids encoded at this position of the gene by some of the mutants can have effects on other stages of the life cycle. The active clone gave rise to virus, but the virus was not wild-type: Replicating DNA genomes were readily detected in infected cells, but no integrated DNAs were found. Thus, this mutation suggests that the region is needed for the reaction. The mutant was furthermore shown to be able to complement a mutant with a lesion at the LTR edge. Coinfection of cells with both viruses resulted in the formation of integrated structures, suggesting that the *pol* gene product was a *trans*-acting element. Low levels of recombinant virus could be detected in these coinfections, but the level was presumably too low to account for the observed complementation.

Could other viral gene products be required for the integration reaction? This is a distinct possibility, if such gene products have multiple functions. The 3' end of the *pol* gene probably encodes the only function specifically required for integration and not needed at any other time. It might well be, however, that reverse transcriptase is needed both for DNA synthesis and for integration; the mutations studied to date are not helpful because no DNA is made. Similarly, *gag* proteins could be involved, but it is difficult to determine whether they are because these proteins are required for the assembly of virions and for very early steps in infection. The role of these proteins in integration, if any, awaits the recovery of very special mutations that block only one of their potentially numerous functions or the development of *in vitro* systems that carry out the recombination reaction.

3.3. Biochemistry of Viral Integration Enzymes

The first hint that an integrase might be present in the infecting virion came from biochemical studies of virion-associated studies. Early analyses revealed the presence of an endonuclease activity in virions of avian retroviruses (Golomb and Grandgenett, 1979; Golomb et al., 1981; Kopchick et al., 1981; Hizi et al., 1982). In the avian system, the pol gene serves to encode several proteins. First, the entire gene encodes a protein of about 92,000 daltons termed beta; a portion of this protein is cleaved to form an N-terminal fragment of about 62,000 daltons termed alpha and a C-terminal phosphorylated fragment of 32,000 daltons termed pp32. The mature reverse transcriptase is a heterodimer of beta and alpha. The purified enzyme was found to exhibit endonuclease activity, and that activity has been characterized (Leis et al., 1983). The activity showed a preference for supercoiled DNA, converting the DNA to nicked forms, and required manganese as cofactor. Genetic evidence that the activity was indeed encoded by the pol gene was provided by a study of two mutants that encode a temperature-sensitive reverse transcriptase, ts335 and ts337. These mutants were also found to encode a temperature-sensitive endonuclease, suggesting that both activities were affected by the lesions and therefore probably encoded by the same gene (Golomb et al., 1981).

The purified beta subunit, and the pp32 protein as well, showed endonuclease activity, while the alpha subunit did not. These results strongly suggested that the endonuclease activity was encoded by the 3' portion of the pol protein in the region common to beta and pp32. Other studies with the pp32 protein showed that it exhibited tight DNA-binding activity, being able to cause the retention of DNA fragments on nitrocellulose filters. Most important, the enzyme showed specific binding to the LTR sequences of the viral genome (Misra et al., 1982). The protein was able to protect a small region at the LTR–LTR junction from DNAse I digestion, and the region was exactly what one would predict for an integrase function. The enzyme also showed endonuclease activity (Grandgenett et al., 1978).

The site of cleavage by the purified reverse transcriptase has been precisely mapped by Skalka and coworkers (Duyk et al., 1983). Single-stranded templates consisting of M13 phage clones of the LTR–LTR junction were treated with the endonuclease activity of the beta subunit, and the position of cleavage was determined by extension of a primer to the new terminus with the Klenow fragment of DNA polymerase I. The single-stranded DNA is an odd template, presumably not the natural one used in vivo, but the enzyme is active on this form. Some cleavage occurred at many sites on the clones, but a preferred site was the LTR–LTR junction. The enzyme could cut both the minus and plus viral strands; the site of cleavage was exactly three bases away from the edge of the viral LTR. Thus, the enzyme apparently cut one base away from the site of joining to the host DNA. Very recently, conditions have been found such that cleavage occurs two bases from the edge; this activity is thus acting at exactly the correct location to account for the reaction seen *in vivo*. These biochemical experiments strongly suggest that the *pol* endonuclease function is involved in integration. Attempts to recreate other steps in the reaction are under way in many laboratories.

There is less information about the endonuclease function in the murine system. A similar activity has been identified in virions, and the activity has been shown to reside in a 45,000-dalton protein. It is now clear that this protein is encoded by the 3' end of the *pol* gene; in the murine system, the *pol* precursor is entirely cleaved so as to separate the reverse transcriptase from the endonuclease function. This domain has been expressed in bacterial cells by at least two groups, and polyvalent antibodies against the protein have been raised in rabbits. Biochemical studies on the activity of the protein may further clarify its role in integration.

4. INSERTIONAL ACTIVATION AND TRANSFORMATION

The insertion of a proviral DNA into the host genome can have profound effects on the expression of adjacent genes. We know from the seminal work of Hayward and co-workers that leukemogenesis by the replication-competent retroviruses is frequently associated with the integration of a provirus near a cellular protooncogene; in these tumors, the expression of the oncogene is dramatically elevated, presumably initiating the transformation process (Hayward *et al.*, 1981). The effect is genetically dominant, in that only one allele needs to be activated. The appearance of tumors is easy to detect, and for these reasons, more examples of insertional gene activation have been studied than of insertional inactivation. We will review several examples of activation of protooncogenes by both avian and murine viruses.

4.1. *c-myc* Locus

The initial understanding of the mechanism of leukemogenesis by the replication-competent leukemia viruses stemmed from the observation that many bursal lymphomas in chickens expressed an unusual class of mRNAs, containing a short stretch of viral sequences at the 5' end and cellular sequences in the remaining 3' part (Neel *et al.*, 1981; Fung *et al.*, 1981). The sizes of the mRNAs usually fell into one of four classes. Analysis of the site of insertion of the proviruses in these tumors revealed a remarkable fact: The proviruses were all inserted upstream of a single gene, the *c-myc* protooncogene previously known from its transduced version in the genome of the oncogenic virus MC29. The insertion had apparently induced the formation of the novel mRNA; since the mRNAs were initiated within the viral long terminal repeat (LTR) at the normal site for transcriptional initiation of viral mRNAs, the process was termed "promoter insertion." The high levels of *c-myc* mRNA, or its aberrant structure, presumably initiated the process that eventually led to the formation of a full-blown tumor.

It was found that virtually all the activating proviruses were incomplete copies of the viral genome and were defective (Fung *et al.*, 1981). The deletions often spanned the 5' end of the genome and removed the 5' LTR; viral proteins were not expressed. Complete proviruses do not normally induce the formation of mRNAs initiated at the 3' LTR, and it was reasonable to assume that loss of the 5' LTR was involved in the production of the downstream transcripts. This hypothesis has been supported by manipulation of artificial constructs; removal of the 5' LTR indeed increases transcripts from the 3' LTR, roughly 3- to 4-fold (Cullen *et al.*, 1984). Alternatively, expression of the viral genes might be opposed by immunoselection in the infected animal.

Analysis of more bursal tumors has revealed more diverse mechanisms of *c-myc* activation (Payne *et al.*, 1981, 1982). Many tumors show proviruses inserted downstream of the gene, or upstream of the gene but oriented away from the gene. In such circumstances, the viral promoter cannot directly initiate *c-myc* expression. The evidence suggests that the virus is activating the gene in a distance- and orientation-independent manner; these are the features of enhancer elements, known to be carried by the viral LTRs. The more general description of the *c-myc* activation is thus via an "enhancer insertion." The mechanism of activation is thus consistent with observed differences in leukemogenicity of various isolates of viruses (Cullen *et al.*, 1983). The subgroup F viruses, causing tumors of different pathologies, integrate near *c-myc* as well as at other sites in at least some tumors (Simon *et al.*, 1984). Another class of virus, the avian reticuloendotheliosis virus, seems to act in a similar way to activate *c-myc* in the majority of the tumors (Noori-Daloii *et al.*, 1981).

Similar events have now been identified in the murine (Steffen, 1984; Corcoran *et al.*, 1984) and feline systems (Neil *et al.*, 1984). The *c-myc* gene is a frequent site for proviral insertion in leukemias of these animals, although it seems not to be as common a target as in avian bursal tumors. Here, too, the importance of the viral enhancer (Luciw *et al.*, 1983) has been supported by extensive studies, and these sequences are clearly important in determining the target tissue for transformation by the leukemia viruses. DNA-fragment exchanges have been made between nonleukemogenic and leukemogenic fragments, and the pathogenicity of the virus segregates with the enhancer-containing fragments (Lenz *et al.*, 1984; DesGroseillers *et al.*, 1984). Exchanges between Friend and Moloney viruses have shown that the respective tissue tropisms for preerythrocytes and T cells are mainly encoded by the enhancers (Chatis *et al.*, 1983, 1984). These tropisms may simply reflect the tendency of the virus to replicate better in particular tissue, or they may reflect the efficiency of the enhancer to activate *c-myc* in that tissue.

We can say very little about the frequency of insertion at the *c-myc* locus during infection; we do not know the number of cells challenged by virus infection, and we do not know whether the mere insertion into the gene is sufficient to induce a tumor. We do know that transformation cannot readily be reproduced by infection of primary cells *in vitro*. Thus, activation may be very rare; more plausibly, secondary mutations at distant loci may be required for the formation of a full-blown tumor. It is now clear that there are often other mutations in the activated *c-myc* loci (Westaway *et al.*, 1984).

4.2. Other Loci

Other tumors have now been found in which host genes other than *c-myc* have served as targets for insertional activation. Perhaps the best characterized of this class are due to insertions at other previously identified protooncogenes. Avian leukosis virus can induce erythroblastoid tumors, different from the bursal lymphomas described above. In these tumors, a different site of integration was discovered: The cellular protooncogene termed *c-erbB* was the common target of insertional activation (Fung *et al.*, 1983). This gene is the source of the active *v-erbB* oncogene on avian erythroblastosis virus; it has recently been identified as the gene that encodes the epidermal growth factor receptor (Downward *et al.*, 1984). The gene is apparently the target for about 70% of the transforming insertions in these tumors. The propensity for this gene to be active in erythroid tumors, even though the routes of activation are diverse, strongly argues for the tissue-specific effects of the gene product.

A third known protooncogene has also been found to be a possible target. The Abelson murine leukemia virus (A-MuLV) is a defective acute transforming virus that carries its own active oncogene, *v-abl*, and when carried into animals by the appropriate helper virus, it normally transforms cells of the B-cell lineage by the direct action of this oncogene. It was found that a different tumor, a "plasmacytoid" lymphosarcoma, was induced at low frequency by the virus; analysis of cells from these tumors often revealed that no A-MuLV proviruses were present at all. Further analysis revealed that the host cellular protooncogene *c-myb* was rearranged (Mushinski *et al.*, 1983) and that the rearrangement was the insertion of the Moloney MuLV (M-MuLV) helper virus into the 5′ side of the gene (Shen-Ong *et al.*, 1984). This gene is the homologue of the active oncogene *v-myb*, present on avian myeloblastosis virus. Thus, these tumors were probably induced by the insertion of the helper component of the virus population and may not have depended on infection by the A-MuLV genomes.

A slightly different agent has been shown to activate expression of a fourth known oncogene in the induction of mouse plasmacytomas. Screening for rearrangements of the known oncogenes revealed that the *c-mos* locus was altered in one such tumor; transcription of this normally silent gene was activated by changes at the 5′ end of the gene (Rechavi *et al.*, 1982). The changes were found to be due to the insertion of a long DNA element flanked by direct repeats, and comparison with known elements revealed that the insertion was a copy of the repeated gene family that encodes the intracisternal A-type particles (IAPs) oriented to initiate transcription away from the *c-mos* gene (Kuff *et al.*, 1983a; Canaani *et al.*, 1983). These elements are related to retroviral genomes by homology and function, but do not encode extracellular virion particles. There are perhaps 1000 copies in the mouse genome, and many cell lines express mRNAs and proteins to form intracellular particles of unknown fate. The evidence is that these elements can transpose to activate other genes very much like a functional virus, and indeed the new elements contain all the hallmarks of a provirus: Terminal base pairs are lost, and host base pairs are duplicated. It is unclear whether the movement was via reverse transcription of an mRNA or via direct DNA transposition. A curious fact is that the two LTR sequences of a given activating element show numerous polymorphisms; movement via an RNA intermediate should make the LTRs identical, since according to current models for reverse transcription, the LTRs are synthesized from unique sequences in the RNA. Thus, either the IAP element moved to *c-mos* via DNA or an extraordinary number of mutations arose in the LTRs after movement via RNA.

Other IAP transpositions have been found to result in insertional mutations. A second myeloma, NSI, often used as a fusion partner in the formation of hybridomas, also contains an IAP element in the *c-mos* locus (Cohen *et al.*, 1983). This IAP is oriented in the same transcriptional direction as the *c-mos* locus. In another study, two variant myeloma cell lines, selected after loss of expression of the κ light-chain gene, were found to have suffered insertions of IAPs into intervening sequences of the κ gene (Hawley *et al.*, 1982; Kuff *et al.*, 1983). These events represent spontaneous transpositions, in the sense that they were not initiated by any exogenous infection. The appearance of the insertions in myelomas, which often express IAP mRNAs, suggests that transposition might require previous IAP expression, but it is not clear whether IAP mRNA is a precursor or whether IAP gene products are involved in catalyzing DNA-mediated transposition.

Can insertions into new, previously unknown loci be detected? The answer is yes. Transformation by the murine mammary tumor virus (MMTV) has been shown to be clearly associated with the insertional activation of discrete loci not identified by any other means as protooncogenes (Nusse and Varmus, 1982). Surveys of mammary tumors of C3H mice showed that most contain several new proviruses scattered at various locations in the genome, only one of which is presumably responsible for an insertional activation event. The presence of these multiple copies makes the detection of common sites for integration very difficult; this problem was overcome by the use of rare tumors containing only one new provirus. Clones of integrated proviruses were obtained, and the flanking host DNA was subcloned. Analysis of DNAs from other tumors, using the flanking DNA as probe, revealed that a high proportion of the tumors contained a provirus in the vicinity of this locus. The locus, termed *int-1*, was cloned from uninfected cells, has been extensively mapped, and is located on chromosome 15 cytologically near *c-myc*. The complete DNA sequence of the coding region has recently been determined (van Ooyen and Nusse, 1984). Tumors, but not normal tissue, express from this locus an mRNA that is presumably important in the transformation event. The insertions are widely scattered in the locus; as in the case of activation of *c-myc*, insertions were found both 5' and 3' to the gene, and in this case the coding regions were uninterrupted. The proviruses are oriented away from the *int-1* gene, suggesting that enhancer insertion is the mechanism of activation (Nusse *et al.*, 1984).

Similar procedures have identified a second locus, termed *int-2*, that is a frequent site of integrated MMTV proviruses in tumors of BR6 mice (Peters *et al.*, 1983; Dickson *et al.*, 1984). This gene has been mapped to mouse chromosome 7 and is therefore not linked to *int-1* (Peters *et al.*, 1984). There is no homology between the two genes, and it seems likely that activation of either of the two can result in the formation of similar tumors. A third group has identified common insertion sites (Etkind and Sarkar, 1983), in support of these more detailed studies.

Analogous results have been obtained in tumors induced by M-MuLV. Although some insertions are in the vicinity of *c-myc* as noted above, most are not; analysis of many tumors has revealed common sites. The tumors used for the studies are derived in rats rather than in mice, to avoid the complications of the numerous endogenous proviruses in mouse cells with homology to M-MuLV. Even in rats, these studies are made difficult by the fact that virtually all tumors have a large number of proviral insertions. Initially, a locus was cloned that was the site of insertion in 5 of 16 tumors tested; this locus was termed *Mlvi-1* (Tsichlis *et al.*, 1983a). The same group soon identified a second locus termed *Mlvi-2* that was a target in 10 of 16 rat tumors (Tsichlis *et al.*, 1983b). The mouse homologue of the latter gene has been identified with the rat probe; the chromosomal location has been determined to be mouse chromosome 15 (Tsichlis *et al.*, 1984). Parallel work defined a locus termed *RMoInt-1* (Lemay and Jolicoeur, 1984) with similar properties, acting as a target in 11 of 20 tumors. The functional properties of these loci are completely unknown.

Tumors in mice have also been studied. In these animals, viruses like M-MuLV undergo recombination events with endogenous proviruses to form recombinant viruses with altered host range and with different properties. It is thought that these new viruses, dubbed mink-cell focus-forming (MCF) viruses, may be the proximal inducers of the leukemias, since new MCF proviruses are frequently found in mouse tumors. One common site for these MCF genomes in mice has been found and termed *Pim-1* (Cuypers *et al.*, 1984). Transcriptional activation of the gene has been demonstrably coupled to the insertion of the provirus. No homology to known oncogenes, or to *Mlvi-1*, *int-1*, or *int-2*, could be found. Activation could be induced by upstream or downstream integration;

furthermore, it could be induced in different tumors by MCFs or by nonrecombinant ecotropic proviruses.

Could these loci simply represent "hot spots" for insertion by the viruses and have no connection with the transformation event? Although this is a formal possibility, it seems highly unlikely. Infection of cells normally does not result in insertion at particular spots, but rather on restriction fragments of many sizes and on many different chromosomes. It is much more likely that the insertion is relevant to tumorigenesis, although the connection is unknown.

The most impressive use of retrovirus insertion as a means of identifying new genes in mammalian cells has been the finding of common target sites in a variety of virally induced tumors, as described above. The detection of new oncogenes by any route is of considerable significance, and the use of provirus tagging to find them is a strong proof of the power of the method. It seems likely that any gene the activation of which can be readily detected (as by tumor formation) can be cloned by this method.

5. INSERTIONAL MUTAGENESIS

The power of insertional mutagenesis in bacteria and the successful identification of oncogenes as targets of proviral insertion have encouraged several groups to utilize retroviruses for the tagging of interesting genes. In fact, no new genes have yet been cloned from infected cell lines by such tagging methods, but proviruses in the germ line of mice have served to identify new genes. Insertions into specific, known genes have been demonstrated, and it is likely that proviruses will be used to make new mutations and find new genes in the near future. We will summarize the information about a number of insertions that have been studied.

5.1. Examples of Germ-Line Insertions

A large body of work by Jaenisch and coworkers has demonstrated the possibility of generating germ-line insertions of retroviral DNAs by exposure of early embryos to Moloney murine leukemia virus (M-MuLV) (Jaenisch, 1976; Jaenisch et al., 1981; Harbers et al., 1981; Jahner et al., 1982). The exposure could be effected by a variety of protocols: Virus could be simply applied to the early embryo; preimplantation-stage embryos could be cocultivated with virus-producing fibroblasts; producer cells could be introduced into a developing blastocyst; or virus could be microinjected into 8-day embryos (Jaenisch et al., 1981). A high proportion of the progeny were mosaic carriers of integrated proviruses, and in many cases the germinal tissue carried these DNAs. The progeny of these animals were also shown to carry the viral DNA, and the presence of the provirus was subsequently transmitted as a simple Mendelian locus. A total of 13 such loci, termed *Mov* loci, were studied in detail. In some cases, the DNA was expressed during the course of development and could induce viremia that spread virus throughout the animal. These animals eventually developed leukemia as though they had received virus by exogenous infection early in life. In other cases, the DNA was not expressed during the life of the animal, and no viremia or disease occurred. These proviruses were either defective and could not encode functional virus or were integrated in locations in the genome that did not allow expression of the viral genes.

The proviruses were always present in only one of the two homologues of a given chromosome, and the initial animals were thus heterozygous for the presence of the provirus. Insertions could be found on any chromosome, including the X chromosome (Stewart *et al.*, 1983). There were no phenotypes associated with the presence of any of the 13 proviruses in heterozygous state except viremia, and thus no dominant mutations had been generated by these insertions. To test for recessive mutations, homozygous animals were generated by mating. Most of the homozygotes were normal, but one of the proviruses, *Mov13*, caused a recessive lethal mutation that resulted in early embryonic death (Jaenisch *et al.*, 1983). Further analysis revealed the reason: The provirus had inserted into a known gene, encoding the alpha 1(I) collagen protein (Schnieke *et al.*, 1983; Harbers *et al.*, 1984). The insertion had occurred at the 5' end of the gene, presumably preventing the formation of a functional mRNA and protein product, and the result of the loss of the protein was a massive breakdown in vessels and connective tissue. The animals died of hemorrhages before day 12 of development (Lohler *et al.*, 1984).

The group of Copeland and Jenkins has been active in mapping the genetic locations of many of the inherited, endogenous proviruses of mice. Many of these proviruses seem to be linked to interesting genes, and in principle the insertions could be responsible for the creation of some of the mutant alleles at these loci. One very old mutation at the *dilute* (*d*) locus was found to be very tightly linked to the single endogenous ecotropic provirus on chromosome 9 in DBA/2J mice (Jenkins *et al.*, 1981; Copeland *et al.*, 1983a). The mutation is a recessive marker, resulting in altered coat coloration in the homozygous state. The provirus was cloned, and DNA flanking the provirus was subcloned to generate a nucleic-acid probe for the locus. A revertant locus, termed d^{+2J}, was recovered from a spontaneous reversion event, and analysis of this locus revealed that the provirus was no longer present; this strongly suggested that the insertion was in fact responsible for the original mutation (Jenkins *et al.*, 1981). Analysis of several independent reversion events showed that an identical change had occurred: The bulk of the provirus was removed, apparently by homologous recombination between the two long terminal repeats (LTRs), leaving exactly one copy of the LTR in place in the locus (Copeland *et al.*, 1983a). Thus, the presence of a single LTR did not disrupt the locus as did the presence of the entire provirus.

A second coat-color mutation has now been shown to be closely linked to an ecotropic provirus (Copeland *et al.*, 1983b). The lethal yellow allele of the *agouti* locus (A^y) was isolated in 1905 and transferred by crosses into several genetic backgrounds; the locus segregated with an endogenous ecotropic virus termed *Emv-15*. No revertants of the mutation have yet been obtained to allow a determination of whether the mutation was directly caused by the provirus or whether it is simply linked to the provirus. This mutation, like the *dilute* mutation, may well be due to a retroviral insertion.

Recently, a few other germ-line mutations have been generated by manipulation of embryos. These mutations were not formed by retroviral infection, but rather by the microinjection of foreign DNAs into mouse embryos, so we will mention these results only briefly. They do constitute examples of insertional inactivation of loci and suggest that the products of such events can be readily recovered at high frequency.

One insertion of a metallothionein–thymidine kinase gene fusion in the germ line of mice resulted in an interesting phenotype (Palmiter *et al.*, 1984). The insertion was genetically transmitted by females to progeny as a heterozygous marker, but never by males (0 of 83 progeny). The insert was autosomal and not on the X or Y chromosome. Further analysis suggested that sperm carrying the allele were inviable and that males therefore never transmitted the marker to progeny, a phenomenon known as transmission distortion. The simplest explanation is that the wild-type locus encodes an essential product that is produced by postmeiotic transcription in spermatids and that the insertion effectively precludes development of half the spermatozoa. The locus is being studied further.

Another well-characterized mutation was generated by microinjection of DNA. Of many transgenic mice expressing a *c-myc* gene construct, one carried a mutation the phenotype of which was apparent in homozygous animals. The mice exhibited syndactyly, a reduction in the number of digits, and a fusion of bones of the outer limbs (P. Leder, personal communication). The phenotype was identical to that of mice carrying recessive homozygous alleles of a gene known as limb deformity (*ld*), and mapping the integrated DNA revealed that the insertion was very near the *ld* locus. Complementation tests showed that the loci were in fact identical. Analysis of the flanking DNA near the insertion should allow the identification of the *ld* locus.

Two other insertions in mice have been found to generate recessive prenatal lethal mutations (Wagner *et al.*, 1983). A set of six transgenic strains were constructed, carrying the human growth hormone gene; two of the six insertions were found to be lethal when homozygous animals were constructed by crosses. The nature of the defect was not determined. When added to all the known insertions in transgenic mice, the results show that recessive lethal mutations are quite common; perhaps one in ten insertions leads to a visible phenotype.

This compares favorably with any other method of mutagenesis of mice, and since preparation of the mutation in this way tags the gene for cloning, it is clear that this is a powerful technique. Obviously, a limitation is that very limited numbers of animals can be prepared, and the nature of the mutations is unselected. During the formation of many transgenic mice, it is likely that many interesting genes will be mutated and cloned by gratuitous insertional mutagenesis.

5.2. Examples of Insertions in Cell Lines

The first directed application of retroviral infection of cell lines to generate insertional mutations was described by Varmus *et al.* (1981). A clonal rat cell line, B31, transformed by Rous sarcoma virus (RSV) and carrying a single RSV provirus, was infected with M-MuLV to generate a small number of new proviral insertions. Analysis of 60 morphologically flat revertants of the transformed cell line revealed two cell lines with insertions of M-MuLV proviruses into the RSV genome. In both cases, the insertion was not directly into the *v-src* transforming gene, but rather into the RSV provirus upstream of *v-src;* the formation of the *v-src* mRNA was prevented by the insertions, and only aberrant RSV mRNAs were found. The RNAs expressed by the two insertionally inactivated loci were quite different, and it seems that the exact site of insertion can affect the processing of the RSV RNAs in very different ways. One of the flat cell lines bearing an insertion gave rise to retransformants at moderate frequency, and analysis revealed that the mechanism was similar to that responsible for reversion of the *dilute* mutation mentioned above: The bulk of the M-MuLV provirus was lost by an apparent homologous recombination between the two LTRs. In one such retransformant, there were additional deletions in the LTR. The other flat line did not readily give rise to retransformants.

Another insertional inactivation has been reported (Wolf and Rotter, 1984) in a tumor line produced by infection with Abelson murine leukemia virus (A-MuLV). Of the hundreds of tumor lines now produced by infection with this virus, one line was found to have very unusual properties. The L12 line, a transformant of the C57/L strain, is an immortal, clonal transformant carrying the A-MuLV provirus, but this line is very poorly tumorigenic when cells are injected into syngeneic animals. Normally, A-MuLV tumor lines are quite aggressive and result in the early death of the animal host. The major antibody response of the animal, before its death, is usually directed against a host protein termed P53, expressed by the tumor; in the case of L12 tumors, there was no response against this protein, and good antibodies could be produced that were reactive against the A-MuLV gene product. The reason for the absence of a response against P53 was straightforward: The L12 tumor line did not express any of this protein. Examination of the line with nucleic acid probes revealed that the normal mRNA encoding P53 was absent and that instead two aberrant mRNAs of larger size were found. The reason for these aberrant mRNAs, in turn, was found to be a rearrangement in the single gene that encodes P53; an inactive pseudogene present in all mice was unaffected. The rearrangement was demonstrated to be the insertion of a complete copy of the M-MuLV genome used as helper virus in the formation of the L12 line into the first intron of the P53 gene. Curiously, only the rearranged gene was present; the gene is autosomal, mapping to chromosome 11, and the other allele should have been still present, since the L12 line has a complete diploid complement of chromosomes. The simplest interpretation is that both homologues carry the identical rearrangement, and perhaps this might have occurred by gene conversion of the normal allele by the mutant allele. In passing, it is noteworthy that the odd biological behavior of the L12 line has since been demonstrated to be directly attributable to the absence of a functional P53 gene.

Our laboratory, in collaboration with that of Dr. Nguyen-Huu, has tested the possibility of inducing mutations in an endogenous host gene by infection with M-MuLV (King et al., 1985). For use as a mutagen, we constructed a variant of M-MuLV, designed to facilitate the detection and cloning of the integrated proviruses (Lobel et al., 1985). A small DNA fragment containing the *Escherichia coli* tyrosine suppressor transfer RNA gene, $SuIII^+$, was isolated and inserted into the LTRs of a cloned copy of the M-MuLV genome. The resulting virus was fully replication-competent and could spread in mouse cells with retention of the inserted sequences. Two such constructs carrying the insert at different locations in the LTR were tested; one insertion in unique sequences near the LTR edge was completely stable, and one in the duplicated enhancer element was slowly lost by homologous recombination between the two elements. Both viruses were stable enough for use as mutagens. Our initial efforts were done with the less stable construct; more recently, we have used the fully stable one. The utility of these constructs is twofold: First, the inserted sequence allows the ready detection of the exogenously added proviruses and the discrimination of these proviruses from all the highly homologous proviruses endogenous to the mouse genome, by the use of an appropriate nucleic acid probe. Second, the new proviruses can be readily cloned in phage vectors bearing amber mutations in essential genes; of a library of phage clones, only those phages containing the $SuIII^+$ gene and therefore the adjacent provirus can replicate to form a plaque in an Su^- bacterial host. Both unintegrated circular DNAs and integrated proviruses could be readily cloned from infected cells by this procedure (Lobel et al., 1985).

To determine whether insertions in a host gene could be detected, we infected an embryonal carcinoma cell line, F9, with the M-MuLV constructs. These cell lines have several properties that make them good candidates for mutagenesis. First, the cells are euploid and karyotypically stable. The frequency of spontaneous mutation is quite low; it is, for example, much lower than that of NIH/3T3 cells or other immortal fibroblastic lines. Second, these cells can be infected with virus to generate exceedingly high numbers of integrated proviruses. The reason for this is that the proviruses are repressed, by unknown means, and no viral proteins are produced (Steward *et al.*, 1982; Niwa *et al.*, 1983; Gautsch and Wilson, 1983); the absence of expression of *env* protein means that the cells do not exhibit the normal superinfection exclusion and therefore do not become resistant to further infection as do most cells. The maximum number of proviruses, as many as 100, was induced by cocultivation of the F9 cells with virus-producing fibroblasts for an extended time.

The target gene that was used was the gene that encodes hypoxanthine-guanine phosphoribosyltransferase (HGPRT). The gene is X-linked and therefore is present in only a single functional copy per diploid cell. It is also one of the few genes for which there is a strong counterselection: The addition of the toxic substrates 8-azaguanine and 6-thioguanine to the medium kills HGPRT$^+$ cells and allows the survival of HGPRT$^-$ mutant cells. Fewer than 1 in 10^7 cells of an uninfected population survived to grow in selective medium, and there was a 10-fold increase in HGPRT$^-$ survivors after infection. Analysis of several of these survivors (4 of 14) revealed that they had indeed suffered an insertion in the *HGPRT* gene. This is a minimum estimate because insertions in much of the gene could not readily be detected by the nucleic acid analyses, and it is likely that most of the rest of the survivors carried an inserted provirus. The mapped insertions were in introns of the *HGPRT* gene.

A very recent study has demonstrated the mutagenesis of an autosomal gene by retroviral infection of diploid cells, dramatically extending the applications of the approach (Frankel *et al.*, 1985). The key to the method is the use of a cell line heterozygous for the gene to be inactivated and of an allele-specific counterselection for the function of the gene. Clonal A-MuLV-transformed cell lines were derived from C57BL/6 × BALB/c F1 mice, which codominantly express both alleles of the β_2-microglobulin gene (*B2m*). These cells were infected with M-MuLV, and mutants lacking the C57BL/6 allele were immuno-selected by treatment with a monoclonal antibody that specifically recognizes the *B2m^b* gene product and not the *B2m^a* gene product, followed by complement lysis. One survivor of 22 tested was shown to contain a retroviral insertion in the *B2m* gene. The insertion, in or near the first exon of the gene, was surprisingly not the infecting M-MuLV genome, but was an A-MuLV genome, mobilized by the helper virus to reintegrate into the genome. Subsequent analysis showed that 2 more of the original 22 survivors carried insertions, in these cases of the M-MuLV genome, in the *B2m* gene.

5.3. Frequencies, Targets, and Prospects

The utility of retroviruses as mutagens and for the tagging of genetic loci depends heavily on the assumption that proviral integration occurs throughout the genome and that targets are selected with approximately equal frequency. It is premature to say much on this point. Our work on mutagenesis of the *HGPRT* locus suggests that insertion in the locus was less frequent than would be predicted by random insertion. The gene is approximately 30 kb long and thus represents about 10^{-5} of the genome; if integration anywhere in the gene can inactivate it, the insertion of 20–50 proviruses per cell should have resulted in the formation of mutations at a frequency of $2–5 \times 10^{-4}$. In fact, the frequency of mutation was about 7×10^{-7}, suggesting that the *HGPRT* locus was effectively a "cold spot" for insertion. Despite this low frequency, insertions could readily be detected. One must be prepared, however, for some loci to be frequently inactivated and for others to be rarely disrupted.

It is probably significant that no insertional inactivations into new, unknown genes have yet been reported. To detect such insertions, it would be helpful to induce insertions at a given locus at a frequency above, or at least close to, that of spontaneous mutations. There are only preliminary data on these points, but the data suggest that frequencies rarely exceed that of spontaneous events. Insertions into *src* described in Section 5.2 (Varmus *et al.*, 1981) occurred in 2 of 60 mutants; the vast majority were point mutants, deletions of the whole target gene, or other rearrangements. Insertions into the *B2m* gene were found in 3 of 22 mutants (Frankel *et al.*, 1985). Other insertions, into the *P53* gene, occurred at unknown frequencies. Only in the case of inactivation of the *HGPRT* gene, when many proviruses were inserted, could the mutation rate be stimulated above background. Unfortunately, the very large number of proviruses would make the detection of the insertions in the gene very difficult, if a probe for the target did not already exist. Probably the best approach will be a compromise. Cells should be infected to produce only a few proviruses, and many mutant cells should be isolated; only a fraction of these mutants will be due to insertional inactivation. These must then be screened for the presence of a common insertion site, as has been successfully done in the detection of transforming loci such as *int-1*.

A number of selection schemes can be used after insertional mutagenesis. Only a few genes will be counterselectable with specific drugs; there must exist a toxic analogue of the normal substrate, in general, to make the method work. Candidates for such toxic analogues are radioactive isomers of substrates, ones that might kill off cells that have successfully utilized and incorporated such compounds. The immunoselection used to isolate mutants in the *B2m* gene (Frankel *et al.*, 1985) is very important, and probably any surface protein can be used as a counterselectable marker in this way. Finally, there is a very important class of genes that may prove to be particularly easy to identify in this way: those genes that normally are responsible for the terminal differentiation of cell lines. If one activates the terminal differentiation of some lines, the vast majority of the cells in the population may cease replicating and die; the survivors are presumably mutants. If infection by a retrovirus can disrupt a gene needed for differentiation and death, the site of integration in these survivors might be that gene. Another selection is morphological transformation. If there are "antitransforming" genes, the normal function of which is to repress transforming functions and prevent aberrant growth, then their disruption would result in transformation. Such events would be readily seen as a transformed focus.

Is there hope of generating insertion mutations in autosomal loci in diploid cells? The difficulties clearly increase. The use of allotypic selections or screens is one important route around this problem, but it is rare that such selections are available. Occasionally, gene-conversion events may transfer an insertion from one allele to the other, thus generating a homozygous cell. The frequency of this event, however, must be multiplied by the marginal frequencies for detection of even single insertions, and the combined frequency is likely to be well below that of other mutations that allow survival in a given selection. In our preliminary studies, we were unable to detect increased rates of mutation in the gene that encodes adenine phosphoribosyltransferase (APRT) on infection by M-MuLV; the *APRT* gene is an autosomal marker. Perhaps the best hope for the detection of insertional mutations is to select for dominant mutations. In essence, if gene activation can be detected, the problem is solved; such activations should in general be dominant. The retroviruses are particularly suited to such activation events. Tumorigenesis by the leukemia viruses is common and exemplifies the ease with which these agents can activate loci; the aberrrant and elevated expression of a gene, if detectable, would readily allow the cloning of that gene.

ACKNOWLEDGMENTS.   I would like to thank John Colicelli and Marian Carlson for critical comments. The author was supported by NIH Grant R01 CA 30488.

REFERENCES

Bingham, P. M., Levis, R., and Rubin, G. M., 1981, Cloning of DNA sequences from the white locus of *D. melanogaster* by a novel and general method, *Cell* 25:693–704.

Boeke, J. D., Garfinkel, D. J., Styles, C. A., and Fink, G. R., 1985, Ty elements transpose through an RNA intermediate, *Cell* 40:491–500.

Boone, L. R., Myer, F. E., Yang, D. M., Ou, C.-Y., Koh, C. K., Roberson, L. E., Tennant, R. W., and Yang, W. K., 1983, Reversal of Fv-1 host range by *in vitro* restriction endonuclease fragment exchange between molecular clones of N-tropic and B-tropic murine leukemia virus genomes, *J. Virol.* 48:110–119.

Bukhari, A. L., Shapiro, J. A., and Adhya, S. L. (eds.), 1977, *DNA Insertion Elements, Plasmids, and Episomes*, Cold Spring Harbor Press, Cold Spring Harbor, New York.

Canaani, E., Dreazen, O., Klar, A., Rechavi, G., Ram, D., Cohen J. B., and Givol, D., 1983, Activation of the C-mos oncogene in a mouse plasmacytoma by insertion of an endogenous intracisternal A-particle genome, *Proc. Natl. Acad. Sci. U.S.A.* 80:7118–7122.

Chatis, P. A., Holland, C. A., Hartley, J. W., Rowe, W. P., and Hopkins, N., 1983, Role for the 3′ end of the genome in determining disease specificity of Friend and Moloney murine leukemia viruses, *Proc. Natl. Acad. Sci. U.S.A.* 80:4408–4411.

Chatis, P. A., Holland, C. A., Silver, J. E., Frederickson, T. N., Hopkins, N., and Hartley, J. W., 1984, A 3′ end fragment encompassing the transcriptional enhancers of nondefective Friend virus confers erythroleukemogenicity on Moloney leukemia virus, *J. Virol.* 52:248–254.

Chinsky, J., and Soeiro, R., 1981, Fv-1 host restriction of Friend leukemia virus: Analysis of unintegrated proviral DNA, *J. Virol.* **40**:45–55.

Chinsky, J., and Soeiro, R., 1982, Studies with aphidicolin on the Fv-1 restriction of Friend murine leukemia virus, *J. Virol.* **43**:182–190.

Chiu, I.-M., Callahan, R., Tronick, S. R., Schlom, J., and Aaronson, S. A., 1984, Major *pol* gene progenitors in the evolution of oncoviruses, *Science* **223**:364–370.

Cohen, J. B., Unger, T., Rechavi, G., Canaani, E., and Givol, D., 1983, Rearrangement of the oncogene c-mos in mouse myeloma NSI and hybridomas, *Nature (London)* **306**:797–799.

Colicelli, J., and Goff, S. P., 1985, Mutants and pseudorevertants of Moloney murine leukemia virus with alterations at the integration site, *Cell* **42**:573–580.

Colicelli, J., and Goff, S. P., 1986, Isolation of a recombinant murine leukemia virus utilizing a new primer tRNA, *J. Virol.* **57**:37–45.

Copeland, N. G., Hutchison, K. W., and Jenkins, N. A., 1983a, Excision of the DBA ecotropic provirus in dilute coat-color revertants of mice occurs by homologous recombination involving the viral LTRs, *Cell* **33**:379–387.

Copeland, N. G., Jenkins, N. A., and Lee, B. K., 1983b, Association of the lethal yellow (Ay) coat color mutation with an ecotropic murine leukemia virus genome, *Proc. Natl. Acad. Sci. U.S.A.* **80**:247–249.

Corcoran, L. M., Adams, J. M., Dunn, A. R., and Cory, S., 1984, Murine T lymphomas in which the cellular myc oncogene has been activated by retroviral insertion, *Cell* **37**:113–122.

Cullen, B. R., Shalka, A. M., and Ju, G., 1983, Endogenous avian retroviruses contain deficient promoter and leader sequences, *Proc. Natl. Acad. Sci. U.S.A.* **80**:2946–2950.

Cullen, B. R., Lomedico, P. T., and Ju, G., 1984, Transcriptional interference in avian retroviruses—implications for the promoter insertion model of leukaemogenesis, *Nature (London)* **307**:241–245.

Cuypers, H. T., Selten, G., Quint, W., Zijlstra, M., Maandag, E. R., Boelens, W., van Wezenbeek, P., Melief, C., and Berns, A., 1984, Murine leukemia virus-induced T cell lymphomagenesis: Integration of proviruses in a distinct chromosomal region, *Cell* **37**:141–150.

Decleve, A., Niwa, O., Gelmann, E., and Kaplan, H. S., 1975, Replication kinetics of N- and B-tropic murine leukemia viruses on permissive and non-permissive cells *in vitro*, *Virology* **65**:320–332.

DesGroseillers, L., and Jolicoeur, P., 1983, Physical mapping of the Fv-1 tropism host range determinant of BALB/c murine leukemia viruses, *J. Virol.* **48**:685–696.

DesGroseillers, L., Villemur, R., and Jolicoeur, P., 1984, The high leukemogenic potential of Gross passage A murine leukemia virus maps in the region of the genome corresponding to the long terminal repeat and to the 3' end of env, *J. Virol.* **47**:24–32.

Dhar, R., McClements, W. L., Enquist, L. W., and Vande Woude, G. F., 1980, Terminally repeated sequences (TRS) of integrated Moloney sarcoma provirus: Nucleotide sequence of TRS and its host and viral junctions, *Proc. Natl. Acad. Sci. U.S.A.* **77**:3937–3941.

Dickson, C., Smith, R., Brookes, S., and Peters, G., 1984, Tumorigenesis by mouse mammary tumor virus: Proviral activation of a cellular gene in the common integration region int-2, *Cell* **37**:529–536.

Donehower, L. A., and Varmus, H. E., 1984, A mutant murine leukemia virus with a single missense codon in *pol* is defective in a function affecting integration, *Proc. Natl. Acad. Sci. U.S.A.* **81**:6461–6465.

Downward, J., Yarden, Y., Mayes, E., Scrace, G., Totty, N., Stockwell, P., Ullrich, A., Schlessinger, J., and Waterfield, M. D., 1984, Close similarity of epidermal growth factor receptor and v-erb-B oncogene protein sequences, *Nature (London)* **307**:521–527.

Dunsmuir, P., Brorein, W. J., Jr., Simon, M. A., and Rubin, G. M., 1980, Insertion of the *Drosophila* transposable element copia generates a 5 base pair duplication, *Cell* **21**:575–579.

Duran-Troise, G., Bassin, R. H., Rein, A., and Gerwin, B. I., 1977, Loss of Fv-1 restriction in BALB/3T3 cells following infection with a single N-tropic murine leukemia virus particle, *Cell* **10**:479–488.

Duyk, G., Leis, J., Longiaru, M., and Skalka, A. M., 1983, Selective cleavage in the avian retroviral long terminal repeat sequence by the endonuclease associated with the alpha–beta form of avian reverse transcriptase, *Proc. Natl. Acad. Sci. U.S.A.* **80:**6745–6749.

Etkind, P. R., and Sarkar, N. H., 1983, Integration of new endogenous mouse mammary tumor virus proviral DNA at common sites in the DNA of mammary tumors of C3Hf mice and hypomethylation of the endogenous mouse mammary tumor virus proviral DNA in C3Hf mammary tumors and spleens, *J. Virol.* **45:**114–123.

Farabaugh, P. J., and Fink, G. R., 1980, Insertion of the eukaryotic transposable element Ty1 creates a 5-base pair duplication, *Nature (London)* **286:**352–355.

Fitts, R., and Temin, H. M., 1983, Cellular DNA surrounding integration sites of an avian retrovirus, *J. Gen. Virol.* **64:**267–274.

Folger, K. R., Wong, E. A., Wahl, G., and Capecchi, M. R., 1982, Patterns of integration of DNA microinjected into cultured mammalian cells: Evidence for homologous recombination between injected plasmid DNA molecules, *Mol. Cell. Biol.* **2:**1372–1387.

Frankel, W., Potter, T. A., Rosenberg, N., Lenz, J., and Rajan, T. V., 1985, Retroviral insertional mutagenesis of a target allele in a heterozygous murine cell line, *Proc. Natl. Acad. Sci. U.S.A.* **82:**6600–6604.

Fung, Y.-K. T., Fadly, A. M., Crittenden, L. B., and Kung, H.-J., 1981, On the mechanism of retrovirus-induced avian lymphoid leukosis: Deletion and integration of the proviruses, *Proc. Natl. Acad. Sci. U.S.A.* **78:**3418–3422.

Fung, Y.-K. T., Lewis, W. G., Crittenden, L. B., and Kung, H.-J., 1983, Activation of the cellular oncogene c-erbB by LTR insertion: Molecular basis for induction of erythroblastosis by avian leukosis virus, *Cell* **33:**357–368.

Gautsch, J. W., and Wilson, M. C., 1983, Delayed *de novo* methylation in teratocarcinoma suggests additional tissue-specific mechanisms for controlling gene expression, *Nature (London)* **301:**32–37.

Gautsch, J. W., Elder, J. H., Schindler, J., Jensen, F. C., and Lerner, R. A., 1978, Structural markers on core protein p30 of murine leukemia virus: Functional correlation with Fv-1 tropism, *Proc. Natl. Acad. Sci. U.S.A.* **75:**4170–4174.

Gautsch, J. W., Elder, J. H., Jensen, F. C., and Lerner, R. A., 1980, *In vitro* construction of a B-tropic virus by recombination: B-tropism is a cryptic phenotype of xenotropic murine retroviruses, *Proc. Natl. Acad. Sci. U.S.A.* **77:**2989–2993.

Gerwin, B. I., Rein, A., Levin, J. G., Bassin, G. H., Benjers, B. M., Kashmiri, S. V. S., Hopkins, D., and O'Neill, B. J., 1979, Mutant of B-tropic murine leukemia virus synthesizing an altered polymerase molecule, *J. Virol.* **31:**741–751.

Gilboa, E., Mitra, S. W., Goff, S. P., and Baltimore, D., 1979, Detailed model of reverse transcription and tests of crucial aspects, *Cell* **18:**93–100.

Gill, R., Heffron, F., and Falkow, S., 1979, Identification of the protein encoded by the transposable element Tn3 which is required for its transposition, *Nature (London)* **282:**797–801.

Goff, S. P., Traktman, P., and Baltimore, D., 1981, Isolation and properties of Moloney murine leukemia virus: Use of a rapid assay for release of virion reverse transcriptase, *J. Virol.* **38:**239–248.

Golomb, M., and Grandgenett, D. P., 1979, Endonuclease activity of purified RNA-directed DNA polymerase from avian myeloblastosis virus, *J. Biol. Chem.* **254:**1606–1613.

Golomb, M., Grandgenett, D. P., and Mason, W., 1981, Virus-coded DNA endonuclease from avian retrovirus, *J. Virol.* **38:**548–555.

Grandgenett, D. P., Vora, A. C., and Schiff, R. D., 1978, A 32,000-dalton nucleic acid binding protein from avian retrovirus cores possesses DNA endonuclease activity, *Virology* **89:**119–132.

Hampe, A., Gobet, M., Even, J., Sherr, C. J., and Galibert, F., 1983, Nucleotide sequences of feline sarcoma virus long terminal repeats and 5′ leaders show extensive homology to those of other mammalian retroviruses, *J. Virol.* **45:**466–472.

Harbers, K., Jahner, D., and Jaenisch, R., 1981, Microinjection of cloned retroviral genomes into mouse zygotes: Integration and expression in the animal, *Nature (London)* **293:**540–542.

Harbers, K., Kuehn, M., Delius, H., and Jaenisch, R., 1984, Insertion of retrovirus into the first intron of alpha 1(I) collagen gene leads to embryonic lethal mutation in mice, *Proc. Natl. Acad. Sci. U.S.A.* **81**:1504–1508.

Hawley, R. G., Shulman, M. J., Murialdo, H., Gibson, D. M., and Hozumi, N., 1982, Mutant immunoglobulin genes have repetitive DNA elements inserted into their intervening sequences, *Proc. Natl. Acad. Sci. U.S.A.* **79**:7425–7429.

Hayward, W. S., Neel, B. G., and Astrin, S. M., 1981, Activation of a cellular oncogene by promoter insertion in ALV-induced lymphoid leukosis, *Nature (London)* **290**:475–480.

Hizi, A., Gazit, A., Guthmann, D., and Yaniv, A., 1982, DNA-processing activities associated with the purified alpha, beta-2, and beta molecular forms of avian sarcoma virus RNA-dependent DNA polymerase, *J. Virol.* **41**:974–981.

Hopkins, N., Schindler, J., and Hynes, R., 1977, Six NB-tropic murine leukemia viruses derived from a B-tropic virus of BALB/c have altered p30, *J. Virol.* **21**:309–318.

Hughes, S. H., Shank, P. R., Spector, D. H., Kung, H.-J., Bishop, J. M., Varmus, H. E., Vogt, P. K., and Breitman, M. L., 1978, Proviruses of avian sarcoma virus are terminally redundant, co-extensive with unintegrated linear DNA and integrated at many sites, *Cell* **15**:1397–1410.

Hughes, S. H., Mutschler, A., Bishop, J. M., and Varmus, H. E., 1981a, A Rous sarcoma virus provirus is flanked by short direct repeats of a cellular DNA sequence present in only one copy prior to integration, *Proc. Natl. Acad. Sci. U.S.A.* **78**:4299–4303.

Hughes, S. H., Vogt, P. K., Stubblefield, E., Bishop, J. M., and Varmus, H. E., 1981b, Integration of avian sarcoma virus DNA in chicken cells, *Virology* **108**:208–221.

Ikenaga, H., and Saigo, S., 1982, Insertion of a movable genetic element, 297, into the TATA box for the H3 histone gene in *Drosophila melanogaster, Proc. Natl. Acad. Sci. U.S.A.* **79**:4143–4147.

Inouye, S., Yuki, S., and Saigo, K., 1984, Sequence-specific insertion of the *Drosophila* transposable genetic element 17.6, *Nature (London)* **310**:332–333.

Jaenisch, R., 1976, Germ line integration and endogenous transmission of the exogenous Moloney murine leukemia virus, *Proc. Natl. Acad. Sci. U.S.A.* **73**:1260–1264.

Jaenisch, R., Jahner, D., Nobis, P., Simon, I., Lohler, J., Harbers, K., and Grotkopp, D., 1981, Chromosomal position and activation of retroviral genomes inserted into the germ line of mice, *Cell* **24**:519–529.

Jaenisch, R., Harbers, K., Schnieke, A., Lohler, J., Chumakov, I., Jahner, D., Grotkopp, D., and Hoffman, E., 1983, Germline integration of Moloney murine leukemia virus at the Mov13 locus leads to recessive lethal mutation and early embryonic death, *Cell* **32**:209–216.

Jahner, D., Stuhlmann, H., Stewart, C. L., Harbers, K., Lohler, J., Simon, I., and Jaenisch, R., 1982, *De novo* methylation and expression of retroviral genomes during mouse embryogenesis, *Nature (London)* **298**:623–628.

Jenkins, N. A., Copeland, N. G., Taylor, B. A., and Lee, B. K., 1981, Dilute (d) coat colour mutation of DBA/2J mice is associated with the site of integration of an ecotropic MuLV genome, *Nature (London)* **293**:370–374.

Jha, K. K., Siniscalco, M., and Ozer, H. L., 1980, Temperature-sensitive mutants of BALB/3T3 cells. III. Hybrids between ts2 and other mouse mutant cells affected in DNA synthesis and correction of ts2 defects by human X chromosome, *Somat. Cell Genet.* **6**:603–614.

Jolicoeur, P., 1979, The Fv-1 gene of the mouse and its control of murine leukemia virus replication, *Curr. Top. Microbiol. Immunol.* **86**:67–122.

Jolicoeur, P., and Baltimore, D., 1976, Effect of Fv-1 gene product on proviral DNA formation and integration in cells infected with murine leukemia viruses, *Proc. Natl. Acad. Sci. U.S.A.* **73**:2236–2240.

Jolicoeur, P., and Rassart, E., 1980, Effect of Fv-1 gene on product on synthesis of linear and supercoiled viral DNA in cells infected with murine leukemia virus, *J. Virol.* **33**:183–195.

King, W., Patel, M. D., Lobel, L. I., Goff, S. P., and Nguyen-Huu, M. C., 1985, Insertion mutagenesis of embryonal carcinoma cells by retroviruses, *Science* **228**:554–558.

Kleckner, N., 1977, Transposable elements in procaryotes, *Cell* **11**:11–23.

Kopchick, J. J., Harless, J., Geisser, B. S., Killam, R., Hewitt, R. R., and Arlinghaus, R. B., 1981, Endonuclease activity associated with Rauscher murine leukemia virus, *J. Virol.* **37**:274–283.

Kriegler, M., and Botchan, M., 1983, Enhanced transformation by a simian virus 40 recombinant virus containing a Harvey murine sarcoma virus long terminal repeat, *Mol. Cell. Biol.* **3**:325–339.

Kuff, E. L., Feenstra, A., Lueders, K., Rechavi, G., Givol, D., and Canaani, E., 1983a, Homology between an endogenous viral LTR and sequences inserted in an activated cellular oncogene, *Nature (London)* **302**:547–548.

Kuff, E. L., Feenstra, A., Lueders, K., Smith, L., Hawley, R., Hozumi, N., and Shulman, M., 1983b, Intracisternal A-particle genes as movable elements in the mouse genome, *Proc. Natl. Acad. Sci. U.S.A.* **80**:1992–1996.

Leis, J., Duyk, G., Johnson, S., Longiaru, M., and Skalka, A., 1983, Mechanism of action of the endonuclease associated with the alpha–beta and beta–beta forms of avian RNA tumor virus reverse transcriptase, *J. Virol.* **45**:727–739.

Lemay, G., and Jolicoeur, P., 1984, Rearrangement of a DNA sequence homologous to a cell–virus junction fragment in several Moloney murine leukemia virus-induced rat thymomas, *Proc. Natl. Acad. Sci. U.S.A.* **81**:38–42.

Lenz, J., Celander, D., Crowther, R. L., Patarca, R., Perkins, D. W., and Haseltine, W. A., 1984, Determination of the leukaemogenicity of a murine retrovirus by sequences within the long terminal repeat, *Nature (London)* **306**:467–470.

Levin, J. G., Hu, S. C., Rein, A., Messer, L. I., and Gerwin, B. I., 1984, Murine leukemia virus mutant with a frameshift in the reverse transcriptase coding region: Implications for pol gene structure, *J. Virol.* **51**:470–478.

Lilly, F., and Pincus, T., 1973, Genetic control of murine viral leukemogenesis, *Adv. Cancer Res.* **17**:231–277.

Lobel, L. I., Patel, M., King, W., Nguyen-Huu, M. C., and Goff, S. P., 1985, Construction and recovery of viable retroviral genomes carrying a bacterial suppressor transfer RNA gene, *Science* **228**:329–332.

Lohler, J., Timpl, R., and Jaenisch, R., 1984, Embryonic lethal mutation in mouse collagen I gene causes rupture of blood vessels and is associated with erythropoietic and mesenchymal cell death, *Cell* **38**:597–607.

Luciw, P. A., Bishop, J. M., Varmus, H. E., and Capecchi, M. R., 1983, Location and function of retroviral and SV-40 sequences that enhance biochemical transformation after microinjection of DNA, *Cell* **33**:705–716.

Luciw, P. A., Opperman, H., Bishop, J. M., and Varmus, H. E., 1984, Integration and expression of several molecular forms of Rous sarcoma virus DNA used for transfection of mouse cells, *Mol. Cell. Biol.* **4**:1260–1269.

Majors, J. E., and Varmus, H. E., 1981, Nucleotide sequences of host–proviral junctions for mouse mammary tumour virus, *Nature (London)* **289**:253–258.

Majors, J. E., and Varmus, H. E., 1983, Nucleotide sequencing of an apparent proviral copy of env mRNA defines determinants of expression of the mouse mammary tumor virus env gene, *J. Virol.* **47**:495–504.

Misra, T. K., Grandgenett, D. P., and Parsons, J. T., 1982, Avian retrovirus pp32 DNA-binding protein. I. Recognition of specific sequences on retrovirus DNA terminal repeats, *J. Virol.* **44**:330–343.

Mizuuchi, K., Weisberg, R., Enquist, L., Mizuuchi, M., Buraczynska, M., Foeller, C., Hsu, P.-L., Ross, W., and Landy, A., 1981, Structure and function of the phage att site: Size, int binding sites, and location of the crossover point, *Cold Spring Harbor Symp. Quant. Biol.* **45**:429–437.

Mount, S. M., and Rubin, G. M., 1985, Complete nucleotide sequence of the *Drosophila* transposable element copia: Homology between copia and retroviral proteins, *Mol. Cell. Biol.* **5**:1630–1638.

Mushinski, J. F., Potter, M., Bauer, S. R., and Reddy, E. R., 1983, DNA rearrangement and altered RNA expression of the c-myb oncogene in mouse plasmacytoid lymphosarcomas, *Science* **220**:795–798.

Neel, B. G., Hayward, W. S., Robinson, H. L., Fang, J., and Astrin, S. M., 1981, Avian leukosis virus-induced tumors have common proviral integration sites and synthesize discrete new RNAs: Oncogenesis by promoter insertion, *Cell* **23**:323–334.

Neil, J. C., Hughes, D., McFarlane, R., Wilkie, N. M., Onions, D. E., Lees, G., and Jarrett, O., 1984, Transduction and rearrangement of the *myc* gene by feline leukaemia virus in naturally-occurring T-cell leukaemias, *Nature (London)* **308**:814–820.

Niwa, O., Yokota, Y., Ishida, H., and Sugahara, T., 1983, Independent mechanism involved in suppression of the Moloney murine leukemia virus genome during differentiation of murine teratocarcinoma cells, *Cell* **32**:1105–1113.

Noori-Daloii, M. R., Swift, R. A., Kung, H. J., Crittenden, L. M., and Witter, R. L., 1981, Specific integration of REV proviruses in avian bursal lymphomas, *Nature (London)* **294**:574–576.

Nusse, R., and Varmus, H. E., 1982, Many tumors induced by the mouse mammary tumor virus contain a provirus integrated in the same region of the host genome, *Cell* **31**:99–109.

Nusse, R., van Ooyen, A., Cox, D., Fung, Y.-K. T., and Varmus, H., 1984, Mode of proviral activation of a putative mammary oncogene (int-1) on mouse chromosome 15, *Nature (London)* **307**:131–136.

O'Hare, K., and Rubin, G. M., 1983, Structures of P transposable elements and their sites of insertion and excision in the *Drosophila melanogaster* genome, *Cell* **34**:25–35.

Ono, M., Toh, H., Miyata, T., and Awaya, T., 1985, Nucleotide sequence of the Syrian hamster intracisternal A-particle gene: Close evolutionary relationship of type A particle gene to types B and D oncovirus genes, *J. Virol.* **55**:387–394.

Ou, C.-Y., Boone, L. R., Koh, C.-K., Tennant, R. W., and Yang, W. K., 1983, Nucleotide sequences of gag–pol regions that determine the Fv-1 host range property of BALB/c N-tropic and B-tropic murine leukemia viruses, *J. Virol.* **48**:779–784.

Palmiter, R. D., Wilkie, T. M., Chen, H. Y., and Brinster, R. L., 1984, Transmission distortion and mosaicism in an unusual transgenic mouse pedigree, *Cell* **36**:869–877.

Panganiban, A. T., 1985, Retroviral DNA integration, *Cell* **42**:5–6.

Panganiban, A. T., and Temin, H. M., 1983, The terminal nucleotides of retrovirus DNA are required for integration but not virus production, *Nature (London)* **306**:155–160.

Panganiban, A. T., and Temin, H. M., 1984a, Circles with two tandem LTRs are precursors to integrated retrovirus DNA, *Cell* **36**:673–679.

Panganiban, A. T., and Temin, H. M., 1984b, The retrovirus *pol* gene encodes a product required for DNA integration: Identification of a retrovirus *int* locus, *Proc. Natl. Acad. Sci. U.S.A.* **81**:7885–7889.

Patarca, R., and Haseltine, W. A., 1984, Sequence similarity among retroviruses, *Nature (London)* **306**:288.

Payne, G. S., Courtneidge, S. A., Crittenden, L. B., Fadley, A. M., Bishop, J. M., and Varmus, H. E., 1981, Analyses of avian leukosis virus DNA and RNA in bursal tumors suggest a novel mechanism for retroviral oncogenesis, *Cell* **23**:311–322.

Payne, G. S., Bishop, J. M., and Varmus, H. E., 1982, Multiple arrangements of viral DNA and an activated host oncogene in bursal lymphomas, *Nature (London)* **295**:209–214.

Peters, G., Brookes, S., Smith, R., and Dickson, C., 1983, Tumorigenesis by mouse mammary tumor virus: Evidence for a common region for provirus integration in mammary tumors, *Cell* **33**:369–377.

Peters, G., Kozak, C., and Dickson, C., 1984, Mouse mammary tumor virus integration regions int-1 and int-2 map on different mouse chromosomes, *Mol. Cell. Biol.* **4**:375–378.

Pincus, T., Hartley, J. W., and Rowe, W. P., 1975, A major genetic locus affecting resistance to infection with murine leukemia viruses. IV. Dose–response relationships in Fv-1 sensitive and resistant cell cultures, *Virology* **65**:333–342.

Rechavi, G., Givol, D., and Canaani, E., 1982, Activation of a cellular oncogene by DNA rearrangement: Possible involvement of an IS-like element, *Nature (London)* **300:**607–610.

Rein, A., Kashmiri, S. V. S., Bassin, R. H., Gerwin, B. I., and Duran-Troise, G., 1976, Phenotypic mixing between N- and B-tropic murine leukemia viruses: Infectious particles with dual sensitivity to Fv-1 restriction, *Cell* **7:**373–379.

Rein, A., Gerwin, B. I., Bassin, R. H., Schwarm, L., and Schildlovsky, G., 1978, A replication-defective variant of Moloney murine leukemia virus. I. Biological characterization, *J. Virol.* **25:**146–156.

Richter, A., Ozer, H. L., DesGroseillers, and Jolicoeur, J., 1984, An X-linked gene affecting mouse cell DNA synthesis also affects production of unintegrated linear and supercoiled DNA of murine leukemia virus, *Mol. Cell. Biol.* **4:**151–159.

Roeder, G. S., and Fink, G. R., 1980, DNA rearrangements associated with a transposable element in yeast, *Cell* **21:**239–249.

Schnieke, A., Harbers, K., and Jaenisch, R., 1983, Embryonic lethal mutation in mice induced by retrovirus insertion into the alpha 1(I) collagen gene, *Nature (London)* **304:**315–320.

Schwartzberg, P., Colicelli, J., and Goff, S. P., 1984, Construction and analysis of deletion mutations in the *pol* gene of Moloney murine leukemia virus: A new viral function required for establishment of the integrated provirus, *Cell* **37:**1043–1052.

Searles, L. L., Jokerst, R. S., Bingham, P. M., Voelker, R. A., and Greenleaf, A. L., 1982, Molecular cloning of sequences from a *Drosophila* RNA polymerase II locus by P element transposon tagging, *Cell* **31:**585–592.

Seiki, M., Hattori, S., Hirayama, Y., and Yoshida, M., 1983, Human adult T-cell leukemia virus: Complete nucleotide sequence of the provirus genome integrated in leukemia cell DNA, *Proc. Natl. Acad. Sci. U.S.A.* **80:**3618–3622.

Shapiro, J. A., 1979, Molecular model for the transposition and replication of bacteriophage mu and other transposable elements, *Proc. Natl. Acad. Sci. U.S.A.* **76:**1933–1937.

Shen-Ong, G. L. C., Potter, M., Mushinski, J. F., Lavu, S., and Reddy, E. P., 1984, Activation of the c-myb locus by viral insertional mutagenesis in plasmacytoid lymphosarcomas, *Science* **226:**1077–1080.

Shimotohno, K., Mizutani, S., and Temin, H. M., 1980, Sequence of retrovirus provirus resembles that of bacterial transposable elements, *Nature (London)* **285:**550–554.

Shinnick, T. M., Lerner, R. A., and Sutcliffe, J. G., 1981, Nucleotide sequence of Moloney murine leukemia virus, *Nature (London)* **293:**543–548.

Shoemaker, C., Goff, S., Gilboa, E., Paskind, M., Mitra, S. W., and Baltimore, D., 1980, Structure of a cloned circular Moloney murine leukemia virus molecule containing an inverted segment: Implications for retrovirus integration, *Proc. Natl. Acad. Sci. U.S.A.* **77:**3932–3936.

Shoemaker, C., Goff, S., Gilboa, E., Paskind, M., Mitra, S. W., and Baltimore, D., 1981a, Structure of cloned retroviral circular DNAs: Implications for virus integration, *Cold Spring Harbor Symp. Quant. Biol.* **45:**711–717.

Shoemaker, C., Hoffmann, J., Goff, S. P., and Baltimore, D., 1981b, Intramolecular integration within Moloney murine leukemia virus DNA, *J. Virol.* **40:**164–172.

Silverman, S. J., and Fink, G. R., 1984, Effects of Ty insertions on HIS4 transcription in *Saccharomyces cerevisiae, Mol. Cell. Biol.* **4:**1246–1251.

Simon, M. C., Smith, R. E., and Hayward, W. S., 1984, Mechanisms of oncogenesis by subgroup F avian leukosis viruses, *J. Virol.* **52:**1–8.

Spradling, A. C., and Rubin, G. M., 1982, Transposition of cloned P elements into *Drosophila* germ line chromosomes, *Science* **218:**341–347.

Steffen, D., 1984, Proviruses are adjacent to *c-myc* in some murine leukemia virus-induced lymphomas, *Proc. Natl. Acad. Sci. U.S.A.* **81:**2097–2101.

Stewart, C. L., Stuhlmann, H., Jahner, D., and Jaenisch, R., 1982, *De novo* methylation, expression and infectivity of retroviral genomes introduced into embryonal carcinoma cells, *Proc. Natl. Acad. Sci. U.S.A.* **79:**4098–4102.

Stewart, C. L., Harbers, K., Jahner, D., and Jaenisch, R., 1983, X chromosome-linked transmission and expression of retroviral genomes microinjected into mouse zygotes, *Science* **221:**760–762.

Swanstrom, R., DeLorbe, W. J., Bishop, J. M., and Varmus, H. E., 1981, Nucleotide sequence of cloned unintegrated avian sarcoma virus DNA: Viral DNA contains direct and inverted repeats similar to those in transposable elements, *Proc. Natl. Acad. Sci. U.S.A.* **78:**124–128.

Temin, H. M., 1981, Structure, variation and synthesis of retrovirus long terminal repeat, *Cell* **27:**1–3.

Toh, H., Hayahida, H., and Miyata, T., 1983, Sequence homology between retroviral reverse transcriptase and putative polymerases of hepatitis B virus and cauliflower mosaic virus, *Nature (London)* **305:**827–829.

Tsichlis, P. N., Strauss, P. G., and Hu, L. F., 1983a, A common region for proviral DNA integration in MoMuLV-induced rat thymic lymphomas, *Nature (London)* **302:**445–449.

Tsichlis, P. N., Hu, L. F., and Strauss, P. G., 1983b, Two common regions for proviral DNA integration in MoMuLV-induced rat thymic lymphomas: Implications for oncogenesis, in: *ICN–UCLA Symposium on Normal and Neoplastic Hematopoiesis* (D. W. Golde and P. A. Marks, eds.), pp. 399–416, Alan R. Liss, New York.

Tsichlis, P. N., Strauss, P. G., and Kozak, C. A., 1984, Cellular DNA region involved in induction of thymic lymphomas (Mlvi-2) maps to mouse chromosome 15, *Mol. Cell. Biol.* **4:**997–1000.

Van Beveran, C., Rands, E., Chattopadhyay, S. K., Lowy, D. R., and Verma, I. M., 1981, Long terminal repeat of murine retroviral DNAs: Sequence analysis, host–proviral junctions, and preintegration site, *J. Virol.* **41:**542–556.

Van Ooyen, A, and Nusse, R., 1984, Structure and nucleotide sequence of the putative mammary oncogene int-1; proviral insertions leave the protein-encoding region intact, *Cell* **39:**233–240.

Varmus, H. E., 1983, Retroviruses, in *Mobile Genetic Elements* (J. A. Shapiro, ed.), pp. 411–503, Academic Press, New York.

Varmus, H. E., Quintell, N., and Oritz, S., 1981, Retroviruses as mutagens: Insertion and excision of a nontransforming provirus alter expression of a resident transforming provirus, *Cell* **25:**23–36.

Wagner, E. F., Covarrubias, L., Stewart, T. A., and Mintz, B., 1983, Prenatal lethalities in mice homozygous for human growth hormone gene sequences integrated in the germ line, *Cell* **35:**647–655.

Wain-Hobson, S., Sonigo, P., Danos, O., Cole, S., and Alizon, M., 1985, Nucleotide sequence of the AIDS virus, LAV *Cell* **40:**9–17.

Westaway, D., Payne, G., and Varmus, H. E., 1984, Deletions and base substitutions in provirally mutated c-myc alleles may contribute to the progression of B-cell tumors, *Proc. Natl. Acad. Sci. U.S.A.* **81**843–847.

Wolf, D., and Rotter, V., 1984, Inactivation of p53 gene expression by an insertion of Moloney murine leukemia virus-like DNA sequences, *Mol. Cell. Biol.* **4:**1402–1410.

Yang, W. K., Kiggans, J. O., Yang, D., Ou, C., Tennant, R. W., Brown, A., and Bassin, R. H., 1980, Synthesis and circularization of N- and B-tropic retroviral DNA in Fv-1 permissive and restrictive mouse cells, *Proc. Natl. Acad. Sci. U.S.A.* **77:**2994–2998.

13

HOMOLOGOUS RECOMBINATION IN MAMMALIAN SOMATIC CELLS

Raju Kucherlapati

1. INTRODUCTION

During the last decade, gene-transfer methods using purified DNA have developed from a novelty to a powerful tool in the study of gene structure–function relationships. Though success with DNA transfer was reported as early as 1962 (Szybalska and Szybalski, 1962), widespread use of this method had to await later developments. On the basis of a method developed by Graham and van der Eb (1973), Wigler *et al.* (1977) and Maitland and McDougall (1977) successfully introduced the herpes simplex virus (HSV) thymidine kinase (*TK*) gene into mouse cells deficient in this enzyme. Wigler and colleagues were also able to identify and isolate from the HSV genome a fragment of DNA that carried the *TK* gene. The calcium phosphate coprecipitation method was augmented by other methods of gene transfer such as DEAE–dextran-mediated transfer (McCutchan and Pagano, 1968; Sussman and Milman, 1984; and Milman and Herzberg, 1981) microinjection into somatic cells (Anderson *et al.*, 1980; Capecchi, 1980) or into embryos (Gordon *et al.*, 1980). Understanding the structure of DNA and RNA viral genomes has permitted the construction of a variety of different vectors for introduction of genetic information into mammalian cells (see Chapters 5 and 6). The availability of these different methods for gene transfer makes it possible to introduce virtually any gene into mammalian cells.

1.1. Fate of Introduced Genetic Information

Extensive investigations regarding the fate of exogenously introduced genetic information have been conducted. Pellicer *et al.* (1978) studied the organ-

Raju Kucherlapati ● Center for Genetics, University of Illinois College of Medicine, Chicago, Illinois 60612.

ization of the HSV *TK* gene introduced into mammalian cells. Using restriction-endonuclease digestion and the Southern-blotting procedure (Southern, 1975), they have shown that the introduced DNA becomes covalently integrated into the cellular DNA. This feature is retained irrespective of the method used for gene transfer. In some cases, depending on the nature of the recipient cell and the nature of the vector, the exogenous sequences may be maintained as autonomously replicating molecules [e.g., simian virus 40 (SV40)-based vectors in monkey cells (see Chapter 5)]. With the exception of material introduced by retrovirus-mediated transfer, the site on the input DNA that is used for integration is random. In the case of retrovirus-mediated transfer, integration is mediated by sequences present in the long terminal repeats. In all cases, the sites in the chromosomal regions that are used for integration are random. Analysis of several integrated copies of the exogenous sequences revealed no significant homology at the sites of recombination. Mammalian cells seem to have efficient mechanisms to mediate nonhomologous recombination between exogenously introduced molecules. Perucho *et al.* (1980b) have shown that plasmids bearing the selectable genes are capable of forming complexes with carrier DNA, and these complexes, which are referred to as "Pekalasomes" (Perucho *et al.*, 1980) or "transgenomes" (Scangos *et al.*, 1981) and are as large as several hundred kilobases, integrate as a unit into cellular DNA. Even when carrier DNA is not used, multiple tandem copies of the exogenously introduced plasmids have been observed (Folger *et al.*, 1982; Scangos *et al.*, 1981).

1.2. Effects of Random Integration

The random integration of exogenous DNA into cellular genomes affects the expression of cellular genes and genes borne by the plasmids. Following the elegant demonstration by Hayward *et al.* (1981), there is a large body of literature showing that retrovirus insertions in or near cellular genes may result in gene activation (see Chapter 12). Such gene activation has been shown to be implicated in the progression toward tumorigenesis. Retrovirus insertion also results in inactivation of genes. In some cases, retroviruses have been used intentionally as insertional mutagenic agents (King *et al.*, 1985); in other cases, insertions at anonymous sites have led to identifiable alteration of the phenotype that helped in identification of the gene that had been inactivated (Jaenisch *et al.*, 1983; Woychik *et al.*, 1985). The exogenously introduced genes are generally, though not always, expressed if they carried the appropriate transcriptional and translational signals. This feature has enabled the identification of sequences that are important for gene transcription, processing of the messenger RNA, and other features that have a role in gene expression.

1.3. Regulation of Gene Expression

Though exogenously introduced genes are expressed in somatic cells, they are not usually regulated properly. For example, a number of investigators have shown that globin genes introduced into fibroblasts are expressed (Wold *et al.*, 1979; Hsiung *et al.*, 1982). Fibroblasts are, of course, not the normal tissue in which globin genes are expressed. Similarly, initial reports of genes microinjected into embryos indicated that the genes are inserted into the germ line in several instances, but no tissue-specific expression was observed (Lacy *et al.*, 1983). Though there are now several reports of tissue-specific gene expression, such expression is not observed in all transgenic mice [for reviews, see Palmiter and Brinster (1985) and Chapter 7].

All these results can be summarized by stating that defined DNA sequences can be readily introduced into mammalian cells, that these sequences become covalently and randomly integrated into the mammalian genome, and that the properly regulated expression of the genes is dependent on intrinsic features of the gene as well as on the site into which it has integrated.

1.4. Rationale for Study of Homologous Recombination

There are two basic reasons for studying homologous recombination in mammalian cells. Homologous recombination between a cellular gene and its counterpart introduced as a plasmid would permit gene modification in a predefined fashion. Such gene modification would be ideal in gene-replacement therapy. In addition, homologous recombination should permit insertional mutagenesis as well as site-directed mutagenesis *in vivo*. These approaches should permit study of gene structure–function relationships.

Genetic recombination is a fundamentally important biological process that plays a key role in generating diversity among species. Though recombination was first discovered in higher eukaryotes, a detailed mechanistic analysis of this process has not been possible. Much of our understanding of the mechanism of homologous recombination comes from study of bacteria, bacteriophage, and fungi. Although the fundamental processes of recombination are likely to be similar in different organisms, it is important to identify the specific enzymes involved in mammalian recombination. In addition, it is expected that the recombination machinery will share enzymes with other key cellular processes such as DNA replication and repair. Thus, understanding recombination may provide important information about enzymes that play a role in DNA replication and repair of damage caused by mutagenic and clastogenic agents.

Study of recombination in mammalian cells is in its infancy, but there are a number of developments that are worthy of report and consideration. This chapter is aimed at describing the current status of studies of recombination between DNA molecules introduced into mammalian cells and how this information has aided in studying recombination between cellular sequences and their homologous sequences introduced by gene-transfer methods. Finally, attempts at unraveling the biochemical aspects of homologous recombination in mammalian cells will also be described.

2. MITOTIC RECOMBINATION

Since it is rather difficult to study meiotic recombination in mammals, early attempts were focused on cultured somatic mammalian cells and mitotic recombination between markers located on homologous chromosomes. These attempts took the form of generating chromosomes carrying mutant markers and bringing together two chromosomes, each having a pair of different markers, into the same environment by cell-fusion methods. One of the chromosomes was allowed to segregate, and recombination was monitored by association of markers that were originally located on different chromosomes. For example, Rosenstraus and Chasin (1978) fused two Chinese hamster cell lines that each had identifiable genetic markers at the hypoxanthine phosphoribosyltransferase ($HPRT$) and glucose-6-phosphate dehydrogenase loci. A recombination event would have been detected by the exchange of these X-linked markers. Despite extensive analysis, Rosenstraus and Chasin were unable to obtain any evidence for homologous recombination. Similar efforts by Tarrant and Holliday (1977) to detect intragenic recombination at the $HPRT$ locus, and by Campbell and Worton (1981) to detect recombination between Chinese hamster chromosome-2-linked markers, yielded negative results.

Recent investigations by Wasmuth and Vock-Hall (1984) and Cavenee et $al.$ (1983, 1985) have changed the situation. Wasmuth and Vock-Hall investigated possible mitotic recombination events between hamster chromosome 2 markers. They utilized the amplified dihydrofolate reductase ($dhfr$) gene on the short arm as a morphological and genetic marker. In addition, a temperature-sensitive mutation leu^{ts} and emetine resistance ($emt\ B^r$) on the long arm of this chromosome are used as additional markers. A chromosome containing the unamplified $dhfr$ gene, leu^{ts} gene, and $emt\ B^r$ gene were brought together by cell hybridization. A product of recombination between the centromere and the leu^{ts} gene would be cells in which the leu^{ts} and $emt\ B^r$ genes are located on the chromosome bearing the amplified $dhfr$ gene. Using morphological, genetic, and additional cell-hybridization experiments, Wasmuth and Vock-Hall were able to show that a homologous recombination of the type described above had indeed occurred. Their calculations indicate that the frequency with which the recombination event occurred is approximately 7×10^{-5}.

A completely independent and unrelated set of experiments by Cavanee and colleagues also provided evidence for mitotic recombination. Human retinoblastomas are the result of deletion or mutation at a locus designated *rb* that is located on human chromosome 13. Cavenee and colleagues have examined a number of independent isolates of retinoblastoma tumor tissues, and in some cases their normal counterparts, by cytological methods, by assaying for a closely linked enzymatic marker, esterase-D, and have determined the fate of polymorphic markers located on either side of the *rb* locus. They have noted that different chromosomal mechanisms could lead to homozygosity at the *rb* locus. In some cases, they observed that the two chromosomes retained polymorphic markers in the proximal region of chromosome 13, but have become homozygous for cell markers distal to the *rb* locus. They interpret these results to indicate that the homozygosity of the *rb* locus and markers distal to it has resulted from a mitotic recombination event. More recent results reported by Cavenee *et al.* (1985) are also consistent with this view. The observations with retinoblastoma cells do not permit the determination of the frequency of mitotic recombination. Taken together, these results indicate that mitotic recombination between homologous mammalian chromosomes occurs, although at relatively low frequencies.

3. RECOMBINATION BETWEEN EXOGENOUSLY INTRODUCED HOMOLOGOUS SEQUENCES

A large amount of information on homologous recombination in mammalian cells is now becoming available from the use of diverse systems. One of the early experiments involved the use of oligomeric SV40 DNA. Wilson and colleagues (Wake and Wison, 1979, 1980) constructed oligomeric SV40 DNA molecules by ligating *Eco*RI-cleaved SV40 DNA derived from two temperature-sensitive mutants. Using appropriate restriction-enzyme cleavages, they were able to generate partial dimeric DNA containing 1.84 genomes. This DNA was introduced into monkey cells. Recombination was assayed by plaque formation. For the SV40 DNA to be packaged, it must be of unit size. Unit-size molecules could be generated by homologous recombination in the oligomeric SV40 DNA. Analysis of the DNA from the resultant plaques indicated that unit-size SV40 molecules were indeed generated. The generation of these molecules was interpreted to be the result of homologous recombination. Similar experiments using other viral molecules also yielded results that are consistent with homologous recombination (Upcroft *et al.*, 1980a,b; Volkert and Young, 1983). The use of viral sequences for recombination has raised the question of whether the recombination reaction is mediated completely by cellular enzymes or whether some part of the reaction is mediated by viral gene products. The fact that defective viral genomes are capable of undergoing recombination eliminated the possibility of viral-gene involvement in the recombination process.

When viral sequences are introduced into competent host cells, the assay for recombination is usually the generation of infective virus. This procedure has many advantages, because relatively rare events can be easily scored and individual viral molecules can be amplified by plaque purification and their DNA isolated and analyzed. An alternative approach that is gaining widespread use is the use of a pair of plasmids carrying mutant or deleted genes with overlapping homologies. De Saint Vincent and Wahl (1983) have used the gene for the multifunctional enzyme abbreviated as CAD for these experiments. In these experiments, they have used a uridine auxotroph of a Chinese hamster cell line as the recipient. This cell line was previously shown to become prototropic for uridine when a complete Syrian hamster CAD gene was introduced by protoplast fusion. To study recombination in the hamster cells, de Saint Vincent and Wahl constructed two plasmids, each carrying a truncated CAD gene. The two sequences carried a 3-kilobase (kb) DNA homology. These two plasmids were first introduced into a rec A⁻ strain of bacteria. The plasmid DNA was introduced into the hamster cells by protoplast fusion. Selection in uridine-deficient medium yielded colonies at frequencies of 10^{-5}–10^{-6}. Examination of DNA from these cell lines revealed that an intact CAD gene was reconstructed and that it arose through reciprocal recombination between the exogenously introduced molecules. Small and Scangos (1983) reported similar results. In this case, they used a 5' deletion mutant of the HSV TK gene as one substrate and a 4-base-pair (bp) deletion mutant within the coding region of the gene as the second substrate. Recombination within a 610-bp region has to occur to yield a wild-type TK gene. Small and Scangos (1983) showed that cotransfer of the two plasmids into tk⁻ mouse L-cells yielded tk⁺ colonies at a rate that is 10% of that obtained with the wild-type gene. Analysis of the DNA of the cell lines again revealed that the reconstruction of an intact TK gene was achieved by homologous recombination. Earlier, Folger et al. (1982) had investigated the fate of DNA injected into mammalian cells with glass micropipettes. The fact that the sequences are found in many cells as head-to-tail tandem concatemers was interpreted to have resulted from homologous recombination of input plasmid sequences. Several investigators have observed head-to-tail concatemerization of transfected DNA in mammalian cells (for a review, see Kucherlapati and Skoultchi, 1985).

The experiments described above clearly indicated that mammalian cells have all the enzymes necessary to mediate homologous recombination between exogenously introduced molecules. These data should not, however, be misconstrued to mean that homologous recombination events are most prevalent. A number of investigators have shown that both homologous and nonhomologous recombination events occur in mammalian cells. Subramanian (1979) has shown that linearized SV40 DNA molecules can become circularized in the cell by end-to-end ligation. He noted that the nature of the ends (flush or overhangs) did not affect the ligation reaction. Wilson *et al.* (1982) introduced pBR322 sequences into nonessential regions of SV40. The molecules were then linearized by cutting within the pBR322 sequences. The linear molecules were introduced into monkey cells. One common feature of all the molecules is that they are larger than the sequences that can be packaged into SV40 capsids. Viable virus can be produced only if the sequences are trimmed and ligated. Wilson and colleagues were able to find that viable plaques resulted from these infections. Analysis of the DNA from the purified virus by restriction-enzyme digestion as well as nucleotide sequencing of the junctions indicated that end-to-end joining is extremely efficient in mammalian cells. Similar results of nonhomologous end-to-end joining were reported by a number of investigators (Winocour and Keshet, 1980; Miller and Temin, 1984; Anderson *et al.*, 1980; Subramani and Berg, 1983). The mechanisms by which this end-to-end joining occurs are currently under investigation.

4. USE OF AUTONOMOUSLY REPLICATING MOLECULES

The use of viral molecules to study recombination has several advantages. The complete viral molecules are capable of undergoing replication in appropriate mammalian cells, resulting in the amplification of the recombinant products. The viral molecules are packaged into virions that can be used to infect susceptible cells that can be scored for foci. Viruses like SV40 have a narrow range in size of viral DNA that can be packaged, so that recombination in larger than unit-size molecules can be studied readily. These features also permit examination of both homologous and nonhomologous events in the same experiment, thus permitting comparisons. Roth and Wilson (1985) reported such experiments. They have introduced a linear molecule containing a 131-bp terminal repeat. Recircularization of this molecule or homologous recombination involving the terminal repeats would result in a circular packageable SV40 molecule. Since the same substrate is used for nonhomologous and homologous events, the relative efficiencies of these two processes can be easily assessed. Results from these experiments indicated that nonhomologous events occur at a level 2–3 times higher than do homologous events. These results are quite encouraging for investigators who are interested in increasing the frequencies of homologous recombination.

An alternative method that utilizes autonomously replicating plasmids to study recombination has been described by two laboratories (Rubnitz and Subramani, 1985; Ayares *et al.*, 1985). Ayares and colleagues studied intermolecular recombination, while Rubnitz and Subramani examined intramolecular recombination. Ayares *et al.* (1985) have used recombination substrates constructed by Kucherlapati *et al.* (1984b). The parental plasmid is a eukaryotic–prokaryotic shuttle vector, pSV2neo (Southern and Berg, 1982). This plasmid carries two markers that are selectable in bacteria. One confers ampicillin resistance and the other resistance to neomycin and kanamycin. Deletions were introduced into the *neo* gene of this plasmid. One deletion resulted in the loss of a 248-bp fragment spanning the 5' end of the gene. A second plasmid has a different deletion that spans the 3' end and is 283 bp long. Ayares and colleagues introduced each of these plasmids separately or both together into monkey COS cells. COS cells are SV40-transformed monkey kidney cells that produce SV40 large tumor antigen in a constitutive fashion (Gluzman, 1981). Any plasmids that contain an SV40 origin of replication (ori) are capable of autonomous replication in these plasmids. The pSV2neo and the deletion derivatives contain an SV40 ori sequence. At 24 hr after transfection, low-molecular-weight DNA was isolated by the method of Hirt (1967) and used for direct analysis by blot hybridization (Southern, 1975) or used to transform *rec A⁻ Escherichia coli*. The frequency of *neo^r/amp^r* bacterial colonies provided information about recombination between the plasmids. Using these plasmids, Ayares *et al*. (1985) reported recombination frequencies of 3–5 × 10^{-3}. Rubnitz and Subramani constructed plasmids that contained truncated but overlapping segments of the *neo* gene into a plasmid that contained the SV40 ori sequence. The truncated sequences were introduced in the same orientation (direct configuration) or in opposite orientation (inverted configuration). These plasmids were introduced into monkey cells, and the low-molecular-weight DNA isolated after transfection was used for direct analysis or for transformation of *rec A⁻* bacteria. These investigators detected as high as 4% recombination. Direct analysis of DNA before bacterial transformation confirmed the view that recombination was mediated by COS-cell enzymes. These two systems have the advantage that a large number of recombination products can be readily examined. The intermolecular reaction studied by Ayares *et al*. (1985) has the additional advantage that the recombination reaction occasionally yielded dimeric molecules that retained the reciprocal products of recombination. Study of such dimeric molecules is somewhat similar to the study of spores in an ascus that results from fungal meiosis. Thus, the COS-cell system along with the use of SV40- "ori"-containing plasmid substrates combines the advantages of a bacteriophage cross and yeast random-spore analysis.

5. FACTORS THAT INFLUENCE HOMOLOGOUS RECOMBINATION

Since nonhomologous events outnumber homologous events in mammalian cells, it is desirable to devise methods that are capable of enhancing homologous recombination. It is known that DNA-damaging agents are capable of inducing higher levels of recombination in bacteria. Kucherlapati *et al*. (1984a) tested the effect of treating input DNA with ultraviolet light and found that it results in a modest increase in the yield of recombinants.

In yeast, homologous recombination between a cellular gene and its plasmid-borne counterpart can be enhanced as much as 1000–3000 times by introduction of double-strand breaks in the input plasmid within the region of homology (Orr-Weaver et al., 1981; Orr-Weaver and Szostak, 1983). The effect of similar breaks on intermolecular recombination was tested by Kucherlapati et al. (1984b), and Lin et al. (1984b) examined intramolecular reactions. Both these groups reported that double-strand breaks enhance homologous recombination. Kucherlapati et al. (1984b) and Song et al. (1985) showed that double-strand breaks within the region of homology on one of the substrates results in an increase of 5- to 80-fold in recombination. Cuts outside the region of homology had no effect on recombination. Linearization of both substrates by cutting at appropriate sites resulted in a greater increase of recombination. Several other investigators have used different substrates and showed that double-strand breaks and gaps are highly recombinogenic in mammalian cells (Ayares et al., 1985; Brenner et al., 1985; Folger et al., 1985; Subramani and Rubnitz, 1985; Wake et al., 1985).

Several investigators studied homology requirements for recombination. Wake et al. (1985) reported that terminal nonhomologies did not affect recombination. Brenner et al. (1985) reported that short heterologous sequences at the sites of breaks did not affect recombination frequency. Rubnitz and Subramani (1984) have systematically investigated the minimum amount of homology required for homologous recombination. In these experiments, they used a plasmid that contained SV40 sequences cloned into pBR322. It also contained repeated sequences at the ends of SV40 sequences in a direct orientation. Recombination between these repeated sequences would yield a wild-type SV40 DNA that can replicate and form plaques on appropriate host cells. Rubnitz and Subramani generated a series of deletions in one of the repeats by limited Bal31 digestion. Each of these plasmids was introduced into monkey cells, and intramolecular recombination was measured by the yield of plaque-forming units per microgram of input DNA. They showed that as little as 14 bp of homology is adequate for recombination, though 165 bp or more was considered optimal. Studies by Brenner et al. (1985) showed that 70 bp of homology is adequate, while Roth and Wilson (1985) find that 130 bp of homology results in efficient recombination. We have studied this aspect in intermolecular recombination and find that as little as 25 bp of homology is adequate for recombination (Ayares et al., 1986). Another feature of the results obtained by Rubnitz and Subramani (1984) is that the recombination frequency is biphasic. They observe a low but detectable level of recombination up to 160 bp of homology, and the rate of increase changes for regions of homology greater than 160 bp. The results obtained from mammalian cells qualitatively parallel those obtained from studies of recombination in bacteria (Watt *et al.*, 1985; Gonda and Radding, 1983; Singer *et al.*, 1982). These reports indicate that as little as 14–16 bp of homology is adequate for recombination, but 50–200 bp is needed for efficient recombination. Nearly 20 bp of homology is needed for stable DNA duplex formation. Thus, it is possible that the lower limits reflect this constraint, while the homology that is required for efficient recombination may reflect the length needed for a stable complex formation between DNA and the recombination enzymes. Thomas (1966) estimated that an approximately 20-bp stretch in *E. coli* would appear only once within the genome, while the sequence has to be much longer to be unique in higher eukaryotes. It is possible that the recombination systems have evolved to accommodate the differences in the complexity of the prokaryotic and eukaryotic genomes.

6. RECOMBINATION BETWEEN "CELLULAR" SEQUENCES AND PLASMID DNA

One of the goals of studying recombination in mammalian cells is to examine the feasibility of gene replacement. For gene replacement to be a reality, it is necessary to establish that recombination between cellular sequences and their counterparts introduced by plasmid transfer can occur. Several strategies have been used to detect such recombination events. One strategy is to introduce a defective copy of a gene by gene-transfer methods and permit it to integrate into the cellular genome. The site of integration would, of course, be random. Cells that retain the foreign DNA sequence in a stable fashion are then used as recipients for a second round of gene transfer using a sequence that shares homology with the integrated DNA. Recombination between the two molecules would be able to generate an intact gene. If the gene under study is selectable and if the original gene cannot undergo spontaneous reversion, homologous recombination events could be selected. There are three possible mechanisms by which the cell would have acquired the selectable phenotype. They are: (1) Reciprocal recombination between the "cellular" sequence and the exogenously introduced sequence. If the exogenous sequence is circular, a single reciprocal recombination event would result in the integration of the input DNA with a concomitant increase in the size of the native gene segment. (2) The defective cellular sequence would become normal through a double crossover event or a nonreciprocal recombi-nation (gene-conversion) event. Under these circumstances, the endogenous se-quence would become normal, and the donor gene containing plasmid might be lost or be integrated elsewhere in the genome by a nonhomologous recombination mechanism. (3) The incoming defective plasmid-borne gene is corrected by mechanisms similar to those described in possibility 2, and the normal copy of the gene would then be integrated at a different site in the genome. Each of these possibilities can be distinguished from the others by a molecular analysis of the cellular DNA.

In one series of experiments, a defective *neo* gene ligated to a selectable gene such as *SV2gpt* was introduced into mouse or hamster cells. *Gpt*[+] colonies were selected in an appropriate medium. The resulting cells were examined for the number of integrated copies of input plasmid. These cells were then used as recipients for a second round of gene transfer with a plasmid carrying a defective *neo* gene. The cells were selected in G418. Only cells that contain a functional *neo* protein are able to survive in this medium. Smith and Berg (1984), Folger *et al.* (1984), and Smithies *et al.* (1984) reported obtaining G418[r] colonies from such experiments. Smith and Berg (1984) state that examination of the DNA from G418[r] cell lines provided evidence for homologous recombination. Folger *et al.* (1984) rescued the integrated plasmids from the primary and secondary cells. Examination of these plasmids revealed that the G418[r] resistant cells are not the result of a homologous recombination event, but a frameshift mutation downstream from the original mutation. How the frameshift mutation could give rise to a functional *neo* gene and protein is not clear. DNA analysis of cells studied by Smithies *et al.* (1984) was not complete enough to draw definitive conclusions about the mechanism of acquisition of G418 resistance. Smith and Berg (1984) allude to information that indicates that a true homologous event has occurred.

Lin *et al.* (1984a, 1985) used a strategy similar to that described above to study recombination. They introduced a mutant HSV *TK* gene into mouse cells. All the cell lines they isolated contained multiple copies of the input plasmid. Several of these cell lines were then transfected with a second mutant *TK* plasmid. Reconstruction of *TK* activity was determined by selection in hypoxanthine–aminopterin–thymidine medium. Two of the ten cell lines tested provided positive results. Examination of the DNA revealed that at least in some instances, a reciprocal recombination event between the cellular *TK* and plasmid-borne *TK* genes has occurred. Estimations of the relative rates of homologous and nonhomologous recombination events by different investigators ranged from a high of 1% to a low of 10^{-5}.

7. RECOMBINATION WITH A CELLULAR SEQUENCE

The experiments involving chromosomal plasmid–plasmid recombination events are very instrumental in developing strategies to study recombination with true cellular sequences. Results from attempts to study recombination with cellular sequences have been published. Goodenow *et al.* (1984) have transfected mouse *tk*[−] cells with a plasmid containing the *TK* gene and a truncated *H2* gene. Using immunological methods, they observed that as high as 10% of the transfectants expressed an antigen that was interpreted to have resulted from a recombination event between the truncated *H2* gene and a cellular gene. However, blot data that would have confirmed this view were not presented. A somewhat similar type of experiment was performed by Yoshie *et al.* (1984), in which they introduced a truncated *HLA* gene into human cells. They too were able to detect the expression of an antigenic determinant that could have resulted from a recombination event. However, analysis of the DNA from these cell lines failed to reveal a homologous recombination event.

Smithies *et al.* (1984, 1985) have devised a scheme to detect the relative frequencies of homologous and nonhomologous recombination events involving a truncated human β-globin gene on a plasmid. The strategy they have utilized is very similar to that routinely used in yeast (Hinnen *et al.*, 1978; Orr-Weaver *et al.*, 1981). Smithies *et al.* (1984, 1985) constructed a plasmid containing the 5' end of the human β-globin gene. This plasmid also contained a bacterial transfer RNA suppressor gene (*supF*) at the 5' end of the truncated globin gene. In addition, the plasmid carried SV2 *neo*, a gene the expression of which can be readily selected for in mammalian cells. This plasmid, referred to as pΔβ117, was introduced into human EJ cells by the calcium phosphate transfection method. At 48 hr later, the cells were harvested, and high-molecular-weight DNA was isolated. The DNA was digested with *Xba*I and introduced into a bacteriophage vector that carries two amber mutations. When this phage is introduced into wild-type bacteria, only those that contain the *supF* gene will be able to grow and form plaques. Each of the plaques would represent a DNA molecule that contained the *supF* gene. In addition, if a homologous recombination event between the input plasmid and the recipient cell globin gene has occurred, the DNA of the phage will also contain a complete and intact β-globin gene. This can be readily assayed by transferring the phage DNA to nitrocellulose filters and hybridizing it with a globin-specific probe that was not represented in the input plasmid. If positive plaques are detectable by this hybridization method, DNA from these phage could be prepared and examined. Smithies and co-workers observed that when DNA from EJ cells transfected with pΔβ117 was analyzed, a small but detectable proportion of the *supF* plaques hybridized to the β-globin second intervening sequence probe. Analysis of the DNA from these phages clearly indicated that the phages do indeed represent homologous recombination events between the plasmid and cellular sequences. It is to be realized that in EJ cells, the globin gene is inactive.

In a second series of experiments, Smithies *et al.* (1985) have used an active globin gene as a target. They have chosen a murine erythroleukemia cell × human fibroblast somatic-cell hybrid that retained the part of the human chromosome 11 that carries the β-globin gene by virtue of its translocation to the X chromosome, which carries a selectable marker in the form of *HPRT*. These hybrid cells were transfected with pΔβ117, and G418[r] colonies were selected. The colonies were pooled into large groups. Each group was expanded, and their

DNA was analyzed by the method described above. Smithies and colleagues observed that nearly 1 in 200 of the *supF*-containing phages also contained an intact globin gene. To confirm that the recombinational events they observed were the result of recombination in the cells, rather than in bacteria, a pool of 300 clones that gave positive results was subcloned and pooled into several groups of 20. DNA from one of these groups of 20 tested positive in the assay. When individual members of this group of 20 were tested, by blot hybridization, 2 of them revealed blot patterns that were identical to those that are expected from a reciprocal recombination event between the cellular and plasmid sequences. Taken together, these experiments show unambiguously that homologous recombination between cellular sequences and plasmid sequences occurs at an appreciable frequency. This method, then, has the potential to permit modification of cellular genes in predefined ways.

Additional evidence for recombination between chromosomal sequences and plasmid sequences was obtained from studies involving introduction of defective viral genomes into cells and recovering intact viral sequences. The sequences that were introduced were of RNA viral or DNA viral origin. In one set of experiments, Schwartzberg *et al.* (1985) have introduced a deletion mutant of Moloney murine leukemia virus into NIH 3T3 cells. The recipient cell population began to produce virions. Virions from one such cell line were used to infect a fresh batch of cells, and the cells were reassayed after several rounds of infection and assayed periodically for viral reverse transcriptase. After 6 weeks of transfer, the cells produced wild-type levels of reverse transcriptase. Analysis of the virions from two of the cloned cell lines from this experiment revealed that the deletion in the original mutant virus was repaired. The region that was repaired was not identical to the wild-type virus. Southern-blot hybridization of mouse cell DNA with a probe that corresponded to the repaired region indicated that it is indeed represented in the mouse genome. The best explanation for these results is that a recombination event between the defective viral genome and a cellular sequence has occurred to give rise to a functional virus. Another experiment of this type was conducted by Shaul *et al.* (1985). In these experiments, a plasmid that contained a defective SV40 viral sequence was introduced into monkey COS cells that carry an origin-defective but intact SV40 virus integrated into the cellular genome. Recombination between the cellular sequences and the plasmid-borne SV40 sequences would result in the generation of intact functional SV40 sequences. This was monitored by plating the viral stock on CV1 cells. Using this method, Shaul *et al.* (1985) were able to obtain a virus. Detailed examination of the virus revealed that it has resulted from recombination between the cellular and plasmid sequences.

Taken together, the results presented in this section and in the previous section indicate that homologous recombination between introduced DNA molecules and their homologous cellular counterparts occurs at a detectable frequency. The next challenge is to enhance the frequency with which these events occur or reduce the rate of nonhomologous events, or both. Higher recombination efficiencies would permit *in vivo* modification of genes with its attending implications for gene therapy and examination of gene function.

8. CELL-FREE SYSTEMS TO STUDY RECOMBINATION

Since it is now known that mammalian somatic cells are capable of mediating homologous recombination events, it may be possible to begin attempts at understanding the mechanisms of this process. The first steps toward reaching this goal have been taken. Kucherlapati *et al.* (1985) have described a cell-free system to study homologous recombination events. Nuclei from human bladder carcinoma cell line EJ were isolated and broken open by sonication. The resulting nuclear extract were treated with high salt, and the mixture was passed through DEAE–Sepharose to remove nucleic acids. The flow-through fractions that were rich in protein were pooled and precipitated with ammonium sulfate and assayed for recombination activity. Two nonoverlapping noncomplementing deletion plasmids of the prokaryotic–eukaryotic shuttle vector pSV2neo [pSV2neo DL and pSV2neo DR (Kucherlapati *et al.*, 1984b)] were coincubated with the extract in the presence of ATP, Mg^{2+}, and all four deoxyribonucleoside triphosphates (dNTPs). The DNA was then used to transform recombination-deficient (*rec A⁻*) *E. coli.* They were able to recover recombinant molecules. Several, if not all, of the recombination events seem to have occurred in the extract because the reaction is dependent on the presence of ATP and Mg^{2+} and is enhanced by the presence of dNTPs. Since individual recombination products can be isolated from bacteria, it is possible to examine the nature of the event that yielded a recombinant molecule. Results from these experiments indicated that double-strand breaks in the region of homology enhance the recombination frequency. In addition, it seems that the site of the breaks is the site of initiation of recombination and that nonreciprocal recombination plays an important role in the generation of wild-type molecules. This *in vitro* system is now permitting genetic analysis of recombination and will constitute the first step in a biochemical analysis of the recombination process. Darby and Blattner (1984) have reported a similar activity from a number of mouse cell extracts. Keene and Ljundquist (1984) have used a strand-invasion assay to detect mammalian recombinase in cell extracts from normal cells and cells from a patient with Bloom's syndrome. They reported a 10-fold enhancement of the recombinase activity in Bloom's cell extracts. All these results are promising a future when it will be possible to isolate and characterize the proteins that are involved in the recombination reaction.

9. SUMMARY AND CONCLUSIONS

It is quite clear that mammalian somatic cells have all the enzymes needed to mediate homologous recombination. Recombination between two exogenously introduced molecules as well as an exogenous molecule and cellular sequences has been detected. Recombination involving cellular sequences occurs at a relatively low frequency. The fact that these types of events occur is encouraging because it may permit manipulation of cellular genes in predefined ways. Such manipulations may in turn have significance for gene therapy and studies of gene regulation. The fact that cell extracts are capable of catalyzing homologous recombination is also useful because it predicts a future in which we may be able to dissect the genetic and biochemical aspects of the mammalian recombination enzymes.

ACKNOWLEDGMENTS. I thank P. Moore, A. Skoultchi, and O. Smithies, who have contributed many ideas for my research program. Individuals from my laboratory, D. Ayares, S. Ehrlich, F. Schwartz, W. Keown, K. Noonan, S. Rauth, L. Chekuri, and D. Ganea, have contributed data and criticism. Original work reported in this chapter was supported by grants from the March of Dimes Birth Defects Foundation and the National Institutes of Health.

REFERENCES

Anderson, W. F., Killos, L., Sanders-Haigh, L., Kretschmer, P. J., and Diaumakos, E. G., 1980, Replication and expression of thymidine kinase and human globin genes microinjected into mouse fibroblasts, *Proc. Natl. Acad. Sci. U.S.A.* **77**:5399–5403.

Ayares, D., Chekuri, L., Song, K. Y., and Kucherlapati, R., 1986, Homology requirements for intermolecular recombination in mammalian cells, *Proc. Natl. Acad. Sci. U.S.A.* **83** (in press).

Ayares, D., Spencer, J., Schwartz, F., Morse, B., and Kucherlapati, R., 1985, Homologous recombination between autonomously replicating plasmids in mammalian cells, *Genetics* **111**:375–388.

Brenner, D. A., Smigocki, A. C., and Camerini-Otero, R. D., 1985, Effects of insertions, deletions and double-strand breaks on homologous recombination in mouse L cells, *Mol. Cell. Biol.* **5**:684–691.

Campbell, C. E., and Worton, R. G., 1981, Segregation of recessive phenotypes in somatic cell hybrids: Role of mitotic recombination, gene inactivation and chromosome disjunction, *Mol. Cell. Biol.* **1**:336–346.

Capecchi, M. R., 1980, High efficiency transformation by direct microinjection of DNA into cultured mammalian cells, *Cell* **22**:479–488.

Cavenee, W. K., Dryza, T. P., Phillips, R., Benedict, W. F., Godbout, R., Gallie, B. L., Strong, L., Murphee, A. L., and White, R. L., 1983, Expression of recessive alleles by chromosomal mechanisms in retinoblastoma, *Nature (London)* **275**:617–623.

Cavenee, W. K., Hansen, M. F., Nordenskjold, M., Kock, E., Maumenee, I., Squire, J. A., Phillips, R. A., and Gallie, B. L., 1985, Genetic origin of mutations predisposing to retinoblastoma, *Science* **228**:501–503.

Darby, V., and Blattner, F., 1984, Homologous recombination catalyzed by mammalian cell extracts *in vitro*, *Science* **226**:1213–1215.

De Saint Vincent, B. R., and Wahl, G. M., 1983, Homologous recombination in mammalian cells mediates the formation of a functional gene from two overlapping gene fragments, *Proc. Natl. Acad. Sci. U.S.A.* **80**:2002–2006.

Folger, K. R., Wong, E. A., Wahl, G., and Capecchi, M. R., 1982, Patterns of integration of DNA microinjected into cultured mammalian cells: Evidence for homologous recombination between plasmid DNA molecules, *Mol. Cell. Biol.* **2**:1372–1387.

Folger, K. R., Thomas, K., and Capecchi, M. R., 1984, Analysis of homologous recombination in cultured mammalian cells, *Cold Spring Harbor Symp. Quant. Biol.* **49**:123–138.

Folger, K. R., Thomas, K., and Capecchi, M. R., 1985, Nonreciprocal exchanges of information between DNA duplexes coinjected into mammalian cell nuclei, *Mol. Cell. Biol.* **5**:52–58.

Gluzman, Y., 1981, SV40 transformed simian cells support the replication of early SV40 mutants, *Cell* **23**:175–182.

Gonda, D. K., and Radding, C. M., 1983, By searching processively RecA protein pairs DNA molecules that share a limited stretch of homology, *Cell* **34**:647–654.

Goodenow, R. S., Stroynowski, I., McMillan, M., Nicolson, M., Eakle, K., Sher, B. T., Davidson, N., and Hood, L., 1984, Expression of complete transplantation antigens by mammalian cells transformed with truncated class I genes, *Nature (London)* **301**:388–394.

Gordon, J. W., Scangos, G. A., Plotkin, D. J., Barbosa, J. A., and Ruddle, F. H., 1980, Genetic transformation of mouse embryos by microinjection of purified DNA, *Proc. Natl. Acad. Sci. U.S.A.* **77**:7380–7384.

Graham, F. L., and van der Eb, A. J., 1973, A new technique for the assay of infectivity of human adenovirus 5 DNA, *Virology* **52**:456–467.

Hayward, W. S., Neel, B. G., and Astrin, S. M., 1981, Activation of a cellular oncogene by promoter insertion in ALV induced lymphoid leukosis, *Nature (London)* **290**:475–480.

Hinnen, A., Hicks, J. B., and Fink, G. R., 1978, Transformation of yeast, *Proc. Natl. Acad. Sci. U.S.A.* **75**:1929–1933.

Hirt, B., 1967, Selective extraction of polyoma DNA from infected mouse cell culture, *J. Mol. Biol.* **36**:365–369.

Hsiung, N., Roginski, R. S., Hanthron, P., Smithies, O., Kucherlapati, R., and Skoultchi, A. I., 1982, Introduction and expression of a fetal human globin gene in mouse fibroblasts, *Mol. Cell. Biol.* **2**:401–411.

Jaenisch, R., Harbers, K., Schnieke, A., Lohler, J., Chumakov, I., Jahner, D., Gorotkop, D., and Hoffmann, E., 1983, Germline integration of Moloney murine leukemia virus at the MOV 13 locus leads to recessive lethal mutation and early embryonic death, *Cell* **32**:209–216.

Keene, K., and Ljungquist, S., 1984, A DNA-recombinogenic activity in human cells, *Nucleic Acids Res.* **12**:3057–3068.

King, W., Patel, M. D., Lobel, L. I., Goff, S. P., and Nguyen-Huu, M. C., 1985, Insertion mutagenesis of embryonal carcinoma cells by retrovirus, *Science* **228**:554–558.

Kucherlapati, R. S., and Skoultchi, A. I., 1985, Introduction of purified genes into mammalian cells, *CRC Rev. Biochem.* **16**:349–379.

Kucherlapati, R. S., Ayares, D., Hannaken, A., Noonan, K., Rauth, S., Spencer, J. M., Wallace, L., and Moore, P. D., 1984a, Homologous recombination in monkey cells and human cell-free extracts, *Cold Spring Harbor Symp. Quant. Biol.* **49**:191–197.

Kucherlapati, R. S., Eves, E. M., Song, K. Y., Morse, B. S., and Smithies, O., 1984b, Homologous recombination between plasmids in mammalian cells can be enhanced by treatment of input DNA, *Proc. Natl. Acad. Sci. U.S.A.* **81**:3153–3157.

Kucherlapati, R. S., Spencer, J., and Moore, P. D., 1985, Homologous recombination catalyzed by human cell extracts, *Mol. Cell. Biol.* **5:**714–720.

Lacy, E., Roberts, S., Evans, E. P., Burtenshaw, M. D., and Constantini, F. D., 1983, A foreign β-globin gene in transgenic mice: Integration at abnormal chromosomal positions and expression in inappropriate tissues, *Cell* **34:**343–358.

Lin, F. L., Sperle, K., and Sternberg, N., 1984a, Homologous recombination in mouse L-cells, *Cold Spring Harbor Symp. Quant. Biol.* **249:**139–149.

Lin, F. L., Sperle, K., and Sternberg, N., 1984b, Model for homologous recombination during transfer of DNA into mouse L-cells: Role for DNA ends in the recombination process, *Mol. Cell. Biol.* **4:**1020–1034.

Lin, F. L., Sperle, K., and Sternberg, N., 1985, Recombination in mouse L-cells between DNA introduced into cells and homologous chromosomal sequences, *Proc. Natl. Acad. Sci. U.S.A.* **82:**1391–1395.

Maitland, N., and McDougall, J. K., 1977, Biochemical transformation of mouse cells by fragments of HSV DNA, *Cell* **11:**233–241.

McCutchan, J. H., and Pagano, J. S., 1968, Enhancement of the infectivity of simian virus 40 deoxyribonucleic acid with diethylaminoethyl–dextran, *J. Natl. Canc. Inst.* **41:**351–357.

Miller, C. L., and Temin, H. M., 1984, High efficiency ligation and recombination of DNA fragments by vertebrate cells, *Science* **220:**606–609.

Milman, G., and Herzberg, M., 1981, Efficient DNA transfection and rapid assay for thymidine kinase activity and viral antigenic determinants, *Somat. Cell Genet.* **7:**161–170.

Orr-Weaver, T. L., and Szostak, J. W., 1983, Yeast recombination: The association between double-strand gap repair and crossing-over, *Proc. Natl. Acad. Sci. U.S.A.* **80:**4417–4421.

Orr-Weaver, T. L., and Szostak, J. W., and Rothstein, R. J., 1981, Yeast transformation: A model system for the study of recombination, *Proc. Natl. Acad. Sci. U.S.A.* **78:**6354–6358.

Palmiter, R. D., and Brinster, R. L., 1985, Transgenic mice, *Cell* **41:**343–345.

Pellicer, A., Wigler, M., Axel, R., and Silverstein, S., 1978, The transfer and stable integration of the HSV thymidine kinase gene into mouse cells, *Cell* **14:**133–141.

Perucho, M., Hanahan, D., Lipsich, L., and Wigler, M., 1980a, Isolation of the chicken thymidine kinase gene by plasmid rescue, *Nature (London)* **285:**207–210.

Perucho, M., Hanahan, D., and Wigler, M., 1980b, Genetic and physical linkage of exogenous sequences in transformed cells, *Cell* **22:**309–317.

Rosenstraus, M. J., and Chasin, L. A., 1978, Separation of linked markers in Chinese hamster cell hybrids—mitotic recombination is not involved, *Genetics* **90:**735–760.

Roth, D. B., and Wilson, J. H., 1985, Relative rates of homologous and non homologous recombination in transfected DNA, *Proc. Natl. Acad. Sci. U.S.A.* **82:**3355–3359.

Rubnitz, J., and Subramani, S., 1984, The minimum amount of homology required for homologous recombination in mammalian cells, *Mol. Cell. Biol.* **4:**2253–2258.

Rubnitz, J., and Subramani, S., 1985, Rapid assay for extrachromosomal homologous recombination in monkey cells, *Mol. Cell. Biol.* **5:**529–544.

Scangos, G., Huttner, K. M., Juricek, D. K., and Ruddle, F. H., 1981, Deoxyribonucleic acid mediated gene transfer in mammalian cells: Molecular analysis of unstable transformants and their progression to stability, *Mol. Cell. Biol.* **1:**111–120.

Schwartzberg, P., Colicelli, J., and Goff, S. P., 1985, Recombination between a defective retrovirus and homologous sequences in host DNA: Reversion by patch repair, *J. Virol.* **53:**719–726.

Shaul, Y., Laub, O., Walker, M. D., and Rutter, W. J., 1985, Homologous recombination between defective virus and a chromosomal sequence in mammalian cells. *Proc. Natl. Acad. Sci. U.S.A.* **82:**2781–3784.

Singer, B. S., Gold, L., Gauss, P., and Doherty, D. H., 1982, Determination of the amount of homology required for recombination in bacteriophage T4, *Cell* **31:**25–33.

Small, J., and Scangos, G., 1983, Recombination during gene transfer into mouse cells can restore the function of deleted genes, *Science* **219**:174–176.

Smith, A. J. H., and Berg, P., 1984, Homologous recombination between defective neo genes in mouse 3T6 cells, *Cold Spring Harbor Symp. Quant. Biol.* **49**:171–181.

Smithies, O., Koralewski, M. A., Song, K. Y., and Kucherlapati, R. S., 1984, Homologous recombination with DNA introduced into mammalian cells, *Cold Spring Harbor Symp. Quant. Biol.* **49**:161–170.

Smithies, O., Gregg, R. G., Boggs, S. S., Koralewski, M. A., and Kucherlapati, R. S., 1985, Insertion of DNA sequences into the human chromosomal beta globin locus via homologous recombination, *Nature (London)* **317**:230–234.

Song, K. Y., Chekuri, L., Rauth, S., Ehrlich, S., and Kucherlapati, R. S., 1985, Effect of double strand breaks on homologous recombination in mammalian cells and extracts, *Mol. Cell. Biol.* **5**:3331–3336.

Southern, E., 1975, Deletion of specific sequences among DNA fragments separated by gel electrophoresis, *J. Mol. Biol.* **98**:503–517.

Southern, P. J., and Berg, P., 1982, Transformation of mammalian cells to antibiotic resistance with a bacterial gene under control of the SV40 early region promoter, *J. Mol. Appl. Genet.* **1**:327–341.

Subramani, S., and Berg, P., 1983, Homologous and nonhomologous recombination in monkey cells, *Mol. Cell. Biol.* **3**:1040–1052.

Subramani, S., and Rubnitz, J., 1985, Recombination events after transient infection and stable integration of DNA into mouse cells, *Mol. Cell. Biol.* **5**:659–666.

Subramanian, K., 1979, Segments of simian virus 40 DNA spanning most of the leader sequence of the major late viral messenger RNA are dispensable, *Proc. Natl. Acad. Sci. U.S.A.* **76**:2556–2560.

Sussman, D. J., and Milman, G., 1984, Short-term high-efficiency expression of transfected DNA, *Mol. Cell. Biol.* **4**:1641–1643.

Szybalska, E. H., and Syzbalski, W., 1962, Genetics of human cell lines. IV. DNA mediated heritable transformation of a biochemical trait, *Proc. Natl. Acad. Sci. U.S.A.* **48**:2026–2034.

Tarrant, G. M., and Holliday, R., 1977, A search for allelic recombination in Chinese hamster cell hybrids, *Mol. Gen. Genet.* **156**:273–279.

Thomas, C. A., 1966, Recombination of DNA molecules, *Prog. Nucleic Acids Res. Mol. Biol.* **5**:315–348.

Upcroft, P., Carter, B., and Kidson, C., 1980a, Analysis of recombination in mammalian cells using SV40 genome segments having homologous overlapping termini, *Nucleic Acids Res.* **8**:2725–2735.

Upcroft, P., Carter, B., and Kidson, C., 1980b, Mammalian cell functions mediating recombination of genetic elements, *Nucleic Acids Res.* **8**:5835–5844.

Volkert, F. C., and Young, C. S. H., 1983, The genetic analysis of recombination using adenovirus overlapping terminal DNA fragments, *Virology* **125**:175–193.

Wake, C. T., and Wilson, J. H., 1979, Simian virus 40 recombinants are produced at high frequency during infection with genetically mixed oligomeric DNA, *Proc. Natl. Acad. Sci. U.S.A.* **76**:2876–2880.

Wake, C. T., and Wilson, J. H., 1980, Defined oligomeric SV40 DNA: A sensitive probe of general recombination in somatic cells, *Cell* **21**:141–148.

Wake, C. T., Vernaleone, F., and Wilson, J. H., 1985, Topological requirements for homologous recombination among DNA molecules transfected into mammalian cells, *Mol. Cell. Biol.* **5**:2080–2089.

Wasmuth, J. J., and Vock Hall, L., 1984, Genetic demonstration of mitotic recombination in cultured Chinese hamster cell hybrids, *Cell* **36**:697–707.

Watt, V. M., Ingles, C. J., Urdea, M. S., and Rutter, W. J., 1985, Homology requirements for recombination in *Escherichia coli*, *Proc. Natl. Acad. Sci. U.S.A.* **82**:4768–4772.

Wigler, M., Silverstein, S., Lee, L. S., Pellicer, A., Cheng, Y. C., and Axel, R., 1977, Transfer of purified herpes virus thymidine kinase gene to cultured mouse cells, *Cell* **11**:223–232.

Wilson, J. H., Berget, P. B., and Pipas, J. M., 1982, Somatic cells efficiently join unrelated DNA segments end to end, *Mol. Cell. Biol.* **2**:1258–1269.

Winocour, E., and Keshet, I., 1980, Indiscriminate recombination in simian virus 40-infected monkey cells, *Proc. Natl. Acad. Sci. U.S.A.* **75**:1929–1933.

Wold, B., Wigler, M., Lacy, E., Maniatis, T., Silverstein, S., and Axel, R., 1979, Introduction and expression of a rabbit β-globin gene in mouse fibroblasts, *Proc. Natl. Acad. Sci. U.S.A.* **76**:5684–5688.

Woychik, R. P., Stewart, T. A., Davis, L. G., D'Eustachio, P., and Leder, D., 1985, An inherited limb deformity created by insertional mutagenesis in a transgenic mouse, *Nature (London)* **318**:36–40.

Yoshie, O., Schmidt, H., Lengyel, P., Reddy, E. S. P., Morgan, W. R., and Weisman, S. M., 1984, Transcripts of human HLA gene fragments lacking the 5'-terminal region in transfected mouse cells. *Proc. Natl. Acad. Sci. U.S.A.* **81**:649–653.

14

INTRACHROMOSOMAL RECOMBINATION IN MAMMALIAN CELLS

Anthea Letsou and R. Michael Liskay

1. INTRODUCTION: METHOD FOR STUDYING RECOMBINATION BETWEEN CHROMOSOMAL SEQUENCES

The technologies of gene transfer and recombinant DNA can be exploited to construct artificial gene duplications of a selectable gene in cultured mammalian cells. These pairs of closely linked genes, harboring different mutations, are stably integrated in the host genome by means of a separate dominant selectable marker and therefore provide an opportunity to systematically study the process of homologous recombination as it occurs between repeated chromosomal sequences in cultured mammalian cells.

In the broadest sense, homologous recombination events can be either reciprocal (crossover) or nonreciprocal (gene conversion) in terms of information transfer. Classically, both types of recombination events have been defined by segregation analyses in fungi, analyses that permit examination of all the participants or products of a given recombination event (for a review, see Orr-Weaver and Szostak, 1985). In the absence of recombination, a cell heterozygous for three linked markers (A, B, and C) and undergoing meiosis produces four spores, each of which harbors parental-type chromosomes (Fig. 1B). Recombinant chromosomes produced by a single reciprocal exchange anywhere between markers A and C in meiosis can be recognized because outside markers A and C are no longer in the parental configuration. All markers (A, B, and C), however, still segregate in a Mendelian fashion (2:2); there is no net gain or loss of information, merely a rearrangement of information (Fig. 1C). In the fungi, non-Mendelian (3:1) segregation of markers is the hallmark of gene conversion. In

Anthea Letsou and R. Michael Liskay ● Departments of Therapeutic Radiology and Human Genetics, Yale University School of Medicine, New Haven, Connecticut 06510.

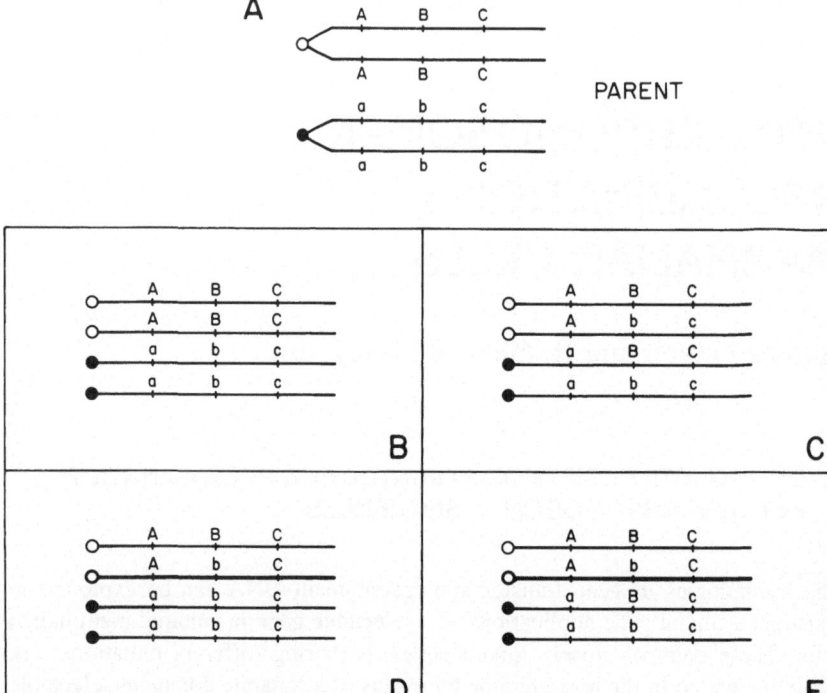

FIGURE 1. Comparison of meiotic products arising in a diploid parent. (●, ○) Centromeres of homologous chromosomes, which define chromosomal origins. (A, B, and C) Linked markers; the cell is genotypically *ABC/abc*. (A) Parental homologous chromosomes are depicted prior to the second meiotic division. (B) Meiotic products in the absence of recombination. (C) Meiotic products following a single reciprocal exchange or crossover in the interval between markers A and B. (D) Meiotic products demonstrating conversion of B to its allelic form b. (E) Meiotic products resulting from a double reciprocal exchange in which one crossover occurs in the interval between markers A and B and a second crossover occurs in the interval between linked markers B and C. Note in (D) and (E) that without the ability to examine all products of a given meiotic event, recombinant chromosomes depicted by the bolder lines are indistinguishable.

the simplest case (i.e., conversion not associated with reciprocal exchange), despite a unidirectional and nonreciprocal transfer of information between homologous chromatids, outside markers remain in the parental configuration (Fig. 1D). Recombinant chromosomes can also be produced by a double reciprocal-exchange event whereby information between two break points is exchanged. Segregation of all markers is 2:2, and outside markers remain in the parental configuration (Fig. 1E). In isolation, therefore, a recombinant chromosome produced by a double reciprocal exchange between chromatids is identical, at the chromosomal level, to a recombinant chromosome produced by a simple gene-conversion event. When one can examine all the products of a single meiosis, however, all types of recombination events are easily distinguishable.

Unfortunately, the elegant types of segregation analyses used to study fungal genetic recombination are not yet amenable to mammalian cell systems. Obvious disadvantages of mammalian systems compared to yeast are our current inability to culture meiotic cells, our inability to easily recover the appropriate mitotic products as pairs, and finally the paucity of suitable genetic and molecular markers. Studies of genetic recombination in cultured mammalian cells are, however, valuable primarily for three reasons. First, the study of intrachromosomal homologous recombination in cultured mammalian cells should be valuable in furthering our practical understanding of the mechanisms by which genetic recombination events proceed. Genomic rearrangements dependent on homologous recombination are postulated to include small circular DNA excision (Calabretta *et al.*, 1982; Reis *et al.*, 1983), gene amplification (Schimke, 1982), and immunoglobulin rearrangements (Brack *et al.*, 1978; Sakano *et al.*, 1981). A role for recombination has been postulated in the etiology of certain childhood tumors as well (e.g., see Leder *et al.*, 1983; Cavenee *et al.*, 1983). Second, the success of programs for gene therapy is contingent on comprehension of the processes of somatic recombination. In theory, the curative power of gene therepy depends on our ability to target incoming wild-type sequences so that they will recombine with inherited mutant alleles of affected individuals. Such interactions should result in replacement or correction, or both, of defective inherited genes, thereby providing cures for at least some inherited disorders. Target-gene modification at an endogenous locus, human β-globin, has been achieved, albeit at very low frequencies (Smithies *et al.*, 1985). Target-gene modification will also be valuable for generating a variety of genome alterations in cultured cells. A final reason for interest in studies of homologous recombination in mammalian cells is that a role for gene conversion is likely in evolution—both in the maintenance of sequence homogeneity and in the generation of variability in multigene families (for a review, see Petes and Klein, 1986).

Simplicity in product recovery and analysis initially made viral systems attractive to investigators of homologous recombination in mammalian cells (e.g., Wake and Wilson, 1979; Volkert and Young, 1983). More recently, gene-transfer techniques coupled with recombinant-DNA technologies have led to the development of additional systems dedicated to detailed studies of extrachromosomal recombination occurring between plasmids in mammalian cells (e.g., de Saint Vincent and Wahl, 1983; Shapira *et al.*, 1983; Small and Scangos, 1983). Difficulties in analysis inherent in both these methods include identification of partners in and timing of recombination events and restriction of analyses to cell lines that are easily transfected. Nonetheless, these studies continue to yield valuable information that lends insight into the processes of genetic recombination, e.g., in defining preferred recombination substrate morphologies (Kucherlapati *et al.*, 1984a; Lin *et al.*, 1984a; Ayares *et al.*, 1985; Brenner *et al.*, 1985; Chakrabarti *et al.*, 1985; Rubnitz and Subramani, 1985; Wake *et al.*, 1985). These extrachromosomal systems, most notable because they can be assayed transiently, thereby speeding up analyses, are dealt with in Chapter 13.

We (Liskay and Stachelek, 1983; Liskay *et al.*, 1984) and other investigators (Lin and Sternberg, 1984; Lin *et al.*, 1984b; Smith and Berg, 1984; Subramani and Rubnitz, 1985) have developed systems that differ from those mentioned above in that these experimental systems have been designed to study *intrachromosomal* recombination. An important advantage of this type of analysis is that both partners are most likely chromosomal sequences, although this inference has yet to be proven. In addition, the participants in the recombination event are known with great certainty and can be examined before as well as, to a certain extent, after the recombination event. The latter asset of the system, however, is not unique; two reports have appeared in the literature (Folger *et al.*, 1984; Kucherlapati *et al.*, 1984b) describing systems that have been carefully designed to permit identification of recombination partners in extrachromosomal recombination.

2. GENERAL SYSTEM

Briefly, the system developed in our laboratory to study homology-dependent intrachromosomal recombination employs plasmids that contain two different mutants of the herpes thymidine kinase (*H-TK*) gene as well as a dominant selectable marker that facilitates introduction and stable integration of the mutant gene duplication into host cells. The experimental plasmid is incapable of extrachromosomal replication and maintenance in the host cells we employ. Our substrate for recombination, therefore, is an artificially constructed gene duplication that consists of two different mutants of the *H-TK* gene and presumably resides in a chromosome. Recombination events between the two defective sequences can be detected by growth in the appropriate selective medium.

As illustrated in Fig. 2, a pair of *H-TK* mutants is inserted into a derivative of the pSV2neo plasmid that carries (1) the neomycin gene of Tn5, which confers G418 resistance to transformed cells and therefore serves as a dominant selectable marker in mammalian cells (Southern and Berg, 1982), and (2) a unique restriction site (*Cla* I) that permits plasmid linearization prior to transfection and consequently integration into the genome in a predictable fashion. Several restriction sites of pSV2neo have been altered to facilitate cloning of the mutants into unique *Bam*HI and *Hind*III sites (Liskay *et al.*, 1984). Both cloned *H-TK* mutants are full length and contain all the necessary 5' and 3' regulatory sequences required for gene expression. In addition, each member of the gene pair harbors a single mutation; the mutations, however, are localized to different sites in the coding region of *H-TK*. The *H-TK* genes can, of course, be oriented in

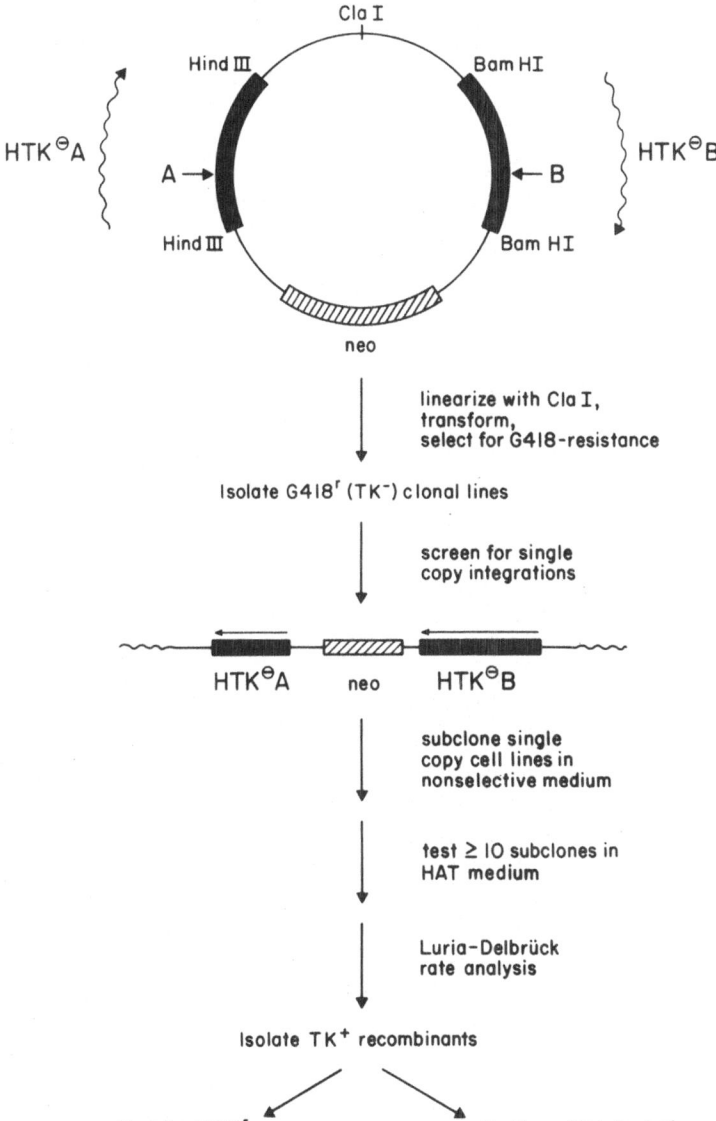

FIGURE 2. General design of the experiments. The plasmid is a derivative of pSV2neo (Southern and Berg, 1982) and contains two different mutants of the *H-TK* gene. (〰➤) Direction of *H-TK* transcription. See the text for details of experimental rationale and methodologies.

the experimental plasmid as either direct or inverted repeats. The studies presented initially employ direct repeats.

We have chosen duplicated *H-TK* sequences to study homologous recombination in cultured mammalian cells for a number of reasons. First, there are numerous available and well-characterized *H-TK* mutants from herpes simplex virus type 1 (HSV-1). (The majority of mutants we have employed are *Xho*I-linker insertions into the coding region of HSV-1 *TK*, a generous gift of Drs. David Zipser and Jesse Kwoh.) Second, *H-TK* genes obtained from different types and/or strains of HSV have been cloned and partially or completely sequenced (Wagner *et al.*, 1981; Kit *et al.*, 1983; Sharp *et al.*, 1983; Swain and Galloway, 1983; Kwoh, personal communication). The availability of cloned *TK* genes from both HSV-1 and HSV-2 permits construction of intertypic *H-TK* gene duplications. As determined from DNA-sequence information, different regions of the two types of *H-TK* genes vary in degrees of homology; therefore, estimates of overall homology requirements for recombination and the effects of various degrees of mismatch on the efficiency of recombination should be possible. In addition, the wealth of silent restriction-site polymorphisms should make practicable accurate assessments of the distribution and lengths of conversion events. Third, appropriate TK⁻ cell lines serve as host cell lines for our analyses. Cultivation of cell lines harboring pairs of different *H-TK* mutants in hypoxanthine–aminopterin–thymidine (HAT) selective medium provides a rapid genetic assay for normal *H-TK* gene expression and consequently provides a rapid genetic assay for recombination events that reconstruct a wild-type *H-TK* gene. Initially, we have employed mouse TK⁻ L cells because they do not revert to the TK⁺ phenotype at a detectable frequency ($<10^{-9}$), and since *H-TK* does not share any known homology with cellular *TK* (Kwoh and Engler, 1984), the great majority of TK⁺ segregants that arise in the duplication-carrying LTK⁻ population should result from the interaction of the duplicated *H-TK* sequences. As will be discussed in Section 4, we can demonstrate that this interaction is indeed homology-dependent, and we can infer the nature of the recombination event (reciprocal or nonreciprocal) on the basis of both molecular and genetic data. Finally, duplicated *H-TK* genes are chosen for these analyses because selections exist for detection of not only the TK⁺ phenotype but also the TK⁻ phenotype. The importance of a means to back-select the TK⁻ phenotype will be addressed in greater detail in Section 9.

3. DERIVING SINGLE-COPY PARENTAL CELL LINES

Our basic experimental protocol is schematically represented in Fig. 2. Plasmid is introduced into TK⁻ mouse L cells by CaPO₄–DNA-mediated gene transfer. Other methods for gene transfer can be used as well (Folger *et al.*, 1984), and we are at present exploring the utility of microinjection and electroporation. Selection for resistance to the drug G418 permits isolation of parent lines in which the *neo* sequence borne by input plasmid is efficiently expressed. DNA isolated from G418^r TK$^-$ transformants is screened by molecular hybridization to determine copy number and integrity of input plasmid. Suitable G418^r TK$^-$ "parent" lines are thereby selected for further analysis—namely, determination by molecular or genetic analyses, or both, of the types of recombination products observed and determination by fluctuation analysis of the rates of recombination.

Analyses are greatly simplified if our studies employ single-copy (or low-copy-number) parent lines. A systematic comparison of transformation conditions indicated that we could optimize our yield of single-copy parent lines by using a transformation protocol that employs a small amount of linearized plasmid (1–100 ng/4 × 10^4 cells) in the presence of excess LTK$^-$ carrier DNA (10–20 μg/4 × 10^4 cells). Candidates for single-copy parent lines are identified by molecular-hybridization analyses. Genomic DNA isolated from G418^r TK$^-$ transformants is digested to completion with *Bam*HI or *Hind*III, electrophoresed in agarose, and transferred to nitrocellulose. Radiolabeled, purified *H-TK* sequences are used to probe these Southern blots. The molecular-hybridization profile of an intact, single-copy parent line should consist of only two bands: Depending on the restriction enzyme employed for analysis, one band migrates at the size of the cloned *Bam*HI or *Hind*III gene; the second band migrates more slowly and contains the second member of the gene pair as well as the junction containing high-molecular-weight host DNA.

The single-copy profile described above is identical to that of integration into a single site of head-to-tail concatemers wherein the unit of amplification contains both plasmid and host sequences. Molecular-hybridization comparisons to single-copy standards are useful for eliminating such multiple-copy lines from our single-copy candidate pool. The finest distinctions, however (e.g., one vs. two copies of input plasmid), require physical analyses of gene copy number in TK$^+$ recombinants. Since recombination events are rare events, it is most probable that in a tandem (e.g. two-copy) G418^r TK$^-$ parent line, TK$^+$ recombinants would arise from a single recombination event. Hence, in a molecular analysis of DNA isolated from TK$^+$ recombinants derived from these tandem parent lines, a single band produced by digestion with *Bam*HI (or *Hind*III) that migrates at the position predicted by the length of *H-TK* sequence inserted into that acceptor site will represent not one, but two, *Bam*HI (or *Hind*III) cloned *H-TK* gene sequences. One gene is wild-type in sequence; the second is mutant. In experiments designed such that mutants are susceptible to enzymatic digestion whereas wild-type gene copies are resistant, one can dissect by molecular-hybridization analyses the single *Bam*HI (or *Hind*III) band into its wild-type and mutant components. Identification of concatemer integrations into single sites is therefore very straightforward when mutant and wild-type alleles are discriminated on the basis of differential restriction-enzyme susceptibilities.

Molecular-hybridization analyses are also employed to confirm that the relative organization of integrated plasmid sequences is as predicted by plasmid linearization prior to transfection. Although restriction sites encoded at or very near sites of plasmid linearization are most frequently, albeit not always, lost on integration, other plasmid restriction sites are useful for depicting post integration plasmid organization. For example, the HSV-1 *TK* gene from strain F encodes a unique *Sac*I site that is useful in many of our putative parent lines for demonstrating that the two *H-TK* genes are linked and gross rearrangements have not accompanied plasmid integration. Routinely, we recover 1–3 single-copy parent lines per 20 primary G418^r transformants screened. Our original studies of intrachromosomal recombination (Liskay and Stachelek, 1983) employed pSV2gpt (Mulligan and Berg, 1981), rather than pSV2neo. For the purpose of generating single-copy lines in mouse L cells, the *neo* system seems highly preferable to the *gpt* (xanthine-guanine phosphoribosyltransferase) system, at least when using *H-TK* mutant pairs.

All single-copy parent lines are next subcloned in nonselective medium, for several reasons:

First, subcloning provides a genetic method for ascertaining stability of the input sequences—i.e., whether the duplication behaves as though it is integrated in a chromosome. We have found that G418^r parent lines shown by molecular hybridization to harbor the *H-TK* gene duplication in association with high-molecular-weight cell DNA can oftentimes give rise to subclones that are G418^s. Subclones sensitive to G418 are evidence that the segregation behavior of the transfected *neo* gene does not always reflect that of an integrated chromosomal sequence. We interpret these observations to indicate that input plasmid has not yet stably integrated into the host genome. Resistance to G418 of at least ten subclones derived in non-selective medium from a single parent line indicates reasonable stability and implies chromosomal segregation behavior of input plasmid sequences.

Second, by definition, subclones originate from a single cell progenitor. It follows, therefore, that by subcloning parent lines, we can be confident that the recombination events we are studying are not events associated with gene transfer. If recombination could occur only between plasmids during gene transfer, then all TK$^+$ cells must accumulate in the parental culture, prior to subcloning. Consequently, randomly selected, single-cell-derived subclones should consist purely of TK$^+$ or purely of TK$^-$ cells. Since all subclones derived from the parent lines, we examine generate TK$^+$ recombinants at low but detectable frequencies (e.g., 10^{-6} events/duplication frequencies), it is certain that the recombination events we are studying are not associated with gene transfer, but rather take place during expansion of the subcloned populations and are consistent with intrachromosomal events.

Third, by subcloning parent lines, we can demonstrate that the recombination events we are studying are random events. Routinely, we analyze at least ten independent subclones from each TK⁻ parent line. The frequency of recombination for each subclone is determined by the number of HAT^r colonies scored per total cell count at the time HAT selection was imposed. The frequency of recombination varies greatly between subclones derived from a common parent. This variability can be as large as two orders of magnitude and is by no means disconcerting. In fact, this variability (or fluctuation) in frequencies demonstrates that the recombination events we are studying are not events induced by the selective media, but rather are random events that can occur any time during the expansion of the subclone (Luria and Delbrück, 1943). Clearly, because frequencies can fluctuate so greatly, accurate measurements of the rates of recombination must not be expressed as frequencies (events/duplication), but rather must be expressed as rates (events/duplication per cell generation), and can be determined by a Luria–Delbrück type of fluctuation analysis as discussed below.

Rates of recombination are determined utilizing a computer program that permits simple calculation by the formula of Luria and Delbrück (Capizzi and Jameson, 1973):

$$r = aN_t \ln(CaN_t)$$

The formula is derived from the statistical distribution of Poisson, and r = average number of recombinants scored in a limited number of subclones; a = rate of recombination/duplication per cell generation; N_t = average number of cells/subclone at the time of HAT selection, and C = number of subclones examined. A most important assumption of the Luria–Delbrück calculation is that the probability of a recombination event is equal at every cell division. Consequently, occurrence of rare recombination events at random cell divisions within defined growth periods of parallel cultures causes frequencies of recombination in these parallel cultures to fluctuate. For example, consider two exponentially expanding cell populations: Frequencies of recombination can greatly differ simply because a single rare recombination event occurs early in the growth period (or cascade of cell division) of one population, thereby producing a large number ("jackpot") of TK⁺ descendants, whereas in a second and parallel population, a single rare recombination event occurs relatively much later in the growth period, so that only a very few TK⁺ L cells are present in the final sample. Clearly, the number of cell divisions is a very important factor in determining the rate (events/duplication per cell generation). In the equation presented above, the number of cell divisions is approximated by N_t because when N_t (final cell number) $\gg N_0$ (initial cell number) the number of cell divisions $(N_t - N_0) = N_t$.

Finally, only on analysis of independent recombinants (i.e., those arising in independent subclones) can we propose with certainty the relative contributions of, for example, reciprocal and nonreciprocal pathways to intrachromosomal recombination in cultured mammalian cells.

4. ANALYSIS OF INTRACHROMOSOMAL RECOMBINATION BETWEEN DIRECT REPEATS

Using the methods already described, we have studied G418^r TK$^-$ cell lines harboring single pairs of linked H-TK mutants, $TK26$ and $TK8$. In addition, control TK$^-$ lines harboring at least one intact copy of only $TK26$ or only $TK8$ have also been studied. The molecular nature of these mutations, which are localized to different sites in the coding region of H-TK, will be discussed in greater detail below. Three G418^r TK$^-$ parent lines were identified, each harboring a single copy of the mutant H-TK gene pair. Presumably, in each line, plasmid integration has occurred at a different genomic site. Prior to subcloning, each line gives rise to approximately 1 TK$^+$ colony per 10^5 cells subjected to selection in HAT medium. To obtain more accurate measurements of recombination rates and to ensure that these are recurring TK$^-$ $\rightarrow$ TK$^+$ events, rather than remnants of events that occur during gene transfer, G418^r TK$^-$ parent lines are subcloned in nonselective medium. Each of at least 10 subclones per original parent is expanded to larger cell populations totaling $1-3 \times 10^7$ cells. At this time, HAT selection is imposed on each subcloned culture. After 10 days in HAT selective medium, the number of TK$^+$ segregants is scored. These raw data are used to determine rates of recombination by the Luria–Delbrück-derived equation discussed in Section 3.

TK$^+$ segregants arise at detectable levels only in TK$^-$ lines that carry linked pairs of H-TK mutants (pJS-3 experimental cell lines). The reversion frequencies of the two genes used in this analysis are less than 10^{-8}. More than 10^9 cells for lines carrying only the $TK8$ mutant gene were plated in HAT selective medium, and not a single TK$^+$ segregant was detected. A total of $3-4 \times 10^7$ cells from each of four lines harboring only H-TK mutant $TK26$ were subjected to selection in HAT; only a single colony was harvested that could *not* be cultivated further in HAT medium. The appearance of stable TK$^+$ segregants therefore requires the homology-dependent interaction of closely linked, repeated H-TK sequences and is consistent with the occurrence of intrachromosomal homologous recombination. The rates of recombination between closely linked sequences sharing 1.8 kilobases (kb) of homology are shown in Table I.

The experimental system has been designed to permit determination of the

Table I. Recombination Rates[a]

Parent line	Copy number	Rate
Experimentals		
pJS-3-10	1	3.5×10^{-6}
pJS-3-3-3	1	4.8×10^{-6}
pJS-3-3-4	1	2.6×10^{-6}
pAL-5-12	2	7.8×10^{-7}
pAL-5-13	2	7.4×10^{-7}
pAL-5-14	2	8.0×10^{-7}
pAL-5-15	2	7.8×10^{-7}
pAL-5-22	1	1.1×10^{-6}
pJS-Al-9	2	6.0×10^{-7}
pJS-Al-18	1	6.1×10^{-7}
pJS-Al-23	1	5.9×10^{-7}

Controls	Number of TK$^+$ segregants/cells plated in HAT
TK8-1	$0/10^8$
TK8-2	$0/10^8$
TK8-3	$0/10^8$
TK26-1	$0/3 \times 10^8$
TK26-2	$1/4 \times 10^{7b}$
TK26-3	$0/4 \times 10^7$
TK26-4	$0/3 \times 10^7$

[a] The table presents the rates of recombination for the experimental cell lines: The pJS-3 lines maintain full-length sequence duplications oriented as direct repeats that share 1.8 kb of homology (see Fig. 3). The pJS-3 lines carry two different mutants (*TK8* and *TK26*). The pAL-5 lines and pJS-Al lines maintain conversion-specific duplications that share only 1.2 kb of homology and are similar in construction to those shown in Fig. 7. The mutant designated *TK8* is the "recipient" in the pAL-5 and pJS-Al experimental lines. Repeats are oriented as direct repeats in pAL-5 lines; in contrast, repeats are oriented as inverted repeats in pJS-Al lines. Rates of recombination for experimental lines are determined by Luria–Delbrück fluctuation analysis and are expressed as TK$^+$ segregants/duplication per cell generation. Also presented are the mean frequencies of TK$^+$ segregation from control cell lines. TK8 control lines harbor only mutant sequences of *TK8*. Similarly, TK26 control lines harbor only mutant sequences of *TK26*.
[b] One colony arose that could not be continuously maintained in HAT medium.

type of recombination event most likely responsible for production of each TK$^+$ recombinant. In the simplest sense, TK$^+$ recombinants can result from either a single reciprocal exchange (crossover) or a nonreciprocal transfer of information (gene conversion) (Fig. 3A). The products of these types of recombination events can be easily distinguished genetically when gene pairs are oriented as direct repeats and flank the *neo* gene (G418^r). As can be seen in Fig. 3A, TK$^+$ segregants arising from a single reciprocal exchange, either within a chromatid or following unequal pairing between sisters, lose the sequences that encode resistance to neomycin and are TK$^+$ but G418^s. Alternatively, TK$^+$ segregants

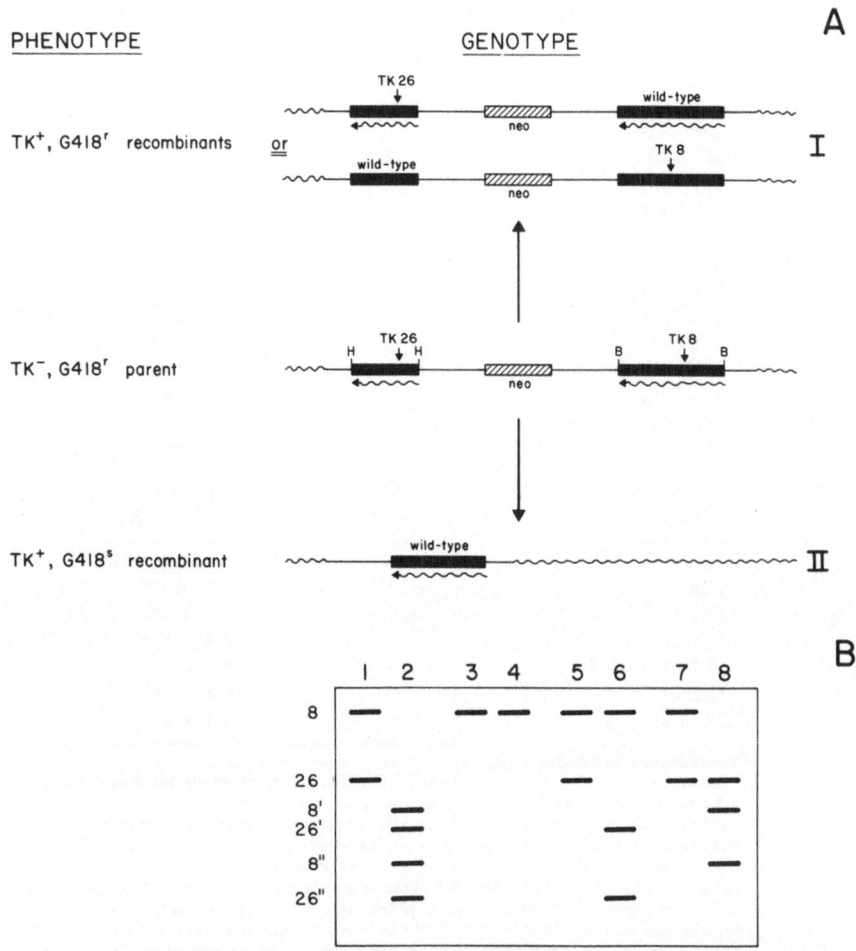

FIGURE 3. (A) Three of the most frequently observed recombination products plus the relative organization of mutant HSV-1 *TK* and *neo* genes in the TK⁻ G418ʳ parent lines. Class I recombinants are most easily explained by a nonreciprocal transfer of wild-type information from one mutant to correct another, a gene conversion. This event can occur within a chromatid or between sisters. Class II recombinants are readily explained by either single reciprocal exchanges within a chromatid or a single exchange between sister chromatids following unequal pairing. *TK8* and *TK26* are designations for two different *Xho*I-linker insertion mutations. (B) Cartoon of Southern-hybridization analysis of TK⁻ parent and three frequently observed recombination products depicted in (A). (1) *Bam*HI/*Hind*III and (2) *Bam*HI/*Hind*III/*Xho*I digestions of the TK⁻ parent. In this parent line, the *TK8* mutant gene resides on a 2.4-kb *Bam*HI fragment and the *TK26* mutant gene resides on a 1.8-kb *Hind*III fragment. The *TK8* and *TK26* mutant genes are both susceptible to cleavage by *Xho*I. (3, 5, 7) *Bam*HI/*Hind*III and (4, 6, 8) *Bam*HI/*Hind*III/*Xho*I digestions of recombinants 1, 2, and 3. Recombinant 1 exhibits a single *Bam*HI/*Hind*III band that is *Xho*I-resistant and is by our definition the wild-type gene. The recoverable product of reciprocal exchange is an *H-TK* gene that now resides

TABLE II. Types of Recombinants[a]

Parent line	G418[r] (conversions)	G418[s] (crossovers)
3-10	19	5
3-3-3	12	2

[a] The table presents the results of genetic analysis of the types of recombination products observed from two different parental lines.

resulting from a conversionlike event retain *neo* sequences and are TK[+] and G418[r]. Hence, we take growth of recombinants in G418 to be synonomous with conversion, whereas failure to survive selection in G418 belies the occurrence of a reciprocal-exchange event (a crossover).

In all, 14 independent TK[+] segregants from parent line 3-3-3 and 24 independent TK[+] segregants from parent line 3-10 were subjected to selection in G418. In both samples, approximately 80% of the segregants were G418[r] and are therefore in accordance with convertants; a minority, approximately 20%, were sensitive to G418—the result predicted if these segregants are produced by a single reciprocal exchange (Table II).

In a similar analysis of homologous recombination that employed chromosomally integrated, duplicated mutants of *neo* in mouse L cells, Smith and Berg (1984) reported an excess of unequal reciprocal sister-exchange events relative to gene-conversion events. The authors are, however, reluctant to speculate on relative frequencies of the two types of events because only a small number of recombinants were examined. It is significant, however, that measurements of recombination frequencies determined for single-copy parent lines in both systems are in good accord.

Our experimental system has been designed so that molecular analyses can be performed in conjunction with the genetic analyses already presented. Not surprisingly, data obtained from the molecular analyses discussed below are entirely consistent with the genetic data.

The *H-TK* mutants used for the initial study, designated *TK26* and *TK8*, are each full length and contain 8-base-pair (bp) synthetic oligonucleotide *XhoI*-linker insertions into the coding region of the HSV-1 strain F *TK* gene (Fig. 3A). These two mutations lie 485 bp apart within the *H-TK* gene, mapping at positions 735 (*TK26*) and 1220 (*TK8*) according to the numbering system of

on a 2.4-kb *Bam*HI/*Hind*III fragment equal in size to the *Bam*HI fragment originally harboring the *TK8* allele. In recombinant 2, (5, 6), the duplication is maintained, and the 1.8-kb *TK26* band is cleaved once by *Xho*I in a manner identical to its parent. The 2.4-kb band, in contrast, is *Xho*I-resistant. The simplest explanation for this blot profile is a *TK8* → wild-type (*wt*) gene conversion. Recombinant 3 (7, 8) retains a 1.8-kb fragment resistant to cleavage by *Xho*I, whereas the *TK8* gene (2.4 kb) is cleaved once by *Xho*I in a fashion identical to the parent. This profile is consistent with a *TK26* → *wt* conversion event.

Wagner *et al.* (1981). These *Xho*I-linker insertions produce nonleaky frameshift mutations, and in each case the newly created *Xho*I site is unique to that mutant *H-TK* gene. *Xho*I does not cleave the wild-type *H-TK* gene or other sequences in the plasmids employed in our studies. Molecular-hybridization analyses therefore permit determinations of the genotypes of *H-TK* sequences before and after TK$^+$ segregation. Susceptibility to cleavage by *Xho*I reveals the presence of the mutation; resistance to cleavage by *Xho*I, on the other hand, is our hallmark of a wild-type gene. By employing such *in vitro* constructed *H-TK* mutants, we are able to present evidence for a wild-type recombinant *H-TK* gene genetically (using survival in HAT selective medium as our criterion) as well as physically (using resistance to cleavage by *Xho*I as our criterion). More specifically, we can demonstrate that the TK$^+$ phenotype dictated by selection and survivability in HAT medium correlates with the presence of a wild-type *H-TK* gene copy that is resistant to cleavage by *Xho*I.

The introduction into the *H-TK* gene of identifiable mutations, such as synthetic oligonucleotide insertion of unique *Xho*I restriction sites, permits determination of the nature of recombination events at the molecular level. The probe employed to screen genomic blots of DNAs prepared from TK$^+$ segregants is specific for *H-TK* sequences, so that in TK$^+$ lines produced by a single reciprocal exchange, the duplication is lost and we should detect a single *H-TK* gene that is resistant to digestion by *Xho*I. TK$^+$ segregants arising from a gene conversion event, on the other hand, should retain the duplication: One *H-TK* sequence is wild-type and resistant to digestion by *Xho*I, and the second has a single mutation and is susceptible to cleavage by *Xho*I. In these studies, wild-type recombinant genes (produced by either pathway) always lack the *Xho*I site, whereas both *H-TK* genes in the parent always restrict with *Xho*I. We can detect three different products of recombination arising in the same TK$^-$ parent line (Fig. 3B): (1) single reciprocal exchanges, (2) conversion of *TK8* to wild-type, and (3) conversions of *TK26* to wild-type.

TK$^+$ segregants could also arise from a double reciprocal-exchange event occurring either within chromatids or between sisters. Using direct repeats as substrates for recombination, we cannot formally distinguish between recombinant chromosomes resulting from gene-conversion events and those resulting from double unequal reciprocal exchanges between sister chromatids (Fig. 4). We argue on the basis of statistical analyses, however, that the recombinant chromosomes we recover result from gene conversions and not from double reciprocal exchanges. First, in our system, relative to events undeniably attributable to single reciprocal exchanges, a 4- to 5-fold excess of events (the major class) are formally consistent with either gene conversion or an unequal double reciprocal exchange between sisters. Although not yet proven, it seems reasonable that the TK$^+$ recombinants comprising the major class are convertants, since it seems unlikely that the frequency of a double reciprocal exchange would

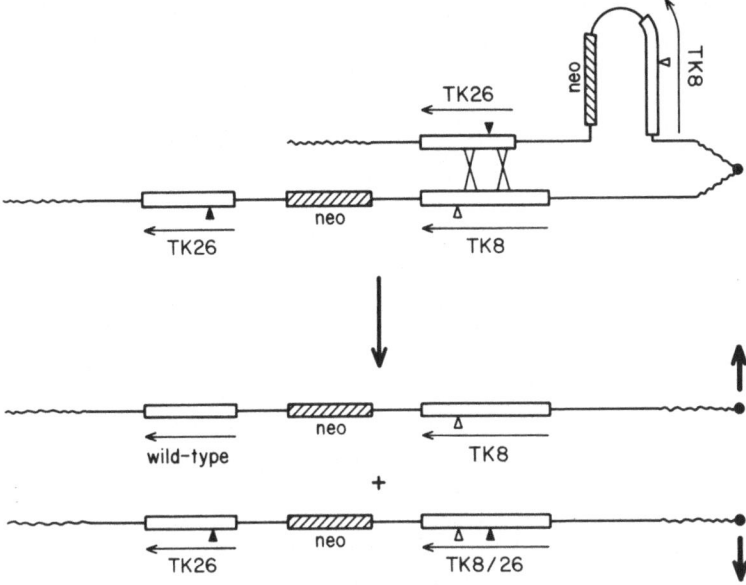

FIGURE 4. Unequal pairing between sister chromatids and the recombinant chromosomes produced by a double unequal reciprocal exchange. When the substrate for recombination consists of duplicated sequences oriented as direct repeats, one recombinant chromosome harbors both a *wt* gene and a single parental-type mutant (*TK8*) (compare with Fig. 3A Class I recombinant). The second recombinant chromosome is not recovered, maintains both the *TK* duplication and *neo* sequences, and contains a new double mutant *TK* allele (*TK8/26*).

be greater than that for a single reciprocal exchange. Second, in conducting these analyses, we have detected no evidence for double reciprocal exchanges occurring within a chromatid. In experiments taking advantage of inverted repeat recombination substrates, however, we have observed single intrachromatid exchanges. Third, in analogous studies initially performed in yeast, the most frequent type of recombination event is gene conversion (Jackson and Fink, 1981, 1985; Klein and Petes, 1981; Klein, 1984).

5. POSITION EFFECTS ON GENE CONVERSION

For the purpose of studying factors that affect gene conversion but are not associated with reciprocal exchange, we have constructed recombination substrates that should be specific for gene conversion. Whereas previously both cloned *H-TK* mutants have been full length, in a "conversion-specific" substrate, the sequence inserted into the *Hind*III acceptor site of our experimental plasmid is a 1.2-kb internal fragment derived from the wild-type *H-TK* gene. This internal sequence lacks essential 5' and 3' information necessary for expression, but spans the mutant site of its partner in all the *H-TK* gene duplications constructed and therefore should serve in all duplications as the *donor* of wild-type information. Consequently, on growth of G418^r TK$^-$ parent lines harboring the conversion-specific recombination substrate in nonselective medium and subsequent selection in HAT medium, TK$^+$ or HAT^r colonies should result from intrachromosomal gene-conversion events. The experimental system is designed such that for any gene-conversion event, the 2.4-kb full-length mutant gene cloned in the *Bam*HI site can be considered the *recipient* of wild-type information from the 1.2-kb donor fragment; single reciprocal-exchange events are nonproductive, resulting in 5' or 3' truncated genes, and the 1.2-kb internally derived *H-TK* gene fragment, because it is not full length and because it is flanked by nonhomology, should not be an efficient substrate for gene conversion. Finally, although double reciprocal exchanges can produce wild-type *H-TK* gene copies, we do not expect, as discussed in Section 4, that these events would contribute significantly to the generation of TK$^+$ segregants.

Five G418^r TK$^-$ parent lines (pAL-5 series) in which one copy, or two copies, of the conversion-specific plasmid have stably integrated were constructed. As determined by molecular-hybridization analyses, the junction fragments in each parent differ. We presume, therefore, that the parent lines originate from independent integration events and that the sites of plasmid integration in each line are different. Consistent with this conclusion are studies that document integration of foreign DNA sequences at many genomic sites (Pellicer *et al.*, 1980). Because our transfection protocol utilizes excess carrier DNA, there remains the unlikely possibility that all integrations have occurred into the same site, but in association with different carrier DNA.

The recipient of wild-type information in each experimental parent line is the *H-TK* 8-bp *Xho*I-linker insertion mutant designated *TK8*. Donor and recipient sequences are oriented as direct repeats. DNAs from a total of 24 independent TK$^+$ recombinants recovered from the 5 parent lines harboring the conversion-specific substrate were analyzed by molecular hybridizations. In all cases, the full-length *Bam*HI allele is resistant to cleavage by *Xho*I and therefore wild-type, whereas in all DNAs isolated from parents, the full-length *Bam*HI recipient gene copy is susceptible to cleavage by *Xho*I and therefore mutant. More specifically, the only difference detected molecularly between TK$^-$ parent and TK$^+$ recombinant is the susceptibility to cleavage by *Xho*I of the full-length *H-TK* gene copy. This single difference is consistent with TK$^+$ segregants resulting from gene-conversion events. In no instance, however, did we detect a *Hind*III donor fragment that had acquired an *Xho*I restriction site, as would be predicted if, for example, the recombination event were a double reciprocal exchange occurring within a chromatid.

Rates of recombination for the pAL-5 lines, expressed as events/duplication per cell generation, are determined by fluctuation analyses as previously described and are presented in Table I. The variability observed in rates among the independent parent lines is considerably less than 2-fold. Clearly, the interpretation is that the position of plasmid integration into the mouse genome does not frequently affect the rate of gene conversion between the duplicated gene sequences. In accordance with this interpretation are our observations in other crosses of little variation in rates of recombination between parent lines that each carry the same duplication. It is worthwhile noting, however, that the integration sites we are able to study are defined and most probably limited by our initial requirement for expression of the neomycin-resistance gene.

The rates of recombination calculated for the parent lines that harbor conversion-specific substrates are less than the rates calculated for the parent lines discussed initially, which harbor full-length *TK26/TK8* gene duplications. We can propose two reasons that are both likely to contribute to this decrease: First, the overall shared homology between input mutant sequences has decreased by approximately 30%—from 1.8 to 1.2 kb; second, one class of recombination event (single reciprocal exchange) is no longer productive. Further discussion of the effects of decreasing homology on the rate of gene conversion follows in Section 6.2.

Primarily for technical reasons, the significance of our failure to detect frequent position effects should be stressed. This lack of position effect demonstrates that recombination rates should most often depend on factors intrinsic to the *H-TK* sequences themselves, rather than on genomic locations.

6. FACTORS THAT AFFECT GENE CONVERSION

As detailed above, the intrachromosomal recombination system can be adapted to study gene-conversion events not associated with reciprocal exchange. These conversion-specific substrates have been exploited to study individual properties of gene conversion and more specifically to (1) determine what effect, if any, the molecular nature of a given mutation has on the efficiency of its conversion to wild-type; (2) determine what effect decreasing shared homology has on conversion rates; and (3) determine what effect inverted orientation of duplicated sequences has on conversion rates.

6.1. Molecular Nature of Mutation

To determine the effects of the molecular nature of mutation on conversion, we have constructed parent cell lines with duplications wherein the donor is, in all cases, the same 1.2-kb internal fragment derived from the wild-type *H-TK* gene as discussed previously. The recipient, however, now carries a variety of different types of molecular mutations, including insertions, deletions, inversions, and base substitutions. Preliminary results suggest that differences in conversion efficiency correlate with the gross size of the mutation. Aberrations 100 bp long are less efficiently converted than those 8 or 1 bp long. Although these analyses are incomplete, we have as yet detected no differences in rates of conversion among mutations that differ in class; there appears to be no difference in the conversion behavior of insertions and deletions of similar sizes.

6.2. Homology Requirements

Much interest has recently arisen in determinations of homology requirements for recombination. Observations by Singer *et al.* (1982) support the hypothesis that bacteriophage T4 requires at least 50 bp of homology for operation of its major recombination pathway. Similarly, *in vitro*, substrates for the *Escherichia coli* RecA pairing reaction can share 151 bp of homology, but fragments that share only 30 bp of homology fail to undergo RecA-promoted homologous pairing efficiently (Gonda and Radding, 1983).

To determine the amount of homology necessary for intrachromosomal gene conversion in mammalian cells, we have tested internal donor fragments that decrease in length from 1.8 to 1.2 kb, 300 bp, 200 bp, and finally 96 bp. In all cases, the donor fragment overlaps the *Xho*I mutant site in the full-length recipient gene. Results obtained thus far indicate that a linear decrease in conversion rate occurs between 1.8 kb and 300 bp, after which we have observed a sharp drop in recombination rate. More specifically, with a 200-bp donor, recombinants have been observed, but at a rate 10 times less than that with a 300-bp donor. Furthermore, in examining more than 10^9 cells from parents that harbor the 90-bp *H-TK* fragment as donor, we have been unable to recover a single TK$^+$ recombinant. We interpret these initial results to indicate that a minimum shared homology of approximately 300 bp is required for "efficient" gene conversion in cultured mammalian cells (Liskay *et al.*, 1986).

The sharp decline in recombination between sequences that share less than 300 bp of homology is consistent with observations made by Rubnitz and Subramani (1984). Using a very sensitive plaque assay to measure frequencies of extrachromosomal recombination in monkey cells as a function of decreasing homology length (5 to 0 kb), a steep drop in recombination frequencies was localized to a length of shared homology between 163 and 214 bp. Reports of equivalent lengths of minimum shared homologies required in both the intra- and extrachromosomal recombination assays suggest that at least some recombination machinery is presumably shared between the pathways for intra- and extrachromosomal recombination.

6.3. Orientation of Repeats

By simply inverting the orientation of the cloned *TK8* mutation with respect to the internal donor fragment, we constructed three independent parent lines carrying duplicated *H-TK* sequences oriented as inverted repeats (pJS-A1 series) and determined rates of recombination for these parent lines. Once again, we observe no variability in conversion rates between parents harboring identical plasmids. Comparison of conversion rates measured between inverted repeats (pJS-A1 lines) to conversion rates measured between the same sequences orientated as direct repeats (pAL-5 lines) reveals no significant differences (see Table I).

7. UTILITY OF RECOMBINATION STUDIES EMPLOYING INVERTED REPEATS

Albeit incomplete, our preliminary evidence suggests that the interactions of inverted repeats are functionally equivalent to the interactions of direct repeats, in terms of both the overall rates of recombination and the proportion of events consistent with conversion vs. that fraction consistent with reciprocal exchange. Most interesting, however, are not analyses of inverted repeats that confirm and extend observations already made employing direct repeats, but rather the additional information that can be garnered from inverted repeats. More specifically, analyses of recombination between inverted repeats should permit discriminations of pathways obscured in the examinations of direct repeats discussed in Section 4. On the basis of classic analyses of homologous recombination in the fungi (Case and Giles, 1958a,b, 1964; Stadler and Towe, 1963; Fogel and Hurst, 1967; Savage and Hastings, 1981), an intimate relationship between gene conversion and reciprocal recombination has been assumed. Furthermore, it has been proposed that the occurrence or nonoccurrence of reciprocal exchanges constitutes two alternative pathways for the resolution of conversion events (Fogel *et al.*, 1978). More recently, however, the postulate that conversion and reciprocal exchange are necessarily associated events has been challenged. Several reports that describe gene-conversion events in yeast not associated with reciprocal exchange have appeared in the literature (Roman and Fabre, 1983; Klar and Strathern, 1984; Klein, 1984; see also Fink and Petes, 1984). Klein (1984) demonstrated the value of inverted repeats for examining conversion and reciprocal exchange in intrachromosomal homologous recombination. A similar approach should be possible in cultured mammalian cells.

If we first consider *only* intrachromatid recombination, three possible classes of events are discernible when gene pairs are examined before and after recombination. Simple gene conversion is straightforward and yields one mutant gene plus a wild-type gene. No other change accompanies a conversion (Fig. 5B). A recombinant chromosome arising from a single reciprocal exchange *not* associated with gene conversion at one of the two mutant sites is identified by inversion of all sequences between the sites of crossover. There is no net gain or loss of information, and one wild-type gene and one double mutant are recovered (Fig. 5C). However, when the crossover, signaled by sequence inversion, is accompanied by conversion at one of the two mutant sites, a single mutant *H-TK* gene, parental in molecular nature, is recovered in addition to the wild-type gene (Fig. 5D).

Because reciprocal sister chromatid events will give rise to aberrant products, we have discussed only reciprocal events occurring within a chromatid. These aberrant products include dicentric chromosomes and acentric fragments plus duplications and deficiencies of flanking chromosomal material (Fig. 6). It does not seem likely that such unstable and unbalanced recombination products would be recovered. In Fig. 6, we have depicted the consequences of sister chromatid reciprocal exchange for only one inverted repeat orientation and one placement of the centromere. Changing these parameters does not circumvent

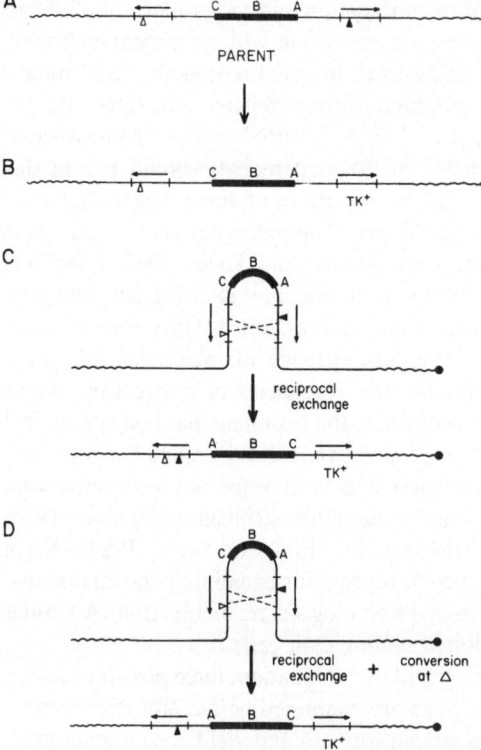

FIGURE 5. (A) Recombination substrate in which duplicated *TK* sequences are oriented as inverted repeats and the direction of transcription is outward. (B) Recombinant chromosome produced by a gene conversion. (C) Recombinant chromosome produced by a single reciprocal exchange (crossover) following intrachromatidal pairing of duplicated *H-TK* sequences. (D) Recombinant chromosome produced by a single reciprocal exchange associated with a conversion of $\Delta \rightarrow wt$. The major difference between recombinant chromosomes in panels (C) and (D) is the single vs. double mutant nature of the mutant *H-TK* allele contained in the duplication.

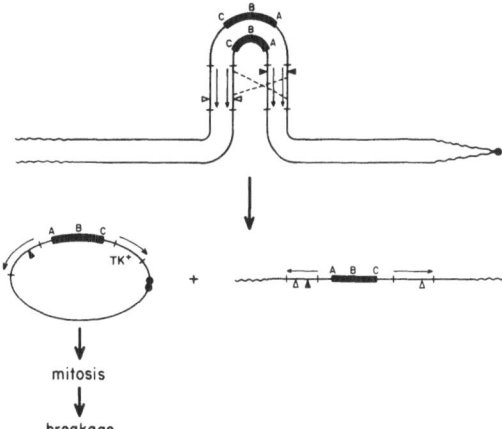

FIGURE 6. Aberrant products (dicentric chromosome and acentric fragment) of reciprocal exchange between sister chromatids when duplicated *TK* sequences are oriented as inverted repeats. It is unlikely that either product would be recoverable.

the problems discussed above, but only affects which undersirable product harbors the wild-type *H-TK* gene.

8. COCONVERSION

A convenient difference between the *H-TK* genes from closely related strains F and 101 of HSV-1 is a silent restriction-site polymorphism that determines susceptibility or resistance to digestion by *Sac*I. This restriction-site polymorphism results from a single base change in the coding region of *H-TK* such that the *H-TK* gene of strain F contains the *Sac*I site whereas the *H-TK* gene of strain 101 does not. Using interstrain *H-TK* sequence duplications as our substrate for recombination (Fig. 7), we first determined that this silent site is always coconverted (22 of 22 events) when we require coconversion of two flanking mutant *Xho*I sites (*TK29* and *TK8*). The rate of coconversion to wild-type of the double mutant, 4.7×10^{-8}, is significantly greater than the rate predicted if conversions at the two *Xho*I insertion sites were truly independent events ($\approx 10^{-12}$). We conclude, therefore, that in each of these events, a single conversion tract, contiguous in nature, encompasses the three sites: *TK8, TK29,* and the *Sac*I silent polymorphism. Second, we demonstrated that this silent site can be coconverted with single selected mutant sites, and the rate of coconversion is inversely proportional to the distance between the silent and selected *Xho*I mutant site, at least when the selected sites are *Xho*I insertions lying 3' of the silent site. A mechanism for gene conversion involving contiguous blocks of DNA differing in length or position or both is consistent with our results. An additional and somewhat surprising finding is that the frequency of coconversion of the

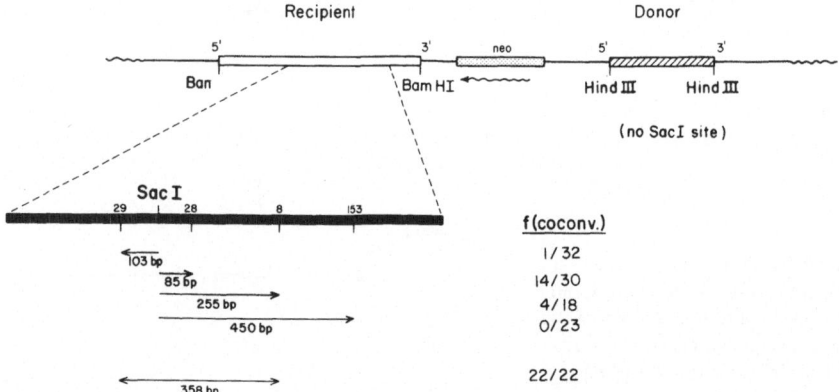

FIGURE 7. Recombination substrates used for coconversion studies. The 1.2-kb donor is derived from *wt TK* sequences of HSV-1 strain 101 and does not encode a restriction site for *Sac*I. The 2.4-kb mutant recipients (*TK29, TK28, TK8, TK153,* or the double mutant *TK29/8*) are all derived from HSV-1 strain F *TK* sequences. HSV-1 *TK* encodes a unique restriction site for *Sac*I by virtue of a single base change difference. *TK* sequences are oriented as direct repeats and flank the *neo* gene. The enlarged recipient sequence is the region of shared homology with the donor. The frequencies of coconversion [f(coconv.)] of the *Sac*I site with the single selected *Xho*I-linker insertion mutations (*29, 28, 8,* and *153*) and the double insertion mutant (T*29/8*) are displayed.

*Sac*I silent site with the *Xho* I insertion mutation *TK29* is significantly less than that observed for mutation *TK28,* even though these two mutations are equidistant from the *Sac*I site. The rates of conversion *per se* at the four different selected (*Xho*I) sites are essentially the same; the unequal distribution in the frequency of coconversion prompts us to suggest that although initiation of conversion events is randomly distributed within the duplication, the propagation of these events may be asymmetric (Liskay and Stachelek, 1986).

9. FIDELITY OF CONVERSION

Identity or nonidentity of converted and parental alleles has been investigated in fungi (for a review, see Orr-Weaver and Szostak, 1985). On the basis of genetic analyses, the converted allele was in all cases identical to the parental allele; i.e., no new mutants were observed. The gene-conversion process therefore exhibits "fidelity." Relevant to the question of fidelity in mammalian cells, we have thus far analyzed a total of 150 independent homologous recombination events leading to the production of a TK$^+$ phenotype. In all but one segregant, the TK$^+$ phenotype correlates with an *H-TK* gene that is now resistant to digestion by the restriction enzyme *Xho*I. These data include the examination of conversion using five different single *Xho*I mutants and one double *Xho*I mutant. We have extended this analysis of conversion fidelity in the examination of TK$^+$ segregants arising from intrachromosomal conversion of a 104-bp deletion. Within this deleted fragment maps a unique *Nru*I restriction site. As determined by molecular-hybridization analyses, conversion to wild-type of five of five *H-TK* mutants carrying the 104-bp deletion was accompanied by recovery of the *Nru*I site.

Although it appears that most conversion events proceed with fidelity, experiments of Folger *et al.* (1984) suggest that some conversion events may be error-prone. In view of these observations, we have undertaken a more rigorous test of fidelity. Determinations of fidelity based on analyses of the TK$^+$ recombinants presented above are biased by our ability to recover only TK$^+$, HAT^r products of recombination. If gene conversion can be an unfaithful process as suggested by Folger and coworkers, we anticipate that an undetermined number of recombination events will be nonproductive. Failure of these TK$^-$ conversion products to survive on cultivation in medium supplemented with HAT precludes their detection. To circumvent bias inherent in a system dependent on positive selection and to more stringently assess the fidelity of recombination, we are at present examining conversion events that yield a change from the TK$^+$ to the TK$^-$ phenotype in cell lines containing a single wild-type *H-TK* gene closely linked to a single defective *H-TK* gene. The molecular lesion responsible for the mutant phenotype of the latter gene copy is an *Xho*I-linker insertion. The strategy of back-selection should permit recovery of both faithful and nonfaithful TK$^-$ recombination products, and molecular analyses should permit distinctions between these two putative classes or recombinants. Faithful events are signaled by susceptibility of both mutant gene copies to digestion by *Xho*I at the same site (i.e., the parent mutant and the convertant mutant are identical). In recombinants arising from nonfaithful interactions, the new mutant gene will not encode the *Xho*I cleavage site diagnostic of fidelity. Eight of ten initial "back-convertants" display a newly mutated gene that contains the same *Xho*I-linker insertion as does its partner sequence.

10. DOUBLE DOMINANT PLASMIDS TO STUDY HOMOLOGOUS RECOMBINATION

"Double dominant plasmids" comprise another class of vector potentially very useful for studying intrachromosomal recombination. Plasmids such as these, employing *gpt* to construct parent lines and duplicated sequences of *neo* for recombination substrates, have been employed by Smith and Berg (1984).

Experiments are currently under way in our laboratory to develop a double dominant plasmid in which *neo* functions for identification of parent lines and duplicated mutant sequences will be derived from the bacterial plasmid-encoded hygromycin-B-phosphotransferase (*hph*) gene (Gritz and Davies, 1983; Rao *et al.*, 1983; Kaster *et al.*, 1984). The *hph* gene confers resistance to the toxic drug hygromycin-B and, like *neo* and *gpt*, is a dominant selectable marker in mammalian cells. Since mutagenesis to construct TK⁻ cell lines is a cumbersome task, difficult and oftentimes impossible, the primary advantage of a system employing double dominant experimental plasmids is obvious. With such plasmids, we can extend our analyses of homologous recombination to many mammalian cell lines or possibly even strains. Double dominant experimental plasmids should allow studies of recombination in a variety of interesting cell lines, e.g., cell lines derived from ultraviolet- and radiation-sensitive human syndromes such as xeroderma pigmentosum, Bloom's syndrome, and Fanconi's anemia.

11. SUMMARY

The system for the analysis of homologous recombination in mammalian cells described herein is notable in design for several reasons. Most important, the substrates for recombination are chromosomal sequences; consequently, the participants in recombination are examined both before and after the recombination event. Our results indicate that duplicated sequences in mammalian cells can undergo intrachromosomal recombination. When these sequences share 1.8 kb of homology, the rate of recombination is 3×10^{-6} events/duplication per cell generation. A mechanism consistent with gene conversion can account for the majority of these events (80%), whereas the remainder (20%) result from a single reciprocal exchange. A second advantage in experimental design is that recombination substrates are easily altered. Direct and inverted orientations of repeats each facilitate analyses of different types of recombination events. Similarly, recombination substrates adapted to be conversion-specific are valuable for examining factors that affect conversion in the absence of a reciprocal exchange. Evidence derived from such substrates suggests that: (1) the efficiency of gene conversion is inversely correlated to the size of molecular lesions (i.e., large mutations convert at a lower rate than small mutations); (2) efficient gene conversion requires a minimum shared homology between 300 and 200 bp long; and (3) conversion tracts appear to be contiguous rather than "patchy" in nature, and these tracts are of variable length or position or both. Finally, we hope to detect mutants that affect recombination by employing additional vectors that will permit analyses of intrachromosomal recombination to be conducted in a variety of interesting cell types.

REFERENCES

Ayares, D., Spencer, R. J., Schwartz, F., Morse, B., and Kucherlapati, R. S., 1985, Homologous recombination between autonomously replicating plasmids in mammalian cells, *Genetics* **111**:375–388.

Brack, C., Hirama, M., Lenhard-Schuller, R., and Tonegawa, S., 1978, A complete immunoglobulin gene is created by somatic recombination, *Cell* **15**:1–14.

Brenner, D. A., Smigocki, A. C., and Camerini-Otero, D., 1985, Effect of insertions, deletions, and double-strand breaks on homologous recombination in mouse L cells, *Mol. Cell. Biol.* **5**:684–691.

Calabretta, B., Robberson, D. L., Barrera-Saldana, H. A., Lambrou, T. P., and Saunders, G. F., 1982, Genome instability in a region of human DNA enriched in Alu repeat sequences, *Nature (London)* **296**:219–225.

Capizzi, R. L., and Jameson, J. W., 1973, A table for the estimation of the spontaneous mutation rate of cells in culture, *Mutat. Res.* **17**:147–148.

Case, M. E., and Giles, N. H., 1958a, Evidence from tetrad analysis for both normal and aberrant recombination between allelic mutants in *Neurospora crassa, Proc. Natl. Acad. Sci. U.S.A* **44**:378–390.

Case, M. E., and Giles, N. H., 1958b, Recombination mechanisms at the pan-2 locus in *Neurospora crassa, Cold Spring Harbor Symp. Quant. Biol.* **23**:119–135.

Case, M. E., and Giles, N. H., 1964, Allelic recombination in *Neurospora:* Tetrad analysis of a three point cross within the pan-2 locus, *Genetics* **49**:529–540.

Cavenee, W. K., Dryja, T. P., Phillips, R. A., Benedict, W. F., Gadbout, R., Gallie, B. L., Murphree, A. L., Strong, L. C., and White, R. L., 1983, Expression of recessive alleles by chromosomal mechanisms in retinoblastoma, *Nature (London)* **305**:779–784.

Chakrabarti, S., Joffe, S., and Seidman, M. M., 1985, Recombination and deletion of sequences in shuttle vector plasmids in mammalian cells, *Mol. Cell. Biol.* **5**:2265–2271.

De Saint Vincent, B. R., and Wahl, G. M., 1983, Homologous recombination in mammalian cells mediates formation of a functional gene from two overlapping gene fragments, *Proc. Natl. Acad. Sci. U.S.A.* **80**:2002–2006.

Fink, G. R., and Petes, T. D., 1984, Gene conversion in the absence of reciprocal recombination, *Nature (London)* **310**:728–729.

Fogel, S., and Hurst, D. D., 1967, Meiotic gene conversion in yeast tetrads and the theory of recombination, *Genetics* **57**:445–481.

Fogel, S., Mortimer, R., Lusnak, K., and Tavares, F., 1978, Meiotic gene conversion: A signal of the basic recombination event in yeast, *Cold Spring Harbor Symp. Quant. Biol.* **43**: 1325–1341.

Folger, K., Thomas, K., and Capecchi, M. R., 1984, Analysis of homologous recombination in cultured mammalian cells, *Cold Spring Harbor Symp. Quant. Biol.* **49**:123–138.

Gonda, D. K., and Radding, C. M., 1983, By searching processively RecA protein pairs DNA molecules that share a limited stretch of homology, *Cell* **34**:647–654.

Gritz, L., and Davies, J., 1983, Plasmid-encoded hygromycin B resistance: The sequence of hygromycin B phosphotransferase gene and its expression in *Escherichia coli* and *Saccharomyces cerevisiae, Gene* **25**:353–358.

Jackson, J. A., and Fink, G. R., 1981, Gene conversion between duplicated genetic elements in yeast, *Nature (London)* **292**:306–311.

Jackson, J. A., and Fink, G. R., 1985, Meiotic recombination between duplicated genetic elements in *Saccharomyces cerevisiae, Genetics* **109**:303–332.

Kaster, K. R., Burgett, S. G., and Ingola, T. D., 1984, Hygromycin B resistance as dominant selectable marker in yeast, *Curr. Genet.* **8**:353–358.

Kit, S., Kit, M., Qavi, H., Trkula, D., and Otsuka, H., 1983, Nucleotide sequence of the herpes simplex virus type 2 (HSV-2) thymidine kinase gene and predicted amino acid sequence of thymidine kinase polypeptide and its comparison with the HSV-1 thymidine kinase gene, *Biochim. Biophys. Acta* **741:**158–170.

Klar, A. J. S., and Strathern, J. N., 1984, Resolution of recombination intermediates generated during yeast mating type switching, *Nature (London)* **310:**744–748.

Klein, H. L., 1984, Lack of association between intrachromosomal gene conversion and reciprocal exchange, *Nature (London)* **310:**748–753.

Klein, H. L., and Petes, T. D., 1981, Intrachromosomal gene conversion in yeast, *Nature (London)* **289:**144–148.

Kucherlapati, R. S., Eves, E. M., Song, K.-Y., Morse, B. S., and Smithies, O., 1984a, Homologous recombination between plasmids in mammalian cells can be enhanced by treatment of input DNA, *Proc. Natl. Acad. Sci. U.S.A.* **81:**3153–3157.

Kucherlapati, R. S., Ayares, D., Hanneken, A., Noonan, K., Rauth, S., Spencer, J. M., Wallace, L., and Moore, P. D., 1984b, Homologous recombination in monkey cells and human cell-free extracts, *Cold Spring Harbor Symp. Quant. Biol.* **49:**191–197.

Kwoh, T. J., and Engler, J. A., 1984, The nucleotide sequence of the chicken thymidine kinase gene and the relationship of its predicted polypeptide to that of the vaccinia virus thymidine kinase, *Nucleic Acids Res.* **12:**3959–3971.

Leder, P., Battey, J., Lenoir, G., Moulding, C., Murphy, W., Potter, H., Stewart, T., and Taub, R., 1983, Translocations among antibody genes in human cancer, *Science* **222:**765–771.

Lin, F.-L., and Sternberg, N., 1984, Homologous recombination between overlapping thymidine kinase gene fragments stably inserted into a mouse cell genome, *Mol. Cell. Biol.* **4:**852–861.

Lin, F.-L., Sperle, K., and Sternberg, N., 1984a, Model for homologous recombination during transfer of DNA into mouse L cells: Role for DNA ends in the recombination process, *Mol. Cell. Biol.* **4:**1020–1034.

Lin, F.-L., Sperle, K., and Sternberg, N., 1984b, Homologous recombination in mouse L cells, *Cold Spring Harbor Symp. Quant. Biol.* **49:**139–149.

Liskay, R. M., and Stachelek, J. L., 1983, Evidence for intrachromosomal gene conversion in cultured mammalian cells, *Cell* **35:**157–165.

Liskay, R. M., and Stachelek, J. L., 1986, Information transfer between duplicated chromosomal sequences in mammalian cells involves contiguous regions of DNA, *Proc. Natl. Acad. Sci. U.S.A.* **83:**1802–1806.

Liskay, R. M., Stachelek, J. L., and Letsou, A., 1984, Homologous recombination between repeated chromosomal sequences in mouse cells, *Cold Spring Harbor Symp. Quant. Biol.* **49:**183–189.

Liskay, R. M., Letsou, A., and Stachelek, J. L., 1986, Homology dependence for gene conversion between duplicated chromosomal sequences in mammalian cells (submitted).

Luria, S. F., and Delbrück, M., 1943, Mutations of bacteria from virus sensitivity to virus resistance, *Genetics* **28:**491–511.

Mulligan, R. C., and Berg., P., 1981, Selection for animal cells that express the *Escherichia coli* gene for xanthine-guanine phosphoribosyl transferase, *Proc. Natl. Acad. Sci. U.S.A.* **78:**2072–2076.

Orr-Weaver, T. L., and Szostak, J. W., 1985, Fungal recombination, *Microbiol. Rev.* **49:**33–58.

Pellicer, A., Robins, D., Wold, B., Sweet, R., Jackson, J., Lowy, I., Roberts, J. M., Sim, G. K., Silverstein, S., and Axel, R., 1980, Altering genotype and phenotype by DNA-mediated gene transfer, *Science,* **209:**1414–1422.

Petes, T., and Klein, H. L., 1986, Genetic analysis of repeated yeast genes, in: *The Recombination of Genetic Material* (K. Brooks Low, ed.), Academic Press, New York.

Rao, R. N., Allen, N. E., Hobbs, J. N., Jr., Alborn, W. E., Kirst, H. A., and Paschal, J. W., 1983, Genetic and enzymatic basis of hygromycin B resistance in *Escherichia coli, Antimicrob. Agents Chemother.* **24:**689–695.

Reis, R. J. S., Lumpkin, C. K., McGill, J. R., Riabowol, K. T., and Goldstein, S., 1983, Extrachromosomal circular copies of an "inter-Alu" unstable sequence in human DNA are amplified during *in vitro* and *in vivo* ageing, *Nature (London)* **301:**394–398.

Roman, H., and Fabre, F., 1983, Gene conversion and associated reciprocal recombination are separable events in vegetative cells of *Saccharomyces cerevisiae, Proc. Natl. Acad. Sci. U.S.A.* **80:**6912–6916.

Rubnitz, J., and Subramani, S., 1984, The minimum amount of homology required for homologous recombination in mammalian cells, *Mol. Cell. Biol.* **4:**2253–2258.

Rubnitz, J., and Subramani, S., 1985, Rapid assay for extrachromosomal homologous recombination in monkey cells, *Mol. Cell. Biol.* **5:**529–537.

Sakano, H., Kurosawa, Y., Weigert, M., and Tonegawa, S., 1981, Identification and nucleotide sequence of a diversity DNA segment (D) of immunoglobulin heavy-chain genes, *Nature (London)* **290:**562–565.

Savage, E. A., and Hastings, P. J., 1981, Marker effects and the nature of the recombination event at the His1 locus of *Saccharomyces cerevisiae, Curr. Genet.* **3:**37–47.

Schimke, R. T., 1982, Summary, in: *Gene Amplification* (R. T. Schimke, ed.), pp. 317–333, Cold Spring Harbor Press, Cold Spring Harbor, New York.

Shapira, G., Stachelek, J. L., Letsou, A., Soodak, L., and Liskay, R. M., 1983, Novel use of synthetic oligonucleotide insertion mutants for the study of homologous recombination in mammalian cells, *Proc. Natl. Acad. Sci. U.S.A.* **80:**4827–4831.

Sharp, J. A., Wagner, M. J., and Summers, W. P., 1983, Transcription of herpes simplex virus genes *in vivo:* Overlap of a later promoter with the 3' end of the early thymidine kinase gene, *J. Virol.* **46:**10–17.

Singer, B. S., Gold, L., Gauss, P., and Doherty, D. H., 1982, Determination of the amount of homology required for recombination in bacteriophage T4, *Cell* **31:**25–33.

Small, J., and Scangos, G., 1983, Recombination during gene transfer into mouse cells can restore the function of deleted genes, *Science* **219:**174–176.

Smith, A. J. H., and Berg, P., 1984, Homologous recombination between defective neo genes in mouse 3T6 cells, *Cold Spring Harbor Symp. Quant. Biol.* **49:**171–181.

Smithies, O., Gregg, R. G., Boggs, S. S., Koralewski, M. A., and Kucherlapati, R. S., 1985, Insertion of DNA sequences into the human β-globin locus by homologous recombination, *Nature (London)* **317:**230–234.

Southern, P., and Berg, P., 1982, Transformation of mammalian cells to antibiotic resistance with a bacterial gene under control of the SV40 early region promoter, *J. Mol. Appl. Genet.* **1:**327–341.

Stadler, D. R., and Towe, A. M., 1963, Recombination of allelic cysteine mutants in *Neurospora, Genetics* **48:**1323–1344.

Subramani, S., and Rubnitz, J., 1985, Recombination events after transient infection and stable integration of DNA into mouse cells, *Mol. Cell. Biol.* **5:**659–666.

Swain, M. A., and Galloway, D. A., 1983, Nucleotide sequence of the herpes simplex virus type 2 thymidine kinase gene, *J. Virol.* **46:**1045–1050.

Volkert, F. C., and Young, C. S. H., 1983, The genetic analysis of recombination, *Virology* **125:**175–193.

Wagner, M. J., Sharp, J. A., and Summers, W. P., 1981, Nucleotide sequences of the thymidine kinase gene of herpes simplex virus type 1, *Proc. Natl. Acad. Sci. U.S.A.* **78:**1441–1445.

Wake, C. T., and Wilson, J. H., 1979, Simian virus 40 recombinants are produced at high frequency during infection with genetically mixed oligomeric DNA, *Proc. Natl. Acad. Sci. U.S.A.* **76:**2876–2880.

Wake, C. T., Vernaleone, F., and Wilson, J. H., 1985, Topological requirements for homologous recombination among DNA molecules transfected into mammalian cells, *Mol. Cell. Biol.* **5:**2080–2089.

15

DEVELOPMENTS LEADING TO HUMAN GENE THERAPY

John W. Belmont and C. Thomas Caskey

1. INTRODUCTION

The development of recombinant-DNA technology has provided an exciting research stimulus to the field of human genetics. The building of the human gene map, the establishment of new linkage methods, the isolation of many disease-related genes, and the application of molecular genetics to the problems of Huntington's chorea and Duchenne muscular dystrophy all indicate the power of this technology. Currently, several groups are exploring the use of these methods for a new therapeutic purpose—development of gene-replacement therapy (Cline, 1982; Anderson, 1981, 1984). The purpose of this review is to examine the elements needed for *in vivo* gene transfer and to assess the potential for human gene therapy.

The ability to deal medically with the more than 3000 known genetic diseases is limited. Prevention of recessive diseases by carrier detection is available for only a few common disorders, e.g., Tay–Sachs and hemoglobin S diseases. Recombinant-DNA prenatal diagnostic techniques are possible for α1-antitrypsin deficiency, hemophilia A and B, Duchenne's muscular dystrophy, Huntington's disease, and a handful of other disorders, but are costly and have restricted availability. A larger number of diseases do not have carrier-detection methods. Thus, unsuspecting couples will continue to discover their genetic risk by the birth of an affected child. Much has been written and discussed regarding prevention of severe genetic disease by prenatal diagnosis and voluntary termination of pregnancy (Milunsky, 1976). However, as many as 50% of at-risk couples elect not to use prenatal diagnosis or not to act on positive results (Adams *et al.*, 1980; Luthy *et al.*, 1980). While dietary therapy has been a success for a restricted group of inborn errors of metabolism, mental retardation or other severe

---

John W. Belmont and C. Thomas Caskey • Institute for Molecular Genetics and Howard Hughes Medical Institute, Baylor College of Medicine, Houston, Texas 77030.

complications or both are often not avoided even with the best clinical care (Matsui *et al.*, 1983). The high frequency of spontaneous mutations in X-linked recessive lethal diseases places an additional limitation on medical management via carrier detection and prenatal diagnosis (Haldane, 1947). In summary, there are a wide variety of social and scientific reasons that prevention and treatment of severe inborn errors of metabolism are insufficient. We will continue to have a medical obligation to the affected child and at present have limited treatment options. Gene-transfer technologies offer some hope for the development of gene therapy for human heritable diseases.

2. GENETIC DISEASES AS MODELS FOR GENE THERAPY

2.1. Constraints on Disease Selection

Development of human gene therapy requires the selection of suitable human disease model systems. The choices of diseases for early investigation reflect our sophistication in understanding the pathophysiology of human metabolic diseases and the inherent technical limitations imposed by available gene transfer methods. We believe that some of the relevant limitations include: (1) the necessity to transfect a readily available tissue; (2) the necessity to transfect that tissue *ex vivo* and place it back into the recipient; (3) the necessity to avoid live virus vector infection; (4) the inability, so far, to closely regulate expression of the transferred gene and the poor expression in the target cells. The diseases chosen for the pathfinding experiments require: (1) knowledge of the exact causes of the disease at the DNA level; (2) normal gene expression in a tissue accessible to *in vitro* manipulation or at least preliminary evidence that expression of the gene in an accessible tissue will improve the disease; (3) the possibility of ameliorating the disease process by relatively low-level expression or unregulated expression.

These limitations are stringent, and even assuming that in the near future functionally specific regulatory elements could be introduced with the gene of interest, the number of diseases that might be treated is small. Diseases that affect structural proteins in the connective tissue and skeleton, diseases that require gene expression in the central nervous system (CNS), diseases that require gene expression at critical early developmental stages, diseases in which the gene product interacts with protein subunits, cofactors, or substrates specific to an inaccessible tissue must all be eliminated from immediate consideration. It is our opinion that only bone marrow and liver offer a realistic opportunity for gene transfer with present vector systems. As we will describe in Section 3.1, preliminary experiments with gene transfer into bone-marrow cells have demonstrated the general feasibility of this approach. To date, no comparable experiments have been performed on normal liver cells, but the easy access, the known regenerative capacity of hepatocytes, the growing experience with liver allografts, and the number of inborn errors of metabolism that affect liver-specific proteins all draw attention and effort to the development of gene-transfer methods for the liver.

Table I is a compilation of metabolic disease genes that have been cloned (Beaudet, 1985). The genes that encode several of the lysosomal enzymes have been included, since limited data on bone-marrow transplantation suggest that gene transfer into bone marrow may be helpful in some of these disorders. Table II is a noncomprehensive listing of other genetic diseases in which molecular cloning of the disease gene has not been reported or in which the metabolic defect is unknown. The diseases listed in Tables I and II are imaginable targets of gene therapy either because the disease is expressed in hematopoietic or hepatic cells or because the pathophysiology is sufficiently well understood to make gene transfer attractive. There might be included in this list an additional group of diseases in which expression of the disease process occurs in many tissues,

TABLE I. Targets for Gene Therapy: Cloned Genes

Hematopoietic
  Adenosine deaminase (immune deficiency)
    α1-Antitrypsin[a,b]
    Carbonic anhydrase II (osteopetrosis)
    Complement—C2, C3, C4, Factor B, and C9[a]
    Fucosidase (fucosidosis)
    α-Galactosidase (Fabry)
    Globin α, β (thalassemia, hemoglobinopathies)
    Glucocerebrosidase (Gaucher)
    β-Glucuronidase (Sly)
    Purine nucleoside phosphorylase (immune deficiency)

Liver
  Arginase (argininemia)
    Argininosuccinic acid lyase (hyperammmonemia)
    Argininosuccinic acid synthetase (citrullinemia)
    Carbamoyl phosphate synthetase (hyperammonemia)
    Dihydrobiopterin reductase[c] (hyperphenylalaninemia)
    Factor VIII, IX (hemophilia)
    LDL receptor (familial hypercholesterolemia)
    Ornithine transcarbamylase (hyperammonemia)
    Phenylalanine hydroxylase (PKU)
    Propionyl CoA carboxylase, α and β subunits (propionic acidemia)
    Pyruvate carboxylase (congenital lactic acidosis)

[a] Circulating blood proteins; site of synthesis is both liver and macrophages.
[b] Expression in hematopoietic cells is not expected to correct hepatic disease.
[c] Significant CNS disease in which gene function in situ may be necessary.

TABLE II. Targets for Gene Therapy: Uncloned Genes or Defect Unknown

Hematopoietic
  Defect known
    Arylsulfatase B (Maroteaux–Lamy)
    Aspartylglycosaminadase
    C3bi receptor
    β-Galactosidase (Morquio B)
    Galactosamine 6-sulfate sulfatase (Morquio A)
    α-L-Iduronidase (Hurler, Scheie)[a]
    α-D-Mannosidase (mannosidosis)[a]
    α-Neuraminadase (sialidosis)[a]
    Iduronate sulfatase (Hunter)[a]

  Defect unknown
    Ataxia–telangiectasia[a]
    Bare lymphocyte syndrome
    Chediak–Higashi syndrome
    Chronic granulomatous disease
    Glanzman's thrombasthenia
    Hermansky–Pudlak syndrome
    Osteopetroses
    Severe combined immune deficiencies
      (autosomal recessive and X-linked)
    Wiskott–Aldrich syndrome
    X-linked agammaglobulinemia

Liver
  Defect known
    Acetoacetyl 3-ketothiolase
    Acyl-CoA dehydrogenases
    Amylo-1,6 glucosidase (GSD Type III)
    Amylo-1,4:1,6 transglucosidase (GSD Type IV)
    ATP:cobalamin adenosyltransferase (methylmalonic acidemia cb1B)
    Branched-chain ketoacid dehydrogenase (maple syrup urine disease)
    Cl inhibitor (hereditary angioedema)
    Cystathionine β-synthetase (homocystinuria)
    Dihydrobiopterin synthetase[a] (hyperphenylalaninemia)
    Electron-transfer flavoprotein (glutaric aciduria Type II)
    Fructose-1 phosphate aldolase (hereditary fructose intolerance)
    Fructose 1,6 diphosphatase
    Fumarylacetoacetase (tyrosinemia Type I)
    Galactokinase
    Galactose 1-phosphate uridyl transferase (galactosemia)
    Glucose 6-phosphatase (GSD Type I)
    3-Hydroxy-3-methylglutaryl-CoA lyase
    Isovaleryl-CoA dehydrogenase (isovaleric acidemia)
    Lipoprotein lipase
    Pyruvate dehydrogenase complex[a]
    Phosphorylase, hepatic (GSD Type VI)
    Phosphorylase B kinase (GSD Type VIII)

TABLE II. (Continued)

Defect unknown
Hyperlysinemia
Hypervalinemia
Hyperleucinemia–isoleucinemia
Methylmalonic acidemia cbl A, C, D
Nonketotic hyperglycinemia[a]
Wilson's disease
Byler's disease
Congenital hepatic fibrosis

[a] Significant CNS disease in which gene functions *in situ* may be necessary.

but in which bone-marrow transplant or liver transplant is known to have therapeutic usefulness.

2.2. Exemplary Model Systems

It is clear that each metabolic disease must be evaluated individually as a candidate for gene transfer, since each deficiency causes injury in a distinctive way. Often, the pathophysiological mechanism may involve tissues and organs far removed from the site of gene expression, but leave the organ in which the gene is specifically expressed largely unaffected. Two general situations must be addressed: (1) diseases in which gene replacement must be directed to an organ that is itself pathologically affected by the disease; and (2) diseases in which gene transfer is targeted to a tissue that is not directly affected by the disease, but in which gene expression favorably changes the course of the disease.

Consideration of several examples clarifies some of the problems faced. Deficiency of arylsulfatase B causes the mucopolysaccharidosis known as Maroteaux–Lamy syndrome (DiFerrante *et al.*, 1974). This disease is characterized by accumulation of dermatan sulfate in various organs, with the most severe problems arising in skeletal and connective tissue. Bone-marrow transplant has been reported to decrease urinary excretion of acid mucopolysaccharide and to significantly improve the skeletal and cardiovascular disease (Krivit *et al.*, 1984; Gasper *et al.*, 1984). Transfer of the arylsulfatase B gene into autologous marrow may be expected to improve the disease if expression is sufficient. A similar example of this type of "long-range" pathogenic effect is provided by familial hypercholesterolemia. Deficiency of low-density lipoprotein (LDL) receptor in the liver leads to massive atherosclerosis and life-threatening coronary-artery disease. Liver transplant is strikingly effective in reducing serum lipids (Bilheimer *et al.*, 1984). Thus, if an adequate method for gene delivery to the liver were possible, transmission of the LDL-receptor gene to liver would be expected to ameliorate the disease.

Several metabolic diseases are caused by deficiencies of enzymes expressed primarily in the liver, but the main pathological feature is the accumulation of circulating metabolites. Several of the clinically most important and common diseases are characterized by neonatal hyperammonemia. In addition to the hyperammonemia, other abnormal metabolites may be found that are themselves toxic, e.g., certain organic acids, or that may be relatively nontoxic, e.g., citrulline. Examples include argininosuccinate synthetase (AS) deficiency (Walser *et al.*, 1977), the organic acidemias propionyl-CoA carboxylase deficiency and methylmalonyl-CoA mutase deficiency (Tanaka and Rosenberg, 1983), and ornithine transcarbamylase (OTC) deficiency (Walser, 1983). AS deficiency is characterized by protein intolerance leading to episodic severe hyperammonemic coma. Death can result from even mild intercurrent infections because of the increased catabolic nitrogen load. Dietary management is partially effective and may not prevent severe episodic illness. Administration of sodium benzoate, arginine, and phenylacetate can provide alternative pathways for nitrogen excretion and lead to stabilization of metabolic status. Thus, provision of a mechanism for clearance of the excess nitrogen load may stabilize the patient. Introduction of the AS gene into bone marrow may be therapeutic, since the substrate, citrulline, is a circulating, freely permeable metabolite. Bone-marrow transplantation might be used to test this hypothesis, but there is little AS activity in blood cells. Alternatively, delivery of the AS gene to the liver would be expected to correct the defect. In contrast, both propionic acidemia and methylmalonic acidemia are characterized by acidosis, hypoglycemia, hyperglycinemia, and hyperammonemia, which are not directly attributable to the metabolic block. The most likely mechanisms for these effects involve secondary inhibition of several mitochondrial systems by the organic acids building up behind the block. It is unclear whether removal of these toxic organic acids from the plasma would completely relieve the secondary mitochondrial effects.

OTC deficiency, on the other hand, may require delivery of the gene to the liver because of the necessity for active carbamoyl phosphate synthetase I activity in the mitochondria to provide carbamoyl phosphate for the synthesis of citrulline (Walser, 1983). Carbamoyl phosphate is not freely diffusible, and thus expression of OTC in a tissue that lacks the ability to adequately synthesize carbamoyl phosphate substrate would likely be unsuccessful in clearing the hyperammonemia. Whether delivery of the OTC gene to bone marrow would allow sufficient ammonia metabolism is an open question. Delivery of the gene to hepatocytes seems preferable.

Finally, it should be noted that with any metabolic disease, the successful treatment of one biochemical parameter may not influence other long-term complications that result directly from the enzyme deficiency. Examples include the successful dietary treatment of the hypoglycemia of Type I glycogen storage disease, only to have late onset of primary hepatocellular carcinoma (Howett *et al.*, 1976), and the successful treatment of the peripheral hyperuricemia of Lesch–Nyhan syndrome by allopurinol with no effect on the CNS disease (Rosenbloom *et al.*, 1967).

Medical genetics is armed with a powerful set of techniques for obtaining cloned human genes. The construction of both genomic and complementary DNA (cDNA) libraries has been greatly improved. The creation of subtraction cDNA (Chien *et al.*, 1984) and genomic libraries (Kunkel *et al.*, 1985) is noteworthy as an example of the rapid technical advances in this area. Screening of recombinant libraries has been facilitated by the availability of rapidly produced synthetic nucleic-acid probes (Suggs *et al.*, 1981; Ullrich *et al.*, 1984) and by the development of expression libraries for immunological (Broome and Gilbert, 1978) and bioassay screening (Wong *et al.*, 1985) protocols. A number of human genes have been cloned, but the task of developing cloned probes for all human genes of medical interest is a large one. For investigators interested in gene-transfer therapy of metabolic diseases, the list of target genes can be narrowed substantially because of the technical limitations inherent in current and envisioned methods of gene transfer.

3. TRANSPLANTATION IN THE TREATMENT OF GENETIC DISEASES

Bone-marrow transplantation provides an important and definitive therapy for several inherited diseases that affect the immune system. In addition, it is seeing wider application in the treatment of other hematological disorders and of some inborn errors of metabolism, particularly the lysosomal storage diseases. Liver transplantation has been used with increasing safety for the treatment of end-stage liver disease secondary to a variety of inherited disorders. Both bone-marrow and liver transplantation provide the opportunity to assess the effectiveness of single-organ correction of a specific enzyme deficiency. In this sense, experience with bone-marrow and liver transplantation provides the necessary test of feasibility of gene-transfer therapy for a number of genetic diseases.

3.1. Bone-Marrow Transplantation

The technique of bone-marrow transplantation is well defined and based on extensive clinical experience in the treatment of neoplastic disease as well as the treatment of severe combined immunodeficiency (SCID) disorders (O'Reilly *et al.*, 1984; Witherspoon 1984). As it became clear that organ transplantation was limited by the major histocompatibility complex, human leukocyte antigen

(*HLA*-matched sib donors were used exclusively. In the last several years, however, increasing experience has been gained with allogeneic bone-marrow transplantation, in which mature T cells are depleted from the donor bone marrow (Prentice *et al.*, 1982; O'Reilly *et al.*, 1983a). Under ideal conditions, this can substantially reduce the risk of graft-vs.-host (GVH) disease, which had previously limited haplo-identical bone-marrow transplantation. However, with both techniques—*HLA*-matched sib and haplo-identical allogeneic marrow—problems with GVH disease continue to play a major role in the long-term morbidity (Sullivan *et al.*, 1984).

Bone-marrow transplantation has been applied to several different classes of inherited disorders. These include immune deficiency diseases, hematological disorders, osteopetrosis, and lysosomal storage diseases. Bone-marrow transplantation is the treatment of choice for SCID and is curative when the reconstitution is complete (Bortin and Rimm, 1977; O'Reilly *et al.*, 1983b). In addition to the SCID, bone-marrow transplantation has been applied extensively in the treatment of the Wiscott–Aldrich syndrome (Parkman *et al.*, 1978; Ochs *et al.*, 1982). In this disorder, both T cells and platelets are affected. Bone-marrow reconstitution fully corrects the deficit in both lineages of cells. Much more limited experience with the treatment of other hematological disorders has been accumulating. This includes selective treatment of patients with β-thalassemia (Thomas *et al.*, 1982; Lucarelli, *et al.*, 1983), sickle-cell anemia (Johnson *et al.*, 1984), Fanconi anemia (Gluckman *et al.*, 1984) and various neutrophil disorders including Kostman's agranulocytosis (Rappeport *et al.*, 1980) and chronic granulomatous disease (Foroonzofar *et al.*, 1977; Rappeport *et al.*, 1982). Because of the availability of other supportive therapeutic modalities for these disorders, bone-marrow transplantation has been applied less aggressively. The associated risks of high morbidity and mortality have inhibited its use except in desperate situations. Morbidity and mortality following bone-marrow transplantation in patients with these hematological disorders is strongly influenced by the selection of patients who have failed other treatments. However, it is known that following successful reconstitution, there is a complete correction of these diseases.

Another example of the application of bone-marrow transplantation to inherited disease is in the treatment of osteopetrosis (Coccia *et al.*, 1980; Sorell *et al.*, 1981). The osteopetroses are a heterogeneous group of disorders that affect osteoclast function and lead to abnormal bone remodeling. Eventually, bone-marrow failure occurs because of encroachment on the bone-marrow space. Recently, one form of osteopetrosis has been attributed to carbonic anhydrase II deficiency (Sly *et al.*, 1985). Osteoclasts are known to be bone-marrow-derived cells. Following bone-marrow transplantation, normal osteoclast function reverses the bone-marrow-cavity encroachment. Gradual improvement in the hematological symptoms and abnormalities of bone formation result.

Finally, limited experience has been obtained in the use of bone-marrow transplantation for a variety of lysosomal storage diseases (O'Reilly *et al.*, 1984; Hobbs *et al.*, 1981; Hugh-Jones *et al.*, 1982). Both clinical and pathological experience suggest a prominent role for the bone-marrow-derived reticuloendothelial cells in the normal metabolism of complex carbohydrates and lipids. Thus, these diseases are often characterized by macrophage–monocyte accumulation of abnormal metabolites. Experience with bone-marrow transplantation in patients suffering from these disorders has been mixed, but preliminary evidence suggests significant clinical improvement in the Maroteaux–Lamy syndrome (Krivit *et al.*, 1984) and in Gaucher's disease (Ginns *et al.*, 1982). However, application of bone-marrow transplantation to storage diseases that affect the CNS has been disappointing, and there is no evidence for either stabilization or reversal of CNS disease (Tutschka *et al.*, 1983; Joss *et al.*, 1982). Clinical experience with bone-marrow transplantation, in summary, indicates that the procedure can be fully curative in diseases that affect cells of the hematopietic lineages. The limited experience with the lysosomal storage diseases demonstrates that diseases that are characterized by the accumulation of abnormal metabolites might also be ameliorated by the transplantation of bone-marrow cells.

Despite the extensive experience in bone-marrow transplantation, the procedure is both costly and dangerous. A limited number of patients have *HLA*-compatible sibling donors available. It requires the close attendance of a highly skilled and experienced transplant team. Even in the most experienced centers, many short-term and long-term complications of the procedure arise. The short-term complications include failure of engraftment, acute infection, and toxicity associated with the use of preparative regimens (Champlin and Gale, 1984; van Bekkum, 1984). Long-term complications include GVH disease, incomplete engraftment leading to incomplete reconstitution of all hematological functions, and the long-term toxicity associated with chronic immunosuppression (Sullivan *et al.*, 1984). The most important of these long-term complications is GVH disease. This is a significant problem both in *HLA*-matched sibling transplantation and in the newer allogeneic bone-marrow transplantation procedures. Immunosuppressive agents, including cyclosporin A, may alter the course of GVH disease, but also have their own toxicities. An increased incidence of lymphoma may be related to use of immunosuppressive agents before and after transplant or may possibly be influenced by the technique of depletion of mature T cells from the bone-marrow allograft (Shearer *et al.*, 1985). It is difficult to estimate the associated risk from bone-marrow transplantation in immunologically normal hosts. Short-term mortality may be limited to 0–10%. However, overall mortality may be as high as 30–50% (O'Reilly *et al.*, 1984). Therefore, application of bone-marrow transplantation has been limited to patients who are suffering end-stage disease. It is intriguing to speculate whether or not earlier application of bone-marrow transplantation might significantly alter the course of a number of the lysosomal storage diseases or other diseases in which there is the necessity to process an external abnormal metabolite. Up to now, physicians have been unwilling to expose patients to the attendant risks of bone-marrow transplantation, but improvement in the technique with lowering complication rates may alter this attitude.

3.2. Hepatic Transplantation

Hepatic transplantation has become feasible in the last several years with the improvement in surgical technique and postoperative management (Starzl *et al.*, 1985; Iwatsuki *et al.*, 1985). Liver transplantation has been applied to the treatment of several genetic diseases. These include α1-antitrypsin deficiency, Wilson's disease, tyrosinemia Type I, galactosemia, glycogen storage disease Types I and IV, LDL–receptor deficiency, hemochromatosis, protoporphyria, Crigler–Najar syndrome Type I, Neiman–Pick syndrome, Byler's disease, sea blue histiocyte syndrome, and congenital hepatic fibrosis (Matas *et al.*, 1978; Scharschmidt, 1984; Vierling, 1984; Starzl *et al.*, 1984; Rolles *et al.*, 1984). In several of these diseases, the liver transplantation has improved the metabolic defect and has been lifesaving. A radical improvement in mortality following hepatic transplantation has occurred since 1980 (Iwatsuki *et al.*, 1985). The single greatest factor in this improvement has been the introduction of cyclosporin A plus prednisone immunosuppression. Improved surgical technique, better tissue typing, *HLA*-matching, more effective donor-organ procurement and preservation, and better patient selection have also contributed. It is worth noting that patients being treated for inherited metabolic diseases have consistently had lower mortality and morbidity compared to the whole group of transplant recipients (Scharschmidt, 1984). Since 1980, the 5-year survival in this restricted subset of patients has been greater than 50% at one major center even though end-stage liver failure was the major indication for transplant. Complications of liver transplantation are numerous and severe, relating mainly to surgical complications, graft rejection, and the generally debilitated state of the recipients. Short-term complications are intraoperative death, infection, hemorrhage, and complications related to immunosuppressive preparative therapy. Long-term complications include chronic graft rejection with concomitant liver failure and the long-term risks associated with immunosuppression, including renal failure and neoplasia.

In summary, organ transplantation, especially bone-marrow and liver transplantation, provides the theoretical and practical test of the idea that introduction of normal genes into selected organs can fully or partially correct inborn errors of metabolism. It can be concluded that diseases that are limited to the bone-marrow-derived lineages can be effectively treated with bone-marrow transplantation. It remains to be seen whether either bone-marrow transplantation or hepatic transplantation will be successful in the management of diseases in which a circulating abnormal metabolite plays a significant role in the pathogenesis of the disease. Experimental use of these treatment modalities provides the important precedent for application of gene-transfer techniques in bone marrow and liver. For any particular genetic disease, it is unlikely that gene transfer would be an effective treatment if preliminary experience with organ transplantation failed to show therapeutic utility.

4. GENE-TRANSFER METHODS

4.1. Minimum Requirements

An adequate delivery system is a key requirement for human gene therapy. Both physical methods and virus-based vector systems are available for *in vitro* eukaryotic gene-transfer applications. Intense efforts aimed at the development of retrovirus vectors are under way in several laboratories. Several requirements for any gene-transfer method for use in gene therapy must be met. These include: (1) high efficiency of transmission, (2) stable replication as either an integrated transgene or an extrachromosomal element, (3) effective expression in the target tissue, and (4) adequate safety over both the acute period of transfer and the life of the treated patient.

Transmission efficiency varies greatly in gene-transfer techniques. Physical techniques such as $CaPO_4$ precipitation have transformation efficiencies in the range of 0.01–0.1% transformants (Corsaro and Pearson, 1981a). This may vary with the cell type (Corsaro and Pearson, 1981b). A new technique, termed "electroporation," that utilizes a high-voltage pulse to allow DNA to enter cells may have a significantly higher efficiency, in the range of 1–10% transformants, but will likely fall short of the efficiency necessary for *in vivo* gene transfer (Potter *et al.*, 1984). In contrast, viral vector systems that utilize either DNA or RNA viruses allow infectious transfer efficiencies ranging from 10 to 100% (Weiss, 1984). In addition to the superior efficiencies of transfer achieved with viral vectors, conditions can be optimized so that on average 1–5 copies of foreign DNA enter individual target cells. Physical methods tend to cause the integration of multiple copies—up to hundreds—as large tandem arrays (Perucho *et al.*, 1980). Neither physical DNA-transfer methods nor any viral vector systems allow targeting of chromosomal position of the inserted DNA. However, with DNA-virus vectors, stable episomal replication is utilized, obviating concerns about insertional mutagenesis or chromosome-position effects.

Following insertion of foreign DNA into target cells, the transgene must be stable to ensure continuation of function. Generally, instability results from recombination–excision events that remove functional components from the transgene. Extrachromosomal recombination mechanisms have been shown to be very active. Some of the vector rearrangement observed in retrovirus systems, and rearrangements following $CaPO_4$ transfer, occur extrachromosomally before the integration of the transgene (Bandyopadhyay et al., 1984; de Saint Vincent and Wahl, 1983; Rubnitz and Subramani, 1985). However, some retroviral constructions are prone to recombinational excision of functional minigenes, demonstrating selective pressure against certain gene configurations (Joyner and Bernstein, 1983a,b; Emerman and Temin, 1984). In addition, following apparently stable insertion, a variable rate of loss or rearrangement of inserted sequences is observed, possibly implying selective pressures resulting from variations at the site of insertion (Jolly et al., 1984; Willis et al., 1984). With all modes of gene delivery, therefore, assessment must be made of the stability of individual constructions, since the parameters of instability have not yet been examined systematically. Instability arising from the insertional site is an uncontrollable factor, so that transmission efficiency must be great enough to numerically avoid this obstacle.

Effective expression of the inserted DNA will be controlled largely by the chromosomal context of insertion and by the particulars of the minigene construction. Effects of chromosomal-insertion site have been well documented in the Drosophilia P element model (O'Hare and Rubin, 1983; Spradling and Rubin, 1983; Hazelrigg et al., 1984). It is clear from less comprehensive studies in mammalian cells that the activity of the surrounding DNA influences expression of the transgene. An additional factor to be considered is the developmental timing of the insertion, since experience with retroviral vector transfer into mouse embryos indicates a high rate of nonexpression of the transferred DNA (Jaenisch et al., 1975, 1981; Stuhlman et al., 1984; Van Der Putten et al., 1985). The extent to which this might affect expression in differentiating cells such as bone-marrow-derived cells is unknown.

Finally, any method of gene transfer must be safe. This includes: (1) no direct cytotoxicity of the transfer and insertion; (2) no viral vector propagation within either the cell or the recipient organism or between the recipient and other organisms; and (c) no transfer to the germinal tissues. The requirement for lack of direct cytotoxicity is a relative concern depending on the target organ. Both bone marrow and liver have substantial regenerative capacity, and individual cell loss could be quite large and still be acceptable if the parameters of transmission efficiency, stability, and expression were optimized. However, gene delivery to nonregenerating cells such as muscle and CNS would require minimization of any direct toxic effects or mutational event in the recipient cells. The propagation of infectious particles carrying the transgene would be considered an unacceptable risk because of the increased chance of infection of germinal tissues, insertional activation of protooncogenes, and the unknown public-health risk of uncontrolled spread to other individuals or animals. Finally, there is the

ethical concern about insertion of foreign DNA into germinal tissues which will
be discussed in Section 6.

4.2. Retrovirus Vectors

The considerations discussed above have strongly pulled investigators into
the design and development of retrovirus vectors. Several recent reviews (Gilboa,
1983; Mulligan, 1983; Bernstein *et al.*, 1985) and Chapter 6 excellently cover
this field. However, several points should be emphasized. The natural retorvirus
life cycle lends itself to high-efficiency single-copy insertion of foreign DNA.
Vectors have been constructed in which the normal *gag, pol,* and *env* genes are
removed, leaving the important *cis*-acting functions of the viral long terminal
repeats (LTRs) and the Ψ packaging sequence. The nucleocapsid proteins, reverse transcriptase, insertional enzymatic machinery, and envelope proteins are
all provided in *trans* either by helper-virus coinfection or by specifically designed
helper-cell lines.

We consider that use of retrovirus vectors for gene transfer entails three
major concerns: (1) virus-particle production, (2) target-cell expression, and (3)
in vivo expression. Vector production requires adequate transcription of the
provirus and efficient packaging of the genomic provirus transcript into infectious
particles. In the actual construction of virus-producing cell lines, we use *in vitro*
selection to identify lines that produce the highest-titer virus. Vector expression
in the target cell, on the other hand, is theoretically separable from that required
for virus production. The differentiated state of the intended target cell is quite
different from the cell lines used in virus production. Assessment of virus titer
is therefore complicated by the fact that conditions for virus-particle enumeration
are set to enhance the sensitivity of detection, but the relative titer on the eventual
target cell may be substantially less. We know very little about the regulation
of transgene expression at the cellular and organismal level. But we can expect
that cell interactions, the use of *in vivo* selection protocols, and possibly autoimmune phenomena will affect the final result.

Expression of the retrovirus vector is a complicated phenomenon with several overlapping facets. Expression of the defective virus in the producer cell
lines may be assessed by a direct measure of gene function, such as ability to
grow under selective conditions, or by product assay. An alternative method of
assessment is measurement of viral-particle production by titration. Viral titers
and expression of the inserted minigene may be different in the spliced vectors,
since virus production is dependent on the production of unspliced genomic
transcripts, while minigene expression may be dependent on the spliced messenger RNA (mRNA) product. Likewise, virus production may be high in a
given producer cell line, indicating good expression in that highly selected clonal

population, but target cells may have poor expression of the minigene. Thus, viral titer is dependent on: (1) the genomic transcription rate, which is affected by copy number, insertional context or chromosomal position, promoter strength, presence of enhancer elements, and splicing regulation; and (2) adequacy of the *trans*-acting components provided by the helper cell or by the coinfecting helper virus.

Initial vector constructions included the 5' splice signal and 3' splice acceptor sites to allow production of both genomic transcripts and a spliced transcript similar to the natural *env* mRNA species (Cepko *et al.*, 1984). Because of the variable transcription of either the genomic or spliced transcript depending on the particulars of the constructions, it was observed that viral titers or expression or both were poor. A general consensus has arisen that unspliced vectors give superior overall results by avoiding the possible competition of genomic transcript vs. spliced transcript production.

Vector contructions that contain the necessary *cis*-acting sequences can be treated as defective retroviruses. Thus, wild-type Moloney murine leukemia virus (M-MuLV) can be coinfected to allow production of both defective and wild-type virus. Using this general approach, very high-titer defective virus production has been achieved (Miller *et al.*, 1983). The obvious disadvantage is the production of wild-type virus and continued propagation of both wild-type and defectives. The same technique may be used to "pseudotype" viruses, changing the displayed envelope protein from ecotropic to amphotropic (Rubin, 1965). Amphotropic viruses have been shown to infect human fibroblasts and lymphoblasts as well as cells from several rodent species (Weiss, 1984). Another approach to production of defective viruses is the construction of helper-cell lines. Mulligan's group has made two packaging cell lines, Ψ_2 and Ψ_{am}, that allow high-titer production of ecotropic and amphotropic viruses, respectively (Mann *et al.*, 1983; Cone and Mulligan, 1984). These cell lines are derived from the transformation of wild-type NIH 3T3 mouse fibroblasts by a Ψ^- M-MuLV-derived plasmid. The plasmid contains intact *gag, pol,* and *env* genes that provide all functions necessary in *trans* for reverse transcription, integration, and packaging, but because they lack the Ψ sequence necessary for the packaging, they are themselves not produced as infectious particles. Recently, Miller *et al.* (1985) have created another amphotropic packaging line called PA-12 that yields slightly higher titers of defective amphotropic virus than Ψ_{am}.

Expression in the target cell is affected primarily by the particulars of the vector construction, including promoter strength, use of enhancer elements, and requirement for splicing. In some constructions, independent promoters are used for a dominant selectable marker (e.g., *neo* or *dhfr*) and the cDNA minigene (Miller *et al.*, 1985). In these constructions, promoter orientation may be a critical variable. This problem is illustrated by the studies of spleen necrosis virus in which the chicken thymidine kinase (*tk*) gene was placed in either the same or opposite transcriptional orientation (Bandyopadhyay and Temin, 1984).

In the same orientation, viral titer was good but *tk* expression was poor. When cloned in the opposite orientation, the *tk* activity increased, but the viral production was inhibited. This "transcriptional interference" is an inconsistent finding, and no general rule of vector construction can be formulated from published results. Relative promoter strengths may play a role, as may constraints on expression imposed by posttranscriptional steps. For example, no examples of expression of polycistronic messages have been described in eukaryotic systems.

The use of natural regulatory elements has been rather limited so far. There have been difficulties with constructions utilizing natural promoter elements for α-globin, β-globin, hypoxanthine phosphoriposyltransferase (HPRT), and immunoglobulin (Emerman and Temin, 1984 and unpublished data). Experience with enhancer elements for the μ H-chain gene, however, indicates that natural *cis*-acting regulatory elements may significantly enhance tissue-specific expression (Dick *et al.*, 1985).

An additional possibility for the enhancement of expression and the increase in viral titers is alteration of copy number in viral-producing cells by amplification. This is possible only for a few genes with available selection protocols. Two promising systems are the methotrexate-resistant dihydrofolate reductase (*dhfr Mtx^r*) gene in which sequential addition of increasing amounts of methotrexate leads to amplification of transfected *dhfr* (Wigler *et al.*, 1980; Kaufman *et al.*, 1985) and adenosine deaminase (*ADA*) markers in which addition of increasing amounts of the suicide inhibitor deoxycoformycin in 11AAU medium can lead to a several-thousand-fold increase in gene copy number (Yeung *et al.*, 1983a,b). There is only limited unpublished experience with these so far in retroviral constructs, but inclusion of either *dhfr* or *ADA* as the dominant selectable marker may allow not only easy selection but also subsequent amplification leading to markedly increased virus production.

Despite the advantages of the current retrovirus vectors, there are significant limitations. Active target-cell DNA synthesis and therefore replication is rquired for efficient virus infection and insertion (Weiss, 1984). This well-documented observation is the probable basis for the finding that 5-fluorouracil (5-FU) enhances the efficiency of hematopoietic stem-cell infection, since stem cells are in general quiescent, noncycling bone-marrow elements (Dick *et al.*, 1985). Any slowly dividing cell or quiescent cell makes a poor target for retroviruses. 5-FU, by killing the rapidly cycling progenitor cells, causes an increased number of stem cells to enter replication and increases the chances for a virus infection event. Wolf and Ruschetti (1985) have shown that pretreatment of animals with phenylhydrazine (induction of hemolytic anemia) allows mice to be infected *in vivo* with replication-defective retroviruses. Presumably, the increased numbers of dividing erythroid progenitors permit this relatively inefficient infection to be observed.

An additional concern is that retroviral insertion into essentially random target genomic DNA sequences (Shimotohno and Temin, 1980) will cause either insertional mutagenesis and disruption of a critical gene or *cis*-activation of a cellular protooncogene (Varmus *et al.*, 1981; King *et al.*, 1985). It is difficult to assess the frequency with which these events might take place, but it is generally felt to be low-risk on the basis of the prolonged active infection necessary for oncogenic transformation following infection by non-*v-onc*-containing retroviruses.

Finally, for a variety of reasons, retroviruses show tissue perferences. All present constructions utilize LTRs from M-MuLV or its relatives that show preference for hematopoietic cells (Lenz *et al.*, 1984). Different viruses may have to be exploited for efficient infection of other tissues, especially liver. Table III summarizes some important elements in the various phases of retrovirus vector expression.

4.3. DNA-Virus Vectors

DNA-virus vector systems have the same advantages of high-efficiency gene transfer and controlled, low cell copy number as do retrovirus vectors when compared to physical methods for gene transfer. Their utility has been hampered by the lack of availability of helper-cell lines that can provide necessary *trans*-acting functions comparable to the Ψ^- packaging lines for retroviruses. An additional problem has been high recombination of defective simian virus 40 (SV40) vectors to make wild-type virus (Karlson *et al.*, 1985).

DNA-virus vector systems are described in detail in Chapter 5. Several systems with particularly useful applications have been developed. These include bovine papilloma virus (BPV) (DiMaio *et al.*, 1982), Epstein–Barr virus (EBV) (Yates *et al.*, 1985), SV40 (Elder *et al.*, 1981), adenovirus virus (Van Doren and Gluzman, 1984), and vaccinia virus vectors (Mackett *et al.*, 1984). The latter is transcribed and replicated in the cytoplasm and has already proven useful as vector for live vaccine applications (Mackett *et al.*, 1982). BPV and EBV vectors are maintained as extrachromosomal elements. Both appear useful for *in vitro* gene-expression applications in cultured cell lines, but their usefulness for *in vivo* gene transfer has not yet been evaluated. SV40 vectors can be maintained as episomes, but stable integration does occur at low frequency. Karlson *et al.*, (1985) explored both SV40 constructions and adenovirus vectors for infection of murine and human hematopoietic cells. An SV40 vector carrying the chloramphenicol acetyltransferase gene successfully infected mouse and human bone-marrow cells, but had a significant problem with recombination of defective viruses to form wild-type virus in further experiments. Much further work will be needed before these vector systems can be applied to *in vivo* gene-transfer experiments. It appears that the efficiency of transfer and the stability may be inferior in DNA-virus vectors compared to retroviruses, but there may

TABLE III. Factors in Facilitating Gene
Transfer

Vector production
  Transcription rate
    Copy number
    Splicing
    Control elements
    Chromosomal position
  Packaging
    Pseudotyping
    Packaging cell lines
      Copy number
      Control elemenets
      Chromosomal position
  Selection
    Titer
    Product expression
    Coselectable marker
    Construction stability

Vector expression
  Titer
    Target cells
    env Gene
    Selection method
  Expression
    Chromosomal position
    Control elements
    Splicing
    Posttranslational modification
    Subunit interactions
    Cofactor requirements
    Substrate requirements
    Subcellular compartmentalization
In vivo expression
  Selection
    Metabolic
    Drug resistance
  Clonal expression
    Polyclonality
    Clonal succession
  Expression in the target
    Differentiated cell type
  Autoimmune phenomena

be some advantage in the expression of introduced genes. In particular, adenovirus vectors may allow introduction of completely independent transcription units including constructs containing introns. The safety of either retrovirus or DNA-virus systems remains to be tested experimentally.

5. EXPERIMENTAL GENE TRANSFER OF HUMAN-DISEASE-RELATED GENES

5.1. Defective Retroviruses Carrying Human *HPRT*

Over the past two years, extensive experience has been gained in the construction of retroviral vectors carrying the human *HPRT* gene. Miller *et al.* (1983) constructed human *HPRT*-containing defective retroviruses using cloned Moloney murine sarcoma virus and M-MuLV LTRs. The 5′ LTR acted as a promoter element and allowed efficient expression of the human *HPRT* in HPRT⁻ rat 208F cells after CaPO₄ coprecipitation. These defective viruses were rescued by coinfection with helper virus. Initial murine ecotropic virus-producing lines achieved titers of approximately 2×10^5 colony-forming units (CFU)/ml with about 10-fold higher titers of helper virus. They further demonstrated that human Lesch–Nyhan fibroblasts and lymphoblasts could be infected with the defective *HPRT*-containing viruses when an amphotropic helper virus was used (1504A). We have likewise constructed human *HPRT*-containing retroviruses and have used both wild-type helper coinfection and helper-cell packaging lines for the production of defective *HPRT*-transducing viruses. We have subcloned the human *HPRT* gene into the SV(X) vector (Cepko *et al.*, 1984) and then transfected into Ψ_2, Ψ_{am}, and PA-12 packaging-cell lines (Fig. 1). Initial efforts focused

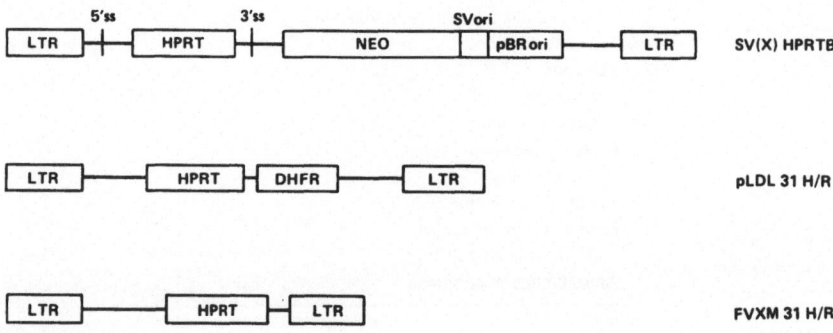

FIGURE 1. *HPRT* retroviral constructions. Human *HPRT* cDNA was ligated to *Bam*HI linkers and subcloned into the unique *Bam*HI sites in SV(X) and pLDL or the unique *Bgl*II site of FVXM.

on the selection of high-titer-producing cell lines using either selection for *neo* gene expression (using the aminoglycoside antibiotic G418) or selection in hypoxanthine–aminopterin–thymidine (HAT) medium of HPRT⁻ target cells. It was found that, on average the titers were poor, ranging from 10^2 to 10^4 CFU/ml. Cell lines formed by infection of packaging-cell lines by these low-titer viruses gave somewhat better results, confirming the observation that expression of foreign sequences in retrovirus vectors is superior following infection compared to CaPO₄ transformation (Hwang and Gilboa, 1984). However, even after extensive screening of individual clonal isolates, titers of *HPRT* retroviruses in this vector were no better than 5×10^4 CFU/ml. Infection of NIH 3T3 mouse fibroblasts with defective retroviruses and then coinfection with wild-type Moloney virus yielded much higher-titer *HPRT* virus production (5×10^5 CFU/ml) as a consequence of wild-type virus infection. It was found that G418-resistance titers and *HPRT* titers were roughly equivalent, indicating that in this spliced vector, the genomic and spliced transcripts were approximately equal functionally. Southern analysis of infected cells demonstrated single-copy insertion. HPRT enzyme expression was approximately 10–40% of wild-type levels. In this construction, the transcription of both the *HPRT* and *neo* genes is promoted by the 5′ viral LTR promoter. An additional construction was made to test the practicality of independent promoter regulation of the *HPRT* gene. Using the SV(X) vector, the natural mouse HPRT promoter and the human cDNA were used as an independent transcription unit. No detectable virus was produced, and this was initially explained by invoking a "transcriptional interference" mechanism (Emerman and Temin, 1984). This seemed to be consistent with the LTR promoter and the natural HPRT promoter interfering with each other. Very recently, a negative regulatory element, possibly a so-called "silencer" (Brand *et al.*, 1985), has been found in the natural promoter, and this may be the explanation for the lack of virus production. We have also used several alternative vectors in the hope of achieving higher-titer virus production. Constructions using the pLDL (Miller *et al.*, 1985) vector and the FVXM (Kriegler *et al.*, 1984) vector have been made. Because the pLDL vector contains a mutant mouse *dhfr Mtx^r* gene, it is possible to select virus-producing lines in methotrexate. The use of the FVXM vector represents a departure from the use of a second dominant-selectable marker. All selection steps are achieved using HAT selection of HPRT⁻ cell lines. Characterization of virus-producing lines using these constructions is under way. It is probable that overall expression can be optimized by the use of the simplest construction. However, an adequate screening method must be available for assessment of viral titer and gene-product expression. Although HPRT deficiency is a severe disease in which limited treatment options are available and even though the molecular biology of the *HPRT* gene is well understood, we do not consider it to be a good model for gene therapy. Because a major pathological feature of the disease is damage to the basal ganglia in the CNS and the high probability that *HPRT* gene expression in these cells is critical to their function, gene delivery to a specific region of the brain appears necessary. No currently envisioned viral-vector system offers any promise for this problem. Research will have to be directed toward vectors that can be targeted to specific areas of the CNS, that work well in nonreplicating cells, and that are noncytopathic.

5.2. Retroviral Vectors Carrying Other Complementary DNAs

Several other human-disease-related genes have been placed into retrovirus expression vectors. These include AS (Wood *et al.*, 1986) and phenylalanine hydroxylase (PH). Ledley *et al.* (1986) have succeeded in using their PH retrovirus-producing Ψ_2 lines to infect a cultured mouse hepatoma line, hepal-a. This is the first example of retrovirus-vector infection of liver cells and is an important step in assessing the feasibility of gene transfer to liver.

5.3. Gene Transfer to Bone-Marrow Progenitors

Several groups have reported success in infecting mouse bone-marrow cells with defective retroviruses (Fig. 2). Joyner *et al.* (1983c) demonstrated transfer of *neo* to murine CFU-GM progenitors by infection of mouse marrow *in vitro*. Supernatants of cell lines that produce defective viruses carrying the *neo* gene were added to primary mouse bone-marrow cultures. The cells were then plated

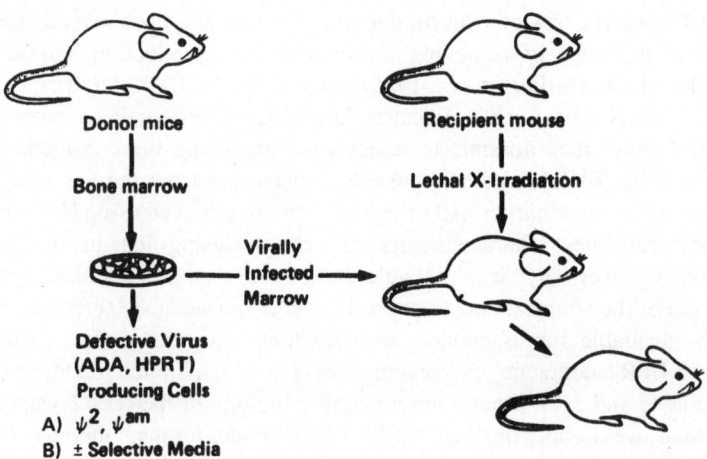

FIGURE 2. Protocol for retroviral transfer of human-disease-related genes into mouse marrow.

in semisolid medium under G418 selection for enumeration of CFU-GMs. The infection efficiency was low, approximately 0.3%. Both Williams *et al.* (1984) and Miller *et al.* (1984) have infected mouse marrow stem cells. In one experiment, mouse bone-marrow cells were infected by cocultivation with a cell line producing human HPRT defective retroviruses. Human HPRT activity was demonstrated in hematopoietic tissue from a single mouse transplanted with the infected marrow. Williams *et al.* (1984) showed that efficiency of infection into spleen colony-forming cells (CFU-S) was about 10–20% using a simple cocultivation protocol. Individual CFU-S colonies were dissected and subjected to Southern analysis. These studies demonstrated clonality and stability on subsequent retransplantation. No assessment of expression efficiency was made. In a more recent study, Dick *et al.* (1985) have demonstrated infection of a primitive stem cell (Sp cell) capable of repopulating all hematopoietic compartments. By making several important technical refinements in the bone-marrow infection protocol, the most important being the addition of a preselection in G418 prior to transplantation, they achieved 100% stem-infection efficiency as assessed by CFU-S Southern analysis and G418 resistance of CFU-GMs *in vitro*. An important aspect of this latter study is the formal demonstration of a primitive stem capable of clonally repopulating the erythroid, myeloid, and lymphoid lineages. The retrovirus gene-transfer system thus provides a powerful analytical tool for the mapping of cell-lineage relationships in the hematopoietic system. This study demonstrates the utility of drug-resistance selection of the bone-marrow stem cells prior to transfer to a recipient animal. Other drug or metabolic selection procedures are under evaluation in several laboratories. The possibility of *in vivo* selection with methotrexate is of special interest. The experiments of the Toronto group (Dick *et al.*, 1985) also illustrate the problem that marrow stem-cell infection is initially polyclonal, and thus variability in the level of transgene expression can be expected to result from the individual infection events. Bone-marrow repopulation of mature blood cells probably operates by a "clonal succession" mechanism; i.e., only a few stem cells at any one time are contributing to the differentiated pool. If less than 100% of the transferred stem cells are infected with retrovirus or there is variability in expression, fluctuations over time in transgene expression can be anticipated. We also have a theoretical concern about autoimmune responses to the gene product. This is known to occur in protein-replacement treatment for Factor VIII deficiency and will need to be evaluated experimentally.

5.4. Adenosine Deaminase Deficiency as a Model for Human Gene Therapy

As Dick *et al.* (1985) note, their experiments set an important precedent for the practicality of gene transfer for therapeutic purposes. Of the genetic diseases that are understood at the molecular level and the relevant genes of which have been cloned, ADA deficiency appears to be the most promising opportunity for early investigation of gene therapy. ADA deficiency is an autosomal recessive disorder that causes an SCID and accounts for approximately 15% of patients with SCID (Kredich and Hershfield, 1983). In the absence of this enzyme, toxic metabolites accumulate, including dATP, which cause selective damage to immature lymphoid cells, especially T cells. Inhibition of ribonucleotide reductase and S-adenosylhomocysteine hydrolase is felt to be important in the toxic effect. Both T cells and B cells are depleted with time in ADA-deficient patients, and death usually results from overwhelming infection. Although enzyme replacement by red-blood-cell transfusion can be used as a temporizing measure, definitive therapy is bone-marrow transplantation (Hirschhorn et al., 1981). The lack of availability of HLA-matched donor marrow, however, has restricted the application of bone-marrow transplant in some cases. ADA is normally expressed at low "housekeeping" levels in most tissues, but is much increased in immature hematopoietic cells, with the highest levels measured in the thymus. The physiological role of these high levels of enzyme is unknown, but appears to be related to the rapid proliferation of these cells. Both human and mouse ADA cDNAs have been cloned (Wiginton et al., 1983; Valerio et al., 1983; Orkin et al., 1983; Yeung et al., 1983c). Recently, the structure of the human gene has been elucidated (Valerio et al., 1984). Analysis of ADA deficiency mutations is under way and indicates that the disease is heterogeneous at the molecular level (Adrian et al., 1984). Both point mutations and larger rearrangements can be expected, on the basis of the molecular analysis of other human mutations.

ADA has attracted much attention over the last several years as a model for gene transfer and its deficiency as a potential model for gene therapy. Two accounts of ADA gene transfer using a retrovirus vector have been published (Friedman, 1985; Valerio et al., 1985). Both studies utilized the SV(X) vector and showed stable transfer of human ADA to mouse target cells. Titers reported by Valerio et al. (1985) reflect the general experience with the SV(X) vector ($\approx 10^4$ CFU/ml). Several labs are making rapid progress in exploiting these ADA retrovirus-producing cell lines for mouse and human bone-marrow infection studies and mouse bone-marrow transplant experiments. We have constructed defective retrovirus vector capable of transducing both Neo and the human ADA cDNA (Fig. 3). This vector, pSVB ADA 211, was used to transfect PA-12 amphotropic helper cells. The resultant defective virus was used to infect the ecotropic helper line $\Psi2$. Helper cell lines formed by infection have generally allowed higher titer virus production than after transfection with plasmid DNA. Highest titer virus production from a clonal isolate has been $\sim 1 \times 10^5$ cfu/ml (G418^r). These replication defective viruses are currently being studied for their ability to infect and be expressed in murine bone marrow cells.

SV(B) ADA 2II

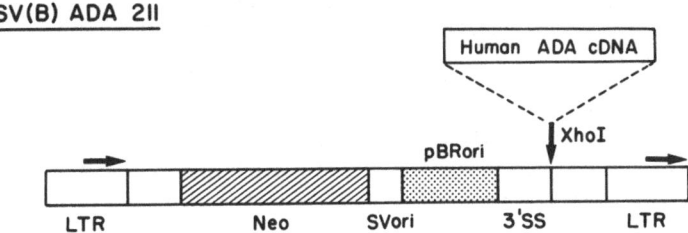

FIGURE 3. *ADA* retrovirus vector. Human *ADA* cDNA was subcloned into the unique *Xho*I site of *pZIP* Neo-SV(B).

Regarding ADA deficiency: (1) It is well understood at the molecular level; (2) experimental evidence from enzyme replacement therapy suggests that relatively low-level and unregulated expression ameliorates the disease; and (3) on the basis of bone-marrow transplant data, it appears that expression in the bone marrow would be curative. It is expected that infection of a defective ADA-encoding retrovirus into autologous marrow stem cells would correct the SCID. The anticipated advantages of this approach when compared to conventional bone-marrow transplantation include: (1) the lack of the necessity for an *HLA*-matched donor; (2) no use of immunosuppression in the posttransplant period; and (3) no possibility of GVH disease. Concerns about the procedure and anticipated complications include: (1) failure of adequate and long-lasting ADA expression; (2) active retroviral infection secondary to recombination of the defective virus with an endogenous defective; and (3) retroviral insertional mutagenesis at the somatic level, possibly leading to oncogenesis. These later concerns will be addressed in animal studies, but definitive answers must await human trials.

6. ETHICAL CONSIDERATIONS

The idea of human gene transfer for treatment of heritable disease has raised concerns about the entrance of medical technology into an unknown and potentially dangerous territory. Treatment of metabolic diseases by gene transfer will require delivery to and expression of genes in somatic tissues. Only two tissues are at present envisioned as realistic targets—hematopoietic and hepatic. Entrance of these transplanted genes into germinal tissue would be highly undesirable for medical reasons. Animal experiments using either embryo microinjection or retrovirus infection demonstrate a high rate of insertional mutagenesis. This risk to future offspring is qualitatively similar to the risk associated with chemotherapy and irradiation for the treatment of cancer. This is especially a concern in the childhood cancers. The comparative quantitative risk between retrovirus infection and chemotherapy is unknown, but we expect that retroviral insertion into germinal tissues would be very infrequent. In addition, an accidental infection of germ cells or a fetus would be easily detectable by prenatal diagnosis using Southern analysis. Somatic gene transfer can be viewed, therefore, as an extension and refinement of allograft techniques currently employed for bone-marrow and liver transplant. This position has been affirmed by a congressional committee (U.S. Congress, 1982), by a Presidential Commission (President's Commission, 1982), and by the National Council of Churches (J. R. Nelson, unpublished).

Germ-line gene transfer, at present, appears to be technically unrealistic. Present techniques for embryo gene transfer are inefficient, do not allow for control of gene copy number, and have a high probability of causing insertional mutagenesis. In addition, there is no realistic way of separating normal from abnormal embryos at the single-cell stage—and if there were, it would make the gene-transfer technology unnecessary. Therefore, the ethical problem of purposeful alteration of the human germ line and gene pool stands outside the medical issue of gene therapy.

7. SUMMARY

Human gene-transfer therapy will be attempted soon. The formidable technical problems have caused attention to be focused on the simplest disease models for early trials. It is expected that ADA deficiency will be treated by gene therapy in the coming months. Given preliminary success with this disorder, a very limited set of the known human genetic disorders will be potential targets for treatment by gene transfer. In the short term, efforts will be limited to diseases in which gene expression in the bone marrow is likely to positively affect the disease. Techniques for *in vivo* infection of the liver by viral vectors need to be developed before human experiments can be proposed. In the interim, both bone-marrow and hepatic transplantation are providing the data base and precedence for the efficacy of gene transfer for the treatment of metabolic diseases. Viral vector systems are improving rapidly, and their efficacy will be a major determinant of success or failure in this technology. The combination of the technologies of gene cloning, viral vector gene transfer, and organ transplantation should allow the establishment of a valuable new tool in the treatment of genetic disease.

REFERENCES

Adams, M. M., Finley, S., Hamsen, H., Jahiel, R., Oakley, G., Sanger, W., Wells, G., and Wertelecki, W., 1980, Utilization of prenatal diagnosis in women 35 years of age and older in the United States, 1977–1978, *Am. J. Obstet. Gynecol.* **139**:673.

Adrian, G. S., Wiginton, D. A., and Hutton, J. J., 1984, Structure of adenosine deaminase mRNA's from normal and adenosine deaminase-deficient human cell lines, *Mol. Cell. Biol.* **4**:1712.

Anderson, W. F., 1981, Gene therapy, *JAMA* **246**:2737.

Anderson, W. F., 1984, Prospects for human gene therapy, *Science* **226**:401.

Bandyopadhyay, P. K., and Temin, H. M., 1984, Expression of complete chicken thymidine kinase gene inserted in a retrovirus vector, *Mol. Cell. Biol.* **4**:749.

Bandyopadhyay, P. K., Watanabe, S., and Temin, H. M., 1984, Recombination of transfected DNA's in vertebrate cells in culture, *Proc. Natl. Acad. Sci. U.S.A.* **81**:3476.

Beaudet, A. L., 1985, Bibliography of cloned human and other selected DNA's, *Am. J. Hum. Genet.* **37**:386.

Bernstein, A., Berger, S., Huszar, D., and Dick, J., 1985, Gene transfer with retrovirus vetors, in: *Genetic Engineering: Principles and Methods*, Vol. 7 (J. K. Setlow and A. Hollaender, eds.) p. 235, Plenum Press, New York.

Bilheimer, D. W., Goldstein, J. L., Grundy, S. M., Starzl, T. E., and Brown, M. S., 1984, Liver transplantation to provide low-density-lipoprotein receptors and lower plasma cholesterol in a child with homozygous familial hypercholesterolemia, *N. Engl. J. Med.* **311**:1658.

Bortin, M. M., and Rimm, A. A., 1977, Severe combined immunodeficiency disease: Characterization of the disease and results of transplantation, *JAMA* **238**:591.

Brand, A. H., Breeden, L., Abraham, J., Sternglanz, R., and Nasmyth, K., 1985, Characterization of a "silencer" in yeast: A DNA sequence with properties opposite to those of a transcriptional enhancer, *Cell* **41**:41.

Broome, S., and Gilbert, W., 1978, Immunological screening method to detect specific translation products, *Proc. Natl. Acad. Sci. U.S.A.* **75**:2746.

Cepko, C., Roberts, B. E., and Mulligan, R., 1984, Construction and application of a highly transmissible murine retrovirus shuttle vector, *Cell* **37**:1053.

Champlin, R. E., and Gale, R. P., 1984, The early complications of bone marrow transplantation, *Semin. Hematol.* **21**:101.

Chien, Y.-H., Becker, D. M., Lindsten, T., Okamura, M., Cohen, D. I., and Davis, M. M., 1984, A third type of murine T-cell receptor gene, *Nature (London)* **312**:31.

Cline, M. J., 1982, Genetic engineering of mammalian cells: Its potential application to genetic diseases of man, *J. Lab. Clin. Med.* **99**:299.

Coccia, P. F., Krivit, W., Cervenka, J., Clawson, C., Kersey, J. H., Sim, T. H., Nesbit, M. E., Ramsey, N. K., Warkentin, P. I., Teitelbaum, S. L., Kahm, A. J., and Brown, D. M., 1980, Successful bone-marrow transplantation for infantile malignant osteopetrosis, *N. Engl. J. Med.* **302**:701.

Cone, R., and Mulligan, R. C., 1984, High efficiency gene transfer into mammalian cells: Generation of helper-free recombinant retrovirus with broad mammalian host range, *Proc. Natl. Acad. Sci. U.S.A.* **81**:6349.

Corsaro, C. M., and Pearson, M. L., 1981a, Enhancing the efficiency of DNA-mediated gene transfer in mammalian cells, *Somat. Cell. Genet.* **7**:603.

Corsaro, C. M., and Pearson, M. L., 1981b, Competence for DNA transfer of oubain resistance and thymidine kinase: Clonal variation in mouse L-cell recipients, *Somat. Cell Genet.* **7**:617.

De Saint Vincent, B. R., and Wahl, G. M., 1983, Homologous recombination in mammalian cells mediates formation of a functional gene from two overlapping gene fragments, *Proc. Natl. Acad. Sci. U.S.A.* **80**:2002.

Dick, J. E., Magli, M. C., Huszar, D., Phillips, R. A., and Bernstein, A., 1985, Introduction of a selectable gene into primitive stem cells capable of long-term reconstitution of the hematopoietic system of W/W^v mice, *Cell* **42**:71.

DiFerrante, N., Hyman, B. H., Klish, W., Donnell, P. V., Nichols, B. L., and Dutton, R. G., 1974, Mucopolysaccharidosis VI (Maroteaux–Lamy disease): Clinical and biochemical study of a mild variant case, *Johns Hopkins Med. J.* **135**:42.

DiMaio, D., Treisman, R., and Maniatis, T., 1982, Bovine papillomavirus vector that propagates as a plasmid in both mouse and bacterial cells, *Proc. Natl. Acad. Sci. U.S.A.* **79**:4030.

Elder, J. T., Spritz, R. A., and Weissman, S., 1981, Simian virus 40 as a eukaryotic cloning vehicle, *Annu. Rev. Genet.* **15**:295.

Emerman, M., and Temin, H., 1984, High frequency deletion in recovered retrovirus vectors containing exogenous DNA with promoters, *J. Virol.* **50**:42.

Foroonzofar, N., Hobbs, J. R., and Hugh-Jones, K., 1977, Bone marrow transplantation from an unrelated donor for chronic granulomatous disease, *Lancet* **1**:210.

Friedman, R., 1985, Expression of human adenosine deaminase using a transmissible murine retrovirus vector system, *Proc. Natl. Acad. Sci. U.S.A.* **82**:703.

Gasper, P. W., Thrall, M. A., Wenger, D. A., Macy, D. W., Ham, L., Dornsife, R. E., McBiles, K., Quackenbush, S. L., Kesel, M. L., Gillette, E. L., and Hover, E. A., 1984, Correction of feline arylsulfatase B deficiency (mucopolysaccharidosis VI) by bone marrow transplantation, *Nature (London)* **312**:467.

Gilboa, E., 1983, Use of retrovirus-derived vectors to introduce and express genes in mammalian cells, in: *Experimental Manipulation of Gene Expression* (M. Inouye, ed.), p. 175, Academic Press, New York.

Ginns, E. I., Rappeport, J., and Brady, R., 1982, Correction of glucocerebrosidase deficiency in Gaucher's disease by bone marrow transplantation, *Blood* **60**:168A.

Gluckman, E., Berger, R., and Dutreix, J., 1984, Bone marrow transplantation for Fanconi anemia, *Semin. Hematol.* **21**:20.

Haldane, J. B. S., 1947, The mutation rate of the gene for hemophilia and its segregation ratios in males and females, *Ann. Eugen.* **13**:262.

Hazelrigg, T., Levis, R., and Rubin, G. M., 1984, Transformation of *white* locus DNA in *Drosophila*: Dosage compensation, *zeste* interaction, and position effects, *Cell* **36**:469.

Hirschhorn, R., Roegner-Maniscalco, V., Kuritsky, L., and Rosen, F. S., 1981, Bone marrow transplantation only partially restores purine metabolites to normal in adenosine deaminase-deficient patients, *J. Clin. Invest.* **68**:1387.

Hobbs, J. R., Hugh-Jones, K., and Barrett, A. J., 1981, Reversal of clinical features of Hurler's disease and biochemical improvement after treatment by bone marrow transplantation, *Lancet* **2**:709.

Howell, R. R., Stevenson, R. E., Ben-Menachem, Y., Philiky, R. L., and Berry, D. H., 1976, Hepatic adenomata in patients with Type I glycogen storage disease (von Gierke's), *JAMA* **236**:1481.

Hugh-Jones, K., Kendra, J., and James, D. C. Q., 1982, Treatment of San Filipo B disease (MPSIIIB) by bone marrow transplant, *Exp. Hematol.* **10**(Suppl.):50.

Hwang, L.-H. S., and Gilboa, E., 1984, Expression of genes introduced into cells by retroviral infection is more efficient than that of genes introduced by DNA transfection, *J. Viol.* **50**:417.

Iwatsuki, S., Shaw, B. W., and Starzl, T. E., 1985, Five-year survival after liver transplantation, *Transplant. Proc.* **17**:259.

Jaenisch, R., Fan, H., and Croker, B., 1975, Infection of preimplantation mouse embryos and of newborn mice with leukemia virus: Tissue distribution of viral DNA and RNA and leukemogenesis in the adult animal, *Proc. Natl. Acad. Sci. U.S.A.* **72**:4008.

Jaenisch, R., Jähner, D., Nobis, P., Simon, I., Löhler, J., Harbers, K., and Grotkopp, D., 1981, Chromosomal position and activation of retroviral genomes inserted into the germ line mice, *Cell* **24**:519.

Johnson, F. L., Look A. T., Gockerman, J., Ruggiero, M. R., Dalla-Pozza, L., and Billings, F., 1984, Bone-marrow transplantaton in a patient with sickle-cell anemia, *N. Engl. J. Med.* **311**:780.

Jolly, D. J., Willis, R., and Friedman, T., 1984, Site dependent frequencies and molecular mechanisms of reversion of single copy introduced HPRT genes in human cells, *Am. J. Hum. Genet.* **36**(Suppl.):141S.

Joss, V., Rogers, T. V., and Hugh-Jones, K., 1982, A bone marrow transplant for metachromatic leukodystrophy, *Exp. Hematol.* **10**(Suppl.):52.

Joyner, A. L., and Bernstein, A., 1983a, Retrovirus transduction: Generation of infectious retroviruses expressing dominant and selectable genes is associated with *in vivo* recombination and deletion events, *Mol. Cell. Biol.* **3**:2180.

Joyner, A. L., and Bernstein, A., 1983b, Retrovirus transduction: Segregation of the viral transforming function and the herpes simplex virus *tk* gene in infectious Friend spleen focus-forming virus thymidine kinase vectors, *Mol. Cell. Biol.* **3**:2191.

Joyner, A., Keller, G., Phillips, R. A., and Bernstein, A., 1983, Retrovirus transfer of a bacterial gene into mouse haematopoietic progenitor cells, *Nature (London)* **305**:556.

Karlson, S., Humphries, R. K., Gluzman, Y., and Nienhuis, A. W., 1985, Transfer of genes into haematopoietic cells using recombinant DNA viruses, *Proc. Natl. Acad. Sci. U.S.A.* **82**:158.

Kaufman, R. J., Wasley, L. C., Spiliates, A. J., Gossels, S. D., Latt, S. A., Larsen, G. R., and Kay, R. M., 1985, Coamplification and coexpression of human tissue-type plasminogen activator and murine dihydrofolate reductase sequences in Chinese hamster ovary cells, *Mol. Cell. Biol.* **5**:1750.

King, W., Patel, M. D., Lobel, L. I., Goff, S. P., and Nguyen-Huu, C., 1985, Insertion mutagenesis of embryonal carcinoma cells by retroviruses, *Science* **228**:554.

Kredich, N., and Hershfield, M., 1983, Immunodeficiency diseases caused by adenosine deaminase deficiency and purine nucleoside phosphorylase deficiency, in: *The Metabolic Basis of Inherited Disease* (J. B. Stanbury, J. B. Wyngaarden, D. S. Fredrickson, J. L. Goldstein, and M. S. Brown, eds.), p. 1157, McGraw-Hill, New York.

Kriegler, M., Perez, C. F., Hardy, C., and Botchan, M., 1984, Transformation mediated by the SV40 T antigens: Separation of the overlapping SV40 early genes with a retroviral vector, *Cell* **38**:483.

Krivit, W., Pierpont, M. E., Ayaz, K., Tsai, M., Ramsay, N., Kersey, J. H., Weisdorf, S., Sibley, R., Snover, D., McGovern, M. M., Schwartz, M. F., and Desnick, R. J., 1984, Bone-marrow transplantation in the Maroteaux–Lamy syndrome (mucopolysaccharidosis Type VI), *N. Engl. J. Med.* **311**:1606.

Kunkel, L. M., Monaco, A. P., Middlesworth, W., Ochs, H. D., and Latt, S. A., 1985, Specific cloning of DNA fragments absent from the DNA of a male patient with an X chromosome deletion, *Proc. Natl. Acad. Sci. U.S.A.* **82**:4778.

Ledley, F. D., Grenett, H. E., McGinnis-Shelnutt, M., and Woo, S. L. C., 1986, Retroviral mediated gene transfer of human phenylalanine hydroxylase into NIH 3T3 and hepatoma cells, *Proc. Natl. Acad. Sci. U.S.A.* **83**:409.

Lenz, J., Celander, D., Crowther, R. L., Patarca, R., Perkins, D. W., and Haseltine, W. A., 1984, Determination of the leukemogenicity of a murine retrovirus by sequences within the long terminal repeat, *Nature (London)* **308**:467.

Lucarelli, G., Izzi, T., and Polchi, P., 1983, Bone marrow transplantation in thalassemia, *Exp. Hematol.* **11**(Suppl.):101.

Luthy, D. A., Emanuel, I., Hoehm, H., Hall, J. G., and Powers, E. K., 1980, Prenatal genetic diagnosis and elective abortion in women over 35: Utilization and relative impact on the birth prevalence of Down syndrome in Washington State, *Am. J. Med. Genet.* **7**:375.

Mackett, M., Smith, G. L., and Moss, B., 1982, Vaccinia virus: A selectable eukaryotic cloning and expression vector, *Proc. Natl. Acad. Sci. U.S.A.* **79**:7415.

Mackett, M., Smith, G. L., and Moss, B., 1984, General method for production and selection of infectious vaccinia virus recombinants expressing foreign genes, *J. Virol.* **49**:857.

Mann, R., Mulligan, R. C., and Baltimore, D., 1983, Construction of a retrovirus packaging mutant and its use to produce helper-free defective retrovirus, *Cell* **33**:153.

Matas, A. J., Desnick, R. J., Najarian, J. S., and Simmons, R. L., 1978, Clinical and experimental transplantation in enzymatic deficiency disease, *Surg. Gynecol. Obstet.* **146**:975.

Matsui, S. M., Mahoney, M. J., and Rosenberg, L. E., 1983, The natural history of the inherited methylmalonic acidemias, *N. Engl. J. Med.* **308**:857.

Miller, A. D., Jolly, D. J., Friedman, T., and Verma, I. M., 1983, A transmissible retrovirus expressing human hypoxanthine phosphoribosyl transferase (HPRT): Gene transfer into cells obtained from humans deficient in HPRT, *Proc. Natl. Acad. Sci. U.S.A.* **80**:4709.

Miller, A. D., Eckner, R. J., Jolly, D. J., Friedman, T., and Verma, I. M., 1984, Expression of a retrovirus encoding human HPRT in mice, *Science* **225**:630.

Miller, A. D., Law, M.-F., and Verma, I. M., 1985, Generation of helper-free amphotropic retroviruses that transduce a dominant-actng, methotrexate-resistant dihydrofolate reductase gene, *Mol. Cell. Biol.* **5**:431.

Milunsky, A., 1976, Prenatal diagnosis of genetic disorders, *N. Engl. J. Med.* **300**:157.

Mulligan, R., 1983, Construction of highly transmissible mammalian cloning vehicles derived from murine retroviruses, in: *Experimental Manipulation of Gene Expression* (M. Inouye, ed.), p. 155, Academic Press, New York.

Ochs, H. D., Lum, L. G., Johnson, F. L., Schiffman, G., Wedgewood, R. J., and Storb, R., 1982, Bone marrow transplantation in the Wiskott–Aldrich syndrome: Complete hematological and immunological reconstitution, *Transplantation* **34**:284.

O'Hare, K., and Rubin, G. M., 1983, Structures of P elements and their sites of insertion and excision in the *Drosophila melanogaster* genome, *Cell* **34**:25.

O'Reilly, R. J., Kirkpatrick, D., and Cunningham-Rundles, S., 1983a, Transplantation for severe combined immunodeficiency using histoincompatible parental marrow fractionated by soybean agglutinin and sheep red blood cells: Experience in six consecutive cases, *Transplant. Proc.* **15**:1431.

O'Reilly, R. J., Kapoor, N., and Kirkpatrick, D., 1983b, Transplantation of hematopoietic cells for lethal congenital immunodeficiencies, *Birth Defects* **19**:129.

O'Reilly, R. J., Brochstein, J., Dinsmore, R., and Kirkpatrick, D., 1984, Marrow transplantation for congenital disorders, *Semin. Hematol.* **21**:188.

Orkin, S. H., Daddona, P. E., Shewach, D. S., Markham, A. F., Bruns, G., Goff, S. C., and Kelley, W. N., 1983, Molecular cloning of human adenosine deaminase gene sequences, *J. Biol. Chem.* **258**:12,753.

Parkman, R., Rappeport, J., Geha, R., Belli, J., Cassady, R., Levey, R., Nathan, D. G., and Rosen, F. S., 1978, Complete correction of the Wiskott–Aldrich syndrome by allogeneic bone-marrow transplantation, *N. Engl. J. Med.* **298**:921.

Perucho, M., Hanahan, D., and Wigler, M., 1980, Genetic and physical linkage of exogenous sequences in transformed cells, *Cell* **22**:309.

Potter, H., Weir, L., and Leder, P., 1984, Enhancer-dependent expression of human κ immunoglobulin genes introduced into mouse pre-B lymphocytes by electroporation, *Proc. Natl. Acad. Sci. U.S.A.* **81**:7161.

Prentice, H. G., Blacklock, H. A., and Janossy, G., 1982, Use of anti-T cell monoclonal antibody OKT3 to prevent acute graft-versus-host disease in allogeneic bone marrow transplantation for leukemia, *Lancet* **1**:700.

President's Commission for the Study of Ethical Problems in Medicine and Biomedical and Behavioral Research, 1982, *Splicing Life*, Government Printing Office, Washington, D.C.

Rappeport, J. M., Parkman, R., Newburger, P., Camitta, B. M., and Chusid, M. J., 1980, Correction of infantile agranulocytosis (Kostman's syndrome) by allogenic bone marrow transplantation, *Am. J. Med.* **68**:605.

Rappeport, J. M., Newburger, P. E., and Goldblum, R. M., 1982, Allogeneic bone marrow transplantation for chronic granulomatous disease, *J. Pediatr.* **101**:952.

Rolles, K., Williams, R., Neuberger, J., and Calne, R., 1984, The Cambridge and King's College Hospital experience of liver transplantation, 1968–1983, *Hepatology* **4**:50S.

Rosenbloom, F. M., Kelley, W. M., Miller, J., Henderson, J. F., and Seegmiller, J. E., 1967, Inherited disorder of purine metabolism: Correlation between central nervous system dysfunction and biochemical defects, *JAMA* **202**:175.

Rubin, H., 1965, Genetic control of cellular susceptibility to pseudotypes of Rous sarcoma virus, *Virology* **26**:270.

Rubnitz, J., and Subramani, S., 1985, Rapid assay for extrachromosomal homologous recombination in monkey cells, *Mol. Cell. Biol.* **5**:529.

Scharschmidt, B. F., 1984, Human liver transplantation: Analysis of data on 540 patients from four centers, *Hepatology* **4**:95S.

Shearer, W. T., Ritz, J., Finegold, M., Guerra, I. C., Rosenblatt, H., Lewis, D. E., Pollack, M. S., Taber, L. H., Sumaya, C. V., Grumet, F. C., Cleary, M. L., Warnke, R., and Sklar, J., 1985, Epstein–Barr virus-associated B-cell proliferations of diverse clonal origins after bone marrow transplantation in a 12-year-old patient with severe combined immunodeficiency, *N. Engl. J. Med.* **312**:1151.

Shimotohno, K., and Temin, H. M., 1980, No apparent nucleotide sequence specificity in cellular DNA juxtaposed to retrovirus proviruses, *Proc. Natl. Acad. Sci. U.S.A.* **77**:7357.

Sly, W. S., Whyte, M. P., Sundaram, V., Tashian, R. E., Hewett-Emmett, D., Guibaud, P., Vainsel, M., Baluarte, H. J., Gruskin, A., and Al-Mosawi, M., 1985, Carbonic anhydrase II deficiency in 12 families with autosomal recessive syndrome of osteopetrosis with renal tubular acidosis and cerebral calcification, *N. Engl. J. Med.* **313**:139.

Sorell, M., Kapoor, N., Kirkpatrick, D., Rosen, J. F., Chaganti, R. S., Lopez, C., Dupont, B., Pollack, M. S., Terrin, B. N., Harris, M. B., Vine, D., Rose, J. S., Goosen, C., Lane, J., Good, R. A., and O'Reilly, R. J., 1981, Marrow transplantation for juvenile osteopetrosis, *Am. J. Med.* **70**:1280.

Spradling, A. C., and Rubin, G. M., 1983, The effect of chromosomal position on the expression of the *Drosophila* xanthine oxidase gene, *Cell* **34**:47.

Starzl, T. E., Iwatsuki, S., Shaw, B. W., Van Thiel, D. H., Gartner, J. C., Zitelli, B. J., Malatack, J. J., and Schade, R. R., 1984, Analysis of liver transplantation, *Hepatology* **4**:47S.

Starzl, T. E., Iwatsuki, S., Shaw, B. W., and Gordon, R. D., 1985, Orthotopic liver transplantation, *Transplant. Proc.* **17**:250.

Stuhlman, H., Cone, R., Mulligan, R. C., and Jaenisch, R., 1984, Introduction of a selectable gene into different animal tissue by a retrovirus recombinant vector, *Proc. Natl. Acad. Sci. U.S.A.* **81**:7151.

Suggs, W. V., Wallace, R. B., Hirose, T., Kawashima, E. H., and Itakura, K., 1981, Use of synthetic oligonucleotides as hybridization probes: Isolation of clones cDNA sequences for human β_2-microglobulin, *Proc. Natl. Acad. Sci. U.S.A.* **78**:6613.

Sullivan, K. M., Deeg, J., Sanders, J. E., Schulman, H. M., Witherspoon, R. P., Doney, K., Applebaum, F. R., Schubert, M. M., Stewart, P., Springmeyer, S., McDonald, G. B., Storb, R., and Thomas, E. D., 1984, Late complications after marrow transplantation, *Semin. Hematol.* **21**:53.

Tanaka, K., and Rosenberg, L. E., 1983, Disorders of propionate and methylmalonate metabolism, in: *The Metabolic Basis of Inherited Disease* (J. B. Stanbury, J. B. Wyngaarden, D. S. Fredrickson, J. L. Goldstein, and M. S. Brown, eds.), p. 474, McGraw-Hill, New York.

Thomas, E. D., Buckner, C. D., and Sanders, J. E., 1982, Marrow transplantation for thalassemia, *Lancet* **2**:227.

Tutschka, P. J., Yeager, A. M., and Moser, H. W., 1983, Bone marrow transplantaton in adrenoleukodystrophy, *Pediatr. Res.* **17**:221A.

Ullrich, A., Berman, C. H., Dull, T. J., Gray, A., and Lee, J. M., 1984, Isolation of the human insulin-like growth factor I gene using a single synthetic DNA probe, *EMBO J.* **3**:361.

United States Congress, 1982, House hearings on human genetic engineering before the subcommittee on investigations and oversight of the Committee on Science and Technology, 97th Congress, 2nd Session, No. 170, Government Printing Office, Washington, D.C.

Valerio, D., Duyvesteyn, M. G. C., Khan, P. M., van Kessel, A. G., de Waard, A., and van der Eb, A. J., 1983, Isolation of cDNA clones for human adenosine deaminase, *Gene* **25**:231.

Valerio, D., Duyvesteyn, M. G. C., and van der Eb, A. J., 1984, Introduction of sequences encoding functional human adenosine deaminase into mouse cells using a retrovirus shuttle system, *Gene* **34**:163.

Valerio, D., Duyvesteyn, M. G. C., Dekker, B. M. M., Weeda, G., Berkvens, T. M., van der Voorn, L., van Ormondt, H., and van der Eb, A. J., 1985, Adenosine deaminase: Character-ization and expression of a gene with a remarkable promoter, *EMBO J.* **4**:437.

Van Bekkum, D., 1984, Conditioning regimens for marrow grafting, *Semin. Hematol.* **21**:81.

Van Der Putten, H., Botteri, F. M., Miller, A. D., Rosenfeld, M. G., Fan, H., Evans, R. M., and Verma, I. M., 1985, Efficient insertion of genes into the mouse germ line via retroviral vectors, *Proc. Natl. Acad. Sci. U.S.A.* **82**:6148.

Van Doren, K., and Gluzman, Y., 1984, Efficient transformation of human fibroblasts by adeno-virus–simian virus 40 recombinants, *Mol. Cell. Biol.* **4**:1653.

Varmus, H. E., Quintrell, N., and Ortiz, S., 1981, Retroviruses as mutagens: Insertion and excision of a nontransforming provirus alter expression of a resident transforming provirus, *Cell* **25**:23.

Vierling, J. M., 1984, Epidemiology and clinical course of liver diseases: Identification of candidates for hepatic transplantation, *Hepatology* **4**:84S.

Walser, M., 1983, Urea cycle disorders and other hereditary hyperammonemic syndromes, in: *The Metabolic Basis of Inherited Disease* (J. B. Stanbury, J. B. Wyngaarden, D. S. Fredrickson, J. L. Goldstein, and M. S. Brown, eds.), p. 402, McGraw-Hill, New York.

Walser, M., Batshaw, M., Sherwood, G., Robinson, B., and Brusilow, S., 1977, Nitrogen me-tabolism in neonatal citrullinemia, *Clin. Sci. Mol. Med.* **53**:173.

Weiss, R., 1984, Experimental biology and assay of retroviruses, in: *RNA Tumor Viruses*, Vol. 1 (R. Weiss, N. Teich, H. Varmus, and J. Coffin, eds.), p. 209, Cold Spring Harbor Press, Cold Spring Harbor, New York.

Wiginton, D. A., Adrian, G. S., Friedman, R. L., Suttle, D. P., and Hutton, J. J., 1983, Cloning of cDNA sequences of human adenosine deaminase, *Proc. Natl. Acad. Sci. U.S.A.* **80**:7481.

Wigler, M., Perucho, M., Kurtz, D., Dana, S., Pellicer, A., Axel, R., and Silverstein, S., 1980, Transformation of mammalian cells with an amplifiable dominant-acting gene, *Proc. Natl. Acad. Sci. U.S.A.* **77**:3576.

Williams, D. A., Lemischka, I. R., Nathan, D. G., and Mulligan, R. C., 1984, Introduction of new genetic material into pluripotent haematopoietic stem cells of the mouse, *Nature (London)* **310**:476.

Willis, R., Jolly, D. J., Miller, A. D., Pleut, M. M., Esty, A. C., Anderson, P. J., Chang, H.-C., Jones, O. W., Seegmiller, J. E., and Friedman, T. F., 1984, Partial phenotypic correction of human Lesch–Nyhan (hypoxanthine-guanine phosphoribosyltransferase-deficient) lympho-blasts with a transmissible retroviral vector, *J. Biol. Chem.* **259**:7842.

Witherspoon, R. P., Lum, L. G., and Storb, R., 1984, Immunologic reconstitution after human marrow grafting, *Semin. Hematol.* **21**:2.

Wolf, L., and Ruschetti, S., 1985, Malignant transformation of erythroid cells *in vivo* by introduction of a nonreplicating retrovirus vector, *Science* **228**:1549.

Wong, G. G., Witek, J. S., Temple, P. A., Wilkens, K. M., Leary, A. C., Luxenberg, D. P., Jones, S. S., Brown, E. L., Kay, R. M., Orr, E. C., Shoemaker, C., Golde, D. W., Kaufman, R. J., Hewick, R. M., Wang, E. A., and Clark, S. C., 1985, Human GM-CSF: Molecular cloning of the complementary DNA and purification of the natural and recombinant proteins, *Science* **228**:810.

Wood, P. A., Herman, G. E., Chao, J. C.-Y., O'Brian, W. E., and Beaudet, A. L., 1986, Retrovirus-mediated gene transfer of argininosuccinic acid synthetase into cultured rodent and human cells, *Cold Spring Harbor Symp. Quant. Biol.* (in press).

Yates, J. L., Warren, N., and Sugden, B., 1985, Stable replication of plasmids derived from Epstein–Barr virus in various mammalian cells, *Nature (London)* **313**:812.

Yeung, C.-Y., Ingolia, D. E., Bobonis, C., Dunbar, B. S., Riser, M. E., Siciliano, M. J., and Kellems, R. E., 1983a, Selective overproduction of adenosine deaminase in cultured mouse cells, *J. Biol. Chem.* **258**:8338.

Yeung, C.-Y., Riser, M. E., Kellems, R. E., and Siciliano, M. J., 1983b, Increased expression of one of two adenosine deaminase alleles in a human choriocarcinoma cell line following selecton with adenine nucleosides, *J. Biol. Chem.* **258**:8330.

Yeung, C.-Y., Frayne, E. G., Al-Ubaid, M. R., Hook, A. G., Ingolia, D. E., Wright, D. A., and Kellems, R. E., 1983c, Amplification and molecular cloning of murine adenosine deaminase gene sequences, *J. Biol. Chem.* **258**:15,179.

INDEX